Aero-Hydrodynamics of Sailing

by the same author

SAILING THEORY AND PRACTICE

_____

AERO-HYDRODYNAMICS OF SAILING

C A MARCHAJ

Dodd, Mead & Company · New York

First published in the United States 1980

1   2   3   4   5   6   7   8   9   10

Library of Congress Cataloging in Publication Data

Marchaj, Czeslaw A
    Aero-hydrodynamics of sailing.

    Includes index.
    1. Sailboats — Hydrodynamics. 2. Sailboats —
Aerodynamics. 3. Sails — Aerodynamics. I. Title.
VM331.M36   1980   623.8′12043   79-27724
ISBN 0-396-07739-0

Contents

PART 2   Basic principles of aero-hydrodynamics: aerofoil and hydrofoil action

List of symbols

A	wetted area of the hull, also lateral area of the fin keel or rudder
AR	aspect ratio
a_o	slope coefficient of lift curve per degree (two-dimensional flow)
B	beam of the hull
b	span of the foil
c	length of the foil chord
CB	centre of buoyancy
C_D	aerodynamic drag coefficient
c_d	section drag coefficient
C_{Do}	minimum profile drag
CE	centre of effort
C_f	friction coefficient
CG	centre of gravity
C_H	aerodynamic heeling force coefficient
C_L	aerodynamic lift force coefficient
C_l	local lift coefficient
c_l	section lift coefficient
CLR	centre of lateral resistance
C_p	pressure coefficient
C_R	aerodynamic driving force coefficient
c_r	chord length at the root of the foil
C_s	side force coefficient (water)
C_T	aerodynamic resultant (total) force coefficient
c_t	chord length at the tip of the foil
C_x	component of C_T along the boat centreline

C_Y	component of C_T perpendicular to centreline
D	aerodynamic drag
D_f	friction drag
D_i	induced drag
D_p	parasite drag
D_r	draft of the hull
D_t	total drag
DWL	designed waterline
F_H	heeling force (air)
F_{lat}	horizontal component of the heeling force
F_R	driving force (air)
F_S	side force (water)
F_T	total aerodynamic force
F_V	vertical downward component of sail force
g	acceleration due to gravity (32.2 ft/sec^2)
H_w	height of the wave
I	height of the foretriangle
J	base of the foretriangle
k_a	admissible roughness height
L	lift, and also length of the hull
LOA	length overall
LWL	load waterline
M_H	heeling moment
M_R	righting moment
O	origin of co-ordinate system
p	static pressure (local)
p_o	standard atmospheric pressure (2116.2 lb/ft^2 = 14.7 lb/in^2)
q	dynamic pressure
R	total hydrodynamic resistance of the hull
RA	righting arm
Re	Reynolds number
R_f	hydrodynamic skin friction
R_h	hydrodynamic resistance due to heel
R_T	total hydrodynamic force on a hull
R_w	wave-making resistance
S	vortex span
S_A	sail area
t	thickness of foil or obstacle, normal to the flow direction
V_A	apparent wind velocity
V_c	circulation velocity
V_{mg}	speed made good to windward
V_o	flow velocity some distance ahead from the foil
V_S	boat speed

V_T	true wind velocity
W	weight of the yacht
α (alpha)	angle of incidence of a foil (geometric)
α_{ef}	effective incidence angle
α_f	angle of incidence of the foresail
α_i	induced angle of incidence
α_{Lo}	incidence at zero lift angle
$\alpha - \alpha_{Lo}$	incidence measured from zero lift angle
α_m	angle of incidence of the mainsail
β (beta)	apparent course between V_A and V_S
$(\beta - \lambda)$	heading angle between V_A and hull centreline
γ (gamma)	true course between V_T and V_S

Courses β, $(\beta - \lambda)$ or γ are measured in relation to apparent or true wind V_A or V_T respectively and, as such, they have nothing to do with courses in navigational sense, such as magnetic course or compass course.

Γ (gamma)	circulation
Δ (delta)	displacement (weight) in long tons or pounds (1 ton = 2240 lb)
Δ_p	differential pressure
δ (delta)	logarithmic decrement, damping
δ_f	angle of trim of the foresail
δ_m	angle of trim of the mainsail
ε_A (epsilon)	aerodynamic drag angle
ε_H	hydrodynamic drag angle
$-\varepsilon$	wash-out (incidence angle decreases towards the foil tip)
$+\varepsilon$	wash-in (incidence angle increases towards the foil tip)
Θ (theta)	angle of heel or rolling angle
λ (lambda)	leeway angle
v (nu)	kinematic viscosity
ρ_A (rho)	mass density of air
ρ_W	mass density of water

There are also some other symbols occasionally used which are explained whenever they are introduced.

Mathematical signs and abbreviations

$\simeq$	approximately equal to...
$>$	greater than...
$<$	less than...
a^n	n^{th} power of a ...
$\sim$	is proportional to...
f( )	is proportional or is a function of the factors given between brackets
∞	infinite length, distance

Author's preface

'There is no virtue in not knowing what can be known'

A. HUXLEY

This present book, a companion volume to my *Sailing Theory and Practice* published originally some twenty years ago, was written with two intentions in mind. Firstly, the aim was to look attentively over the progress and developments in sailing practices, design features and underlying theories which have come along during the last two decades; secondly, to describe as vividly and straightforwardly as possible some basic concepts of aerodynamics and hydrodynamics which are essential as the fundamental tools for inquiry into the technology of the modern sailing vessel.

A deliberate effort has been made to present the subject in such a way that it is readable and comprehensible to the scientifically inclined layman with an inquiring mind. Even those who have not yet been actively concerned with the theoretical and practical problems of sailing should have little difficulty in following most of the discussion and subsequently to reason for themselves about the whys and wherefores of sailing yacht behaviour, tuning, performance, etc. In so far as they are discussed in this book, the theories have shown good agreement with experimental facts; thus the ideas presented will be of use to readers whose primary interest lies in applying these ideas to concrete practical problems.

In a similar manner to *Sailing Theory and Practice*, pictures and graphical representation are extensively used. It is hoped that through these means, rather than through the use of words and mathematics, one can bridge the communication gulf more easily. Such a visual display not only demonstrates a given property or concept more clearly, but it reinforces the text and conveys a great deal that words cannot express adequately.

The whole text leans heavily on the intuitive approach and the pictorial examples emphasize the physical meaning of unfamiliar concepts and terms introduced. Following the view of a prominent expert on applied mathematics that '...the purpose of computing is insight not numbers', almost all quantitative statements are given in the form of simple graphs rather than equations.

This book does not render *Sailing Theory and Practice* obsolete, although one may find that the earlier volume is not up to date in some respects. My present view on some aspects of sailing theory such as, for example, the interaction between the two sails, has changed substantially. This is inevitable. Theories are fallible, and fortunately, or otherwise, depending on one's attitude and expectation, the structure of sailing theory, like any other knowledge, is neither rigid nor static but is bound to continually evolve and expand as new observations and facts come to light.

One must, however, realize–and this is particularly directed to those who value the hard facts and practice most–that science, to quote I B Conant, '...advances not by the accumulation of new facts...but by the continuous development of new and fruitful concepts'. In other words, all the hard facts we observe–and this applies to the whole variety of human experience–speak to us through interpreters which are the theories we all apply, consciously or not, while trying to understand the physical world or events. This is the reason why not everyone emerges with the same viewpoint after encountering the same facts.

Thus sailing theory, like any other concept or interpretation of the facts of life, is not self-terminating, but can always be modified or improved with continually accumulating knowledge. In this sense the present book should by no means be regarded as a set of dogmas. It should serve to suggest ideas and to stimulate thought rather than to provide any definite answer to the ever increasing number of practical problems sailing men have to cope with.

In this respect, certainty does not appear to be a virtue of any science. Unknown or hidden variables, not perceptible at first, may always emerge with time and challenging inconsistency in any theory will become too evident to be ignored. If one agrees on that, one must also agree with K Popper when he says: '...we do not start from observations but always from problems–from practical problems or from a theory which has run into difficulties; that is to say which has raised, and disappointed, some expectations...Thus we may say that our knowledge grows as we proceed from old problems to new problems by means of conjectures and refutations, by refutation of our theories or, more generally, of our expectations.'

In one way or another many people contributed to this book and I gratefully acknowledge my debt to all authors whose names are given in the list of references. In particular, I would like to express my indebtedness to the late Thomas Tanner, whose work helped me on many occasions and who was always ready to assist with friendly, impartial advice. My warmest thanks also go to H Barkla, A Gentry, H Glauert, P V MacKinnon and A M O Smith. Although their work greatly inspired my thoughts I am not saddling them with any responsibility for the views my book now contains.

I should like to record my appreciation of the unfailing help and advice generously given by J Flewitt of Southampton University who read the first draft of this book. I want to thank also B Hayman and J Driscoll of *Yachting World*, P Cook of *Yachts and Yachting*, G C Comer of Lands' End Publishing Corporation, Dr G Corbellini, Beken of Cowes Ltd, The British Hovercraft Corporation Ltd, The Society of Naval Architects and Marine Engineers, and ICI Fibres Ltd for their kind permission to use their photographs.

Finally, I am greatly indebted to Jeremy Howard-Williams, my editor, for advising me about the manuscript and for his effort in correcting my foreign English. Although he waited with extraordinary patience and tact for a result, he held me firmly to my commitment.

Because of the recent change being made from the British to the SI Metric measurement system, one may expect some criticism at the retention of the time-honoured and familiar terms such as knots, displacement/length ratios, etc, when presenting the results of experiments or calculations. The only excuse for my attitude in this respect is that there will unavoidably be a period when both systems are in use together. And since my book is addressed to the general reader rather than to the scientist I believe that an indiscriminate or dogmatic introduction of the new SI Metric system would only cause an unnecessary confusion. Those who might be interested in conversion of British Units into the Metric ones can easily do it with the help of the 'Table of Dimensions and Units' included as an appendix at the end of the book.

Southampton, November 1978   C A Marchaj

To Jana and Martin in remembrance of good days.

PART 1

Fundamental factors governing yacht performance

'Those who fall in love with practice without science are like a sailor who steers a ship without a helm or compass, and who never can be certain whither he is going.'

LEONARDO DA VINCI

Introduction:
The nature of the problem

'The history of technique and engineering testifies to the irresistible urge of humanity towards increasing the speed of locomotion. Means of locomotion on the ground, on the surface of and within water, through the air and, perhaps, through empty space, compete in an ever growing effort towards higher velocities. Obviously there are limitations for every type of locomotion. At a certain speed any particular type becomes so inefficient and uneconomical that it is unable to compete with other more appropriate types.' Thus argued G Gabrielli and Th Karman in their famous paper *What Price Speed?*

F T Marinetti, the founder and leader of Italian futurism, in his *Futurist Manifesto* published in 1916, developed his idea of the new religion of speed. The following extract is evidence of the remarkable sensitivity of Marinetti as an artist who was capable of giving expression to a powerful trend that was hardly discernible at his time.

'Speed having as its essence the intuitive synthesis of every force in movement, is naturally pure. Slowness, having as its essence the rational analysis of every exhaustion in repose, is naturally unclean. After the destruction of the antique good and the antique evil, we create a new good: speed, and a new evil: slowness.

Speed is the synthesis of every courage in action, and is aggressive and war-like. Slowness is the analysis of every stagnant prudence, and is passive and pacifistic. The intoxication of great speeds in cars is nothing but the joy of feeling oneself fused with the only divinity. Sportsmen are the first converts to this religion; then comes destruction of houses and cities, to make way for great meeting places for cars and planes.'

No doubt the peculiar fascination and exhilaration of high speed under sail was, and still is, a powerful emotive drive to stir man's creative imagination and desire to build and sail faster and faster craft. Progress to higher speed is not always, however, a continuous process, and has been achieved in a rather spasmodic fashion through the last 100 years. Right now the majority of sailing men seem inspired by dreams of ultimate speed under sail; the Transatlantic Single Handed Race and the John Player World Sailing Speed Record are just two extreme examples of the competitive spirit which dominates the sailing scene.

For no immediately apparent reasons, this interest in high speed sailing has driven people to discuss new concepts of high performance sailing machines, build them and finally sail almost unthinkable, spider-like sea monsters bordering on pure fantasy.

Concentrating on the competitive and high speed aspect of sailing boats, we may divide existing and anticipated sailing craft into five categories, as follows:

1. Light, flat bottomed *skimming* forms (dinghies, scows).
2. Heavy displacement forms (heavy conventional ballasted yachts).
3. Multihulls (catamarans, trimarans, proas).
4. Sailing hydrofoils.
5. Other, various, craft using sail for propulsion (land yachts, ice boats, surf-boards, skimmers).

What factors limit performance in each of these categories? What price is paid for speed? What has been achieved? What are the prospects for further improvement? To explore ways and means of sailing faster we shall review the basic factors and underlying principles that govern the behaviour and limit the performance of a variety of sailing craft, ranging from heavy displacement yachts ploughing troughs in the water, to sailing hydrofoils, ice boats, and other modern craft which sail unsupported by buoyancy forces and are therefore not subject to any wave drag barrier.

Since sailing boats are not constant cruising speed vehicles such as aircraft, but operate in a variety of wind velocities, ranging from calm to gale, and on various courses relative to the wind, one should not expect that any simple set of criteria can successfully be applied to judge the merits of rigs or hulls.

The choice of a rig or the concept of a hull to match a rig must necessarily be a compromise, depending largely on what one is trying to achieve: to improve a boat's performance on a triangular course, or on an arbitrary course when racing offshore, to beat some absolute speed record in sheltered water, to cross the Atlantic in the shortest possible time singlehanded, etc, etc.

However crude and difficult to determine, a set of evaluation criteria must be established or agreed upon, in order to estimate the quality of a sailing boat and the eventual progress made. With no criteria it is difficult to make any sensible judgement concerning the excellence of a design or development. The criteria by which the merits of a particular type of boat are evaluated will of course vary with the particular design aims of the vessel and, strangely enough, with time. This point requires some clarification.

There is probably no better way to start a violent argument than to ask what constitutes a modern, high performance offshore cruiser-racer. We might agree that the following characteristics or requirements, as written in Table 1.1, are important:

Table 1.1

---

1. Habitability or space for living quarters (convenient accommodation with good head-room etc).
2. Stability, ballast ratio (power to carry sail).
3. Speed (to windward and on other courses; all-round performance).
4. Dryness (adequate reserve buoyancy at bow and stern).
5. Sea-keeping behaviour or easy motion (large anti-rolling inertia and small pitching inertia to prevent deep plunging).
6. Controllability and ease of handling (quality of balance, steering and course-keeping ability).
7. Seaworthiness (strong, durable and water-tight construction; boat's longevity).

However, when one is asked to list those characteristics in order of merit, they become immediately a pretty subject to debate for at least two reasons. Firstly, some of the requirements are incompatible; secondly, people's concepts of the ideal or dream yacht are contrasting and of an emotional rather than a rational character, and therefore highly arguable.

The characteristics 1–7 are interchangeable, at least some of them, and it is the problem of the owner and/or yacht designer to decide just how far a loss in one characteristic is justified by a gain in some other characteristics. No one can design a boat which incorporates all the features 1–7, developed to full satisfaction; compromise is unavoidable.

Performance

As far as performance in terms of speed is concerned, conventional ballasted yachts have reached a stage near to finality, which is strictly conditioned by the fundamental principles and factors governing hull behaviour. It is a characteristic of all so-called displacement type yachts, in which the lateral stability needed to carry sails is provided by heavy metal keels, that their resistance due to wave-making takes a sharp upturn as soon as the relative speed $V_s/\sqrt{L}$ (i.e. boat speed V_s in relation to the hull length L) exceeds only a little more than unity. This fact depicted in Fig 1.17 *New York 32* puts an effective brake on the maximum speed which rarely exceeds $1.4\sqrt{L}$. It has been understood for a long time that one cannot hope to continue indefinitely the speed improvements of the displacement type of yacht.

The basic speed-affecting factors or parameters which every yacht designer must confront when considering a new boat are given in Table 1.2.

Table 1.2

1. Waterline length of the hull (L).
2. Sail area (S_A).
3. Displacement (Δ).
4. Wetted area of the hull (A).
5. Stability or power to carry sails effectively.
6. Prismatic coefficient, which measures the distribution of immersed volume along the length of the hull.
7. Sail area/displacement $\left(\dfrac{S_A}{\Delta}\right)$ ratio.
8. Sail area/wetted area $\left(\dfrac{S_A}{A}\right)$ ratio.
9. Displacement/length $\left[\dfrac{\Delta}{(L/100)^3}\right]$ ratio.

Depending on the emotionally or rationally selected sets of requirements given in Table 1.1, and the factors presented in Table 1.2, different types of boat will result. And those which, like fashion, are generally appreciated today may be rejected tomorrow and rediscovered happily after a lapse of time.

If one wishes to have a fast racer, then the requirement concerning living quarters might be defined as some wit put forward: '...the best accommodation for an ocean-racer is a hull empty but for a load of hay. It gives the best ballast ratio; the crew can sleep in it and eat it; at the end of a race it can be mucked out ready for a fresh lot next time.' Such an approach, however extreme, is almost accepted by go-fast fanatics and this explains why more and more boats competing in international races are, in fact, stripped out shells with their living quarters looking like huge sail-bins; they are organized ruthlessly to be functional and efficient in carrying and operating a large wardrobe of sails for all conditions.

A man who likes racing and also cruising may put emphasis on requirements 1, 5 and 7, in Table 1.1, and still be quite happy if his dream yacht has only a reasonable expectation of winning. For him, convenient accommodation at a cost of 100 lb more displacement may be valued more highly than the better stability which might be achieved by shifting the same 100 lb downwards to form ballast.

The intense competition, or rather preoccupation with speed, which dominates today, permits no half-way approach. The designer's as well as the builder's reputation is made by the racing success of their creations. Inevitably, the design conflicts between features 1–7 listed in Table 1.1 are nowadays almost always resolved in favour of high performance, but a price must be paid in one way or another.

A Forces and geometry of sailing to windward

Let us limit, for the time being, our attention to the windward leg, generally regarded as the most important sailing course. This course, more than any other, intensifies the conflict between the aerodynamic efficiency of a rig and the hydrodynamic efficiency of a hull, together with its resistance and stability. The most obvious manifestation of this conflict in the traditional mono-hull yacht is that between resistance against stability (i.e. narrowness and lightness, giving an easily driven hull) and beam and weight (giving power to stand up to the sail forces). Yacht performance, particularly when sailing to windward, is in fact a complicated game of hull resistance, the driving power of the sail and stability (Ref 1.1)*. The other conflicting factors, very acute in recent times, are wetted surface of the hull, against steering efficiency and lightness of the hull, against strength, seaworthiness and habitability.

Before the relative influence of those factors on a boat's performance can be considered in some detail, a glance at the sailing mechanism of a simple dinghy, sailed nearly upright, would be appropriate. Although a dinghy may seem a far cry from a heavy displacement yacht, it will be seen that it may increase our understanding of sailing by virtue of its simplicity. One may even justifiably claim that big yachts, including the most sophisticated 12-Metres, are nothing but big dinghies. Once the principles or secrets governing their behaviour have been understood they are likely to be of general application to all sailing craft. When necessary, the peculiarities of heavy displacement craft will also be explored and analysed.

* See *References and Notes* at the end of Part 1.

Fig 1.1   Equilibrium of forces and moments in steady-state sailing conditions.

F_R–Driving force
F_{Hlat}–Horizontal Heeling Force
F_V–Vertical Aerod Force
M_{PA}–Trimming Moment
M_H–Heeling Moment
M_{YW}–Yawing Moment

} Air

R–Water Resistance
F_{Slat}–Horizontal Side Force
F_{VW}–Vertical Hydrod Force
M_{PW}–Trimming Moment
M_R–Righting Moment
M_{YL}–Yawing Moment

} Water

W–Weight of the boat
Δ–Displacement of the boat

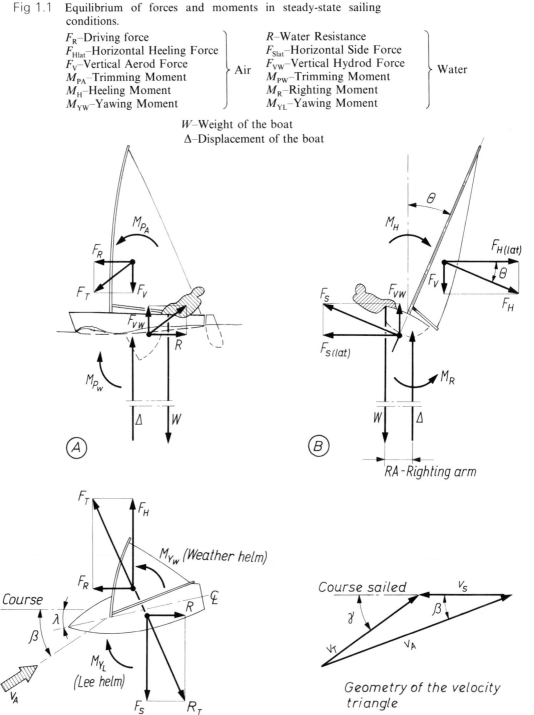

Aerodynamic and hydrodynamic forces

Figure 1.1 illustrates various components of aerodynamic and hydrodynamic forces and their relative position, as well as associated moments, affecting a Finn-type dinghy propelled by a una rig and sailing steadily to windward.

They can be written down as follows:

$$
\left.
\begin{array}{lll}
1. & F_R & = R \\
2. & F_H \text{ lat} & = F_S \text{ lat} \\
3. & F_V & = F_{VW}
\end{array}
\right\} \text{ Forces}
$$

$$
\left.
\begin{array}{lll}
4. & M_{PA} & = M_{PW} \\
5. & M_H & = M_R \\
6. & M_{YW} & = M_{YL}
\end{array}
\right\} \text{ Moments}
$$

The special case of the dinghy in which movable crew weight is sufficient to keep the boat nearly upright implies that, in order to establish the boat's performance at various wind velocities, it is a good approximation to consider only forces acting horizontally (Ref 1.2); this is depicted in Fig 1.2. We can represent the wind action, or aerodynamic forces, on that part of the boat which is above the waterline, as in Fig 1.2A, i.e. from a bird's eye view. The resultant force F_T, labelled 'Total Aerodynamic Force', which goes through the CE (Centre of Effort), arises due to the action of the apparent wind V_A, and includes the aerodynamic forces on the hull, mast and rigging, as well as those developed by the sail alone. The two components, labelled Cross Wind Force or Lift L, and Drag D, should be considered as equivalent to a single total aerodynamic force F_T. These two components L and D can be measured in the wind tunnel, drag D being measured in the same direction as the apparent wind, and the lift L at right angles to it. The two other components in which every sailor is directly interested, labelled Driving Force F_R and Heeling Force F_H, can also be considered as the equivalent of the same single aerodynamic total force F_T. The driving force F_R, shown in the direction of the course sailed, propels the boat; the heeling or capsizing force F_H, at right angles to the former, is responsible for drift and heel.

The essential requirement of the sail is to produce a driving force component F_R, and this it cannot do, except on a dead run, without at the same time producing a heeling force F_H, and the driving force attained is proportional to this heeling force. In the close-hauled condition the driving force F_R is roughly one-fourth to one-third of the heeling force F_H. In other words, every pound of driving force generated on the sail is accompanied by three to four pounds heeling force that the yacht must withstand and absorb by virtue of her stability. By analogy, the heeling force and associated heel can be regarded as the throttle in a motor boat: heel it less and you go faster.

Looking at a sail as an aerofoil, or a lift-producing device, one may regard drag D as a price paid for the lift L. The angle ε_A, between the lift L and the total aerodynamic force F_T, may serve as an index of aerodynamic efficiency of the sail. If

Fig 1.2   Aerodynamic forces on a yacht rig sailed nearly upright resolved in two different ways: lift or cross-wind force L and drag D components as measured in the wind tunnel; driving force F_R and heeling force F_H components which control the boat's behaviour.

Hydrodynamic forces developed on a yacht hull moving through water with an angle of yaw.

Equilibrium of aerodynamic and hydrodynamic forces. When the total aerodynamic force F_T and total hydrodynamic force R_T are equal and opposite no acceleration takes place and the boat is in equilibrium and steady motion.

If wind velocity increases the aerodynamic forces also increase and the boat will accelerate; conversely, if the wind velocity decreases the boat will decelerate until a new equilibrium of forces is again established.

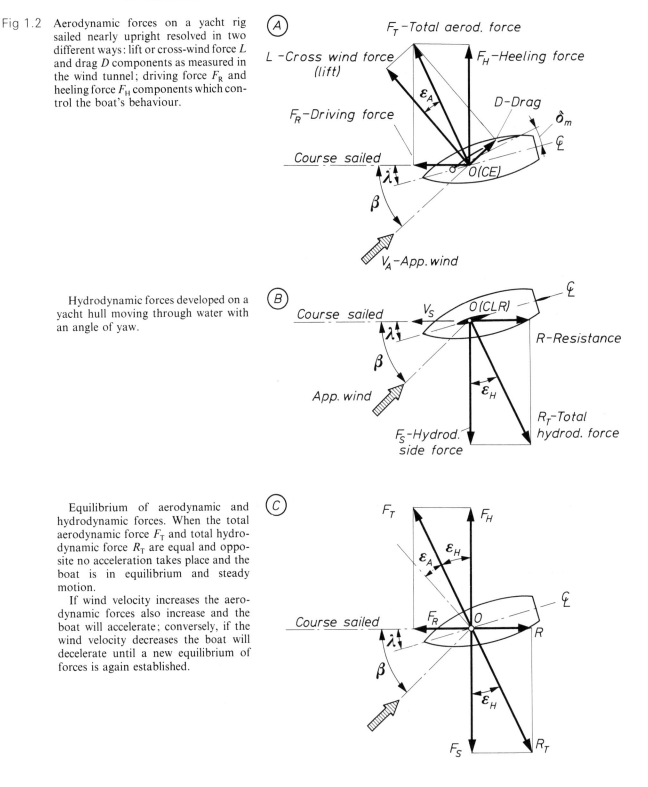

the drag D could somehow be made smaller without altering the value of lift L, the sail would be more efficient in windward work. The total force F_T would then be inclined more forward and the driving force F_R would be a larger fraction of the undesirable heeling or capsizing force F_H, which the hull must withstand.

One may easily find that the higher the L/D ratio, the smaller is ε_A angle. In mathematical terms, it can be expressed by:

$$\cot \varepsilon_A = \frac{L}{D}$$

Similarly, Fig 1.2B illustrates the hydrodynamic forces on the hull moving at velocity V_s through the water, with an angle of yaw or leeway λ. The underwater part of the hull may be regarded as a hydrofoil producing a hydrodynamic side force F_S and a resistance R. These two components F_S and R are measurable in the towing tank, and their effect can be represented by a single hydrodynamic force R_T, which goes through the Centre of Lateral Resistance (CLR) of the hull. The hydrodynamic forces depend on the boat's velocity V_S, its heel, and the leeway angle λ. Generation of a hydrodynamic force F_S may lead to considerable increase in the hull resistance R over that of the upright hull of zero leeway. This increase in resistance is an additional price one has to pay for the driving force F_R produced by the sail.

In a similar manner as in the case of the sail, the angle ε_H defined by:

$$\cot \varepsilon_H = \frac{F_S}{R}$$

reflects the hydrodynamic efficiency of the hull in generating side force F_S. It is obvious that the efficiency of the underwater part of the hull, regarded as a hydrofoil, lies in its ability to produce the necessary side force F_S, at the least resistance R for a particular speed.

In order to determine the relationship between the wind and water forces, we might invoke Newton's law of motion which, when applied to our case, states that if a sailing craft is to proceed at a steady speed in a straight course, the resultants of air and water forces must balance each other. The combined action of wind and water forces is shown in Fig 1.2C, which implies not only that the resultant aerodynamic and hydrodynamic forces F_T and R_T are equal in magnitude and opposite in direction, but that the equivalent system of components are also equal and opposite. From this figure it becomes apparent that the sail-driving force F_R is used to overcome the hull resistance R, while the unwanted but unavoidable sail-heeling force F_H is balanced out by the hydrodynamic side force F_S produced by the underwater part of the hull.

The aero- and hydrodynamic forces developed due to a yacht's motion relative to air and water, and which control her behaviour, are directly related. To illustrate the point, the effectiveness of a rig in driving the boat is directly related to the trim of the sail relative to the wind and to the hull. As the boat begins to move towards the wind, the aerodynamic forces generated on the sail determine the instantaneous leeway angle. While the hull accelerates under the action of continually changing sail forces, the leeway angle also changes, determining the instantaneous hydrodynamic forces developed on the hull. By virtue of a feed-back, existing between the sail and hull forces, the varying hull velocity and hull attitude, regarded as parameters, modify continually the sail attitude and the apparent wind velocity until equilibrium of air and water forces is reached. From now on, the boat proceeds with constant velocity, provided the wind velocity is constant. If the true wind increases the boat will accelerate, and if it decreases the boat will decelerate until a new equilibrium of forces is once again established. The feed-back between aero- and hydrodynamic forces is partly automatic, independent of the helmsman's will, and may partly be intentional whenever the helmsman changes the sail trim or applies rudder action. (Note 1.3).

Under the conditions shown in Fig 1.2 the motion of a boat sailed nearly upright is controlled by nine basic variables which can be divided as in Table 1.3:

Table 1.3

Geometry of sailing velocity triangle Fig 1.1D	Sail aerodynamics	Hull hydrodynamics
V_S	V_A	V_S
V_A	β, δ_m	λ
β	F_R, F_H	R, F_S

If we know how the aerodynamic forces F_R and F_H vary with V_A, β and δ_m, and the hydrodynamic forces R and F_S vary with V_S and λ, we may answer the essential question: at which particular wind velocity V_A and boat velocity V_S will the aerodynamic and hydrodynamic forces balance each other? We may therefore predict, with reasonable accuracy, how fast a yacht may sail at various wind velocities and courses β relative to the apparent wind V_A. Hence, we can estimate the speed performance of a given boat at given wind V_T.

An interesting relationship follows from Fig 1.2C, to quote the original words of F W Lanchester, who in 1907 made an outstanding, although too advanced by his contemporaries' standards, contribution to aerodynamics:

'...the problem of sailing yacht mechanics resolves itself into an aerofoil combination in which the aerofoil acting in the air (a sail spread) and that acting under water (the keel, fin, or dagger plate) mutually supply each other's reaction.

The result of this supposition is evidently that the minimum angle at which the boat can shape its course relatively to the wind is the sum of the under and above water gliding angles.' (Ref 1.4)

Introducing contemporary sailing terminology, 'the gliding angles' are equivalent to the aerodynamic and hydrodynamic drag angles ε_A and ε_H respectively.

Expressing Lanchester's idea in mathematical terms, one may write:

$$\beta = \varepsilon_A + \varepsilon_H \qquad \text{Eq 1.1}$$

where

$$\varepsilon_A = \cot^{-1}\left(\frac{L}{D}\right) \quad \text{or} \quad \cot \varepsilon_A = \frac{L}{D}$$

$$\varepsilon_H = \cot^{-1}\left(\frac{F_s}{R}\right) \quad \text{or} \quad \cot \varepsilon_H = \frac{F_s}{R}$$

The expression $\cot \varepsilon_A = L/D$ is equivalent to the statement 'ε_A is the angle whose cotangent is L/D'; this is often abbreviated:

$$\varepsilon_A = \cot^{-1}\left(\frac{L}{D}\right)$$

where '$\cot^{-1}$' means, literally, 'the angle whose cotangent is'; '$\cot^{-1}$' must not be interpreted as the -1 power of $\cot \varepsilon_A$.

Table 1.4 below gives values of ε_A and ε_H in degrees for various L/D and F_s/R ratios. Thus, the small ε_A or ε_H angles correspond to high L/D or F_s/R ratios respectively.

Table 1.4

ε_A or ε_H	L/D or F_s/R	ε_A or ε_H	L/D or F_s/R	ε_A or ε_H	L/D or F_s/R
4°	14.30	11°	5.14	18°	3.08
5°	11.43	12°	4.70	19°	2.90
6°	9.51	13°	4.33	20°	2.75
7°	8.14	14°	4.01	21°	2.60
8°	7.12	15°	3.73	22°	2.47
9°	6.31	16°	3.48	23°	2.36
10°	5.67	17°	3.27	24°	2.25

See also the more extensive Table 1.4A.

It is possible to derive from Eq 1.1 a number of practical conclusions when interpreting the wind tunnel and towing tank experiments which are particularly important both for competitive sailing and yacht designing. For example, it is obvious that an increase in the hull side-force/resistance F_s/R ratio, either by increasing F_s or by decreasing R *throughout the range of leeway angles* which is equivalent to a decrement in ε_H, will improve the potential windward ability of any sailing craft.

TABLE 1.4A
NATURAL COTANGENTS
N.B.—Subtract mean differences

	0′	6′	12′	18′	24′	30′	36′	42′	48′	54′	mean differences				
											1′	2′	3′	4′	5′
0°	∞	573.0	286.5	191.0	143.2	114.6	95.49	81.85	71.62	63.66					
1	57.29	52.08	47.74	44.07	40.92	38.19	35.80	33.69	31.82	30.14					
2	28.64	27.2/	26.03	24.90	23.86	22.90	22.02	21.20	20.45	19.74					
3	19.08	18.46	17.89	17.34	16.83	16.35	15.89	15.46	15.06	14.67					
4	14.30	13.95	13.62	13.30	13.00	12.71	12.43	12.16	11.91	11.66					
5	11.43	11.20	10.99	10.78	10.58	10.39	10.20	10.02	9.845	9.677					
6	9.5144	3572	2052	0579	9152	7769	6427	5126	3863	2636					
7	8.1443	0285	9158	8062	6996	5958	4947	3962	3002	2066					
8	7.1154	0264	9395	8548	7720	6912	6122	5350	4596	3859					
9	6.3138	2432	1742	1066	0405	9758	9124	8502	7894	7297					
10°	5.6713	6140	5578	5026	4486	3955	3435	2924	2422	1929	*mean* differences not sufficiently accurate				
11	5.1446	0970	0504	0045	9594	9152	8716	8288	7867	7453					
12	4.7046	6646	6252	5864	5483	5107	4737	4374	4015	3662					
13	4.3315	2972	2635	2303	1976	1653	1335	1022	0713	0408					
14	4.0108	9812	9520	9232	8947	8667	8391	8118	7848	7583					
15	3.7321	7062	6806	6554	6305	6059	5816	5576	5339	5105					
16	3.4874	4646	4420	4197	3977	3759	3544	3332	3122	2914					
17	3.2709	2506	2305	2106	1910	1716	1524	1334	1146	0961					
18	3.0777	0595	0415	0237	0061	9887	9714	9544	9375	9208					
19	2.9042	8878	8716	8556	8397	8239	8083	7929	7776	7625					
20°	2.7475	7326	7179	7034	6889	6746	6605'	6464	6325	6187					
21	2.6051	5916	5782	5649	5517	5386	5257	5129	5002	4876					
22	2.4751	4627	4504	4383	4262	4142	4023	3906	3789	3673					
23	2.3559	3445	3332	3220	3109	2998	2889	2781	2673	2566					
24	2.2460	2355	2251	2148	2045	1943	1842	1742	1642	1543	17	34	51	68	85
25	2.1445	1348	1251	1155	1060	0965	0872	0778	0686	0594	16	31	47	63	78
26	2.0503	0413	0323	0233	0145	0057	9970	9883	9797	9711	15	29	44	58	73
27	1.9626	9542	9458	9375	9292	9210	9128	9047	8967	8887	14	27	41	55	68
28	1.8807	8728	8650	8572	8495	8418	8341	8265	8190	8115	13	26	38	51	64
29	1.8040	7966	7893	7820	7747	7675	7603	7532	7461	7391	12	24	36	48	60
30°	1.7321	7251	7182	7113	7045	6977	6909	6842	6775	6709	11	23	34	45	56
31	1.6643	6577	6512	6447	6383	6319	6255	6191	6128	6066	11	21	32	43	53
32	1.6003	5941	5880	5818	5757	5697	5637	5577	5517	5458	10	20	30	40	50
33	1.5399	5340	5282	5224	5166	5108	5051	4994	4938	4882	10	19	29	38	48
34	1.4826	4770	4715	4659	4605	4550	4496	4442	4388	4335	9	18	27	36	45
35	1.4281	4229	4176	4124	4071	4019	3968	3916	3865	3814	9	17	26	34	43
36	1.3764	3713	3663	3613	3564	3514	3465	3416	3367	3319	8	16	25	33	41
37	1.3270	3222	3175	3127	3079	3032	2985	2938	2892	2846	8	16	24	31	39
38	1.2799	2753	2708	2662	2617	2572	2527	2482	2437	2393	8	15	23	30	38
39	1.2349	2305	2261	2218	2174	2131	2088	2045	2002	1960	7	14	22	29	36
40°	1.1918	1875	1833	1792	1750	1708	1667	1626	1585	1544	7	14	21	28	34
41	1.1504	1463	1423	1383	1343	1303	1263	1224	1184	1145	7	13	20	27	33
42	1.1106	1067	1028	0990	0951	0913	0875	0837	0799	0761	6	13	19	25	32
43	1.0724	0686	0649	0612	0575	0538	0501	0464	0428	0392	6	12	18	25	31
44	1.0355	0319	0283	0247	0212	0176	0141	0105	0070	0035	6	12	18	24	30

The same argument applies to sail efficiency; higher L/D or F_R/F_H ratio and hence smaller ε_A over *the operational range of incidence angles of a rig* give better windward performance.

In order to appreciate the potential influence of the sail characteristics on a yacht's behaviour let us consider the geometrical relationship presented in Fig 1.2C. When sailing a boat it is possible, by gradually luffing into the wind and reducing the β angle, to reach a condition where the yacht loses forward motion and only drifts to leeward; this will occur when the total force F_T acts perpendicular to the course sailed, i.e. when the driving force component F_R disappears. This therefore represents a limiting value for β when beating to windward. From the foregoing argument one can infer that the angle β, between the course sailed and the apparent wind direction, depends to a large extent on the L/D ratio. It can be anticipated that for a given course sailed β, and constant total aerodynamic force F_T, the driving component F_R increases when the L/D ratio increases. In general, a small value for ε_A or big value for the L/D ratio is desirable, as it is an obvious factor in improving a boat's ability for close-hauled work. Nevertheless, as we will see later, the L/D maximum has a limited application as a criterion of sail efficiency in the case of conventional water-borne craft. It is, however a factor of primary importance in the case of fast sailing craft such as ice and land yachts.

From Eq 1.1 and Fig 1.2C one may further infer that:

1. the closest possible angle β to the apparent wind is the sum of the minimum values of the two drag angles ε_A and ε_H,
2. the effective driving force F_R is equal to the total aerodynamic force $F_T \times$ sine of the hydrodynamic drag angle ε_H.

This second conclusion may seem a little strange at first, as one naturally reflects that the aerodynamic effectiveness of the rig is directly related to the drag angle ε_H at which the underwater part of the hull operates.

The first proposition seems to be rather attractive, but may easily be misleading. In fact, people have tried to develop the concept of a 'most efficient' yacht which took origin from Eq 1.1. An example given below will illustrate the risk and disappointment involved in jumping too far ahead in interpreting this equation.

The Ten Degree Yacht

It has been argued (see Publications 56 and 61 of the *Amateur Yacht Research Society*) that

'by carrying the design of both sails and hulls to the utmost extreme the drag angles of each can be reduced to 3°. This would produce a yacht which would sail at 6° from the apparent wind. From a research point of view, the concentration of effort on a series of hulls in the test tank to produce all degrees of hull drag angle, from 10° downwards, would be well worth while. At the same time, workers with wind tunnels could be trying out various sails to see how low they can reduce the sail and windage drag angle.'

'When all this work has been completed,' continues the author of *The Most Efficient Yacht*,

'and the sailing characteristics of the resultant yachts found, I think that the very close-winded yacht will be slower to windward than those which sail freer. My guess as to the *best yacht performance to windward is that it will occur when both hull and sail drag angles are 5°, thus making a "Ten Degree Yacht".*'

Is such a concept sound?

The basic theorem involved $\beta = \varepsilon_A + \varepsilon_H$ and the geometrical relations concerning the forces acting on the 'efficient yacht' are depicted in Fig 1.3. For the sake of clarity in representing the very small angles ε_A and ε_H, while reducing the size of the

Fig 1.3 Hypothetical concept of the Ten Degree Yacht.

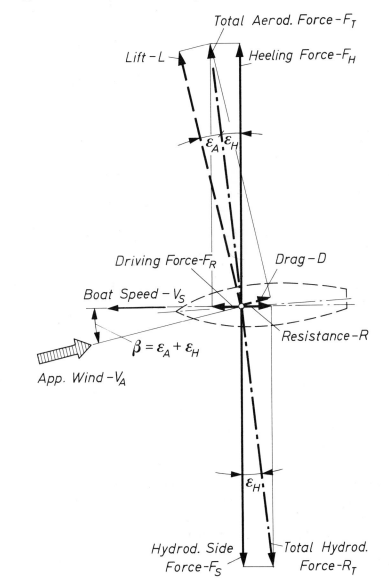

graph, the yacht shown from a bird's eye view sails actually at $\beta = 13°$ instead of 10°. Such a departure from the concept of 'Ten Degree Yacht' does not, of course, invalidate arguments against the whole idea.

In very light winds, ignoring restriction due to stability, one may assume that there would be virtually no limit to the height and size of the sails and therefore it would be quite possible to design a high aspect ratio rig which could secure a desirable maximum L/D ratio of the order of 11.4, i.e. drag angle $\varepsilon_{A\,min} = 5°$ (Table 1.4).

In a range of relatively low boat speeds, while hull motion is dominated by frictional resistance (which varies approximately as the square of boat speed), the high aspect ratio fin might ensure a desirable minimum drag angle $\varepsilon_{H\,min}$ of the order of 5.0°. However, this small hydrodynamic angle of drag ε_H is bound to increase steadily with boat speed for, with increasing velocity, wave drag can no longer be ignored. This increase in ε_H is inevitable because the total resistance R, including wave drag, develops at a rate greater than the square of the boat speed, while at the same time, the hydrodynamic side force F_S varies in relation to the square of the boat speed.

To substantiate the last argument, let us look at Fig 1.4 which illustrates the hull forces by means of a series of polar curves, showing how side force F_S and resistance R vary with leeway λ for a given boat speed V_S. This set of curves for various speeds ranging from 2.0 to 6.0 knots obtained from Ref 1.5 represents the hydrodynamic characteristics of the full scale International 10 sq m Canoe hull. The co-ordinates (or axes, vertical and horizontal) of the polar diagram are side force F_S and resistance R, while the leeway angles λ are inscribed along the curves. The ratio of side force F_S to resistance R is represented by the slope of the line drawn from the origin 0, to any point on the selected curve. Evidently, the maximum value of the ratio F_S/R equivalent to the minimum value of ε_H occurs when this line becomes tangential to the selected curve for given speed V_S. The graph of F_S versus R is called a polar curve because a vector from the origin 0 to any point on the curves represents magnitude and direction of the resultant force, provided that the co-ordinates are plotted to the same scale.

The two vectors plotted as thick broken lines illustrate the resultant hydrodynamic forces developed at the two different boat speeds $V_S = 2.0$ and 4.0 knots, but at the same leeway angle $\lambda = 5°$. One may find that when V_S increases twofold the side force increases fourfold but the associated resistance is six times greater; the relevant drag angles ε_H are therefore different.

By drawing a tangent line to each curve, for a given V_S, one may estimate the minimum drag angle ε_H. As already mentioned, ε_A increases gradually with boat speed due to the increasing contribution of the wave-making resistance to the total resistance of the hull. The wave resistance increases proportionally to the higher powers of V_S than the friction resistance, extending to powers 3, 4, and even more, depending largely on the displacement/length ratio $\Delta/(L/100)^3$ of the boat and her speed/length ratio $V_S/\sqrt{L}$. Table 1.5 demonstrates this trend towards higher ε_H as boat speed gradually increases.

Fig 1.4   Set of curves: resistance R versus side force F_S at various boat
speeds V_S ranging from 2–6 knots (International 10 sq m Canoe).

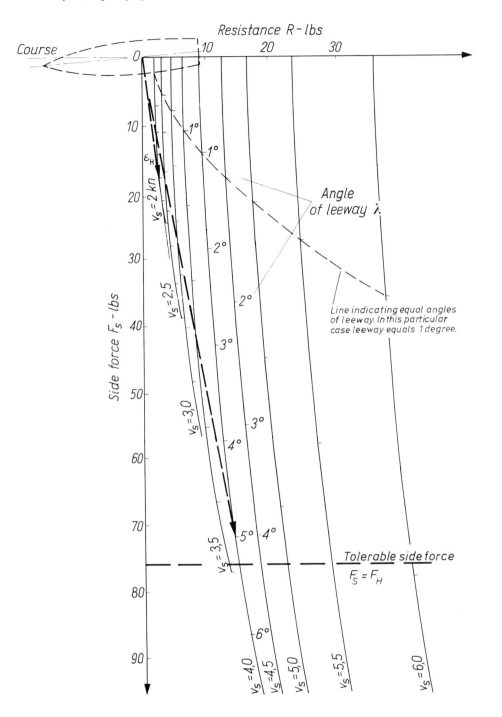

Table 1.5
(International Canoe hull)

V_S in knots	$\varepsilon_{H\,min}$	F_S/R
2.0	ab. 8.0°	ab. 7.0
3.5	9.3°	6.1
4.0	10.8°	5.3
4.5	12.0°	4.6
5.0	14.0°	4.0
5.5	17.5°	3.2

The International Canoe, the upright resistance characteristics of which are shown in Fig 1.5, is a light-displacement type of boat of an exceptionally low displacement/length ratio $\Delta/(L/100)^3$ of about 41.0. She is equipped with an efficient streamlined-section centreboard of high aspect ratio, hence the observed rate of resistance growth with leeway is rather slow if compared with that of a heavy, long-keeled displacement boat.

The variations of ε_A (or L/D) and ε_H (or F_S/R) for a range of constant values of course sailed β, are illustrated in Fig 1.6. The general conclusion one may draw from it is of some immediate practical importance as far as the 'Ten Degree Yacht' concept is concerned. Thus it is evident that if the ε_H angle has increased, as it must if the boat velocity has increased, then the ε_A angle must inevitably be reduced if a given course sailed β is to be maintained. However, if as already stated in the 'Ten Degree Yacht' concept, both hull and sail minimum drag angles, 5° each, are to be maintained, it practically means that the boat cannot be sailed at $\beta = 10°$ in a wind exceeding a certain strictly limited velocity, in fact a very low one.

This lack of freedom in adjusting arbitrarily ε_A and ε_H angles to suit the ever-changing wind conditions can be apprehended from Fig 1.6 by considering the meaning of a circular mark on the curve $\beta = 10°$. From this it follows that since the hydrodynamic drag angle ε_H cannot be maintained at its minimum value of 5° while boat speed increases, and $\varepsilon_A = 5°$ is the minimum available aerodynamic drag angle, the β angle must automatically increase above 10°. The Ten Degree Yacht concept, as presented, is therefore a fiction unable to operate in reality, unless by some means the hull can be lifted out of the water–air interface and no wave drag penalty is paid for it, as in the case of ice and land yachts and also perhaps some futuristic vehicles driven by sails.

With realistic hull and sail data we may answer the question: what is the range of possible variations of drag angles ε_A and ε_H in true sailing conditions? Referring to the hull data of the International Canoe, as represented in Fig 1.4 and Table 1.5, we may find that minimum ε_H at a low boat speed of $V_S = 2.0$ knots is in the order of 8°, corresponding to $F_S/R = 7.0$; this is marked on the horizontal axis of Fig 1.6. Trying to drive this hull using a Finn-type sail, we find that the minimum available ε_A of such a rig (when the rig is trimmed to the minimum ε_A) is about 10.0°, corresponding to a

Fig 1.5   Resistance curve of International 10 sq m Canoe with no leeway
and no heel.

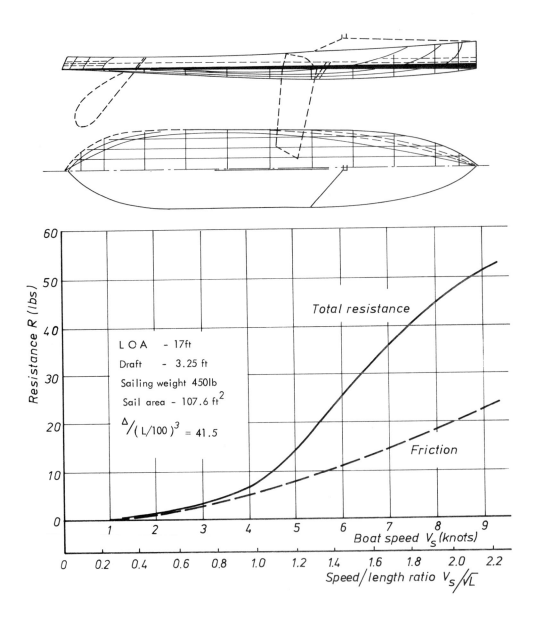

LOA   – 17ft

Draft   – 3.25 ft

Sailing weight 450lb

Sail area – 107.6 ft^2

$\Delta/(L/100)^3 = 41.5$

Fig 1.6   Variation of aerodynamic and hydrodynamic drag angles (ε_A and ε_H) for β angles ranging from 10° to 32°.

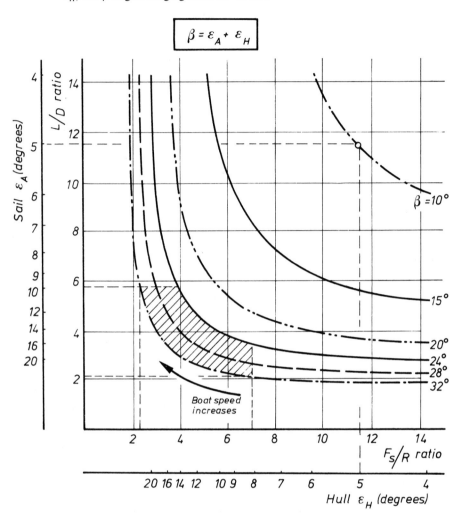

L/D ratio below 6.0; this limit is marked in Fig 1.6 on the vertical axis.

The hatched zone between the hull- and sail-limiting drag angles ε_H and ε_A in Fig 1.6 illustrates the likely conditions for sailing to windward at β angles ranging from 24° to 32°. Taking $\beta = 28°$ constant as a fairly representative course to windward, one may find that at low boat speeds, when ε_H is about 8°, the relevant sail drag angle ε_A at which the rig operates must necessarily be possibly as large as 20°, corresponding to a small L/D ratio of the order of 2.5 only. Vice versa, when boat speed increases and ε_H becomes larger and larger, then the associated aerodynamic drag angle ε_A must progressively be decreased, i.e. the sail must be trimmed in such a way that the L/D ratio becomes progressively higher. This is a matter of elementary geometry.

Fig 1.7   Variation of ε_A and ε_H at different boat velocities and constant β angle.

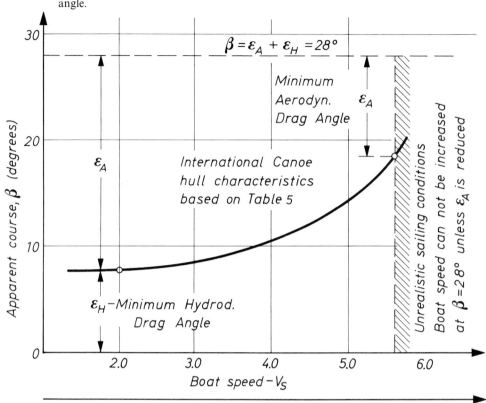

Figure 1.7 based on Table 1.5 represents graphically in a simplified manner (assuming that the centreboard is working at its most efficient leeway angle λ) the variation of ε_A and ε_H angles when the boat speed V_S increases but the course sailed β relative to the apparent wind remains constant at 28°. It is evident that the β angle can only remain unchanged provided that sail is trimmed in such a way that the aerodynamic angle ε_A is supplemental to the hydrodynamic angle ε_H to give a total of 28°. However, as boat speed increases, sooner or later the attainable ε_A minimum is reached, beyond which the boat cannot possibly be sailed faster at this particular course $\beta = 28°$. The hatched zone of Fig 1.7 indicates these unrealistic conditions imposed by the available ε_A minimum that cannot suit the rapidly rising value of ε_H.

Another conclusion from Fig 1.7 is that the realistic, achievable β angle is appreciably greater than the sum of the minimum values of ε_A and ε_H. It disposes of the argument concerning the advantage of separately making ε_A and ε_H minima, expecting that in real sailing conditions both minima can at the same time be used. And since ε_A and ε_H cannot in practice have minimum values simultaneously then it is reasonable to ask whether the optimum windward performance is related to the minimum value that β can attain in any given wind velocity.

Fig 1.8   Definition of V_{mg} (Speed made good to windward).

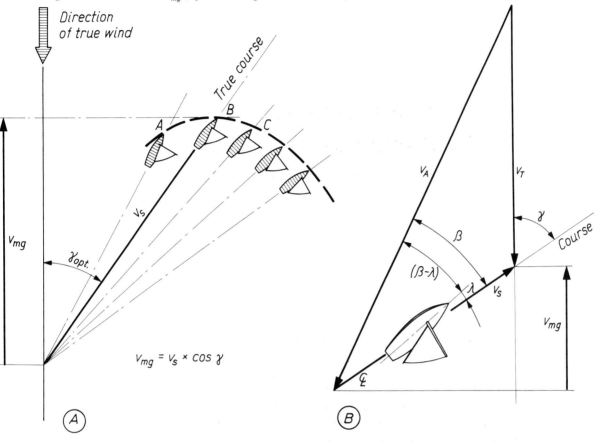

Speed made good to windward

In order to answer this question we must define more precisely what is meant by optimum windward performance. At the present time it is commonly agreed that the performance of a sailing yacht in close-hauled conditions is ultimately measured by the distance it has travelled directly to windward in a given time. This is usually referred to as the 'speed made good to windward' V_{mg} which should be a maximum at each true wind velocity V_T.

Figure 1.8 illustrates in a pictorial way the definition of V_{mg} and also the geometry of velocity vectors in close-hauled conditions. Noteworthy is the fact that speed made good to windward is the composite product of boat speed V_S, and the true sailing course γ

$$V_{mg} = V_S \times \cos \gamma \qquad \text{Eq 1.2}$$

One may deduce from Fig 1.8 that *merely pointing high is not a merit in itself.* Although boat A sails closer to the wind than boat B, her V_{mg} is lower than boat B's.

B Sail/hull interaction in light and strong winds

Graphical analysis of the interaction between the International Canoe hull shown in Fig 1.5 and a Finn-type sail (photo 1.5B)–let us call this case the Canoe-Finn Dinghy–should elucidate further the changing interrelation and feedback between the sail and hull in two different winds–light and strong; this is shown in Fig 1.9 which refers to the close-hauled condition. The whole drawing is divided into four parts: the two sketches on the top right side of Fig 1.9 depict the polar diagrams of sail coefficients C_L and C_D, and there are also two sets of the C_L and C_D coefficients actually employed in the two different sailing conditions, labelled *Light wind* and *Strong wind*. Readers who have followed the author's work on sailing theory, such as, for example, presented in Ref 1.5, will be familiar with the diagrams in Fig 1.9. In a way, these sail polar diagrams are similar to the hull polar curves already shown in Fig 1.4. The co-ordinates of the sail polar diagram are the lift and drag coefficients C_L and C_D, while the angles of sail incidence α are inscribed along the curve. A vector drawn from the origin O to any point on the polar curve represents the magnitude and direction of the resultant coefficient, and an arrow at the end of the vector indicates also the angle of sail incidence α at which this particular force coefficient is obtained. The maximum L/D ratio, equivalent to ε_A minimum, is given by the tangent line drawn from origin O to the polar curve. The origin of the polar diagrams O coincides, in these two sketches, with the Centre of Effort (CE) of the sail; and the silhouette of a hull, seen from a bird's eye view, indicates the boat attitude, i.e. course sailed β relative to the apparent wind V_A.

The polar curves plotted are based on the results of a wind tunnel test on a Finn sail which are described and analysed in Part 3, section D1. The sail coefficients C_L

Fig 1.9   Sail and hull characteristics of the Canoe-Finn dinghy.

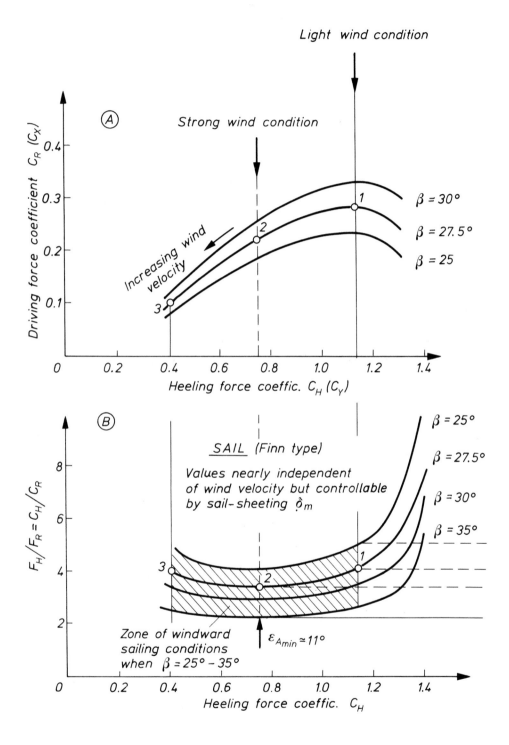

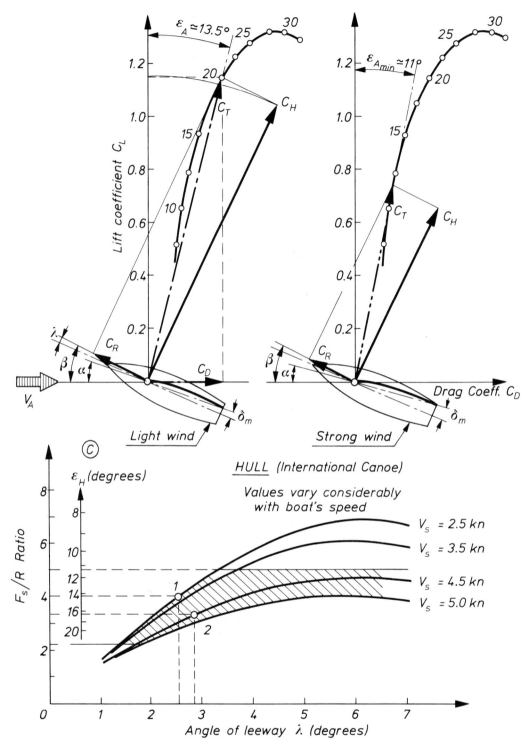

Fig 1.9D   Variation of optimum course sailed β.

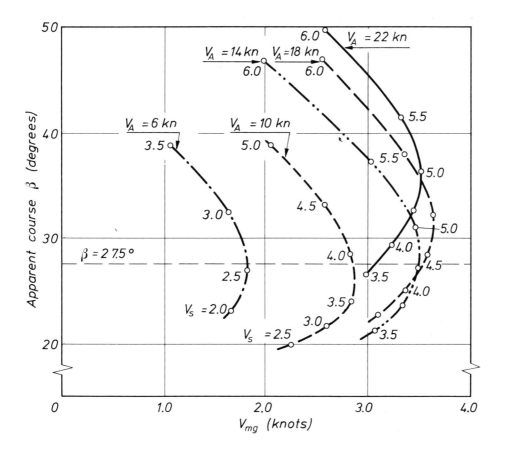

Fig 1.9E   Variation of optimum sheeting angle δ_m.

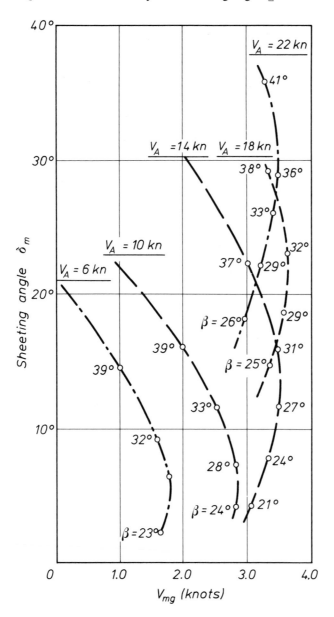

Fig 1.9F   Variation of optimum true course γ.

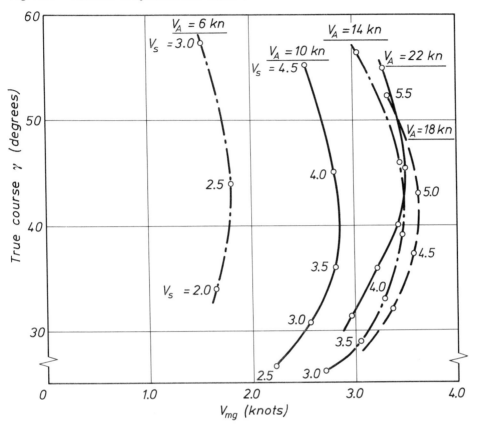

and C_D, given in the form of polar curves, were subsequently used to calculate the relevant values of C_R, C_H and also the C_H/C_R ratio for three or four different courses sailed ($\beta = 25.0°$, $27.5°$, $30.0°$ and $35.0°$); they are plotted in parts A and B of Fig 1.9. Supplementary Fig 1.10 demonstrates the correlation between the C_L and C_D coefficients and the driving and heeling force coefficients C_R and C_H respectively. The relevant equations relating lift L and drag D to driving force F_R and heeling force F_H are also incorporated with Fig 1.10. Attention is invited to the definitions of the symbols F_R, C_R, F_H and C_H; they represent horizontal force components or their coefficients, measured along and perpendicular to the course sailed β respectively. Since the leeway angle λ may not be known beforehand, the driving and heeling forces F_R and F_H or their coefficients C_R and C_H cannot be calculated. Instead, one may quote force components measured parallel and perpendicular to the boat heading, i.e. the centre line of the hull (see Fig 3.14, Part 3); it is common practice to call those force components F_X and F_Y. They can be calculated in a similar manner to F_R and F_H by putting heading angle β-λ instead of apparent course β. The heading angle β-λ is the angle between the centre line of the boat and the direction of the apparent wind V_A. If the leeway angle is small, as in fact it is, the differences between

Fig 1.10   Definition of aerodynamic forces and angles in the close-hauled condition. For the sake of clarity the leeway angle λ is greatly exaggerated.

$$C_R = C_L \sin \beta - C_D \cos \beta \qquad C_H = C_L \cos \beta + C_D \sin \beta$$

$$F_R = L \sin \beta - D \cos \beta \qquad F_H = L \cos \beta + D \sin \beta$$

or   or

$$C_R = C_T \sin \varepsilon_H \qquad\qquad C_H = C_T \cos \varepsilon_H$$

$$F_R = F_T \sin \varepsilon_H \qquad\qquad F_H = F_T \cos \varepsilon_H$$

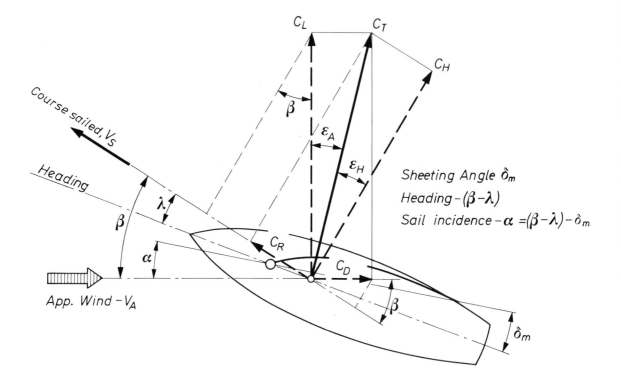

the F_X, F_Y and F_R, F_H are small enough to ignore if the object of the exercise is only a *qualitative comparison*.

In what follows, no differentiation has been made between the F_X, F_Y system and F_R, F_H system, and more familiar terms such as driving force F_R and heeling force F_H are used. Referring again to Fig 1.9, Part A presents the sail characteristics of a Finn-type rig expressed in terms of driving force coefficient: C_R plotted against the heeling force coefficient C_H, for three different β angles of 25.0°, 27.5° and 30.0°. Part B gives the variation of Heeling/Driving force ratios (F_H/F_R) plotted against the Heeling Force Coefficient C_H as in Part A. Part C gives the hull characteristics, expressed in terms of Side Force/Resistance F_S/R ratios plotted against leeway angle λ for four selected boat speeds, $V_S = 2.5$, 3.5, 4.5 and 5.0 knots.

Seeking some qualitative information as to how the sail and hull-centreboard combination interact, we shall neglect for the time being the influence of other factors of secondary importance on yacht behaviour, assuming that:

a. A light helmsman (of all-up weight about 180 lb) by sitting on the weather rail and leaning back can balance the heeling force $F_H = 76$ lb, which is the maximum heeling force that may be tolerated (Figs 1.4 and 1.22). Otherwise, as the boat heels beyond some critical heel angle (which may or may not be upright) both drag angles ε_A and ε_H increase: at first slowly, then more rapidly. The influence of heel, complicating unnecessarily our preliminary investigations, will be ignored and it is assumed that the heel angle, of the order of 10°, is kept independent of wind strength.

b. The sail characteristics, as described by the polar diagram (right side top of Fig 1.9) are also independent of apparent wind velocity, i.e. the mast-sail combination is rigid and the sail camber, as well as its distribution and twist, are constant.

The helmsman can only change the sheeting angle δ_m or corresponding angle of incidence α relative to the apparent wind, but not the shape of sail or its area.

On the above assumptions, supplementary Figs 1.9D, E and F were prepared employing the graphical method of performance prediction described in *Sailing Theory and Practice* (Ref 1.5). They show how apparent course β, sheeting angle δ_m, true course γ, boat speed V_S, apparent wind V_A, and speed made good V_{mg} are mutually interrelated in different sailing conditions.

Thus, Fig 1.9D illustrates how variation of β affects V_{mg} at different apparent wind velocities V_A. Performance predictions are given for $V_A = 6$ knots (10 ft/sec), 10 knots (17 ft/sec), 14 knots (24 ft/sec), 18 knots (30 ft/sec) and 22 knots (38 ft/sec). Along the curves plotted are inscribed the boat speeds V_S. Note how quickly one can shift from the best performance if the β angle is a few degrees off optimum.

Sheeting angle

Figure 1.9E shows how the optimum sheeting angle δ_m varies for various ranges of V_A. Each curve for a particular V_A has on it a point where V_{mg} is a maximum; values of β are inscribed along the curves.

Finally, Fig 1.9F demonstrates the interrelation between the true course γ and V_{mg}, and the boat speed values V_S are inscribed along the curves.

Since the sail area ($S_A = 108$ sq ft in the example discussed) will not be reduced by reefing when the wind increases, the tolerable heeling force, F_H, can only be kept within the limit assumed (up to 76 lb) by adjusting the angle of sail incidence α, i.e. the sheeting angle δ_m. In other words, by changing the heeling force coefficient C_H, the heeling moment is adjusted to match the available righting moment. The heeling force given by Eq 1.3 is:

$$F_H = C_H \times q \times S_A \qquad \text{Eq 1.3}$$
$$= C_H \times 0.00119 \times V_A^2 \times S_A$$

where q is the dynamic pressure due to apparent wind action (see Table 2.3), and can be expressed in lb/ft^2 by (see Table 2.1):

$$q = \frac{\rho_A \times V_A^2}{2} = 0.00119\, V_A^2$$

Hence the tolerable heeling force coefficient:

$$C_H = \frac{F_H}{0.00119 \times S_A \times V_A^2} \qquad \text{Eq 1.4}$$

Putting $F_H = 76$ lb and $S_A = 108$ sq ft into Eq 1.4 we obtain:

$$C_H = \frac{76}{0.00119 \times 108 \times V_A^2} = \frac{592}{V_A^2} \qquad \text{Eq 1.4A}$$

Table 1.6 gives calculated values of the tolerable heeling force coefficient C_H, and relevant approximate incidence α and sheeting angles δ_m, for various values of V_A, ranging from 10 to 38 ft/sec (6.0 to 22.5 knots) corresponding to true wind variation from about force 2 to 5 on the Beaufort scale. Coefficients C_H are tentatively calculated regardless of whether the rig in question may or may not produce them:

Table 1.6

V_A (ft/sec)	V_A (knots)	V_A^2	q	Tolerable C_H	$\alpha°$	$\delta_m°$
10	5.9	100	0.12	5.92	ab. 20	ab. 4
14	8.3	196	0.23	3.02	20	4
17	10.1	289	0.34	2.05	20	4
20	11.8	400	0.48	1.48	20	4
22	13.0	484	0.58	1.21	20	4
23	13.6	529	0.63	1.13	20	4
24	14.2	576	0.69	1.03	20	4
28	16.6	782	0.93	0.76	12	12
30	17.8	900	1.07	0.66	11	13
34	20.1	1156	1.37	0.51	8	16
38	22.5	1444	1.72	0.41	5	19

In Table 1.6 are also given the values of wind dynamic pressure q (lb/sq ft) from which one may find that, in the range of apparent wind speeds indicated, the sail experiences minimum and maximum dynamic pressures differing by a factor of about 14, from 0.12 to 1.7 lb/sq ft of the sail area. One may be puzzled to learn that those dynamic pressures that affect the sail, are a minute fraction of the atmospheric pressure, which is of the order of 2116 lb/sq ft (standard atmosphere). As shown in Table 1.6, the heeling force coefficients C_H are inversely proportional to the dynamic pressure q, i.e. their product is constant as indicated by Eq 1.4A.

What is needed in light wind is the sheeting angle for which the largest driving force coefficient C_R is generated. This is usually associated with large values of C_L and C_H, as depicted in the sketch labelled 'Light wind' in Fig 1.9. It demonstrates, in the form of vectors, respective sail coefficients C_L, C_D and C_T, as well as C_H and C_R. It will be seen in the sketch that the C_H coefficient, almost equal to C_L, is about 1.13. This value is marked on the horizontal axis in Fig 1.9A. The vertical line going through $C_H = 1.13$ and a circlet marked 1 on the C_R versus C_H curve labelled $\beta = 27.5°$ helps to find the corresponding driving force coefficient $C_R = 0.28$ on the vertical axis. It is at the maximum which can be produced by this particular sail at the predetermined β angle. This maximum C_R coefficient is achieved by sheeting the sail well in and maintaining the incidence angle α at about 20°. This corresponds to a sheeting angle δ_m of about 5°, and this value can be found in Fig 1.9E on the curve marked $V_A = 6$ knots relevant to light winds. At any other β angle the maximum C_R will of course be different, requiring a different incidence angle α.

Heeling forces

The heeling/driving force ratio F_H/F_R (equal to C_H/C_R ratio), corresponding to $C_{R\,max}$ at $\beta = 27.5°$, is to be found in Fig 1.9B by dropping a vertical line from Fig 1.9A which intersects at point 1 the relevant curve marked $\beta = 27.5°$. This F_H/F_R ratio is about 4.0, i.e. F_H is about 4 times greater than F_R.

Referring to Table 1.6, it will be seen that when V_A is about 10 ft/sec (light weather conditions) the heeling force coefficient C_H which might be tolerated, bearing in mind the available stability, is pretty high, almost 6.0. However, such a high C_H coefficient associated with high C_R coefficient cannot possibly be generated by this particular Finn-type sail, nor by any other conceivable practical soft sail either. This is beyond the potential capability of any realistic sail, even of a wing equipped with high-lift devices, such as that shown for example in Photo 1.1, illustrating a C-class cat with a revolutionary rigid sail-wing with flaps. In fact, $C_{L\,max}$ recorded in the course of full scale as well as wind tunnel tests on soft, conventional sails, is below 2.0.

Since, as shown in Fig 1.9A, the biggest heeling force coefficient generated by the sail ($C_H = 1.13$) is well below the theoretically tolerable one which is about 6.0, one may say that, in light winds, our Canoe-Finn boat is undercanvassed, i.e. she is not making use of her full stability. In winds where $V_A = 10$ ft/sec she might carry about 5 times that sail area and still be sailed upright.

Photo 1.1 Chris Wilson's *Miss Nylex*. She is a cat with revolutionary, rigid
sail-wing. The much talked about 'zap flaps' are visible. The
total weight of the rig was less than 150 lb; built of balsa-wood
frames and covered with 2 oz Terylene sailcloth. Photograph
reproduced with kind persmission of *Yachting World*.

Hull responses

Let us digress, for a while, from sail to hull responses. The horizontal line intersecting the curve marked $\beta = 27.5$ in point 1, Fig 1.9B, and extended through the set of curves representing hull properties in Fig 1.9C, facilitates an estimation of hull response to given sail input. It can be seen in supplementary Fig 1.9D that, when V_A is about 6 knots (10 ft/sec), V_S will be of the order of 2.5 knots. At this speed, the leeway angle λ marked in Fig 1.9C by circlet 1 is about 2.5°, and the corresponding drag angle ε_H is about 14°. When this value of ε_H is added to the sail drag angle ε_A, which is about 13.5°, as indicated in the sketch labelled 'light wind', the resulting $\beta = 27.5°$, as it should be.

Returning to Table 1.6 one may infer that the boat in question remains undercanvassed up to the critical apparent wind velocity V_A about 23 ft/sec (13.6 knots) at which the tolerable C_H coefficient reaches the value of 1.13. This is actually the value that a given sail is capable of producing. Therefore one may say that, from this particular wind speed onwards, the boat carries just the right sail plan for her stability, but to a certain point only.

When V_A increases beyond 23 ft/sec the incidence angle α which is about 20° (corresponding to sheeting angle $\delta_m = 4°$) and held constant up to this critical velocity, should gradually be decreased in order to lower the heeling coefficient C_H, if a tolerable heeling force $F_H = 76$ lb is to be maintained.

The sketch labelled 'strong wind' in Fig 1.9 illustrates force coefficients C_H, C_R and C_T, when the sail operates at $\varepsilon_{A\ min} = 11°$, corresponding to maximum available L/D ratio of this particular rig. This would suit the demands imposed by V_A of about 28 ft/sec. In such circumstances, $C_H = 0.72$ and $C_R = 0.22$. Both coefficients are smaller than the previous ones, employed in light winds, but give the lowest possible F_H/F_R ratio of about 3.3 for the selected $\beta = 27.5°$. It means that, in a certain range of incidence angles α close to the angle at which L/D max occurs, the price paid for the driving force F_R in terms of harmful heeling force F_H is the lowest one. The heeling force is about 3.3 times greater than the driving force, as shown in Fig 1.9A and B by the circlet 2.

In attempting to adjust a rig for heavy winds a practical deduction would be that the sail shape should be modified in such a way that the highest possible L/D ratio in the range of applicable sheeting angles is achieved. We will see in Part 3, when discussing Finn test results, that a gradual flattening of the sail up to the utmost drum-like membrane, together with a reduction of angle α, is more appropriate action than spilling the wind and flogging over-full canvas. The latter sailing routine is fighting for survival rather than efficient racing.

The hull response in higher winds is indicated in Fig 1.9C by the circlet 2. As expected, the hydrodynamic drag angle $\varepsilon_H = 16.5°$ is higher than that in the case of light winds. The boat reaches V_S just above 4.5 knots, which can be interpolated from Fig 1.9D, curves marked $V_A = 14$ knots (24 ft/sec) and $V_A = 18$ knots (30.0 ft/sec).

If the wind continues to increase and, for example, V_A reaches about 22 knots

(38 ft/sec), the heeling force coefficient must be reduced by further increasing the sheeting angle δ_m which may cause partial flogging of the sail. The tolerable C_H value, given in Table 1.6, is now only 0.41. If the head of the sail is allowed to flog, there will not only be a drastic reduction in driving force, but also a rapid, undesirable increase in F_H/F_R ratio, shown in Fig 1.9B by the circlet 3. In such conditions the boat will slow down. The actual reduction in speed from about 4.5 to 3.7 knots can be estimated from Fig 1.9D by comparing curves marked V_A = 18 knots (30 ft/sec) and V_A = 22 knots (38 ft/sec). It will be seen that due to insufficient available stability, only a fraction of the potential driving power of the sail can be used. One may say that in such circumstances the boat becomes overpowered, or overcanvassed. Fast sailing is therefore largely a matter of stability.

Some conclusions just derived from Fig 1.9 and based on analysis of sail–hull interaction at $\beta = 27.5°$, are in a qualitative sense applicable to other courses of β under close-hauled conditions. Some other, as yet unrecorded, conclusions concerning yacht behaviour are left to those crossword-minded hardy spirits who might feel happy to unravel them.

In the case of a dinghy, the tolerable heeling force F_H is largely determined by the sheer sailing weight of the crew, including the additional ballast carried in the form of water stored in wet clothing, or even lead, and crew ability of sitting out more or less uncomfortably, for a sufficiently long period of time, demanded on the windward leg. An obvious relationship between the aerodynamic characteristics of the sail, the available righting moment, and resulting boat's performance, explains why the technique of wearing soaking wet, almost grotesque, sweatshirts by some gladiator-type competitors has become routine in all international classes. It has been established by practical observation that a Finn helmsman should weigh 230–250 lb to be competitive in winds over 35 ft/sec (20 knots). In one of the 1969 Finn Gold Cup races the leading helmsman wore 100 lb of wet clothing! Since then and in spite of the fact that 'wet sweaters', or heavy garments, as a stability aid is still a controversial issue amongst the IYRU rule-makers and competitors, an additional paragraph has been introduced to the IYRU rule 22, Shifting Ballast, which now reads:

'22.3 CLOTHING AND EQUIPMENT
- (a) A competitor shall not wear or carry any clothing or equipment for the purpose of increasing his weight.
- (b) A class which desires to make an exception to rule 22.3(a) may so prescribe in its class rules. However, unless a lesser weight is prescribed in the class rules, the total weight of clothing and equipment worn or carried by the competitor shall not be capable of exceeding 15 kg when saturated with water.

  For the purposes of this rule, water pockets or compartments in the clothing and equipment of a competitor shall be permitted unless otherwise prescribed in the class rules. The weight of water in pockets or compartments shall be included in the total weight.'

C Centreboard or fin keel efficiency

The hull action is somewhat less simple to analyse than that of the sail.

For although it may be regarded as a hydrofoil, in so far as it develops a hydrodynamic force to balance the sail force, the hull and its appendages are burdened with some other onerous duties, which are not easy to recognize at first glance even if unsteady sailing conditions are ignored.

For example, from Fig 1.9C one may infer that, in the working range of leeway angles $\lambda = 2 - 3°$, the centreboard-hull combination is operating well below the attainable maximum of F_S/R ratio, which is about 7 when $V_S = 2.5$ knots. The boat is actually sailing at an F_S/R of about 4, and such a response is not a feature peculiar to the International Canoe, but is typical of most sailing craft, including keel-boats, except perhaps those modern craft with much reduced appendage wetted surface (those with shark-fin keels, for instance).

Figures 1.11 and 1.12 illustrate hydrodynamic forces on a 6-Metre hull at $V_S = 5.9$ knots (Ref 1.6). Under the action of a heeling force $F_H = 465$ lb the hull heels to 20° and the total resistance $R = 141$ lb, giving $F_S/R = 3.3$. The implication of Fig 1.11 is that the side force F_S generated by the hull at $\lambda = 3°$, is much too small to develop the maximum hypothetical $F_S/R = 4.3$, of which the hull form is inherently capable, as in the case of the International Canoe hull in Fig 1.9C. More than twice the actual side force F_S (or heeling force F_H) would be needed to develop this maximum attainable F_S/R ratio. Since a 6-Metre boat already has nearly three-quarters of its total weight concentrated in lead ballast at the bottom of the keel, there is no chance of doubling lateral stability and hence the side force F_S. Is the low angle of leeway that both the 6-Metre and International Canoe experience evidence that their appendages have excessive lateral area, and therefore that too much

Fig 1.11   Forces on 6-Metre hull at $V_S = 5.9$ knots with 20° heel angle.
Total resistance $R = R_{Upright} + R_{Heel} + R_{Induced}$
$$R = 112 \quad\quad + 4 \quad\quad + 25 = 141\text{ lb}$$

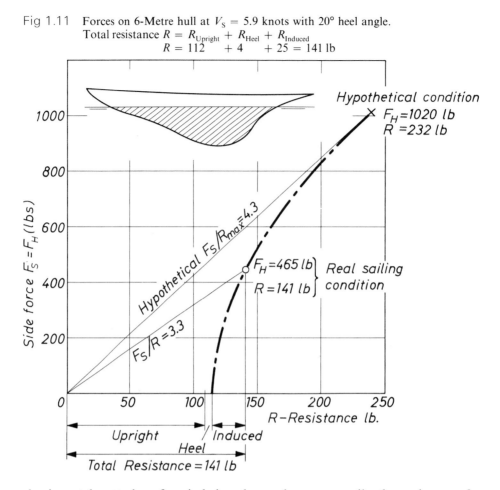

detrimental wetted surface is being dragged unnecessarily through water?

To touch this subject briefly, it looks as if the centreboard, or keel proper in the case of a displacement type of boat, may not be exploited fully as efficient lift-producing devices if the angle of leeway, in normal sailing conditions, is below that at which F_S/R max occurs. Such feelings have been supported by the following sequence of related logical statements, given by Bruce in Ref 1.7:

    a. The leeway angle of hull appendages is identical with the angle of incidence of a symmetrical foil, and therefore the hydrodynamic side force is equivalent to lift (Fig 1.12).
    b. There exists an angle of incidence which produces an optimum F_S/R ratio for a foil or hull/foil combination.
    c. The highest F_S/R ratio corresponds to (L/D) max and to the smallest hydrodynamic drag angle ε_H in Eq 1.1.
    d. Therefore a hull leeway angle exists which will produce the highest pointing of the hull's course, in respect to the apparent wind direction.

Fig 1.12   A 6-Metre yacht. Equilibrium of forces in the close-hauled
            sailing condition, $V_T = 12$ knots.
            LWL–23.5 ft
            Beam–6.5 ft
            Draft–5.4 ft
            Displacement–9400 lb
            Sail Area–600 sq ft
            Lateral Area (hull)–70 sq ft
            Angle of heel–20°

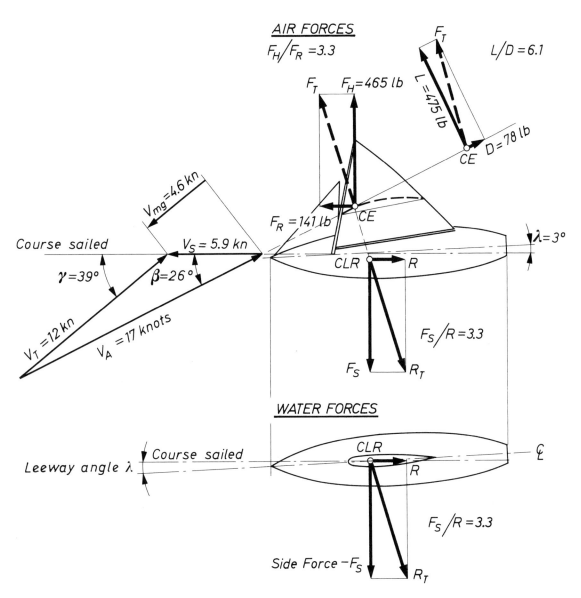

Let us compare the characteristics of several foils of NACA sections, shown in Fig 1.13, of aspect ratio 6 and Reynolds Number of about three millions (3.0×10^6). One may notice in Table 1.7 that all incidence angles for (L/D) max are within the range of 4–6°. For a lower aspect ratio, say 3, this range of incidence angles is about 1° higher, i.e. 5–7°.

Table 1.7

Foil	(L/D) max	Incidence degrees for (L/D) max	C_L at (L/D) max	$C_{L\,max}$	Incidence degrees for $C_{L\,max}$	C_D at (L/D) max
Flat plate	about 10	4–5	—	0.72	20–25	
NACA 0006	24	4	0.30	0.88	16	0.0125
0009	23	5	0.35	1.27	18	0.0152
0012	22	5	0.35	1.53	22	0.0159
0015	21	5	0.35	1.53	22	0.0167
0018	20	6	0.35	1.50	23	0.0175

Based on Ref 1.8.

As we shall see, thickness of a section has relatively small effect on the rate of C_L growth with incidence. Since drag increases slightly with thickness, it affects to a small extent values of (L/D) max. However, this higher drag of the thicker foil is generously offset by its higher $C_{L\,max}$.

Aspect ratio

For reasons which will be explained in the following chapters, the aspect ratio AR (a term which describes the planform proportion of a foil, as looked at from the side) has a profound influence on the magnitude of lift generated at a given angle of incidence. This is shown in Fig. 1.14 which facilitates a quick estimate of side force F_S (lift) generated at speed = 3.0 knots, by a series of foils of any symmetrical NACA section, shown earlier in Fig. 1.13, that have the same lateral area A = 4.0 sq ft but different aspect ratios AR ranging from 1 to 6.

A circlet and an arrow in Fig 1.14 indicate a side force produced by the centreboard of an International Canoe. The geometric aspect ratio of her centreboard shown in Fig 1.5 is about 2.75, as determined in accordance with conventional calculation:

$$AR = \frac{\text{Span}^2}{\text{Lateral Area}} = \frac{3.28^2}{3.92} = 2.75$$

If the root section of the centreboard attached to the hull is sealed so that there is no flow over the root, the so-called effective aspect ratio is about twice the geometric

Fig 1.13 Lift and drag characteristics of NACA 0015 section of AR
= 6.0. Reynolds Number Re = 3.2×10^6 (Ref 1.8). (c.p.–
Centre of Pressure.)

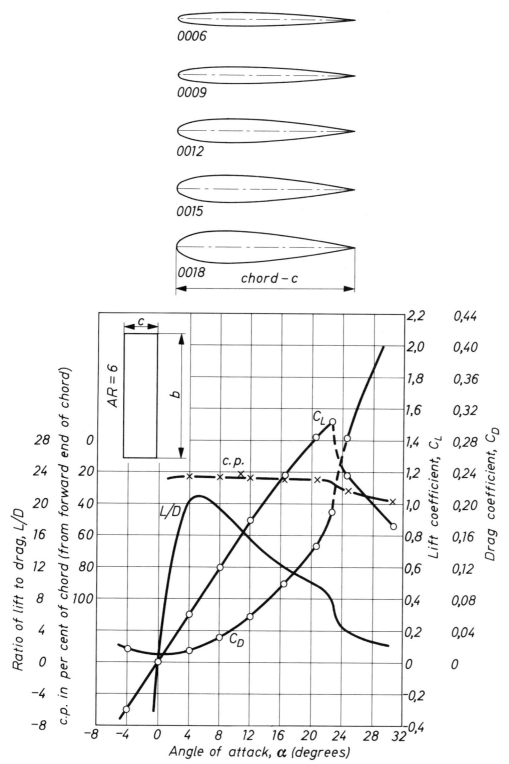

Fig 1.14   Side force F_s (or Lift) curves of NACA 0009 foil of lateral area
$A = 4.0$ sq ft and of various aspect ratios AR.
Speed $V = 3.0$ knots (5.07 ft/s).
For fresh water $F_S = L = 0.97 \times C_L \times A \times V^2$
$= 0.97 \times C_L \times 4.0 \times 5.07^2$
$\simeq 100\, C_L$ (in lb.)

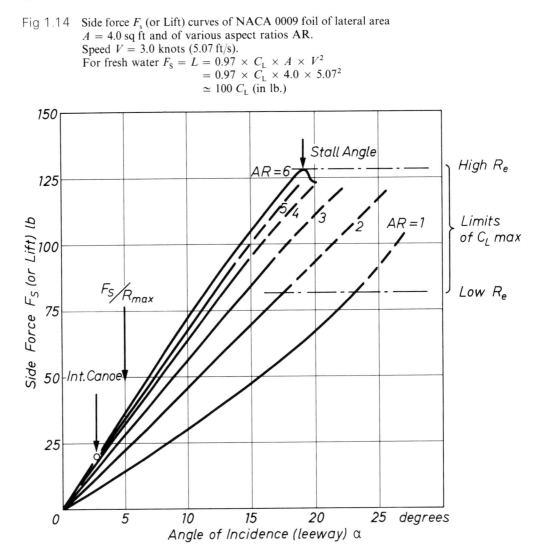

aspect ratio, i.e. in the case of the International Canoe the effective AR would be about 5.5. The actual maximum value of the side force $F_{S\,max}$ depends appreciably upon the Reynolds Number, and this fact shown in Fig 1.14 is of some practical consequence as far as yacht behaviour in unsteady sailing conditions is concerned.

From Fig 1.14 it becomes evident that the International Canoe has a large potential reserve for generating hydrodynamic side forces, but such forces do not seem to be used in normal sailing conditions. Does it mean that the area of her centreboard is unnecessarily large?

Following reasoning already quoted from Ref 1.7, and assuming that boat speed is the primary objective, it seems apparent that a competent designer should adjust the size of his chosen underwater appendages so that a maximum side force/resistance F_S/R ratio is achieved in the range of expected boat speeds. This appears reasonable since one may expect to gain something, in terms of boat speed, by providing the required lateral hydrodynamic force with the least possible drag, which is associated with a small wetted surface. These arguments are however only partly true and apply to steady sailing conditions. If reduction in lateral area of appendages is taken too far it may bring disappointing, if not disastrous, results in unsteady sailing conditions, i.e. when rolling or tacking in strong gusty winds and rough seas, or even, as we shall see, in light winds. The steering and close-windedness deficiencies, which are directly coupled and observed in the case of the International Tempest class and other modern boats, in some weather and sea conditions, can be attributed to the small lateral area of their appendages. Not infrequently, this is well below the adequate area suggested by the statistical analysis of successful boats, and usually expressed in terms of minimum fin keel area/sail area ratio. Apparently, there were good reasons behind these statistical recommendations based on past experiences.

One of the reasons is as follows: hydrodynamic side force F_S is generated mainly due to fin action, the hull itself contributing little to it (Ref 1.5). A marked increase in leeway, which would enable the fin keel to work at best side force/drag ratio, is bound to increase noticeably the hull wave-making resistance. In an attempt to reduce this additional resistance due to crabwise motion of the hull, fin keels with tabs, or rotating fins with incidence adjustable in relation to the hull centreline, were developed.

Evidently, there are two conflicting requirements, set forth by Barkla in Ref. 1.1, namely: when the lateral area is increased in order to reduce the angle of leeway, the wetted surface increases and the resulting greater frictional resistance, dominant at low speed, may outweigh possible gains at higher speed when wave-making resistance dominates. It seems that the size of the fin or centreboard, masterpieces of empirical development, established statistically as about 4 per cent of the sail area, is the minimum consistent with its function of generating sufficient side force in a variety of sailing conditions, in steady as well as unsteady motion.

12-Metre boats

It was suggested in Ref 1.7 that the rather large lateral plan area and wetted surface of older 12-Metre hulls could be reduced by at least one-third; this would result in lower resistance. Such a conclusion appears to agree with aeronautical practice which confirms that one of the best ways of increasing speed is to reduce the wing area or wetted area. Figure 1.15 illustrates the evolution of 12-Metre yachts in this respect from 1958 to 1974. The modern Twelves have far less wetted area, mainly due to smaller appendages, and they are faster, but as has been reported by Ficker (*Intrepid*'s helmsman: Ref 1.9)

Fig 1.15   Evolution of 12-Metre class hull.

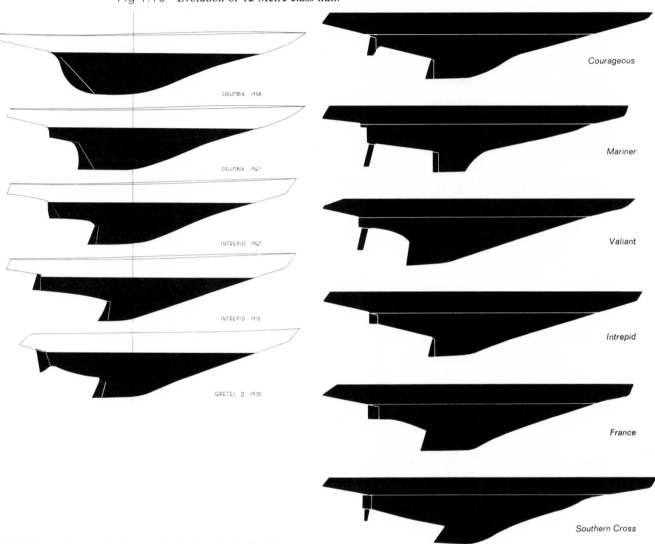

'the present breed of 12-Metres is very difficult to steer and keep "in the groove". *Intrepid*'s biggest difficulties were experienced when tacking in light weather. It was not easy at all to get her moving again on the wind and to regain the speed of the previous tack. Every combination of tacking technique and sail trim were tried without success.'

The directional stability of these boats has deteriorated by the standards of older 12-Metres; to quote an experience recorded in Ref. 1.10: '...When I took the helm of one 12-Metre on a moderate day on Long Island Sound I had to fix the bow on a

Photo 1.2   View of *Norsaga* showing some instrumentation.

Mast for
water speed log

Mast for wind speed
and direction

Yaw transducer Camera

point on the shore. I couldn't steer the boat otherwise. It was yawing 10° either side of the mean track and needed constant control.'

When, in similar conditions, the author of this book was given an opportunity to steer the old 12-Metre *Norsaga* (Photo 1.2) with a conventional long keel, it was rather with a feeling of amazement to discover that the boat steered herself.

These steering deficiencies are probably an unavoidable price one has to pay for the reduced wetted area of appendages: fin, keel, and separate rudder. On the other hand, they can partly result from inefficient action of the rudder operating in the wake of the bustle, a device which is nowadays a common feature of almost all contemporary, high performance offshore racers.

Experience has shown that the bustle, when properly designed, may be beneficial in reducing wave-drag. However, if badly shaped it may cause flow separation, which in turn has an adverse effect on the performance of the rudder.

The same kind of disease is bound to afflict the modern breed of cruiser-racers, for which the 12-Metres are pointers to progress. 'Too radical a break'–to quote Olin Stephens–

'from the line of thinking you have been following could bring about a surprising result. Unpleasantly surprising, that is. There is still so much to learn from all the refinements that are possible in so many different applications, that you have to try just a few more ideas each time and hope that they do take you forward.'

The why's and wherefore's of these modern yachts' misbehaviours will be examined in the chapter dealing with unsteady sailing conditions.

D Optimum course to windward

The optimum γ angle in close-hauled conditions is affected by three things. Apart from the two factors already mentioned, namely, the aerodynamic drag angle ε_A and the hydrodynamic drag angle ε_H, it also depends on the rate of resistance build-up as boat speed increases.

In other words, the optimum course γ relative to the true wind is seen to depend largely on the so-called power law index of the boat's resistance. This can be written in a crude form as:

$$\text{Resistance} = C \times V_s^n \qquad\qquad \text{Eq 1.5}$$

where

      C = variable coefficient depending on hull form, its attitude etc.

      n = power law index

This index n is close to 2.0 at low speeds, when skin friction predominates.

Resistance

A glance at Fig 1.5, showing the resistance characteristics of a light displacement International Canoe in upright condition, reveals that when boat speed is doubled, say from 2 to 4 knots, resistance increases about fourfold. Since forces developed by sails are proportional to wind velocity squared, one may expect that in a low speed regime boat speed would increase roughly in direct proportion to the wind velocity. When boat speed increases further and further, the power law index rises gradually to 3, 4, or even more, depending on displacement/length ratio, $\Delta/(L/100)^3$ of the boat in question.

H Barkla has shown (Refs 1.1 and 1.11) that the ratio of V_{mg} to V_T, which may also be regarded as a measure of yacht performance, can be expressed in terms of γ and β:

$$\frac{V_{mg}}{V_T} = \frac{\cot \gamma}{\cot (\gamma - \beta) - \cot \gamma} \qquad \text{Eq 1.6}$$

Fig 1.16 Diagram for calculating yacht performance in close-hauled conditions.

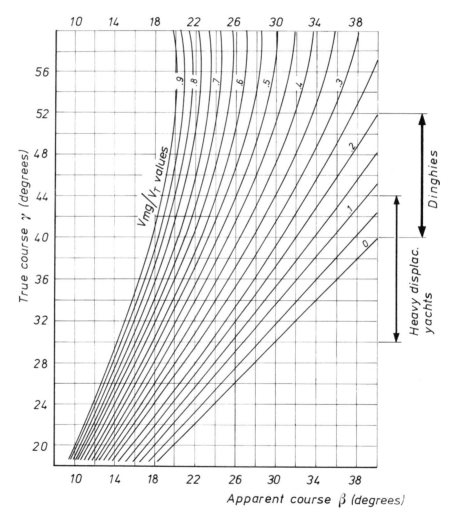

since

$$\beta = \varepsilon_A + \varepsilon_H$$

then

$$\frac{V_{mg}}{V_T} = \frac{\cot \gamma}{\cot (\gamma - \varepsilon_A - \varepsilon_H) - \cot \gamma} \qquad \text{Eq 1.7}$$

Fig 1.16 illustrates the variation of V_{mg}/V_T ratio with γ and β. From it the relative merits of different yacht types, in terms of V_{mg}/V_T ratio, can be directly assessed with a knowledge of the angles between the course sailed and the true wind γ and apparent wind β. For heavy keel boats, representative values of V_{mg}/V_T vary from 0.3 (strong winds) to 0.6 (light winds) with corresponding γ angle relative to the true wind direction within a range of 30–44°.

For light displacement dinghies γ angles are larger, in between 40–52°. Why? Does it mean that heavy keel boats are superior in getting to windward? Certainly not. The closer-windedness of the keel yacht simply implies that the build-up of resistance with speed is so sharp that it does not pay to sail faster and further off the wind, whereas for a dinghy it does. Since the only thing that matters in close-hauled work is the attainable V_{mg} given by Eq 1.2, a closer-winded boat may lose to another, faster one, with a better, less steep resistance/speed relation.

To estimate the true resistance characteristics of different hull forms it is desirable to eliminate, as far as possible, the effects of size. This may be accomplished by comparing values of resistance per ton, R/Δ_T, sometimes called specific resistance, plotted against speed length ratio $V_S/\sqrt{L}$. Figure 1.17 represents comparative 'specific resistance' curves of four different hull forms, which also have different displacement/length ratios $\Delta/(L/100)^3$. It should be remembered that the displacement/length ratio describes, in a way, the load put on a given length L of the hull (Note 1.12). In contrast with very light displacement craft, such as the International Canoe, or the A-Class Scow, the specific resistance curve of the heavily loaded NY 32, a representative of the displacement form, shoots upwards steeply when the hull approaches its so-called displacement speed, at which the wave barrier makes any further speed increase practically impossible.

The remarkable flattening of resistance curves of the 'Scow' and International Canoe, observed at higher speeds, is mainly due to the great reduction of wave-making by virtue of their hull forms and lightness. A flat-bottomed, lightly loaded hull may be lifted substantially at a certain speed, so that it tends to plane or skim over the water surface, instead of ploughing through it.

Between these two distinctive hull forms, a displacement form such as the NY 32 and lifting or skimming form, such as the A Class Scow, representing two different and contrasting approaches to the sailing yacht, there are all sorts of intermediate, occasionally semi-planing craft, embodying the full range of compromise between sheer speed and the other desirable attributes which a 'perfect boat' should possess.

Fig 1.17   Resistance characteristics of different hull forms.

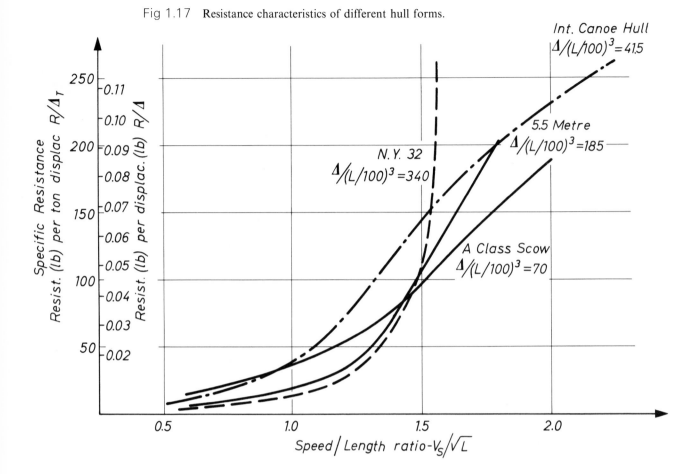

The 5.5-Metre class, with displacement/length ratio about 185, represents an intermediate type of boat (see remarks about the division of sailing yachts–Note 1.13).

The main difference between various hull forms demonstrated in Fig 1.17 lies in their rate of build-up of specific resistance with speed, and is well reflected in the optimum course γ relative to the true wind; this is illustrated in Fig 1.18. A conclusion one may derive from it confirms Barkla's findings as stated in Ref 1.11:

> 'While it is true that *two similar hulls* with different rigs may have different best course angles γ, in which case the smaller course angle indicates the better rig and probably, though not inevitably, the faster boat, greater speed to windward and a closer course do follow inevitably from an increase of keel efficiency. But when we are comparing *different types of boat*, the best course angle γ ceases to have any validity as a criterion. The owner of a keel yacht has no right to feel superior to the dinghy owner on the grounds that the dinghy points best at, say, 42° while his best is 35°.'

Fig 1.18 True course angles γ for best V_{mg}; NY 32 data taken from TM 85 part II (Davidson Laboratory, USA), Finn-Canoe data calculated by the author.

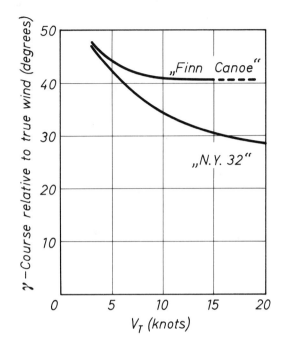

Figure 1.19 demonstrating the performance characteristics of five boats of various displacement/length ratios including skimming forms (Photos 1.3, 1.4 and 1.5) in terms of $V_{mg}/\sqrt{L}$, so the size effects are eliminated, may be used to support perhaps an unexpected conclusion that the dinghy man could feel entitled to claim superiority. The best γ angle for his boat is greater than that of a keel boat; thus indicating a lower power law index of hull resistance. Therefore he may take advantage of this by sailing further off the wind and hence faster, in light and moderate winds, with better resulting $V_{mg}/\sqrt{L}$ ratio than that of the heavy racer of displacement form which is pointing higher.

The $V_{mg}/\sqrt{L}$ ratio attainable by the light-displacement, skimming Scow sailed in smooth water, exceeds 1.1 while the modern 12-Metre hardly reaches this value. No available data exist concerning Flying Dutchman performance in various wind velocities, but almost certainly it is better in terms of $V_{mg}/\sqrt{L}$ than that of our mythical Canoe-Finn hybrid dinghy, where the hull length was taken as 17 ft, unusually long by modern dinghy standards. Such a length in denominator reduces substantially the $V_{mg}/\sqrt{L}$ value. One may expect that the performance curve of the relatively shorter Flying Dutchman should be bodily shifted to the right relative to the Canoe-Finn curve. It would mean that the Flying Dutchman is a better

Fig 1.19 Performance curves of different boats compared in terms of $V_{mg}/\sqrt{L}$ ratio. Calculated curves are based on data obtained from the following references: NY 32–Davidson Laboratory TM 85 1948; *Intrepid*–Ref 1.10; A-Class Scow–Dav Lab Rep 133; 5.5-Metre–Performance Trials of the 5.5-Metre yacht *Yeoman*, NPL Report 1955; Canoe-Finn–Author's calculations–see Fig 1.23; Tornado curve–see section G–High Speed Sailing.

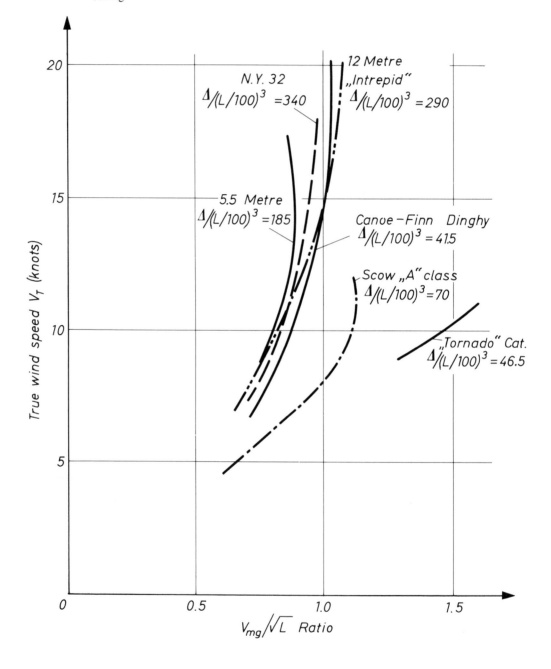

Photo 1.3   *Intrepid*, America's Cup defender 1970.

Photo 1.4   *A-Class Scow* competing in the 'One of a Kind Regatta'–1966. Heeled to leeward intentionally in order to reduce the wetted surface of the hull.

LOA–38 ft.
Sail Area S_A–557 sq ft.
Wetted Surface A–ab 200 sq ft at 0° Heel
A–ab 155 sq ft at 30° Heel
Displacement about 1.27 ton.

Photo 1.5A   International 10 sq m Canoe
LOA about 17 ft.
Sailing weight about 450 lb (0.2 ton).
Sail area S_A–107 sq ft.
Wetted Surface A–about 51 sq ft.

Photo 1.5B   International Finn
LOA = 14.75 ft.
Sailing weight–about 500 lb.
Sail Area S_A–about 108 sq ft.
In the example labelled 'Canoe-Finn' the boat's performance was calculated on the assumption that the Canoe hull was driven by the Finn-type sail. Canoe hull characteristics are given in Figs 1.4 and 1.5. Characteristics of one of the Finn sails are presented in Fig 1.9.

performer to windward than the 12-Metre, over possibly the whole range of recorded wind velocities.

Comparing the two curves representing performances of a rather conventional NY 32 and the 12-Metre *Intrepid*, which can be taken as the epitome of current progress in yacht designing, one may be surprised to learn that the 12-Metre, though superior in stronger winds, can be beaten by the NY 32 in light and moderate winds, provided of course that performance is compared in terms of $V_{mg}/\sqrt{L}$. This example evidently demonstrates the fact that it is very difficult indeed to improve heavy displacement yacht performance over the whole range of sailing conditions.

Heeding the warning that all comparisons are odious and generalizations untrue, we may accept a certain risk while analysing further the virtues of different types of boat, in an attempt to find out what is 'general in what is particular and what is permanent in what is transitory'. From Figs 1.17 and 1.19 one may infer, for example, that since dinghies or scows are fast and they are light-displacement, everything light-displacement is fast. Unfortunately, such a conclusion would be wrong.

Examining carefully Fig 1.17, we should notice that in the low speed regime, light-displacement craft have much higher specific resistance, 3–4 times higher than displacement forms. It reflects the square-cubic law involved which, translated into common language, states that big, heavy yachts have relatively less wetted area (i.e. lower wetted area/displacement ratio A/Δ) than small, light yachts. It is due to the fact that displacement Δ increases in proportion to the cube of linear dimensions of the yacht, while the wetted surface A increases with the square of those dimensions (see Fig A.1, Appendix).

Table 1.8

	NY 32	A Scow	Finn-Canoe
1. Displacement Δ in tons	11.38	1.27	0.2
2. Length LWL in ft	32.26	26.3	17.0
3. Sail Area S_A in sq ft	950	557	107
4. Wetted Surface A in sq ft	378	200 (upright) 155 (heeled)	51
5. S_A/Δ in sq ft/ton (see Ref 1.14)	92	440	535
6. S_A/A	2.5	2.8 3.6 (heeled)	2.1
7. $\Delta/(L/100)^3$	340	70	41.5
8. A/Δ in sq ft/tons	33	157 122 (heeled)	255

Sail area

Such a trend is well displayed in line eight of Table 1.8, giving particulars of the boats in question. There are also given in lines 6 and 5 the two other important ratios, namely sail area/wetted surface (S_A/A) and sail area/displacement (S_A/Δ). These two ratios are measures of the driving power available to the boats. The ratio S_A/A governs at low speeds, while S_A/Δ governs most of the time.

From statistical analysis it emerges that, in order to secure a reasonable performance in light winds, the S_A/A ratio should be in the range 2.0–2.5. The sail area is the dominant speed-producing factor in the low speed regime. However, not in every case of light-displacement craft is this crucial, light weather criterion, (S_A/A ratio) large enough, and this explains the occasionally observed superiority of heavy-displacement racers over the light-displacement ones, both having similar length, in very light airs or drifting conditions.

Shifting ballast

If a light-displacement boat has a sufficiently high S_A/A ratio, the higher values of S_A/Δ may be expected and the particulars of the exceptionally light-displacement Scow and Finn-Canoe, given in Table 1.8, clearly demonstrate this. By NY 32 standards, these boats are enormously overpowered, carrying 5–6 times more canvas per ton displacement. Under these circumstances, enough lateral stability, or power to carry relatively large sails, can only be provided by shifting a sufficiently heavy crew to the weather rail, on to a sliding seat or on a trapeze, otherwise the performance potentials of those boats cannot fully be realized.

For example, the weight of the Flying Dutchman is about 400 lb. The best all-round combination for a crew is a lightweight skipper and a tall, heavy, but obedient fellow on a trapeze. They should weigh as a team 330–370 lb (Photo 1.6). The crew contributes about 45 per cent of the all-up sailing weight whilst the displacement/length ratio is about 55! In these conditions, enormous power to carry a lot of canvas can be provided by shifting the crew to the weather rail, and a man on a trapeze serves as an effective and cheap alternative to beam and ballast.

As pointed out by Dr Davidson in Ref 1.6, the key to these light-displacement craft is that they are essentially small. They are small enough for the weight of their crew to form a sizeable proportion of their total displacement. Light racing dinghies with their crew shifted as far to windward as possible can take or absorb, at the tolerable angle of heel, a heeling force in the order of $\frac{1}{5}$ to $\frac{1}{6}$ of the total weight of the boat, W; trapezes and sliding seats may bring the limiting heeling force to about $\frac{1}{4}$ or even $\frac{1}{3}$ W. For comparison, the 6-Metre and 5.5-Metre, both displacement forms, heel to 20° under the action of a heeling force which is about $\frac{1}{20}$ (i.e. 5 per cent) of the yacht's weight (Figs 1.12 and 1.20B). Lack of sufficient stability in strong winds and too low a S_A/A ratio in light winds explain why the 5.5-Metre light-displacement boat has such a poor close-hauled performance in comparison with NY 32, and this

Photo 1.6 Crew weight for Flying Dutchman appears to be best in the range 330–370 lb: a 150 lb helmsman and a tall 180–200 lb crew.

Fig 1.20 Performance data of a 6-Metre boat in close-hauled conditions.

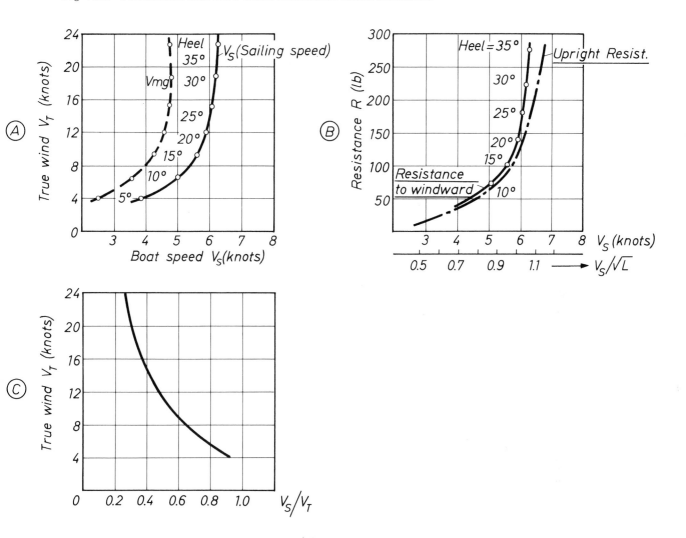

is shown by the relevant curves of $V_{mg}/\sqrt{L}$, in Fig 1.19. That near religious chant we hear occasionally:

> 'Half the Displacement
> Half the Sail Area
> Twice the Speed'

is nothing but a deceit.

No doubt, when reaching under favourable conditions in strong winds, the light displacement craft may show bursts of speed. However, to enjoy really high speed sailing one must have a boat that incorporates in her design features lightness or low $\Delta/(L/100)^3$ ratio, combined with large S_A/Δ ratio, which automatically demands high stability. The 5.5-Metre is too heavy to use her crew weight as the crucial stability factor, which is so effectively employed in light and fast dinghies.

'It would be idle to attempt to argue the relative merits of heavy displacement boats against those of the particular light displacement craft which have been considered here. It is simply a matter of how much one chooses to emphasize maximum speed for its own sake and how much one cares to sacrifice in the way of sea-keeping, and so-on, in order to get it.'

These remarks expressed by Dr Davidson some years ago, together with the results presented in Figs 1.17 and 1.19, can be summed up: in sailing as in other matters, the promise of something for nothing rarely works.

Fig 1.21 Relationship between the optimum true course γ and sail (ε_A) and hull (resistance) characteristics.

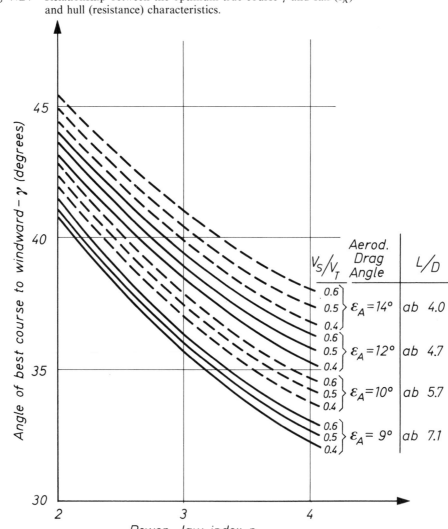

Figure 1.21, taken from Ref 1.11, gives an insight into the relative influence of some important factors, such as: power law index n in Eq 1.5, ε_A and V_S/V_T ratio on the best course sailed γ to windward. First, the γ angle varies much with the actual rate of resistance build-up, i.e. power law index n which, as a factor of primary importance, depends entirely, as shown in Fig 1.17, on the underwater form of the hull and the load put on its length, i.e. displacement/length ratio. Secondly, the best course γ varies with the aerodynamic drag angle ε_A, or the lift/drag ratio L/D, which describes the aerodynamic properties of the rig. Finally, γ varies with the boat speed/true wind speed ratio V_S/V_T, which depends on all the design factors involved in a given type of sailing craft.

The significance of the V_S/V_T ratio can be appreciated by considering its variation, taking as an example a 6-Metre boat (a displacement form) sailing close-hauled in increasing wind. Fig 1.20A, B, C, illustrates the sailing speeds V_S and V_{mg}, the upright and heeled resistance and also V_S/V_T variation over the range of true wind speeds up to 24 knots (Ref 1.6). Such a set of diagrams supplementing Fig 1.12 provides a better insight into a boat's behaviour than does Fig 1.12 alone.

The attainable boat's velocity V_S, which is at its maximum at 25–30° of heel, is strictly limited by the wave drag barrier. The situation is in a way analogous to the supersonic aircraft at the so-called sound barrier, trying to fly faster than Mach 1; only an adequate reserve of driving power makes it possible. A conventional, heavy-displacement yacht cannot possibly develop enough power to surpass the wave-drag barrier. Sailing with an increasingly greater angle of heel, which is commonplace for keel yachts, causes an additional resistance build-up entirely due to heel. Moreover, for the same reason, namely stability deficiency, the large driving forces which might be harnessed in strong winds cannot be developed. Consequently, the V_S/V_T ratio, at which the displacement type of boat operates, becomes progressively smaller when the wind speed increases. A typical 'vicious circle' is established which precludes high speed sailing. In fact, in average winds a displacement type of boat sails about half of the true wind speed, i.e. $V_S/V_T = 0.4 - 0.5$. There are, however, types of sailing craft discussed in chapter H 'Land and Hard Water Sailing Craft', based on different design principles, that may approach $V_S = 5 \times$ true wind speed, i.e. $V_S/V_T = 5.0$.

E Stability effect on performance

Common sense whispers furtively that really to improve a yacht's performance one must be able to reduce resistance. This blinding glimpse of the obvious, accepted as an axiom in ship science, is of limited value as far as the sailing yacht is concerned, except perhaps in the case of a yacht sailing upright and dead before the wind. One may prove an apparently paradoxical point that a boat's performance can be improved when its hull resistance is increased. After all, a sailing yacht is not just a ship but rather an aeroplane-ship hybrid.

We are already aware of the fact that as the boat heels beyond some critical heel angle (which may or may not be upright) both drag angles ε_A and ε_H increase, at first slowly then more rapidly. At more or less the same rate at which the drag angles increase, both the driving efficiency of the rig and the hydrodynamic efficiency of the hull deteriorate and so does the boat's performance. One may rightly infer that in stronger winds, in which the highest speeds may be attained, stability becomes the supreme merit of any kind of boat. Experience with dinghies supports this statement, for it is known that for a given wind speed a boat can be sailed closer to the wind or faster, the more upright she is kept. Moreover, it is also known that the performance of a keel boat to windward is at its best if the heeling angle is not allowed to exceed some critical angle.

Stability

Some quantitative assessment of the influence of stability on a boat's performance will illustrate better the whole problem. To make the presentation simple let us

Fig 1.22   Equilibrium of forces in close-hauled sailing. Definition of
symbols.

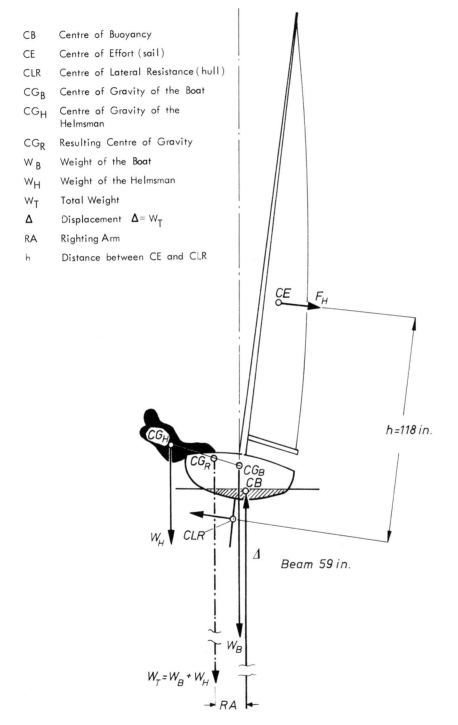

CB   Centre of Buoyancy

CE   Centre of Effort (sail)

CLR   Centre of Lateral Resistance (hull)

CG_B Centre of Gravity of the Boat

CG_H Centre of Gravity of the Helmsman

CG_R Resulting Centre of Gravity

W_B Weight of the Boat

W_H Weight of the Helmsman

W_T Total Weight

Δ Displacement $\Delta = W_T$

RA   Righting Arm

h   Distance between CE and CLR

CE F_H

$h = 118$ in.

CG_H

CG_R CG_B
CB

W_H CLR

Δ Beam 59 in.

W_B

$W_T = W_B + W_H$

$\rightarrow RA \leftarrow$

analyse the Canoe-Finn dinghy type shown in Photo 1.5A, B and Fig 1.22 sailed to windward. The righting moment, which is a product of the total weight of the boat W_T and the righting arm RA, will depend upon the amount of leeward shift of the centre of buoyancy C_B, relative to the resulting centre of gravity of the boat CG_R. The position of the latter will depend largely on the weight of the helmsman and his ability to sit outside the gunwale. Two helmsmen of different weight, say 180 lb and

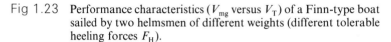

Fig 1.23   Performance characteristics (V_{mg} versus V_T) of a Finn-type boat sailed by two helmsmen of different weights (different tolerable heeling forces F_H).

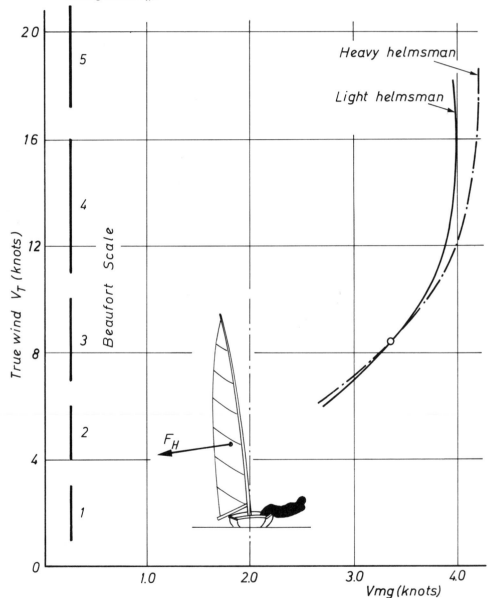

230 lb (including wet sweaters), will obviously produce different righting moments, so that the tolerable heeling force F_H will also be different.

To a reasonable degree of accuracy one may calculate that these tolerable heeling forces will be 95 lb for the heavy helmsman and 76 lb for the light one. Assuming that the weight of the boat W_B is 320 lb, and adding the helmsman's weight W_H, we find that the total weights of the boat in sailing conditions will be 550 lb and 500 lb, for

Fig 1.23A   Effect of helmsman weight on V_{mg} at two different wind speeds V_T.

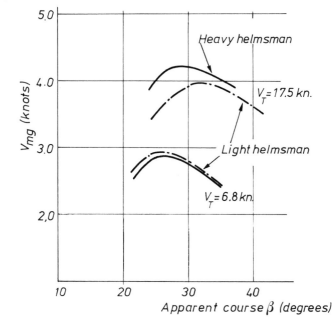

heavy and light helmsmen respectively. In the first case the boat is 10 per cent heavier than in the second case, and will consequently experience higher hull resistance. Generally, resistance increases in proportion to displacement, thus the boat with the heavier helmsman will have 10 per cent more drag than that with the lighter man. The heavy helmsman pays, therefore, a certain penalty in terms of increased drag in exchange for greater stability, expressed in this case as a tolerable heeling force F_H. The effect of this, in terms of speed made good to windward V_{mg} for various wind velocities, is shown in Fig 1.23. Calculations were performed applying a graphical method described in Ref 1.5.

The two performance curves demonstrate that in winds above 9 knots, or force 3 on the Beaufort scale, the advantage of having better stability outweighs the penalty of higher hull resistance. In winds below 8 knots the heavy helmsman is handicapped, but the deterioration of the performance of his boat, in terms of V_{mg}, is very small–1.4 per cent only against 6.4 per cent gain in V_{mg} recorded in stronger winds. Figure 1.23A illustrates in some detail how V_{mg} changes with β angle, i.e. the angle between the course sailed and the apparent wind direction. The two sets of curves refer to two different wind speeds; one set refers to $V_T = 6.8$ knots when the heavy helmsman is handicapped, and the second to $V_T = 17.5$ knots when the light helmsman is handicapped. For races sailed in winds exceeding 8–9 knots, heavy helmsmen are better off than light ones, certainly on the all-important windward legs. On reaching legs, in marginal planing conditions, the light helmsman may enjoy certain advantages when he can plane and a heavy helmsman cannot. In strong winds, when everybody planes, the difference in speed (and in elapsed time) on the triangular course may become primarily the difference in speed on the windward leg.

One should notice that the V_{mg} curve illustrating the best potential performance of the light helmsman in Fig 1.23 bends to the left, towards lower values of V_{mg} for wind speeds above 16 knots. This deterioration in performance is due to the fact that, above a certain critical wind velocity, the sail has to be spilled in order to keep the heeling force down to the tolerable magnitude of 76 lb. In such a condition the sail works very ineffectively; by analogy, it can be compared with an engine firing on only three or perhaps four of its six cylinders. Under similar conditions, the sail of the boat in its heavier state, with its higher stability, works like an engine firing on five of its six cylinders.

That is why the Finn, an Olympic class with a sail area of about 110 sq ft, is in fact a boat for heavy 'tough guys' weighing 200–210 lb. The Finn is supposed to be a strictly one design class, anyway closer to an ideal one design concept than any other Olympic class. It is thus intended to give all competitors an equal chance to match their talents and skill as helmsmen. To make this possible in each of the last Olympiads, identical Finns have been supplied by the host nation to all participants.

This prompts the questions, assuming that all relevant factors but weight are equal, including the helmsman's brains, what is really measured in the Olympic Finn class competition? Are the results merely indicative of stability, i.e. of sheer weight of a human body? Does the concept of the one design make any sense?

The Finn is a weight-sensitive boat and, other factors being equal, relatively small differences in helmsman weight will be reflected in speed performance. In this respect the Laser is even more sensitive; it is a smaller and lighter boat than the Finn, and therefore the effect of ratio of helmsman weight to the total weight of the boat is bound to be more conspicuous in terms of performance. On the other hand, heavier boats such as, say, the Star will not practically be affected in their performance figures by 20–30 lb difference in crew weight.

Since weight sensitivity in relation to boat performance is rather acute, and is integral with the boat lightness, it prompts another question–should the Olympic classes be selected from rather heavy, ponderous and perhaps ballasted boats in order to eliminate, as much as possible, the sheer weight sensitivity from competition? The other alternative would be to divide competitors into weight classes, say, light, medium and heavy, to give them a better chance to compete on equal terms.

Stability is what a sailing machine must have if she wants to go fast. The significance of stability has only been appreciated and incorporated in rating rules

Fig 1.24   Dixon Kemp's plank-on-edge yacht, designed about 1880. An
angle of heel of 15° from horizontal, not the vertical, when
sailing to windward, was not uncommon. This particular type of
boat is a product of the YRA rule in which Beam, B, was severely
taxed. The rule was:

$$\frac{L + B^2 \times B}{1730} = \text{tons}$$

As might have been anticipated, the boat built to this '1730
Rule', as it was often called, had a long, narrow, heavily
ballasted, deep-bodied, wall-sided hull, possessing small initial
stability, and no great speed, considering the sail area employed
to drive it. This rule, nevertheless, '...governed first-class racing
in British waters from 1880 to 1886.'

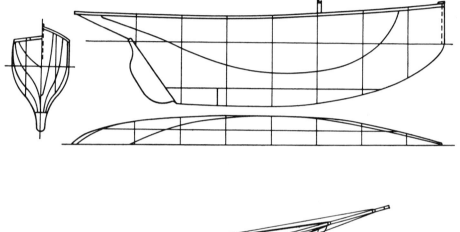

relatively late in the twentieth century. Let us look at Fig. 1.24 which illustrates
a deep and narrow type of hull of extreme displacement form, ploughing through
the green waters in the 1890's. Basically she is slab-sided with no hollow at the
garboards. Stability, provided by heavy ballast of about 60 per cent of displacement,
was pretty poor by the standard of contemporary beamy hulls, which had great
stability due to form. It is therefore not surprising that these boats, unable to stand
up to their canvas, sailed at excessive angles of heel and were notoriously wet. Uffa
Fox who sailed those extreme craft, referred to them in *Thoughts on Yachts and
Yachting* as: 'more like submarines than sailing boats!'.

Development

*An angle of heel of 15° from horizontal, not the vertical, when sailing to windward was
not uncommon* and the story told by H Benham in his book, *The Last Stronghold of*

Sail, bears out this initial stability deficiency in the description (Ref 1.15):

'...with nice, fine weather, when up off the island [Mersea Island] we saw a vessel under way hove down on her beam ends. We looked again and thought there must be a hurricane coming. We started getting sail off the brigantine in double quick time, and then she came by, still rail under and still next to no wind.'

The development of this extremely narrow type of yacht was largely a product of the existing rating rule, which taxed beam *B* heavily. The YRA rule was

$$\frac{(L + B)^2 \times B}{1730} = \text{Rating (in Tons)}.$$

Adopted in 1886, a new YRA rule suggested by Dixon Kemp was based on length and sail area alone, as follows:

$$\frac{L \times S_A}{6000}$$

There was no restriction on beam, therefore no advantage to be gained in terms of lower rating, by building the plank-on-edge yachts. Subsequently, a rather sudden and astonishing change 'from slow to fast' sailing yachts was observed.

According to Heckstall-Smith (Ref 1.16) the *Britannia* designed by G L Watson (shown in Fig 1.25) could be taken as the best example of the change in the type of sailing yacht that then took place: '...with her advent the old slow type died and the new fast type was born.' Trying to determine the discovery at the root of the change, Heckstall-Smith says:

'Scientists may object to the word "discovery" but the evolution of type from 1890 to 1893 was so rapid that I may venture to use it. The *Britannia* and her contemporaries were built to skim over the waves and waters, and not to plough through them. That, in every language, was the discovery.'

Fig 1.25

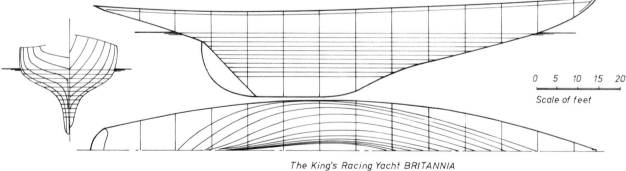

The King's Racing Yacht BRITANNIA

The lines from a drawing by G.Watson, 1893

Fig 1.26   Sand-bagger *Susie S*
LOA–27.25 ft (8.30 m)
Beam–11.0 ft (3.34 m)
Sail Area–about 1500 sq ft.

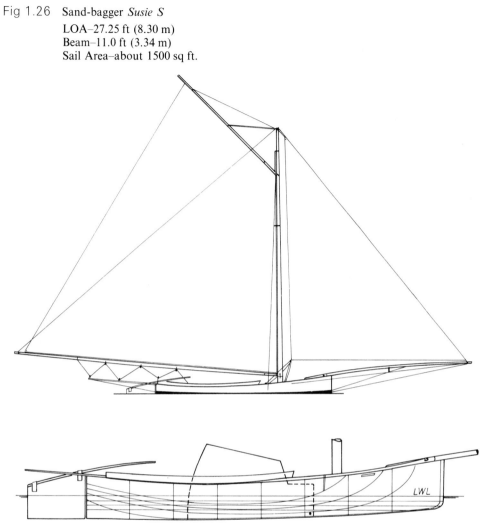

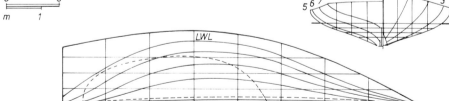

It may sound strange to our generation but Victorian nautical gentlemen ('true to the tradition of the stiff upper lip, and not adept in the language of the feelings', as Phillips-Birt says in his book *An Eye for a Yacht*) universally condemned the new, almost 'modern' by our standard, vessels as being 'hideous machines'! Herreshoff's *Gloriana* was said to have received an even worse reception in the USA.

Sail area, displacement and stability are the three factors that most affect the comparative speeds of boats of the same waterline length. Once length is selected as the only basis for handicap, and no limitations are put on other speed-governing factors, extraordinary sailing machines and rule cheaters are bound to develop.

So-called sand-baggers, popular at the end of the 19th century, are good examples of such an extreme development. Those dish-like surface-skimming, exceptionally fast craft were the product of an early American so-called length or mean-length measurement rule. Sail area, stability and displacement were used to the utmost, with one single purpose–maximum speed; an example is shown in Fig 1.26 (Ref 1.17).

Immense sails could only be carried by means of great beam and shifting ballast. The balast, usually in the form of two dozen sand bags weighing about 45 lb each, supplemented by a crew of up to seventeen made up of waterside toughs, could produce, when moved to the weather rail, an additional righting moment of about 18,000 ft/lb (crew weight about 3400 lb and ballast about 1100 lb, making a total 4500 lb, at a righting arm of say 4 ft). This calculation, applicable to the upright attitude, does not take into account the increased righting moment due to form when the boat progressively heels. Referring to high initial stability as a necessary condition for fast sailing, F Herreshoff expressed a view that: '...while the multihulled craft score high in their sailing ability, I am not sure that with shifting ballast the single hulled craft cannot be made to equal their performance in anything but strong winds and perhaps a running wind on the quarter.'

The idea of acquiring large stability by means of shifting ballast and great width became popular again, and contemporary fast dinghies have exploited this principle to the utmost, though the movable ballast consists only of her crew. Figure 1.27 shows a modern version of a sand-bagger, a boat belonging to one of the celebrated *Pen Duick* family of Eric Tabarly. She is not a sand-bagger in the true sense; there are no sand bags on her deck, or tough guys to shift them from one side to the other, but the design principles are the same, reflected in her lines and method of increasing initial stability. Most of her ballast, in the form of water, is put into side-tanks, so that it may be pumped out to windward to give a large initial righting moment before she heels. Referring to her hull sections, there is a chine just above the waterline. This reduces the wetted area in the upright position and ensures a rapid increase of form stability when the boat is heeled some degrees.

Fig 1.27   *Pen Duick V*–designed by Michel Bigoin.

LOA–10.67 m (35 ft)
LWL–9.10 m (29 ft 10 in)
Beam–3.45 m (11 ft 4 in)
Draft–2.30 m (7 ft 6 in)
Displacement–3200/3700 kg (7050/8150 lb)
Ballast–400 kg (880 lb) lead; 500 kg (1100 lb) water.

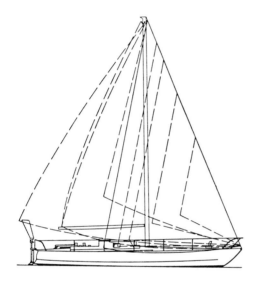

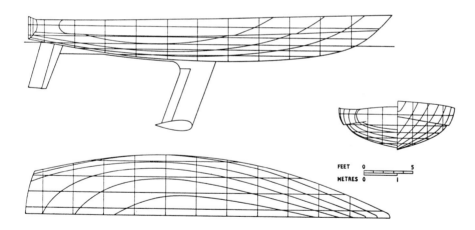

FEET 0   5
METRES 0   1

F All-round performance

Figure 1.28 adopted from Ref 1.18 gives a good overall picture of the performance of a ballasted racing yacht. The three so-called speed polar curves indicate the estimated speeds of a modern 12-Metre yacht sailing in calm water, on courses ranging from close-hauled to running, in true wind strengths of 7, 12 and 20 knots; the yacht is carrying genoa or spinnaker, as appropriate.

The curves are in reasonable agreement with observations of actual 12-Metre performance, as well as with the performance expected from tank tests of hulls and wind-tunnel tests of sails. However, as both full-size observations and predictions from model tests are subject to error, these curves must only be regarded as approximate. In open water, the stronger winds would cause considerable waves and this would lead to a marked reduction in V_{mg}, and would also affect to some extent the speeds on other headings.

The numbers marked alongside the semicircle indicate the course relative to the true wind direction V_T; there are also some figures alongside the speed polar curves which indicate the heading angle $(\beta-\lambda)$ between the centreline of the yacht and the direction of the apparent wind which is felt on the yacht. This angle would be indicated by a wind-vane, if it could be positioned where the airflow would not be affected by the proximity of the sail.

The optimum angle γ when sailing to windward, is seen to be a function of true wind velocity V_T, as shown earlier in Fig 1.18. In light winds the γ angle is greater than for the moderate breeze case. Experienced racing sailors already have a knowledge of what seems to be the optimum sailing course for each type of boat. It can be assumed that a prospective new boat will not go fastest to windward when

Fig 1.28   Performance polar diagram of a 12-Metre yacht (geometry of the velocity triangle given in Fig 1.8B).

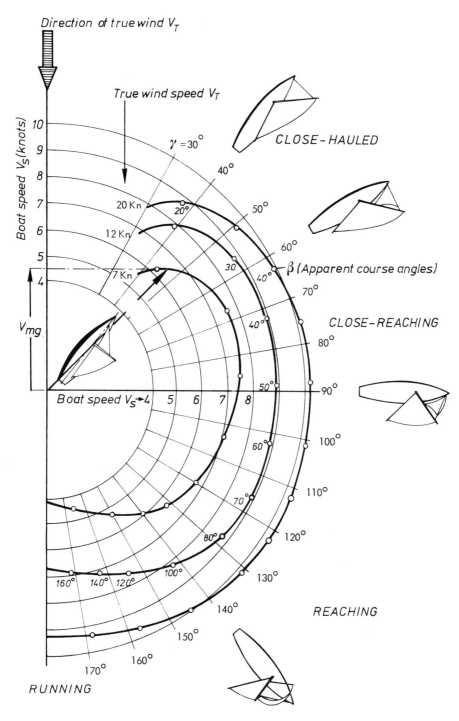

sailed similarly to another boat of a different type. Boats, depending on wind and sea conditions, may respond in a unique, peculiar way when being coaxed to show their best. Interesting remarks about this particular feature in boats' behaviour can be found in Vol. I of *Yachting–The Badminton Library* which served as a modern encyclopaedia for keen 19th century gentlemen sailors.

> 'Some women–I speak it with all respect–bear being "squeezed" and "pinched", they almost seem to like it, at any rate they don't cry out; whereas others will cry out immediately and vigorously. So will yachts. The more you squeeze one vessel, the more you pinch her, the more she seems to enjoy it. Squeeze another, pinch her into the wind, and she lies down and calls out at once. The difference between vessels in this respect is quite funny, and essentially feminine.'

The form of the speed polar curves in Fig 1.28 is somewhat dependent on the fact that the 12-Metre is a yacht of rather narrow beam, rather high displacement/length ratio $(\Delta/(L/100)^3 =$ about 300) and high ballast ratio. Hulls of lower displacement/length ratio, more easily driven at high speeds, would travel faster in broad reaching conditions and would show a hollow in the running part of the polar curve in higher winds; multihulls are extreme examples of this behaviour. For other displacement yachts the polar curves would be of broadly similar form to those for the 12-Metre. For smaller yachts, having hull and rig characteristics similar to those of the 12-Metre, the polar curves giving the approximate performance might be obtained from those for the 12-Metre, by reducing V_S and V_T values in proportion to the square root of the waterline length.

Figure 1.29, based partly on data presented in Ref 1.10, depicts the development of the 12-Metre class during the last 32 years. The two curves, referring to the best yachts *Vim* (1938) and *Intrepid* (1970) representing relevant periods, illustrate the progress made in terms of V_{mg} at various wind velocities. It is noticeable that the performance of 12-Metres has not improved much in light winds, the main differences in the tank tests showing up in wind velocities above 9 knots; V_{mg} improvement at $V_T = 20$ knots is about 12 per cent. Looking at the table in Fig 1.29 one may deduce that this improvement can be attributed mainly to the very much higher stability of *Intrepid* due to higher displacement and ballast ratio, as compared with *Vim*, also to her longer waterline and smaller wave-drag, which most probably was reduced by adding a bustle to the afterbody (see Fig 1.15).

Bearing in mind the enormous cost involved (*Intrepid* is reputed to have cost about $1,000,000) in research work, testing, developing and building of sixteen 12-Metres in various countries during this period of time, the progress made has been painfully slow.

Without great risk, one may say that any further progress in higher $V_{mg}/\sqrt{L}$ that may take place in the future will be even slower than before. The recently introduced amendments to the 12-Metre rule, prohibiting expensive composite materials incorporating fibres of carbon, boron etc. (not to mention gold keels!), the use of

Fig 1.29   Optimum performance curves V_{mg} versus V_T of two 12-Metre boats. After many years of development any genuine and measurable improvement in 12-Metre hull shape is so unlikely that, according to some American sources, there seems little sense in spending money to develop a better hull. As a matter of fact, the average recorded differences in elapsed time between the American contenders in 1977 were about one minute around the 24·5 mile triangular course.

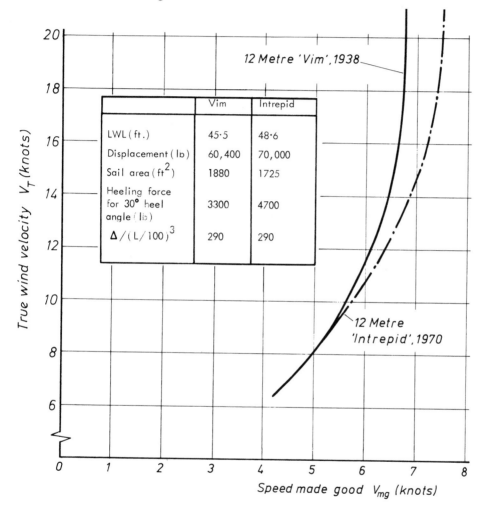

	Vim	Intrepid
LWL (ft.)	45·5	48·6
Displacement (lb)	60,400	70,000
Sail area (ft^2)	1880	1725
Heeling force for 30° heel angle (lb)	3300	4700
$\Delta/(L/100)^3$	290	290

which might save weight or reduce the size of scantlings, make future progress exceedingly difficult.

It seems that, from the designer's point of view, conventional ballasted yachts are approaching certain limits of $V_{mg}/\sqrt{L}$. It is strictly conditioned by the geometry of yachts, which are controlled by existing rules in operation. They reflect in a way, the philosophy established by yachtsmen at the end of the last century, that *a good sound yacht should aim at the best compromise between seaworthiness, habitability, safeness and speed.* Those requirements are conflicting, and it appears that in the case of 12-Metres seaworthiness and course-keeping ability have already been sacrificed to a large extent for the sake of higher speed.

America's Cup

The stipulation in the original deed of gift that the America's Cup challenger must cross the ocean on her own bottom was intended to make sure that the challenger was a seaworthy and ocean-going yacht. Could the contemporary 12-Metre crew accept without hesitation the original stipulation?

The most interesting fact perhaps from the history of the America's Cup is that concerning the schooner *America* herself. She almost certainly was the only yacht to be guaranteed by her builder Mr W H Brown, who undertook 'to build a schooner that should outsail any other vessel at home or abroad, and has agreed to make the purchase of her contingent upon her success.' Who would dare to take a similar risk today?

As matters stand at the moment, races between top helmsmen sailing boats built to strict measurements rules are often lost or won on time margins of a few seconds. An improvement of as little as 2 per cent in performance, in terms of V_{mg}, can be regarded as quite dramatic and very difficult to achieve on the designer's part, bearing in mind that such variables as crew expertise, sails, unsteady winds, waves, sea conditions, etc. also matter a lot, and they can hardly be investigated in the wing tank or wind tunnel.

It has been said that no more esoteric work can ever fall to a naval architect than the designing of America's Cup challengers and defenders. The renowned C P Burgess, making comments some time ago, told the Society of Naval Architects that

'the modern America's Cup racer bears not the slightest resemblance to any useful craft in the world, and she does not even contribute to the development of yachting as a true sport apart from the satisfaction of an illogical national vanity. But having damned them, I must confess to an absorbing interest in the problems set by these extraordinary craft. They have the fascination of sin.'

Waterline

It has already been mentioned that, when comparing the performance of different types of boat or boats of different size, one should realize that it is not speed alone which counts, but speed in relation to hull length L or relative speed $V_S/\sqrt{L}$. According to a fundamental principle of hydrodynamics, the speed V_s of a boat is proportional to the square root of her waterline length L, i.e. $V_s \sim \sqrt{L}$. This refers to the well known fact that mere increase of size, with no change of other design features, will increase sailing speeds roughly in proportion to the square root of the increase of length L. This means, in simpler language, that a longer and

Photo 1.7A,B When the middle part of the hull is in one huge wave trough with the two crests close to each end, a displacement type of yacht experiences a kind of resistance barrier.

obviously more expensive boat is expected to be faster than a shorter one. For instance, a boat of 100 ft LWL, identical in hull design to one of 25 ft LWL, should be able to sail at about double the maximum speed of the smaller one. Even if the hulls are not similar in the strict geometrical sense, but are merely of the same general displacement form, they experience a similar sharp increase in resistance when speed/length ratio $V_S/\sqrt{L}$ approaches 1.3. This dramatic increase in resistance compared earlier with the so-called sound barrier for supersonic aircraft, is due to the peculiar wave-pattern generated by the hull itself. Photographs 1.7A and B illustrate the physical reason for this. When the middle part of the hull is in one huge wave trough with the two crests close to each end, a displacement type of yacht experiences a kind of resistance barrier, which practically puts a limit on the attainable speed. Sailing downwind in strong winds, when the sail area carried is limited by what the structure can stand, the so-called 'sailing' or 'effective length' of the hull is virtually the only speed-limiting factor. In practice, heavy-displacement yachts can only attain V_S about $1.5\sqrt{L}$ (called sometimes 'hull-speed limit') in the most favourable reaching conditions. Photo 1.7B demonstrates convincingly that even a streamlined body such as a hydrofoil, while surface piercing at $V_S/\sqrt{L}$ ratio of about 1.2, produces a conspicuous wave system with consequent high wave drag. In general, the smaller the stern wave, the lower is the wave resistance, and hence different stern waves produced by similar boats, at the same speed/length ratio $V_S/\sqrt{L}$, may serve as a rough indication of their hull efficiencies.

Figure 1.30 shows boundaries of constant values of $V_S/\sqrt{L}$ which separate the potential performance of a variety of sailing craft depending on their length. Thus the mean speed of heavy displacement cruiser-racers, in average weather conditions, usually does not exceed $V_S = 0.9\sqrt{L}$ (V_S in knots, L in feet). It would result in $V_S = 5.0$ knots for a 30-footer, and $V_S = 7.0$ knots for a 60-footer.

The maximum speed, attainable occasionally in most favourable conditions, is unlikely to exceed $1.5\sqrt{L}$, i.e. about 8 knots for a 30-footer. In the same conditions, a light-displacement monohull cruiser of the same length on the waterline shown in Photo 1.8 may reach $2.0\sqrt{L}$, i.e. 11.0 knots.

It has been claimed that racing dinghies, such as the Flying Dutchman (an olympic class shown in Photo 1.6), may attain the speed of 14.5 knots, which yields a speed/length ratio $V_S\sqrt{L}$ of about 3.5. Another olympic class, the Finn (Photo 1.9), with a less effective lever for the available crew weight (just toe straps) is not as fast for her waterline as the Flying Dutchman; her relevant $V_S/\sqrt{L}$ ratio is below 3.5.

The International 10 sq m Canoe depicted in Photo 1.5 with the more effective sliding seat may reach speed V_S about $4.0\sqrt{L}$.

Modern catamarans, such as the Tornado, the fastest in 'B' division, with almost a square foot of sail for each pound of her weight (see Fig 1.33), may occasionally exceed $V_S = 4.5\sqrt{L}$. A catamaran sailing with such a high speed in rough conditions, may easily bury her lee bow in the rising slope of the next wave and be in

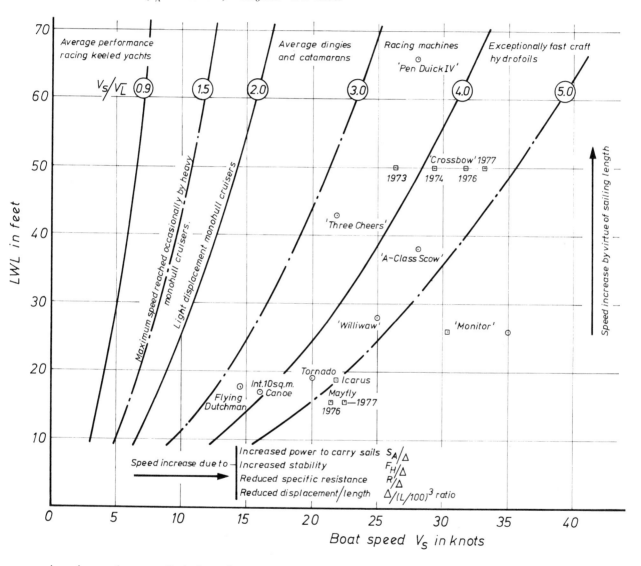

Fig 1.30 Potential speed performance of various types of sailing craft.
World sailing craft records ratified by the Royal Yachting
Association in 1978.
Class
Open–*Crossbow II*–33.8 knots
B–(S_A = 21.84–27.88 m²) *Icarus*–22.2 knots
A–(S_A = 13.94–21.84 m²) *Mayfly*–23.0 knots
10m²–(S_A = 10.00 m²) *Windglider*–19.1 knots

imminent danger of pitch-poling as shown in Photo 1.10.

Some records of the top speeds, marked in Fig 1.30 by circlets, have been given
unofficially, therefore they can only serve as an indication of the potential capacity
of a given type of sailing craft. Other records, such as the *Crossbow* official speed

Photo 1.8   Below–*New World*–designed by J Spencer, New Zealand. An impressive light-displacement cruiser-racer of displacement/length ratio $\Delta/(L/100)^3$ about 90. A huge sail area can be deployed for downwind sailing; in a similar manner as shown in the photo, which depicts *Prospect of Whitby* under all available canvas steaming downwind in the Sydney–Hobart race. Full mainsail, big boy, tallboy and spinnaker at work.

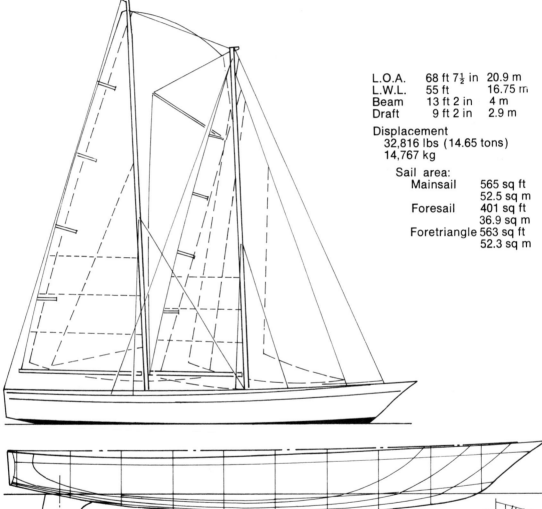

L.O.A.	68 ft 7½ in	20.9 m
L.W.L.	55 ft	16.75 m
Beam	13 ft 2 in	4 m
Draft	9 ft 2 in	2.9 m

Displacement
  32,816 lbs (14.65 tons)
  14,767 kg

  Sail area:

Mainsail	565 sq ft	
	52.5 sq m	
Foresail	401 sq ft	
	36.9 sq m	
Foretriangle	563 sq ft	
	52.3 sq m	

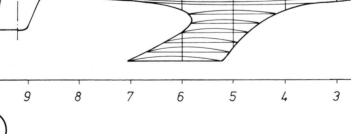

10   9   8   7   6   5   4   3   2   1

(A)

Photo 1.9 International Finn Class.
Boats sailing close to maximum speed.

record to which we will refer later, are marked by little squares. They indicate the maximum speed attained as measured officially.

The significance of length as a major speed-producing factor was recognized by rating rule-makers relatively early in yachting history, and since then they have been desperately trying to find out a fairly good correlation between the 'sailing length', the actual waterline length (LWL), and the rating. This is not an easy task.

The obvious conflict of interest dividing rule-makers and rule-breakers, i.e. yacht designers, cannot be avoided. It manifests itself dramatically in an apparent tendency, observed in all categories of cruiser-racer, towards longer boats than anticipated by rule-makers. In principle, the rating rule intention is to relate the actual waterline length to rating in such a way that rating is reflected more or less accurately by the LWL or vice versa.

Yacht designers, always looking for the proverbial loophole in the rule, are succeeding so well in manipulating design and measurement factors that, however sophisticated and tight the rule seems to be, the underrated rule-cheater may always be produced.

The wave barrier produced by a heavy displacement hull at high speed can be regarded as a trap from which displacement hulls cannot practically escape. Since, generally, the wave resistance increases in proportion to displacement, one may expect that by reducing displacement and developing so-called light displacement yachts, of lower displacement/length ratio $\Delta/(L/100)^3$, higher speeds can be attained. This is partly true–it does apply to reaching, but may not work in close-hauled conditions, as shown elsewhere.

Figure 1.31 illustrates in a qualitative sense, incorporating the most essential

Photo 1.10   The lee-bow burying, with imminent possibility of being pitch-poled is still a great problem for all catamarans sailing in strong winds. Sufficient torsional stiffness is another problem.

Fig 1.31   Relationship between the specific resistance R/Δ and displacement/length ratio ranging from 400 to 25 including fully submerged submarine experiencing no wave drag. The submarine becomes superior only when the wave-making resistance of the surface boats becomes sufficiently great. In other words, with no wave-making resistance, the submarine can beat any surface boat of $\Delta/(L/100)^3$ ratio greater than 50 if the $V_s\sqrt{L}$ ratio exceeds 1.7.

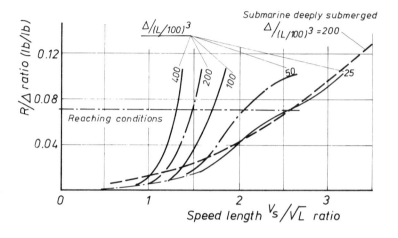

factors, to what extent the maximum speed increases while displacement/length ratio $\Delta/(L/100)^3$ is gradually reduced from 400, a representative ratio for a conventional cruiser, to 200, 100, and finally 25 which defines exceptionally light craft. The five curves marked 400, 200, 100, 50 and 25, adopted with some modification from Ref 1.19, show the specific resistance R/Δ, i.e. Resistance/Displacement ratio, as a function of the Speed/Length ratio $V_S/\sqrt{L}$.

The displacement/length ratio $\Delta/(L/100)^3$ which has been defined as a measure of a load put on a given length L, can also be defined as an index of the 'slenderness' of the submerged form of a hull which, together with the so-called prismatic coefficient (Note 1.20), are the variables which have a large influence on the specific resistance R/Δ.

Just what is meant by slenderness?

Sail-carrying ability and the necessary driving force are a function of hull stability due to form and ballast. It is perhaps self-evident that the slender form, which will sail fastest giving some forward driving force from the sails, has the least ability to withstand the heeling force. Therefore, for the sake of better stability, a certain fatness in hull form, and the drag penalty for it, must be tolerated. Let us assume that, in broad reaching conditions when the heeling force F_H is negligible and hence stability is of secondary importance, the sails are capable of developing a driving force F_R in the order of 0.07Δ, i.e. 7 per cent of the boat displacement Δ. Such a large driving force, although extreme, is still within a realistic limit of some craft, provided the wind is sufficiently strong. On the assumption that in steady sailing condition F_R = R and therefore $F_R/\Delta = R/\Delta = 0.07$, one can find from Fig 1.31 that the speed/length ratios $V_S/\sqrt{L}$ for five gradually lighter boats of $\Delta/(L/100)^3 = 400, 200, 100, 50$ and 25, would be of the order 1.3, 1.5, 1.7, 2.1 and 2.6 respectively. Evidently, in reaching conditions, the light displacement forms seem much superior.

However, in close-hauled conditions and strong winds, the order of merit is somewhat reversed. If heavy ballast and a certain fatness in hull form are the only means of developing sufficient stability, as in the case of ballasted cruiser-racers, one has to accept a penalty for it in terms of drag, in order to gain enough power to carry sail. Since an increase in ballast causes an increase in both hull resistance and in driving power of sails, it is rather obvious that there must be a certain optimum displacement/length ratio $\Delta/(L/100)^3$, when the gains in power to carry sail effectively are just balanced by the losses in terms of hull resistance. It appears that, as far as cruiser-racers are concerned, the displacement/length ratio $\Delta/(L/100)^3$ in the range 300–350 seems to be an optimum for the best all-round performance.

By manipulating the major design variables given in Table 1.2, it is possible to construct yachts having very different properties for the same rating: boats for strong winds as well as boats for light winds and boats for particular courses, etc. However, it is extremely difficult to improve the performance of a yacht over the whole range of true wind velocities and the whole range of courses relative to wind direction unless, as has been done to some extent in the case of modern racers, stripping out their interior enables a higher ballast/displacement ratio to be achieved. As a matter of fact, not only are racers nowadays designed for local wind and sea conditions, but also the particular time correction system which supplements the rating formula in operation has to be taken into account.

G High speed sailing

Figure 1.32 provides a basis for judging the specific resistance R/Δ of two different hull forms of the same length/beam ratio $L/B = 4.5$:

1. A displacement round-bilge form reflecting the ballasted yacht hull.
2. A hard-chine skimming form, suitable for light displacement dinghies (Ref 1.21).

Fig 1.32 Specific resistance of two different hull forms (L/B–length/beam ratio).

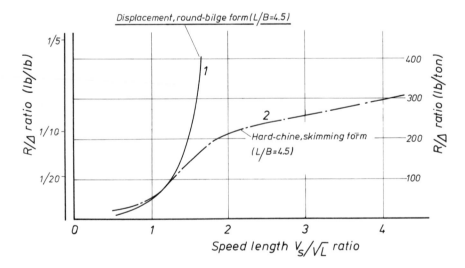

Photo 1.11   Fighting the boat up.
              Getting caught under the boat in a life-jacket can be
              dangerous.
              Drawing from French Magazine *Bateaux*–article by J
              Dumet.

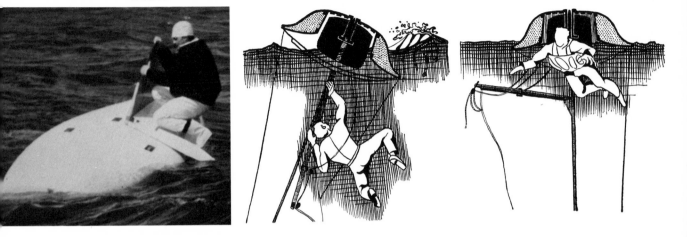

The two curves representing specific resistance in lbs per ton, for a hull of $\Delta/(L/100)^3$ = about 150, clearly demonstrate the high speed potential of a skimming form. Beginning from a speed/length ratio $V_S/\sqrt{L}$ exceeding 2.0, the specific resistance R/Δ becomes nearly proportional to boat speed, and builds up very slowly compared with the resistance experienced by a displacement form.

Provided that the wind is strong enough, a speed/length ratio of the order of 4.0 or even more is quite feasible for highly sensitive planing forms, descendants of those magnificent skimming dishes developed by the end of the 19th century. The only snag is that they are capsizable. For this reason, the fathers of world sailing from the IYRU, anxious to cotton wool the international high speed sailing racing machines and the people who sail them, recommend that 'rescue launches in the ratio of one rescue boat for every 15 starters are required from half an hour before the start'. Somehow, somebody must pay a price for speed, and the *requirements for speed and seaworthiness appear to be fundamentally incompatible*.

Paraphrasing H Saunders' idea (Ref 1.22), one does not need discerning eyes to discover that the civilization of man, his anatomy, as well as sailing boats, are intended to function normally when right side up, corresponding to what might be termed the natural or customary position. Once the sail and centreboard have exchanged the media in which they normally operate, i.e. a boat has turned turtle, as shown in Photo 1.11, it is very difficult to regain the customary position without outside assistance. This applies particularly to catamarans. Beware of getting caught under the boat or inside the cabin in a life-jacket or being swept away from your craft. Even those already expert in the art of capsizing may find it worth reading some information concerning safety, capsizing and self-rescue action, incorporated in Note 1.23.

Multihulls

Multihulls' potential capability of reaching high speed under sail is much greater than that of monohulls for the same sailing length. They are able to carry the large sail area necessary for high speed, by countering heeling moment with the inherent stability of widely separated hulls, which may then be of a fine, slender form and therefore easy to drive. By ballasting the windward hull an almost unlimited righting moment can be provided; speed limit is practically restricted only by the strength of

Fig 1.33A   Tornado's predicted and measured performance.
     B   Polar diagram V_S/V_T based on full-scale experiments (faired data).
       Particulars of Tornado tests:

LWL–19.2 ft
Weight (boat) 400 lb
Weight (crew) 340 lb
Total weight 740 lb (Δ = 0.33 ton)
$\Delta/(L/100)^3 = 46.5$
$S_A = 235$ sq ft (mainsail + jib)
$S_A/\Delta = 710$ sq ft/ton

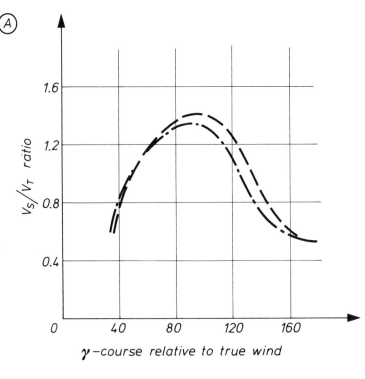

γ –course relative to true wind

—·— theory (Bradfield)

——— faired experim. data line

the hull's structure, and of rigging, and of course the state of the sea. After all, for every type of craft, however cleverly and strongly built, there are wind and sea conditions in which survival becomes the first priority, and it is commonly known that safety then depends upon yielding to the sea and not standing up against it.

There are few data available concerning full-scale performance of catamarans on various courses relative to the true wind. Figure 1.33 presents the unique results of limited tests done by Bradfield (Ref 1.24) on the well known, high performance, International Tornado. The tests were conducted in order to compare the

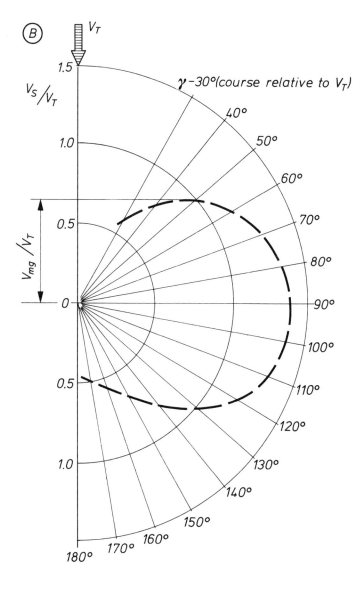

performance of the hydrofoil (Ref 1.25) with that of the multihull, assuming that any daysailer that is competitive with the Tornado is competitive with 95 per cent of present-day fast sailing craft. The test gave also an opportunity to compare the actual full-scale performance with calculations based on Bradfield's theory, described in Ref 1.26 and plotted in Fig 1.33A.

The Tornado was sailed in flat water, at wind speeds ranging from 10 to 15 mph. The maximum boat speed V_S recorded was 19.2 mph in a 13.9 mph breeze, on a course angle γ about 90° relative to the true wind V_T. It gives a V_S/V_T ratio = 1.38, which means that the boat was easily exceeding the true wind speed V_T. Figure 1.33A shows the faired experimental data compared with the performance predicted by theory. The performance is expressed in terms of the ratio: catamaran speed V_S/true wind speed V_T plotted against the course sailed γ, relative to true wind direction. This ratio can be regarded as a measure of the catamaran's efficiency as far as high speed performance is concerned. When the data of Fig 1.33A are crossplotted in the form of the familiar polar diagram in Fig 1.33B, it comes out that the best tacking angle upwind is approximately 50° to the true wind, which is consistent with practice; and the best tacking angle downwind is about 130°.

It is interesting to notice that, within the approximations of Bradfield's theory, which sacrifices some theoretical sophistication and/or breadth of application in the hope of gaining simplicity and perspective, the true speed ratio V_S/V_T on any selected heading is found to be independent of the true wind speed. A relatively simple algebraic solution offered in Ref 1.26 is undoubtedly of great engineering value, as well as of tactical and tuning importance to racing helmsmen.

It is evident from Fig 1.33A that the theory underestimates the measured performance on reaching, but substantial agreement is apparent. It has been concluded that the discrepancy is due to underestimating of the rig characteristics. In fact, the Tornado rig data have not been available and the wind tunnel Dragon rig data of Ref 1.27 were used instead. The full-scale results would indicate that the Tornado rig is an unusually clean and powerful one; anyway more efficient aerodynamically than the Dragon rig, as tested in the wind tunnel.

The V_{mg}/V_T ratio for the Tornado, seen directly from the polar plot of Fig 1.33B, is of the order of 0.64. The speed made good to windward V_{mg}, expressed in terms of $V_{mg}/\sqrt{L}$, is given in Fig 1.19. It facilitates a direct comparison with other sailing craft already discussed; evidently the Tornado is a very good windward performer. She introduces, no doubt, a new quality into the racing scene, being perhaps the right answer to those high speed sailing oriented enthusiasts.

As already mentioned, in Bradfield's dimensionless theory the speed ratio V_S/V_T is shown to be a universal performance criterion, virtually independent, at least in a certain range of true wind velocities, of the true wind speed. This finding is reflected reasonably well in the World Multihull Championship 1968 Records. Figure 1.34 shows, for instance, the results of *Whiplash*, one of the outstanding C-Class catamarans. Assuming certain unavoidable deviation from the best potential performance, on account of crew ability to coax the boat into her superior

Fig 1.34

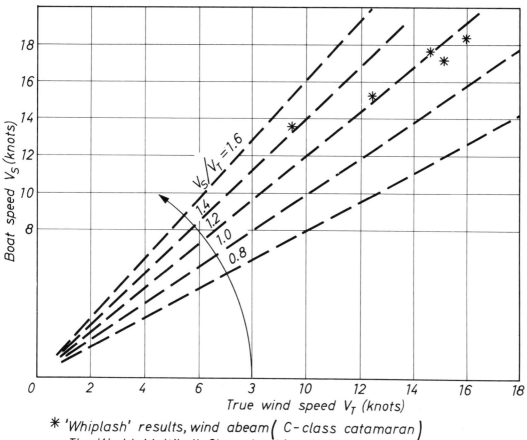

* 'Whiplash' results, wind abeam (C-class catamaran)
  The World Multihull Championships-1968.

performance, measurement uncertainties, unsteadiness of wind, sea conditions etc., the recorded results of trials at wind abeam are pretty close to the straight line labelled $V_S/V_T = 1.2$. Anyway, the observed deterioration of the catamaran's performance (in terms of V_S/V_T ratio) with wind speed V_T, is much less dramatic, almost negligible, when compared with that of keeled yacht performance variation, as depicted in Fig 1.20C. The plot combining V_S, V_T and V_S/V_T, as in Fig 1.34, can be used to compare the relative efficiencies of various multihull configurations.

Transatlantic races

In 1972, multihulls celebrated two outstanding achievements, the first place in the *Observer* Single-Handed Transatlantic Race, and first in the RYA/John Player World Sailing Speed Record.

Figure 1.35 and Photo 1.12 illustrate *Manureva* (ex-*Pen Duick IV*), the winner of

Fig 1.35   LOA–67.0 ft
Beam–35.0 ft
Displacement–about 7.0 ton.

After four years of testing and development *Pen Duick IV*, sailed by A Colas (France), won the 1972 Single-handed Transatlantic Race in 20 days 15 hr. Probably the ugliest as well as the most ruthlessly efficient racing machine and one of the fastest sea-going trimarans in the world.

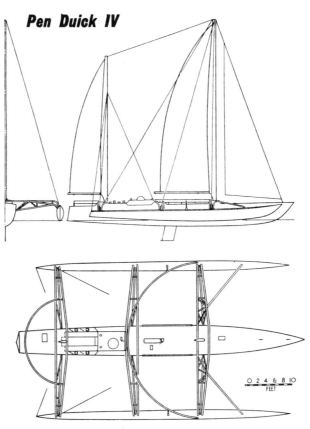

the Transatlantic Race, sailed by A Colas. Built initially by Tabarly for the 1968 race, she became, after four years of testing and development, one of the fastest sea-going trimarans in the world. Regarded by some of the smaller competitors as the most hateful of monsters, *Pen Duick IV* was probably the ugliest as well as the most ruthlessly efficient racing machine in existence at that time. Her bridge structure, shown in Photo 1.12, resembled the innards of an oil refinery, but it was functional and it was this only and not the beauty which mattered.

According to the International Hydrofoil and Multihull Society, there was nothing new in the design concept of *Pen Duick IV* (Ref 1.28). The idea of a central hull connected to the floats with lightweight lattice arms was, they say, a direct copy from the *Trifoil* and *Triform* class of trimarans, developed here nearly twenty years

Photo 1.12   Bridge structure of *Manureva* resembled the innards of an
             oil-refinery, but it was functional (*Yachting World*).

ago. The Society even had their own terminology for this configuration: they called
it the MEROLOA principle, which stands for Minimum Element Resistance Open
Lattice Outrigger Arms.

   For comparison, Photo 1.13 reveals some details of one of the rivals of *Pen Duick
IV*, the outstanding Dick Newick proa *Cheers* (American entry in the 1968 OSTAR,
sailed by Follett), described by Macalpine-Downie before the race as 'original,
inventive, wholly convincing and enormously attractive, ...she is innocent of all but
bare essentials. Inside she is dead white, naked and is tight as a teacup.'

   To qualify for the race *Cheers* sailed across the Atlantic, single-handed, in 29 days.
However, for several perfectly good reasons, the Committee decided that she was
potentially dangerous and could not be accepted. After some hesitation, the verdict

Fig 1.36 *Three Cheers*, designed by R Newick. The solid wing deck is rather unusual in a racing trimaran, but the designer feels that the danger of being overturned due to deck windage when heeled can be reduced to an acceptable level by the prudent handling that any fast vehicle requires.

L.O.A.	14.02 m	46 ft 0 in
L.W.L.	13.18 m	43 ft 3 in
Beam	8.23 m	27 ft 0 in
Draft	0.91 m/	3 ft/5½ ft
	1.68 m	
Displacement	3180 kg	7000 lb
Sail Area	77 sq m/	830 sq ft/
	132 sq m	1420 sq ft

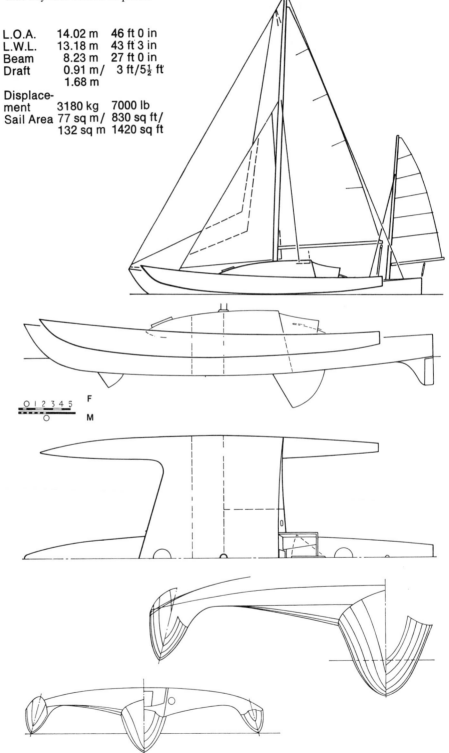

O 1 2 3 4 5   F
O   M

Photo 1.13   *Cheers*, an American entry in the 1968 OSTAR.
Dick Newick's proa, sailed by T Follett, had been described by some commentators before the race as 'original, inventive, wholly convincing and enormously attractive... She is innocent of all but bare essentials. Inside she is dead white naked and is tight as a teacup...' When she finished third in the race, the policeman on watch on Rhode Island commented: 'Gee, he must be nuts to sail that thing.'

LOA–40 ft.
LWL–30 ft.
Displacement–1.34 tons.
Sail Area–340 sq ft.

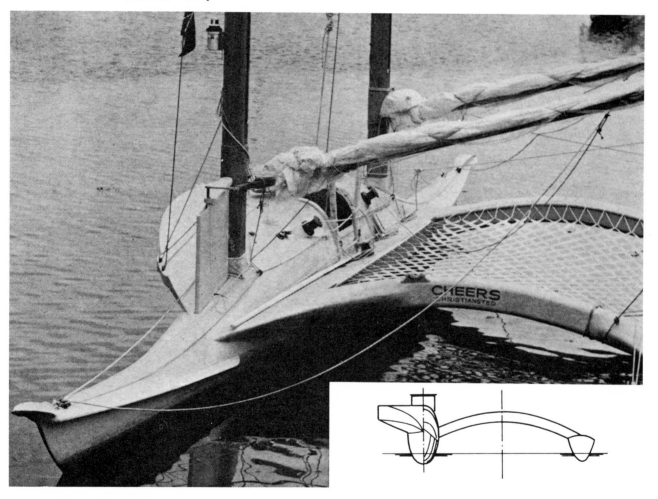

was reversed. She sailed an effortless race and came in a very good third, several days ahead of the next multihuller (in this race *Pen Duick* withdrew at an early stage).

The same team that produced *Cheers* entered the 1972 race in a trimaran, shown in Fig 1.36 and called *Three Cheers*. The designer R Newick said of his boat:

Fig 1.37   *Vendredi 13* (French entry, 1972). D Carter's extreme approach to the sailing machine; in theory longer boats are likely to sail fast by virtue of sheer length, but prevailing wind and sea conditions also matter. Luck with weather plays a large part in the Single-handed Transatlantic Race. *Vendredi 13* was regarded as just about the longest, narrowest, lightest and shallowest boat that anyone could ever imagine.

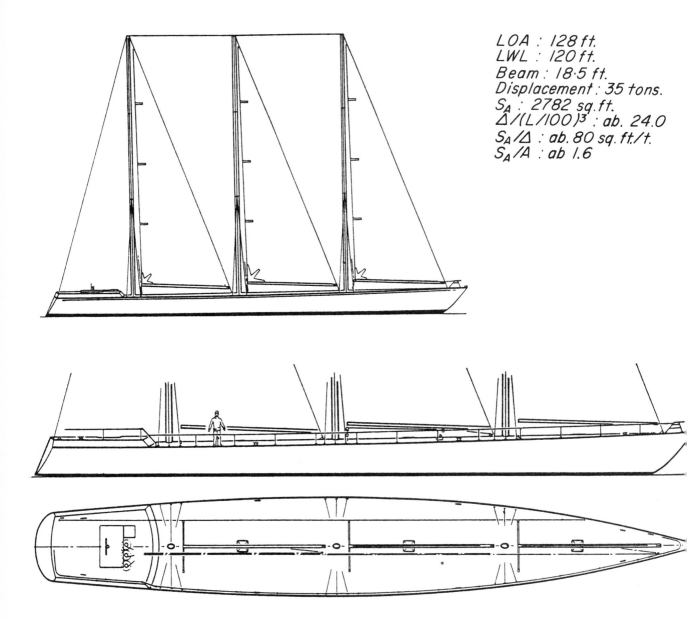

LOA : 128 ft.
LWL : 120 ft.
Beam : 18·5 ft.
Displacement : 35 tons.
S_A : 2782 sq.ft.
$\Delta/(L/100)^3$: ab. 24.0
S_A/Δ : ab. 80 sq. ft./t.
S_A/A : ab 1.6

'Her 7000 lb racing weight is not especially light because I feel that the power to carry sail is important in an offshore windward race. The ketch rig with two headsails, coupled with the well spread lateral plane using a forward centreboard gives quite good directional stability in a wide range of conditions. The solid wing-deck is not now usual in racing trimarans. One of its principal advantages is the ability to distribute stresses over wider areas. It also embraces accommodation and enables headsails to be sheeted exactly as needed, plus making for a drier craft at high speeds. The overturning moment of its windage when heeled is an adverse factor that can be reduced to acceptable levels by its small area, good shape and the prudent handling that any fast vehicle requires.'

It has been claimed that on the day of launching a speed above 22 knots was achieved, which makes $V_S/\sqrt{L}$ about 3.4; there were fourteen people aboard at that time. She finished fifth in the 1972 Race, crossing the Atlantic in 27 days 11 hours.

Another French entry in the 1972 Transatlantic Race, *Vendredi 13* (sailed by Y Terlain who came second, 16 hours behind the winner), is depicted in Fig 1.37. Leading for a long while, Terlain might have won the race if he had had more luck with winds in the closing stage. *Vendredi 13*, described as a logical monster of 120 ft LWL, designed by Dick Carter, was 80 per cent longer than the next biggest boat. This gave her a tremendous potential speed advantage over her competitors. Unfortunately prevailing winds were not strong enough to expose the supreme merit of her sheer sailing length.

In his nostalgic article *My Big Boat and Me* (Ref 1.29) Terlain says: 'I was hoping to turn a page of sailing history, to prove that a big monohull could be faster than the best quality, best tested of the multihulls in the world at this time. I think we turned only half a page of history. One cannot say that multihulls are faster, it's just a matter of sea conditions.' This is rather a confession of faith–very difficult to prove in the special conditions of the *Observer* Single-Handed Transatlantic Race.

There should be no doubt that if both catamaran and monohull are developed to the extreme length, and they are both expertly sailed by equally determined tough crews in similar weather conditions, the catamaran is bound to be faster, particularly when both craft are of the same length. The already existing knowledge about factors affecting performance, as well as speed records incorporated in Fig 1.30, clearly support such a logical conclusion, applicable specially in the case of an excessively long monohull sailed single-handed. The boat must then necessarily be undercanvassed in order to maintain some sort of command over the driving power plant–thus resulting in a small S_A/A ratio. Hence the misery of being 'glued to the water in light winds' cannot be avoided. Consequently *Vendredi 13*, more than any conventional light displacement craft, suffered deficiency of driving power. Both the S_A/A ratio which governs in light winds, as well as the S_A/Δ ratio which governs most of the time, the first being about 1.6, and the second about 80.0, are well below the required values securing a satisfactory performance in a variety of weather conditions (Table 1.8). Trying to reconcile excessive length with single-handedness, one has to pay a penalty in terms of increased wetted surface, reduced sail area, and

Fig 1.38   Alain Colas's giant *Club Méditerranée*.
          Approximate dimensions:

LOA–236 ft (72.2 m)
LWL–208 ft (63.3 m)
Beam–31.5 ft (9.6 m)
Displacement–250 tons
S_A–10,700 sq ft (1000 sq m)

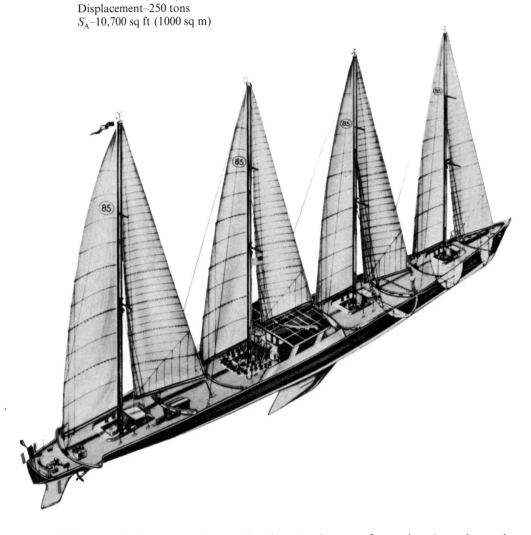

less efficient rig of three staysails working in a tandem configuration (aerodynami-cally ineffective). All these factors help to explain why it was so difficult to coax the boat's head through the wind under force 4, as Terlain complained.

Exactly the same arguments apply to another French entry–Alain Colas' giant *Club Méditerranée*, almost twice as big as *Vendredi 13* and shown in Fig 1.38. This four-masted schooner designed to win the OSTAR race to Newport, Rhode Island, in 1976 is based upon the *Vendredi 13* concept, in which overwhelming emphasis is put on hull length as a dominant speed-producing factor. Although the estimated speed capability of *Club Méditerranée* of about 22 knots, based on hull length alone, gave her enormous theoretical advantages, she failed the OSTAR test, coming 7 hours 28 minutes behind Tabarly's 73 ft ketch *Pen Duick VI*. Her average speed was about 5.2 knots, which in terms of speed/length ratio gives $V_s/\sqrt{L} = (5.2/\sqrt{208} = 0.36$ only.

Certainly, as in every race, much depends on wind and sea conditions. Nothing knocks average speed down more quickly than a couple of days in the windless wilderness of the Atlantic. A map in Fig 1.39, taken from Alan Watts' analysis in Ref 1.30, well illustrates the strategic dilemma facing every sailor participating in OSTAR. Choosing the short, direct Northern route from Ireland to America, one has to accept headwinds almost all the way for about 3000 miles. The Southern route of the old sailing ships, which curves down between 20° and 50° latitude, is much longer (about 4000 miles), but there the happy mariner may find everything in his favour–a wind from about the beam and a mobile sea surface entrained by that wind. Unfortunately this route skirts the zone of unpredictable weather and dreadful calms in the so-called Horse Latitudes. How each sailor/boat combination meets the ever changing wind and sea conditions is of enormous significance. In order to control their luck with the weather *Vendredi 13* and *Club Méditerranée* were equipped with shipboard weather map facsimile machines.

Those two monsters were built for the same specific purpose, according to the 'no holds barred' rule. If man dares to handle such huge boats, are larger boats feasible for single-handed sailing? Bearing in mind the differences in length of the competing boats in the 1976 event–between 23 and 236 ft–one may wonder what is the purpose of this whole exercise in terms of fair play?

Most of the competitors who crossed the starting line in the OSTAR events were probably driven by the wildly romantic expectation underlying Blondie Hasler's concept of this race, that of the lonely man and the sea. But not all of them.

Is the scientific and technological progress which is spreading rapidly in our world, liable to corrupt the soul of Sailing Man? Hasler described *Vendredi 13* as 'a fascinating experiment in the wrong direction'. And the *Yachting Monthly* commentator pointed out:

'It is with regret that one suggests the need in future to limit the overall length of yachts eligible, but with the 128 ft of *Vendredi 13* it must be clear that *we have reached well beyond the limit of desirable development in this direction*. For it is frightening to contemplate such a boat, dwarfing the one man crew and inherently difficult to manoeuvre, while capable of very high speeds, sailing in the crowded waters at either end of the race' (Ref 1.28).

Apart from this obvious and directly practical point, one may ask what is the purpose of pursuing existing knowledge to its extremes? What new can be learned from such a costly experiment like *Vendredi 13* or *Club Méditerranée*, which is not already known or could not easily be predicted with some imagination and a little arithmetic?

Fig 1.39   Atlantic routes as seen by a meteorologist (based on material from *Yachting World*).

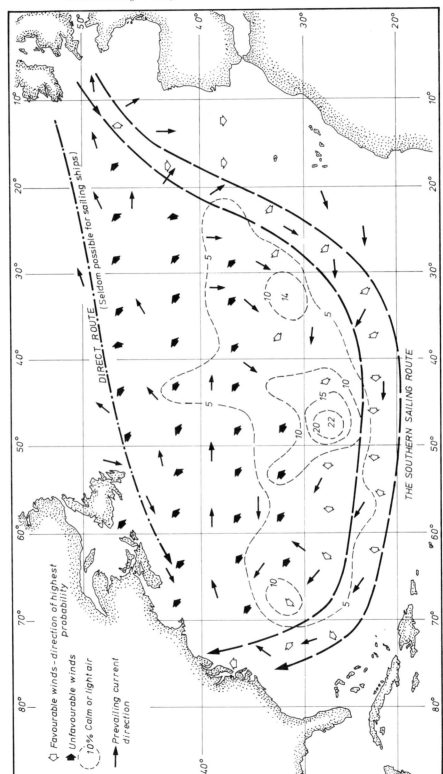

It seems that, in our restless society stirred by its ever rising expectations and its desire for a 'kick', monsters are fashionable. And no doubt it is easier to produce freaks than to create harmonious, well-balanced masterpieces of lasting value.

World speed records

Photo 1.14 and Fig 1.40 illustrate another spectacular hair-raising monster, *Crossbow*, the world sailing speed record holder. The initial speed of 26.3 knots, officially recorded in 1972 and since raised to over 30, was not a shattering success, but rather the beginning of an entirely new form of competition. It started during the week of September 30–October 8, 1972, sponsored by the Royal Yachting Association and John Player, the cigarette manufacturers, who offered £1000 for the first prize and a further £2000 for the highest speed reached anywhere in the world before the end of the year. The object of the competition for the John Player World Speed

Photo 1.14   *Crossbow*, the official world sailing speed record holder 1972–75, at Portland Harbour, England. The *Crossbow*'s speed over the half-kilometre course seemed to prove that she sailed faster when the gondola was just clear of the water. That is where the three men on the plank came in.
  The picture is reproduced with kind permission of *Yachting World*.

Fig 1.40 *Crossbow*–designed by R Macalpine-Downie. Nicknamed 'The Beast' for obvious reasons, this pencil-slim, one-tack proa was designed for ultimate speed under sail.

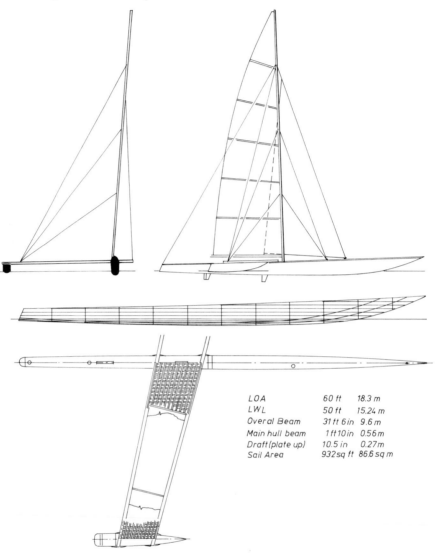

LOA	60 ft	18.3 m
LWL	50 ft	15.24 m
Overal Beam	31 ft 6 in	9.6 m
Main hull beam	1 ft 10 in	0.56 m
Draft (plate up)	10.5 in	0.27 m
Sail Area	932 sq ft	86.6 sq m

Record Trophy, awarded annually, is to sail as fast as possible in a straight line at any angle to the wind along a 500 metre course.

As originally designed, the 50 ft long, pencil-slim, one-tack proa *Crossbow* with a 30 ft outrigger on one side and four men walking to and fro along the narrow trampolin of canvas (to provide just enough righting moment of some 1500 lb on a 25 ft arm, for about 930 sq ft of sail area), won the trophy convincingly. The speed of 26.3 knots recorded in 1972 in winds of about 19 knots makes V_S/V_T about 1.5, well below her anticipated top speed.

Fig 1.41   *Crossbow II.*

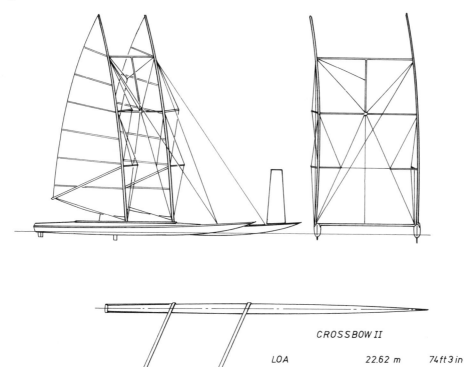

CROSSBOW II		
LOA	22.62 m	74 ft 3 in
Beam	9.06 m	29 ft 9 in
LOA (hull)	18.67 m	61 ft 3 in
Beam (hull)	0.57 m	1 ft 10.5 in
LWL (hull)	15.89 m	52 ft 1.5 in
Sail area	130 sq m	1400 sq ft

To be asked to design such an entirely new type of contestant with literally no restriction is not something which every designer would relish. Macalpine-Downie explains some of his approaches, as follows: 'Accommodation, convenience, adaptability, marketability, all count for nothing; even seaworthiness and manoeuvrability are irrelevant except in so far as they threaten to limit her flat-out, wound-up terminal speed' (Ref 1.31). The designer claims that *Crossbow* is *the most extreme conventional boat* ever built. In a way she is, for she did not make use of hydrofoils in first trials, but in 1973 the outrigger was equipped with a small hydrofoil whose purpose was to prevent it from touching the water at high speed and save the crew from running up and down the trampolin. Subsequently, for the 1976 contest, the boat was developed further and the new *Crossbow II*, which reached the speed of 31.8 knots, is a 60 ft twin-hulled, twin-masted vessel with staggered hulls. As shown in Fig 1.41, the lee rig is ahead of the windward one. The idea of splitting

Photo 1.15   *Icarus* (1972) with all metal foils (one set) and steering by transom-hung inverted 'T' rudder/foils. Retraction is by rotating the foil forward on its mounting.

Breeze of about force 3. Fully foil-borne *Icarus* (1970): one of the tentative steps using two sets of foils. The helmsman on a trapeze operated steerable front foils, but control was rather poor.

the rig into two was to lower the centre of effort and to achieve a more controllable sail plan than *Crossbow I*. It was reported by the *Yachting World*, 1976, that there was a doubt about the air rudder arrangement; the point being that the tiny water rudder on the windward hull will lift out, on occasions, so that the air rudder, which is an aerofoil section to give a fraction of lift, will take over.

Hydrofoils

Her closest rival *Icarus*, shown in Photo 1.15 and Fig 1.42, achieved 21.6 knots in 1972. This is a Tornado cat, equipped with a hydrofoil system developed by the J Grogono team (Ref 1.32). The sailing hydrofoil is likely only to be foil-borne over a relatively narrow range of sea and wind conditions, being sluggish or unmanageable off the foils (the 15 knots wind velocity and the preferably flat water that are needed to get the average sailing hydrofoil foil-borne, do not occur every day). The *Icarus* development was thus restricted to the standard Tornado catamaran. Foils are

Fig 1.42   *Icarus*'s sets of foils drawn to scale as evolved during the period 1969–1972. Dotted parts movable.

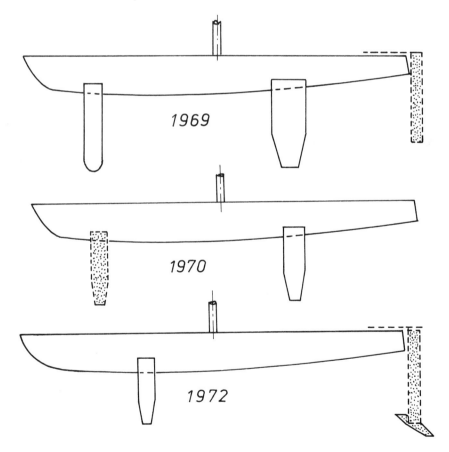

Photo 1.16   An 'ogival section' and a method of machining the metal foil.

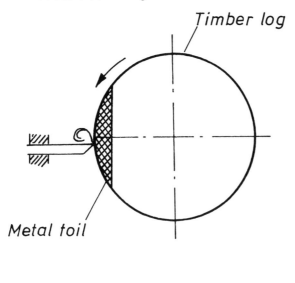

detachable, easily fitted on suitable days, but the boat may be 'in class' on other occasions. As shown in Photo 1.15 and Fig 1.42, the foils, initially constructed in wood, have been developed into all-metal foils, with a less complicated configuration, not so susceptible to mechanical failure.

The design principle of *Icarus* is that of the aeroplane configuration: the main dynamic lift is provided by so-called *surface-piercing foils* at the front, and the rear foils, submerged inverted 'T' rudder foils, merely follow the line set by the front foils. For further details Refs 1.33, 1.34 and 1.35 may be consulted. For some theoretical reasons as well as practical ones, the lifting foils used are of the so-called 'ogival section', shown in Photo 1.16. The attached sketch illustrates schematically the process of machining. One of the surfaces of the ogival section is of constant, circular radius, another one is flat, thus enabling an economical use of man-hours and also of material (usually aluminium alloys, sometimes reinforced by carbon fibre).

Hansford's *Mayfly*, shown in Photo 1.17, smaller than *Icarus* and using a similar foil configuration, was regarded as the most successful, in some ways, of the foilboats competing at Weymouth in 1972. She rises onto her foils readily and remains on them constantly over 10 knots. However, both those most successful boats suffer serious control problems in choppy water and stronger winds. Choppy sea for these sailing hydrofoils was summed up by D Pelly, who had a trial sail in *Mayfly*...

Photo 1.17　Hansford's *Mayfly*, regarded as the most successful, in some ways, of the foil-boats competing at Weymouth in 1972. She rises onto her foils readily and remains on them constantly in winds over 10 knots. Although her speed of 16.4 knots in a wind of about 16 knots was creditable for a boat of only 15 ft it does not represent the best that she can do.

'as the speed rises she rides higher and higher until there is only about six or nine inches of the lifting foils left in the water. This is OK when the foils are going through a nine-inch wave crest but not so good when they come to a nine-inch trough. Bereft of their supporting medium, the foils have no choice but to let go and the boat bellyflops into the water. Smacking straight into a wave crest at 16 knots is not too kind to a 15-footer...I made just two runs when the wind was between 15 and 20 knots. On both the boat felt distinctly out of control...I agreed that the water was too rough and we put the boat away' (Ref 1.36).

Similarly, *Icarus* found the Weymouth chop not particularly to her liking and suffered from fairly severe control problems, resulting from unpredictable ventilation of the inverted 'T' rudder foils (Note 1.37). This caused a high speed capsize, fortunately with little damage.

Speeds achieved by these leading hydrofoil sailing craft, *Icarus* and *Mayfly*, during the 1972–1976 Trials are still below the *Monitor*'s speed record of 30.4 knots, established in 1956 and marked in Fig 1.30. Further details are given in Note 1.38, though *Crossbow II* achieved 33 knots in 1977.

Another outstanding sailing hydrofoil worth mentioning is the ocean voyager *Williwaw*, designed by D Keiper and presented in Photo 1.18. This 31 ft long hydrofoil trimaran (28 ft LWL) is capable of reaching 25 knots while foil-borne. She has already demonstrated good all-round performance in cruising off the California coast to the Hawaiian Islands. It has been reported in Ref 1.39 that some 13 knots of wind are required for her to become fully foil-borne. Lacking that wind, the boat can operate as an efficient trimaran by retracting the foils, as depicted in Photo 1.18. The design approach on *Williwaw* was to use four constant-geometry, retractable hydrofoil units, consisting of bow foil, steerable stern foil and ladder form foils on each side of the centre of effort. The trimaran hull form seems to provide excellent fastening points for this particular four-foil system, which is the subject of both US and British patents.

The system is described by her designer:

'For take-off, the bow foil has excess lifting area, so that the bow rises higher than the stern, augmenting take-off lift. The craft tends to level off at higher speed, bringing lift coefficients closer to high speed optimum. At lower speed, the system is a four foil symmetric system, but by take-off speed it becomes a three foil asymmetric system.'

Using 6 in chord elements, all the foils are trusses made up of lifting elements and struts, and are designed to withstand a hydrodynamic load of one ton/sq ft.

The apparent ability of *Williwaw* to operate in relatively rough sea and strong winds is demonstrated in this account by her designer:

'When *Williwaw* started broaching and getting uncomfortable with retracted foils (about 100 miles off the California coast), sails were dropped temporarily

Photo 1.18   Left–*Williwaw* with all foils retracted.
          Right–*Williwaw* doing 20 knots with five people aboard plus
          a couple of hundred pounds of water and food. Wind
          required for flying is about 12 knots.

LOA–31.3 ft.
LWL–28.0 ft.
Beam overall–15 ft.
Sail Area S_A–380 sq ft.

and all foils were set. The wind was force 4 from the North, but increased to force 5 and 6 later in the day. Seas ran 10–12 ft. We found that heading East we could get *Williwaw* flying, and once flying we could head off onto a broad reach or quarter and continue flying on the larger waves. Sometimes, we could fly for half a mile at a time, until our leeward pontoon bow would slice a steep wavelet and bring us back to the half foil-borne state. We were flying 20 knots or more diagonally across wave faces, sometimes heeled 50° on a wave face, but never sliding down. My crew member was a surfer and he delighted in keeping *Williwaw* howling across these steep wave faces. The foils behaved well all the time.'

–so *Williwaw* seems to be competitive with existing trimarans of her size.

The idea of water-borne craft of low resistance, supported by wings which fly just under or penetrating the water surface, has attracted the attention of inventors for over 80 years. This interest is mainly due to:

1. the higher speed potential of hydrofoils in comparison with other craft,
2. the hydrofoil's potential ability to maintain a comfortable ride and speed in comparatively severe sea conditions.

Those two inducements, which apparently did not appeal to past generations except for military application, suddenly became irresistible to contemporary mentality. This gave great impetus to the development of many successful, powered hydrofoils, and recently to sailing hydrofoils.

A lengthy paper by P Crewe (Ref 1.40) gives valuable insight into some theoretical and practical problems concerning hydrofoil craft. Although his historical review does not take sails into account as a propulsion device, it contains some information applicable to sailing hydrofoils too. The unsteadiness of the driving power from the sail makes the aero-hydrofoil problems more difficult, but certainly the basic hydrodynamic phenomena as well as structural worries are alike.

After a long period of experimentation it has been proved that powered hydrofoils can, as predicted, operate satisfactorily in comparatively rough seas (up to 10 ft waves, depending of course on boat size), and also at high speed on sheltered and inland waters. The *Williwaw* experiment confirms this. The problems of great practical importance are still those associated with dynamic stability, structural reliability, ventilation and cavitation in the case of speeds upwards of 40 knots.

A hull equipped with hydrofoils receives an upward lift which increases in proportion to the boat's speed squared. In consequence, as the craft accelerates, the foils increase their lift unloading the hull, until at take-off speed the entire weight is on the foils and the buoyancy or dynamic lift of the hull itself is reduced to zero. This feature favours a relatively conventional light-displacement hull rather than one of semi-planing, hard-chine type of higher resistance in the lower speed regime. As soon as foil-borne speed is reached and the hull is lifted completely above the water the performance of a hydrofoil craft is significantly improved. The larger the surface of the foils, the lower will be the take-off speed. However, the drag of the hydrofoils themselves will also be relatively greater.

Figure 1.43 facilitates in a qualitative sense a comparison between the characteristics of a semi-planing form such as a hard-chine hull and a hydrofoil. As in Fig 1.32 the curves are plotted in terms of specific resistance R/Δ (resistance in lb, R, divided by displacement in tons Δ) against speed/length ratio $V_S/\sqrt{L}$, therefore they can be applied approximately to any size of craft. Compared with a naked hull, the foils deployed increase substantially the resistance of a hydrofoil at low speeds and impair its performance quite drastically. Configurations embodying fully submerged foils systems are shown to have hump propulsive thrust requirements at take-off speed, but once this has been passed, hydrofoils offer a substantial

Fig 1.43 Comparison between conventional planing craft (hard-chine boat from Fig 1.32) and a hydrofoil craft (Refs 1.25 and 1.40). Resistance characteristics of hydrofoil craft presented qualitatively depend to a large extent upon foil area, cleanness of their supporting structure and foil efficiency.

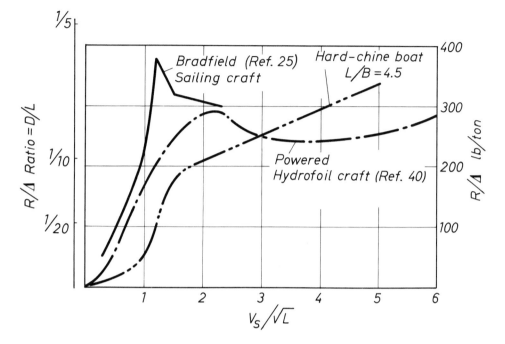

reduction in resistance for a certain range of speed–length ratios. Since at low and medium speeds the hydrofoil is inferior to the normal boat, retractable foils seem to be the only solution to the problem of overcoming such an inferiority. Hydrodynamic lift and a potential for reaching higher speeds are dearly paid for in terms of additional drag which is composed of several parts. There is drag associated directly with lift generation (induced drag), friction drag contribution, the strut's parasite drag, drag due to interaction between the foil's systems and hull, spray drag, etc.

Spray drag, or surface interference drag, occurs at the points where struts or foils pierce the water surface. It results from a complex combination of effects involving ventilation (air entrainment). Most trouble in hydrofoil take-off is caused by the peculiar resistance hump as shown in Fig 1.43. Characteristically, the driving force required to make the craft fully foil-borne reaches its maximum as take-off speed is approached. Then, as the hull clears the water, thrust requirements drop to a minimum value at a speed above that of take-off and then climb again. The surface-piercing V form, and ladder type foil configurations have in general smoother take-off with little resistance hump as compared to the fully submerged foil types.

A certain margin of driving force available over resistance, particularly in undulating seas, is absolutely vital, since this is the accelerating force, the amount of which determines how quickly and in what distance the craft will reach its flying attitude. Similarly, the foils must be capable of producing a certain excess of hydrodynamic lift over weight to provide for the vertical acceleration needed to lift the boat to her fully foil-borne attitude. The large drag penalty associated with foils operating below the fully foil-borne condition explains why for a time (during the 1972 Weymouth Trials) it looked as if the performance of *Icarus*, the Tornado cat equipped with hydrofoils, might not be better than the normal Tornado, but eventually she made a run two knots faster than the Tornado's best.

Lift-off on foils cannot be achieved in the particular case, as illustrated by the relevant resistance curve in Fig 1.43, unless the propulsive thrust F_R is about 300 lb per ton of displacement. In the case of sailing hydrofoils (which have severe limitations in available driving force in light and moderate winds) the submerged foils, and other parasitic structural parts, can contribute so much to the excess of hydrodynamic resistance before lift-off, that the craft may not reach sufficient speed to become fully foil-borne. The vital question then arises: what is the optimum area and configuration of hydrofoils for a given craft? Trying to answer this problem, R Baker defines in Refs 1.43 and 1.44 four basic principles:

1. There is a unique optimum lift-off speed V_{lo} for any given hydrofoil craft. This speed, which can be approximately computed from relatively simple formula, is dependent primarily on the boat's weight W and the hydrodynamic Lift/Drag ratio $(L/D)_h$ (formulae for optimum speed and optimum hydrofoil area are given in Ref 1.43). The hydrofoil craft, like any planing craft, can be characterized by the Lift/Drag ratio (Fig 1.44B), which depends on: the actual load on the hydrofoils, the $(V_S/V_{lo})^2$ ratio, and the craft speed V_S.
2. It is of no value to employ hydrofoils having a maximum area $A_{h\,max}$, different from that associated with the optimum lift-off speed V_{lo}.
3. Before a hydrofoil craft reaches the optimum lift-off speed there is nothing to be gained, in terms of speed, by the use of hydrofoils in combination with partially elevated craft, unless there are appreciable parasitic submerged areas independent of hydrofoil submersion.
4. After a hydrofoil craft has reached the optimum lift-off speed, there is no value to be gained by the use of hydrofoils in combination with a partially elevated craft, unless there is a significant variation in $(L/D)_h$ ratio due to hydrofoil loading.

Foil control

Turning to the problem of longitudinal response, or dynamic trim of the hydrofoil craft relative to the water surface, the ideal foils should lift the hull gently above the water surface and carry it docilely within the predetermined range of speeds without any tendency to bury or skip at high speed. To achieve this goal, some precise control over the vertical limits within which the foils are constrained to operate becomes of great importance. The instantaneous lift generated, the vertical

Fig 1.44   Simplified diagrams of forces acting on hydrofoil craft in steady motion.

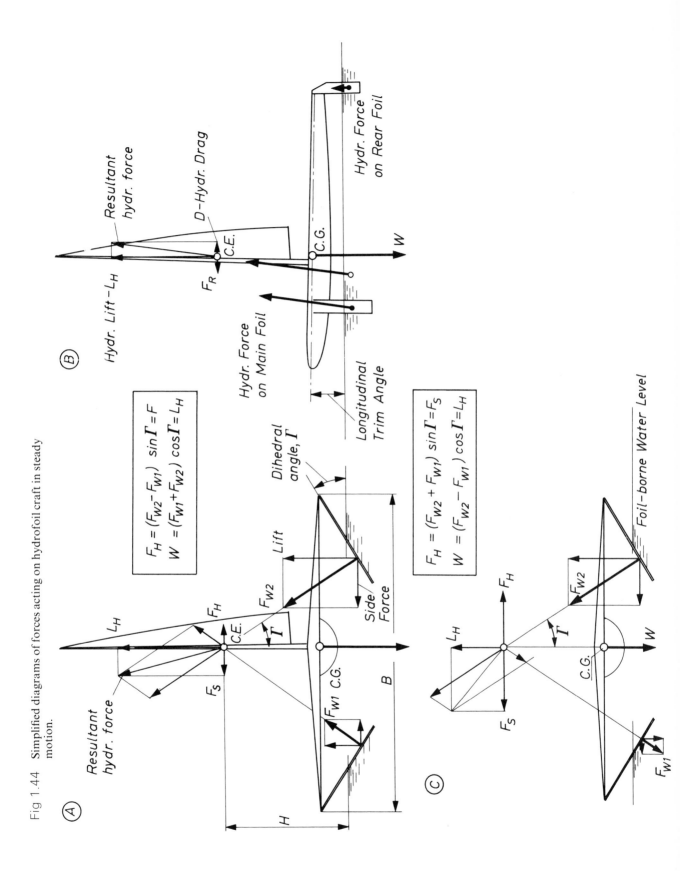

acceleration, the ever important hydrodynamic Lift/Drag ratio $(L/D)_h$, the sea-keeping ability and the other performance characteristics, will depend directly on this control effectiveness. Such an altitude control, which in fact means lift control, can hardly be achieved by manual action of the crew, particularly in confused seas. The human sensory capacity to respond precisely in a repetitive manner compatible with the ever-changing wave configuration is limited on account of the mental and physical fatigue familiar to all people. Human ability to carry on a number of simultaneous activities when a time lag is involved is also limited, compared to that of automatic systems, which have no such limitation. Some type of automatic control, in which the routine tasks have been built in, is therefore essential and must be incorporated in the foil system to operate independently of the crew. Otherwise, a major potential advantage of the hydrofoil craft, that is its ability to ignore waves which are smaller than its hull clearance, can be lost. The foil-borne longitudinal response can be of two kinds (Fig 1.45). One is called platforming, when a sufficiently large craft has enough hull clearance so the mean altitude can be maintained, regardless of the surface contour. Another type of longitudinal response, contouring, is proper to small boats. They must contour up and down the slopes of larger waves and this may incur large vertical accelerations. An intermediate response occurs when the hull just misses the crest and the foils just remain immersed at the troughs, then an occasional slamming cannot be avoided. For further information concerning longitudinal stability Refs 1.40, 1.42 or 1.47 might be consulted.

Fig 1.45   Platforming and contouring modes. In the case of small craft, such as sailing hydrofoils, an intermediate response is what can be expected. If the craft is not sufficiently large for its hull-clearance to exceed the maximum wave height, an occasional and severe slam (depending on sea condition) is inevitable. It can be assumed that the vertical accelerations are virtually independent of wave height in waves larger than 1.5 times the hull-clearance and wave steepness is then the governing factor (Fig 1.42).

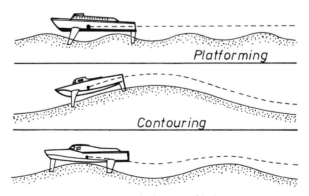

Platforming

Contouring

Intermediate response

There are two basic approaches to the lift control:

1. A rigid surface-piercing system, as exemplified by *Monitor* or *Icarus*, in which hydrodynamic lift is controlled by the change of foil area with depth of immersion.
2. The fully submerged foils, in which the lift generated is controlled by adjusting the angle of incidence, flap deflection, or by introducing atmospheric air into the low pressure (suction) side of the foil. Such a concept of an automatically controlled craft is schematically represented in Fig 1.46, which shows C Hook's idea of a sailing hydrofoil vessel (*Hydrofin* type) in which foils are primarily controlled by input signals of water level, provided by mechanical sensors so arranged as to provide information both in pitch and roll (Ref 1.46).

Fig 1.46   C Hook's idea of a Hydrofoil Sailing Vessel (Hydrofin type) in which the foils are: '…primarily controlled by input signals of water level provided by mechanical sensors so arranged as to provide information both in pitch and roll. In addition to this the pilot is provided with a control column and steering wheel and by moving this fore and aft, he can shift the zero position of the relative setting of sensor and foil to control flight attitude. By rocking his control column sideways he can feed in a difference signal that applies banking' (Ref 1.46).

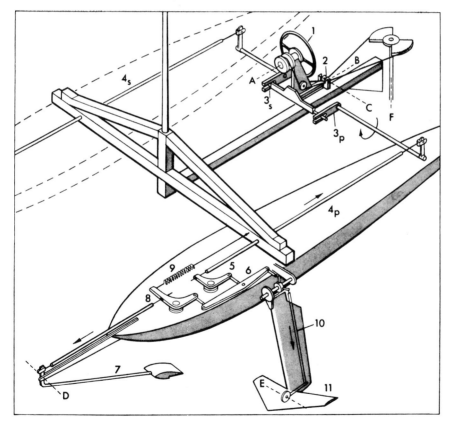

The first system of fixed hydrofoils, shown in Photos 1.17, 1.18 and Fig 1.44, is arranged by using a ladder foil configuration, or single foils set at large dihedral angles, so that their immersed area varies with the altitude of the craft above the water surface. For any speed, there is an equilibrium waterline when the craft weight is balanced by the lift generated on the submerged foil area. The reserve foil area is available above water to become effective immediately the foil unit enters the face of a wave. Because the foil elements necessarily pierce the water surface a drag penalty is involved but, having no moving parts, this simple and reliable system provides the most straightforward control of longitudinal trim of hydrofoil craft. It also provides inherent stability in pitch. This system thus possesses an automatic lift control feature without the employment of any mechanism or special sensing devices.

The second system of lift control employs fully submerged foils and a variety of sensors and gyroscopes to obtain continuous signals of attitude relative to the oncoming waves and craft altitude. These activate electro-hydraulic 'black boxes', which in turn control the deflection of flaps or the incidence angle of an all-flying element of the foil system. The fully submerged foil system is potentially more efficient than the piercing foil in terms of minimum drag, but it entails complete reliance on the electronic and hydraulic systems of control and on some moving mechanical parts under water (Ref 1.42 and 1.47). This is certainly not a field for amateurs, anyway for the time being. Within this fixed foil system, there is a promising possibility, reported in Prof Schuster's paper (Ref 1.48), of controlling lift easily by the air supply into the low pressure side of the foil. The upper surface of a foil is provided with one or two rows of outlet openings distributed spanwise. The openings are connected to a duct arranged inside the foil which in turn is connected by a hollow strut to the free atmosphere. The air that is sucked out of the low pressure side of the foils, is controlled by a valve. Lift decreases in proportion to the quantity of air fed into the boundary layer. Such a system eliminates moving parts.

Lateral stability

Lateral stability and dynamic response in rolling is another problem of crucial importance in the case of hydrofoil craft using sails for propulsion. The basic stability requirement is that for steady motion:

1. The vector sum of all the external forces acting upon the aero–hydro system must be zero, and
2. The sum of the moments of all the forces acting about any single axis is zero.

If a balance of forces and moments is reached, all the forces when resolved into the two convenient components, vertical and horizontal, should pass through a single point, the centre of effort CE, of the hydrodynamic force, as shown in Fig 1.44 (Refs 1.49 and 1.50). Looking at the sketch 1.44A, that represents in a simplified manner the hydrodynamic action of a pair of single foils of surface-piercing type tilted relatively to the water surface at a dihedral angle Γ, one may notice that foils are capable of producing both the lift equal to the weight of the craft, as well as keel effect (side force) to meet the aerodynamic heeling force F_H.

Foils generating differential lift–more lift on the lee foil than on the weather side (due to leeway which increases the angle of incidence of the lee foil as compared with that of the weather foil)–are also the source of stabilizing moment. However, the limit to the stability moment of such a foil configuration is rather low. As Barkla noticed (Ref 1.49), to enjoy really high speed sailing by exploiting foils properly the rig must be able to develop and the whole structure sustain a heeling force F_H, comparable with the all-up weight of the vessel W. To satisfy this requirement the hydrofoil craft would need to have a beam of about two times greater than the height of the centre of effort above the water; not a desirable feature from the standpoint of weight, strength and stiffness of the whole structure.

The beam/height ratio depends on the dihedral angle, but this angle is limited to a relatively narrow range in the vicinity of 40–45°, if foils have to generate the dynamic lift component of the same order as side force component.

The differential lift on the two foils can be augmented with subsequent increase of stability, if the boat is allowed to heel until an obvious limit to the stability is reached when the lift of the weather foil is reduced to zero. The righting moment can also be increased by shifting the crew to weather as is demonstrated in Photo 1.15, where use is made of a trapeze.

By setting the weather foil at negative incidence, as depicted in Fig 1.44C, there is theoretically no limit to the heeling force F_H, and therefore to the driving force F_R, that may be generated by the sails. Practically, a F_H/W ratio of the order of 1.0–an important requirement facilitating high speed sailing–is quite conceivable. However, since one foil is acting against the other (positive and negative lifts being produced) a certain penalty in terms of higher induced drag must be paid. Apart from that the structural strength to sustain large loadings has to be paid for in weight, but those penalties, it is believed, could be offset by the advantages due to the raised limit of the attainable driving force.

Of course, the foils can be arranged in different ways: the ladder system, for example, in which the ladder rungs produce mainly the lift to support the weight of the craft, while vertically orientated struts generate the side-force. In fact, innumerable variations of foil configuration are possible to suit particular demands. Some of the current hydrofoil configurations employed by powered hydrofoil craft are shown in Fig 1.47.

The rig

A radically different solution to the stability problem, which may appeal to some adventurous spirits, is given in Fig 1.48. It makes use of the pair of tilted sails to bring the centre of effort closer to the water level (Ref 1.51). This solution refers in a way to Barkla's idea: the freely pivoted aerofoil unit suggested in 1951 (Ref 1.50 and sketched also in Fig 1.48). The following is the inventor's comment: 'Although

Fig 1.47 Current hydrofoil configurations employed by powered hydrofoil-craft. The position of the main-foils relative to the centre of gravity, CG, shown in the first line, has considerable effect upon seakeeping behaviour. Innumerable variations in foil configuration are possible to suit particular demands (Ref 1.40).

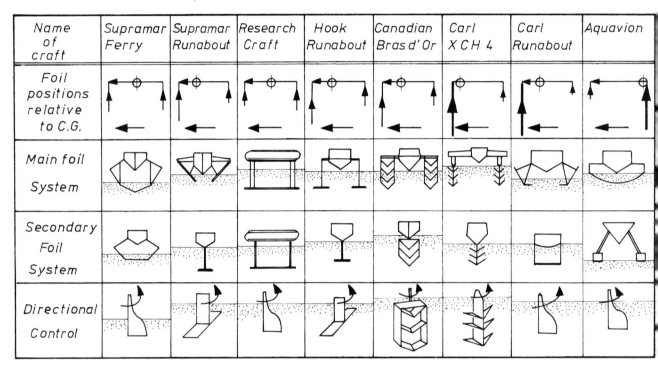

Name of craft	Supramar Ferry	Supramar Runabout	Research Craft	Hook Runabout	Canadian Bras d'Or	Carl X CH 4	Carl Runabout	Aquavion
Foil positions relative to C.G.								
Main foil System								
Secondary Foil System								
Directional Control								

the hull and foil system in this case would be more compact, the rig would spread far beyond it. It would thus be just as vulnerable and unpopular as the first arrangement' (i.e. as depicted in Fig 1.44A).

There are also other arguments against the sloping sails configuration, namely, a sail heeled to 30° incurs an obvious disadvantage over that working at a more or less upright position (Ref 1.5), unfavourable interference effects near the top of such sails should also be expected; both will result in substantial reduction of the driving force/heeling force ratio. To compensate those effects more sail area would be needed, thus imposing two additional penalties: an increase in weight in an already heavy rig, together with an undesirable upward shift of the vertical position of the centre of gravity.

Wing sails

In an attempt to harness more driving power from the wind, at possibly low heeling moment, various novel rig concepts have been developed, including multiple, rigid

Fig 1.48   Freely pivoted aerofoil unit suggested by H Barkla in 1951 (Ref
            50). One of the radical solutions to the stability problem–the two
            tilted sails bring the centre of effort closer to the water level (Ref
            1.51). Figure 1.49 depicts another solution to the same problem.

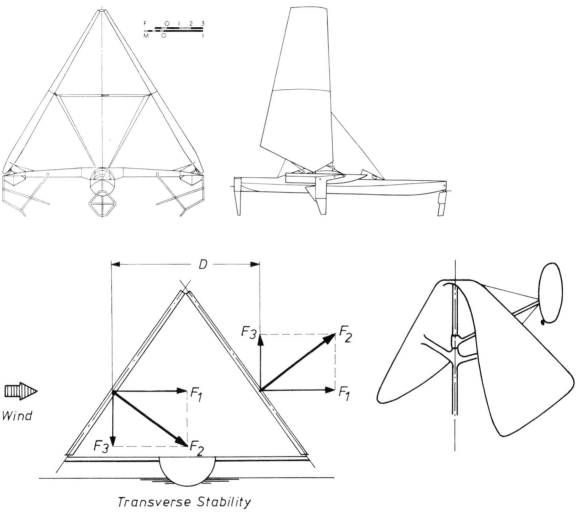

Transverse Stability
Stabilising moment = $F_3 \times D$

wing sails. One of them is schematically presented in Fig 1.49 (adopted from Baker-
Douglas paper, Ref 1.44). Photo 1.19 shows one of the more controversial
contemporary multihulls, *Planesail*, driven by a cascade of four rigid aerofoils.

The essential feature of *Planesail* is the freely pivoted multiplane rig suggested by
Barkla in 1951 (Ref 1.52). The basic principle of such a free-rotating rigid rig is, at
first sight, simple and promising. A small trailing or tail foil controls the angle of
incidence of the main driving wing-sail unit. While the tail-vane is in neutral

Fig 1.49   Novel concepts have been added to sailing technology. Here, multiple rigid wing sails (Ref 1.44).

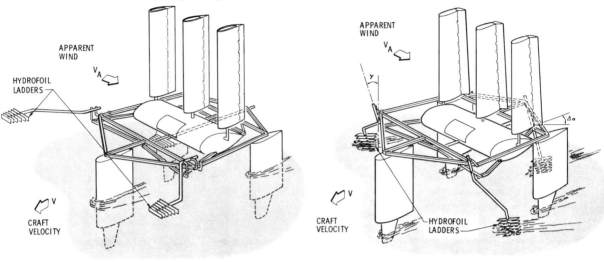

Three-Airfoil Craft With Trimaran Hull Form
(Float-Borne With Hydrofoils Retracted)

Three-Airfoil Craft With Trimaran Hull Form
(Hydrofoil-Borne With Hydrofoils Extended)

position, that is in line with the symmetrical aerofoils, the whole rig behaves like a weather-cock, i.e. there is no lift force generated, but a relatively small amount of drag is transferred to the boat through the pivot bearing. As soon as the tail-vane is set by the helmsman at a certain angle to the rest of the rig, the main rig also takes a certain incidence relative to the prevailing wind and, unlike an ordinary rig, automatically adjusts itself to the wind without further attention. The generated aerodynamic force propels the craft in the direction controlled by an ordinary rudder.

The group of ex-aircraft engineers who developed *Planesail* claimed that they have completely re-thought the concept of ordinary 'soft cloth and wet strings' sailing boat in order to produce:

1. A boat that can be as easy to sail as a motor-car is to drive, and
2. An inherently safe boat under all reasonable conditions, including a gale on the open sea (a very ambitious plan).

The practical execution of this idea of a fast, revolutionary craft has met, like so many inventions in yachting, formidable difficulties. The whole project has gained no acceptance as yet and initially foundered. However, one may foresee that, provided inflation of the monetary system does not get out of hand, we may well witness in the not too distant future, some extraordinary or even bizarre sea monsters along the lines suggested in Fig 1.49 and Photo 1.19. In fact, during the 1974 Speed Trials the five-wing-sailed proa *Clifton Flasher* reached 22.4 knots (Ref 1.45).

Photo 1.19   J Walker's *Planesail*.

LWL–29.0 ft.
Displacement about 1.3 tons.
Sail Area S_A–280 sq ft.

One of the most controversial contemporary multihulls, developed by a group of ex-aircraft engineers, who claimed to have completely rethought the concept of the 'soft cloth and wet strings sailing boat'. The craft is driven by four rigid foils. The small trailing foil controls the angle of incidence of the driving wing sails.

(Photo by J S Biscoe, Burnham-on-Sea.)

Fig 1.50   Planforms of wing sails applied with some success in the past
(Ref 1.53). From left to right:
    Utne's Boat (about 1941). The peculiar cigar-shaped body
protruding from the wing leading edge held a weight to balance
the sail so that its angle of incidence would not be affected by
heeling or rolling. The problem of dynamic stability of the rigid
aerofoil system is unknown to sailors using conventional rig.
    Blackburn's Boat (1962)
    Fekete and Newman's Boat
Basic wing sail area of 30 sq ft can be increased to 54 sq ft (1964).

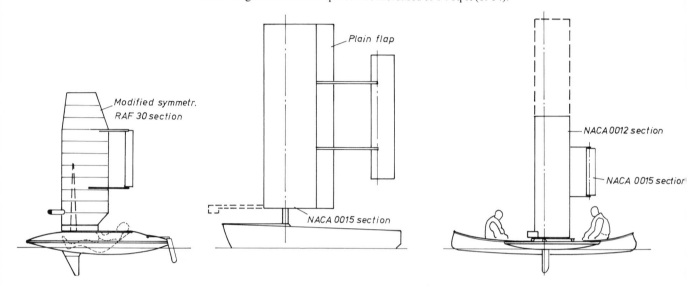

Recent developments in multiplane rigs have been stimulated by earlier attempts
to improve yacht performance, substituting a single rigid wing sail for an ordinary
soft sail. Most wing sail driven craft designed since the late 1930s have been built on
an amateur basis and therefore they range from the freaks to the few which are based
on sound principles. In Fig 1.50 there are shown the planforms of three wing sails,
developed with some success in the past.

What is the theoretical background justifying the use of rigid aerofoils in place of
conventional sails? Are the latter destined to be replaced by wing sails? As we will
see in following chapters, the wing sail can hardly compete with soft sails at low
wind speeds, when the maximum lift coefficient C_L is the factor of primary
importance that limits the performance of the boat. Symmetrical sections, unless
equipped with fairly elaborate flaps, are rather poor devices for producing high lift
coefficients.

For example, the maximum lift coefficient $C_{L\,max}$ of a wing sail rig of symmetrical
section NACA 0012 is of the order of 0.8. Figure 1.51 illustrates the results of wind
tunnel testing on Fekete and Newman's wing sail (one-eighth scale model), shown
in Fig 1.50. The curves of lift coefficient C_L versus incidence angle α (based on
NACA Rep 586) applicable to low Reynolds Number Re, appropriate to realistic

Fig 1.51 Lift coefficient C_L versus incidence angle for three different foil
AR (Ref 1.53).

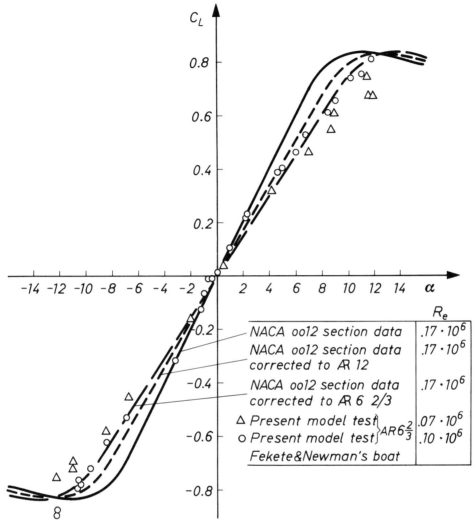

sailing conditions and corrected for aspect ratio are compared with the experimental
data at two Reynolds Numbers; the agreement is pretty good. At similar Re the soft
Bermuda sloop type rig can generate a $C_{L\,max}$ of some 1.9 or even more, and is
certainly less troublesome in handling and maintenance. Soft sails can easily be
hoisted, lowered in emergency and stored in a relatively small space inside the hull
while at moorings.

Since the boat must normally be able to sail with the wind on either side of the sail,
it implies the choice of a symmetrical section for the wing sail as an obvious, least
complicated solution. The $C_{L\,max}$ coefficient of a heavily cambered rigid aerofoil can
be as high as that of a good soft sail, but it requires what may be a sophisticated

mechanism for reversing and varying the camber if the rig is intended to work on both tacks.

At higher wind velocities, and particularly in the case of fast sailing boats and ice boats that make their own wind, i.e. they sail faster than the true wind, the rigid wing sail offers great advantage over the conventional sail. In such a condition, the large lift coefficient cannot be used. What is really needed is a high L/D ratio and the difference between one sail and another lies in the comparison of their drag coefficients at the maximum usable lift coefficient. In this respect the rigid aerofoil is superior and unbeatable by any soft sail. Besides, the rigid wing sail facilitates a fine control of incidence angle, and therefore precise control over the forces generated. At low angle of incidence and small lift coefficients that involve allowing the soft sail to flap, the comparison is even more in favour of the rigid aerofoil. However, it is rather difficult to get something for nothing; apart from handling difficulties, the rigid aerofoils are heavy. Even the simplest symmetrical wing sail is at least twice as heavy as a normal rig. As a fragile structure, the wing sail is awkward to store when not in use, and erecting it in windy conditions involves serious risk. Several good wings have been ruined while at moorings.

MacKinnon, who experimented with the wing sail, writes:

'In theory, the wing can be left up indefinitely with the boat at rest; either chocked up ashore, or afloat. This is because the wing is free to weather-cock and its drag is very low indeed. In practice, this was realised with one complication which I ought to have foreseen. With the boat afloat and head to wind, there was a tendency towards rhythmic rolling with oscillation of the wing from side to side. This effect is quite a serious problem, as otherwise the wing could safely be left up when the boat is not being sailed. In theory, it is curable by mass balancing the wing so that, for example, if the boat is rolled a little to starboard, the trailing edge of the wing moves to starboard and not to port as is the natural tendency' (Ref 1.54).

The problem of dynamic stability of the self-trimming wing sail rig is not a simple one and requires attention. Further insight into it is given in Refs 1.53 and 1.55.

Other solutions

Yet another radical 'sailing' craft, which perhaps has nothing in common with sail propulsion except that it uses wind energy, is shown in Fig 1.52. Such a concept, if workable in practice, makes it possible to drive a vehicle straight into the wind. A fascinating proposition! An analysis of such a man-carrying vehicle, accelerating from zero speed up to a speed greater than the wind speed, is given by Bauer (Ref 1.56). Both land-vehicles as well as water-borne craft have been considered. The practicability of travelling directly to weather has been demonstrated and a land-vehicle built on this principle has made sustained runs of the order of 40 sec in a wind of about 12 mph, with the vehicle speed estimated to be about 2 mph faster than the

Fig 1.52   The catamaran and windmill-propeller combination still awaits attention. Such a concept if workable in practice makes it possible to drive a hull straight into the wind and sail faster than the true wind. A fascinating proposition. Drawing taken from Barkla's paper (Ref 1.1).

wind. For the water-borne craft, the anticipated V_S/V_T ratio is of the order of 1.5, i.e the craft may reach a speed V_S about 50 per cent higher than the true wind velocity V_T. The natural extension would be to construct a vehicle capable of travelling in any desired direction.

A new concept in sailing a boat stripped to bare essentials is shown in Fig 1.53. It combines the pleasure of creating speed using wind power, with surfing. This type of boat, called a sailboard, represents a unique departure in the sport of sailing, facilitating true 'man against the sea' situations; it is a very demanding and exhausting form of physical recreation. The sailboard is ridden in a standing position and controlled by a hand-held sail assembly only (no rudder). The mast is stepped on the board through a fully articulated universal joint but is otherwise unsupported. Figure 1.53 demonstrates, in a self-explanatory manner, gybing and tacking manoeuvres while riding a simpler form of sailboard in a sitting position. The gybing and tacking are initiated by dipping the sail forward or aft, respectively.

Several types of sailboard have been built and tested with varying degrees of success. They are described, in some detail, in Ref 1.57 together with operational theory and instructions.

Reverting to Fig 1.31, there is a curve marked 'submarine', laid over the five curves representing the specific resistance of hulls of various displacement/length ratios. This curve refers to a submarine deeply submerged, of a displacement/length ratio 200. Let us now assume that in conditions in which the sail is capable of providing its maximum driving force, i.e. while reaching, the driving force actually

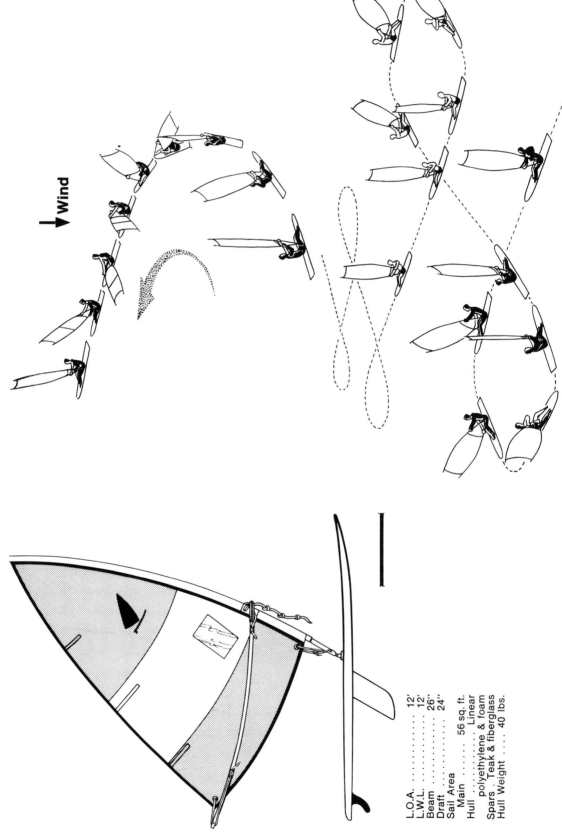

Fig 1.53   A new concept in sailing—a fast boat stripped to the bare essentials. Windsurfer includes a patented sail/spar system and a board of moulded polyethylene for durability, foam-filled for safety.

Sailboarding: tacking and gybing manoeuvres.

Wind

L.O.A.	12'
L.W.L.	12'
Beam	26"
Draft	24"
Sail Area Main	56 sq. ft.
Hull	Linear polyethylene & foam
Spars	Teak & fiberglass
Hull Weight	40 lbs.

Fig 1.54   Sailing skimmer propelled by wing sail. The aerofoil system,
lifting the hull completely above the water, promises, theoretically, great speed sensation.

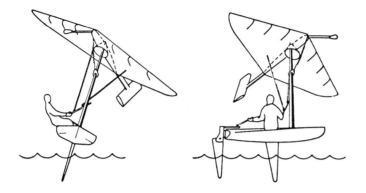

developed F_R, which must be equal to hull resistance R, is 7 per cent of the boat's displacement, i.e. $R/\Delta = 0.07$. This is marked by the horizontal line labelled 'reaching conditions'. It is evident that the submarine, making no waves, can beat any of the surface boats of displacement length ratio $\Delta/(L/100)^3$ greater than 50, and at any speed/length ratio $V_s/\sqrt{L}$ greater than about 1.7; provided that the thrust or driving force developed by the submarine engines, expressed in terms of driving force/displacement F_R/Δ ratio, is the same for all craft.

This perhaps unexpected conclusion indicates drastically how big the wave drag can be and how inefficient, in terms of drag and energy expenditure, are the surface hulls. However if, as illustrated in Fig 1.31, the surface boats are allowed to go to a minimum displacement/length ratio of 25, there is practically a toss-up between the two different types of water-borne vehicles–the surface-planing form versus submarine.

There are two immediate practical conclusions. In order to sail fast, by virtue of drastically reducing wave-drag, one must either submerge the hull well below the water surface, or lift it above the water. The first conclusion, a go-down concept, in fact a sailing submarine propelled by sails, has not been produced as yet (anyway to the writer's knowledge) but, who knows in our progressive world?

The second conclusion, a lift-up concept, is developing quite rapidly. The hydrofoil craft just described may sail fast simply because they obviate wave drag. Figure 1.54 adopted from Ref 1.58 illustrates a sailing skimmer, propelled by a wing sail which also generates sufficient lift to keep the hull above the water; a project bordering on pure fantasy, it is nevertheless analytically correct. This is not an entirely new project. Many people have been developing in dreams such a concept and have even published details of an inclined sail partially lifting the hull and facilitating fast sailing. This craft, shown in Photo 1.20, was called by some the kite-rig or umbrella-rig. An interesting account of an experiment performed with the umbrella-rig was reported in *Yachts and Yachting* (July 17, 1959) by J Rowland:

Photo 1.20   An umbrella rig, photographed in 1895 (Beken). It is believed
that this idea of a non-capsizable boat stems from Polynesian
craft. The basic principle appears to be a rather simple but
practical solution–how to control such a rig in strong winds
is another matter.

...'Then I put the rig on a 15 ft Snipe hull and eventually gave that a real breakdown test by sailing it through a near hurricane that blew the whole rig away and landed it on the lawn of my place, a quarter of a mile away, without even capsizing the boat. Before that happened, she nearly drowned me with spray and I got to laughing so hard that I was weak.'

The Snipe of about 500 lb weight was not perhaps destined to fly. However, taking advantage of recent developments in lightweight, high-strength materials, a much lighter craft could be built. No doubt, the aerofoil system lifting the hull completely above the water, as sketched in Fig 1.54, promises theoretically greater gains than those manifested by the Snipe-umbrella-rig experiment, but a workable skimmer is still unknown. The main drawback of this concept is that it cannot produce a sensation of speed in light winds or when running before the wind. What else might the human imagination invent? What might the next step be?

When high speed fever descended on sailing men at the end of the 19th century and new skimming forms with platinum keels were dreamt about, the famous English yacht designer G L Watson made the prophetic remark that the time would come when, '...I hope, we won't care for sailing in such a sluggish element as the water. I firmly believe that some day the air will become as easily traversed as the earth and the ocean.' Certainly, the air–water interface is not the best place for really high speed. If the aeroplane had not been invented by the Wright Brothers, it would surely be discovered in the very near future by sailors!

H Land and hard-water sailing craft

The origins of land-sailing craft are not as fascinating as those of water·borne sailing vessels, nevertheless they have a long tradition which can be traced to the 6th century. About that time, 'Kaotshang Wu-Shu succeeded in making a wind-driven carriage which could carry thirty men, and in a single day could travel several hundred "li"' (Ref 1.59).

After this record there is a long silence, until just on a thousand years later when Simeon Stevin, the great Dutch mathematician, constructed a sailing carriage of which there exists some historical evidence. Around 1600, Prince Maurice of Nassau invited several ambassadors and distinguished guests to appreciate the performance of these sail-driven land vehicles. Two such craft had been made, a larger and a smaller, and both succeeded in accomplishing the distance between Scheveningen and Pelten, along the beach, in less than two hours, though it took fourteen hours to walk. Photo 1.21 shows a contemporary print (by de Gheyn) of the fleet of these craft; 24 persons were carried by the larger vehicle. These sailing carriages were almost certainly inspired by the stories of the sailing vehicles of China which were prevalent in Europe during the previous century. One of the pictures in Photo 1.21 shows a contemporary artist's impression of Chinese craft incorporated in Mercator's Atlas of about 1613.

Photo 1.22 depicts a modern, but not necessarily top-performance, descendant of those magnificent Dutch chariots of the early 17th century.

Photo 1.21A Sketch of a sailing wheelbarrow, the sail assisting animal traction, and a vignette of an imaginary Chinese land-sailing carriage–from *Mercator's Atlas* of about 1613. The sailing wheelbarrow is an admirable device still widely used in China, notably in the coastal provinces.

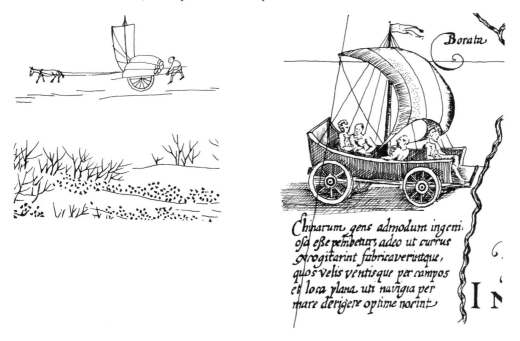

Photo 1.21B Contemporary print by de Gheyn shows the fleet of successful sailing carriages constructed by Simon Stevin *c*. 1600. These land yachts were inspired by stories of the sailing craft of China prevalent in Europe during the previous century.

Photo 1.22   Windbuggy Class (Bob D'Olivo Photo, Ref 1.60).

Ice yachts

The basic difference between water-borne craft, kept afloat mainly by Archimedes' buoyancy forces, and 'dry' sailing vessels (land and ice yachts) is that the latter do not experience such an enormous drag barrier as the former. In consequence, they can sail much faster than the true wind, bringing the apparent wind so far forward that, whether sailing up or down wind, the course is always close-hauled. Some strange or even mysterious stories circulate amongst enthusiasts about the extraordinary property of ice boats in making their own wind (Note 1.62).

To grasp the fundamental principles underlying the behaviour of land and ice yachts, which are very much alike, let us analyse an 'idealized yacht', the hull of which is capable of producing side force without any drag penalty. An ice boat sliding over the 'hard water' with a constant, very small friction resistance, below 2 per cent of its total weight (Photo 1.23), is a best approximation to such an ideal yacht.

Apparent wind

Figure 1.55 shows in several sketches (A, B, C, D, and E) the idealized ice boat moving from rest along a course $\gamma = 90°$ (square reach) from the true wind V_T; as we will see later, this course is not the fastest, but it facilitates the explanation. Beginning from the top sketch A, as soon as the sail is sheeted in the boat accelerates and, provided that the sheets are being pulled harder and harder to match the changing apparent wind direction, the boat speed V_S gradually increases. This is reflected in increasing V_S/V_T ratios, taking values 1, 2, 3, 4, as depicted in the relevant velocity triangles. The apparent wind V_A heads gradually and the β angle decreases, until the sail is close-hauled on the beat, attaining the condition when $V_S/V_T = 4$ (sketch E) and the resultant aerodynamic force F_T becomes perpendicular to the direction of motion, or V_S vector. From now on, there is no net driving force that might accelerate the boat further, i.e. a balance of driving and resisting forces is reached and the boat will proceed in steady motion. Once the sail has reached its best L/D ratio with the wind steady, any further acceleration on this particular course $\gamma = 90°$ is impossible. Reasoning in another way, any further increase in the speed of the boat V_S would bring the apparent wind V_A closer ahead with subsequent feathering of the sail overhead and unavoidable reduction of aerodynamic force; as a result, the boat would decelerate. Such a deceleration with the sail flogging violently may happen in practice in gusty winds, when suddenly the true wind V_T drops but the speed of the boat V_S does not, due to the inertia of the vehicle.

A glance back at the sketches in Fig 1.55 and the table attached will reveal that there is a close relationship between β and V_S/V_T ratio. The smaller the β angle, the higher is the V_S/V_T ratio, i.e. the higher is the attainable boat's speed. An immediate question arises as to how small this β angle might be.

This can be answered by recalling Eq 1.1 ($\beta = \varepsilon_A + \varepsilon_H$) and Lanchester's words from chapter A: 'the minimum angle (β) at which the boat can shape its course relative to the wind is the sum of the under and above water gliding angles' (drag angles ε_H and ε_A respectively). Since the runners of our idealized ice-yacht are presumed to experience no resistance, $\varepsilon_H = 0$ and hence $\beta_{min} = \varepsilon_{A\,min}$. Bearing in mind that cot $\varepsilon_A = L/D$, we may write that:

$$\beta_{min} = \varepsilon_{A\,min} = \cot^{-1}(L/D)_{max} \qquad \text{Eq 1.8}$$

and this indicates that the rig efficiency reflected in attainable L/D_{max} is a factor of primary importance as far as fast sailing is concerned.

From Fig 1.55 one may immediately deduce (and it can be proved, Note 1.61) that

Fig 1.55 Velocity triangles of an ice boat accelerating from stationary
position until $V_S = 4V_T$. Glider shown in sketch F can be
regarded as a close relative.

V_S/V_T	V_A/V_T	$\beta°$	Aer. Force Ratio $(V_A/V_T)^2$
1	1.4	45°	2.0
2	2.2	26.5°	4.8
3	3.2	18.5°	10.0
4	4.1	14.0°	17.0

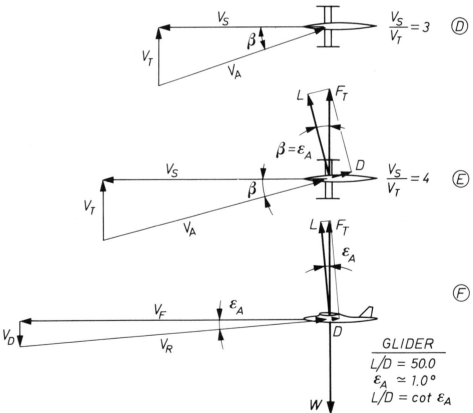

Photo 1.23   Top left and right:
   Ice boats of DN class, probably the most popular type of
craft that can also be transformed into a land yacht, as
illustrated in photo below and Fig 1.57. The parasite drag of
the boat's structure appears to be pretty high.

on the square reach, when the true course angle $\gamma = 90°$, the ratio of the boat's speed V_S to true wind speed V_T:

$$\frac{V_S}{V_T} = \cot \beta \qquad \text{Eq 1.9}$$

or

$$V_S = V_T \times \cot \beta \qquad \text{Eq 1.10}$$

Making use of Eq 1.8 one can write Eq 1.10 in a simpler form:

$$V_S = V_T \times \frac{L}{D} \qquad \text{Eq 1.10}$$

and this explains why ice yachts, land yachts and all other fast sailing craft call for the highest obtainable L/D ratio in the rig.

The table and velocity vectors diagrams in sketch (E) of Fig 1.55 show how a true wind speed V_T of, let us say, 10 knots on a square reach at the start, when added to the reverse of boat speed $V_S = 40$ knots, produces 41 knots close-hauled resultant, or apparent, wind V_A. The total aerodynamic force F_T is a function of the apparent wind V_A and increases in proportion to its velocity squared (V_A^2). An ice boat reaching the ratio of $V_S/V_T = 4.0$, with the associated ratio of $V_A/V_T = 4.1$ develops an aerodynamic force proportional to $(V_A/V_T)^2 = 4.1^2$, which is 17 times greater than that developed at the start (compare sketches A and E, together with the table in Fig 1.55); let us bring these ideas together.

Assuming that an ice boat of total sail area $S_A = 70$ sq ft is capable of generating a L/D ratio of 4.0, while the total force coefficient $C_T = 0.7$, and the true wind velocity $V_T = 20$ ft/sec (11.8 knots), we may calculate the total aerodynamic force F_T, using the familiar formula:

$$F_T = 0.00119 \times C_T \times S_A \times V_A^2$$

At the start, when the boat is stationary, the wind which determines the aerodynamic force is the true wind $V_T = 20$ ft/sec, hence:

$$F_T = 0.00119 \times 0.7 \times 70 \times 20^2$$
$$= 0.00585 \times 400 = 23.3 \text{ lb}$$

When the boat reaches her ultimate speed, i.e. $V_S/V_T = 4.0$, the total force F_T increases 17 times, therefore:

$$F_T = 23.3 \times 17 = 400 \text{ lb}$$

If the initial true wind $V_T = 30$ ft/sec (about 18 knots, which is equivalent to a moderate or gentle breeze) the sail will develop at the start, a total force of only:

$$F_T = 0.00585 \times 30^2 = 0.00585 \times 900 = \text{about } 50 \text{ lb}$$

but at the ultimate speed V_S, when $V_S/V_T = 4.0$, the total force would reach:

$$F_T = 50 \times 17 = 850 \text{ lb!}$$

if, of course, stability would allow such a state to be achieved.

The apparent wind V_A would then be about 123 ft/sec = 73 knots. At such a tearing wind speed, equivalent to a hurricane, the sail and the rig must withstand enormous loads, unusually high by the standard of ordinary water-borne craft. The sail, susceptible to deformation under exceptionally large stresses, must be made of very heavy canvas to preserve a flat draft, almost as flat as a drum, and a twist as low as possible.

Stability

After rig efficiency, stability is the second factor that puts a limit on attainable maximum speed of any ice yacht. In this respect they are not different from any other type of sailing craft. Like catamarans, they are capable of developing a large stabilizing moment by spreading their runners or skates widely on the ice surface. The further the lee runner is from the centre of gravity, the more power to carry sail is available, i.e. the craft can operate efficiently in stronger winds. However, there are limits to the distance over which the structure may be spread. The weight and strength of the connecting components impose one limit; another limit is imposed by handling characteristics, depicted and partly described in Fig 1.56, adopted from Ref 1.63. Although this figure relates to a land yacht, the stability principle illustrated is applicable to ice boats and, in a way, to multihull boats as well; craft which are shorter than they are wide tend to be very wild when running. In a seaway, the older type of catamarans were notorious for their pitch-poling–a tendency to somersault when the bow stuck into the back of a wave (see Photo 1.10).

Ice boats manifest something similar, being subject to what is called in ice boat parlance the awesome 'flicker', or a flat spin occurring when the steering runner is lifted up and the craft loses its grip on the ice surface.

Speed potential

If we still ignore runner friction, assuming that the ice boat is devoid of contact with 'hard water' and operates in one medium–the air only–we may look at the ice boat as a close relative to the glider. Its velocity triangle, shown in Fig 1.55F, is similar. The resultant velocity V_R is a vector sum of the two velocities–forward velocity V_F, and downward velocity V_D, and similarly, as in the case of an ice boat, the ratio:

$$\frac{V_F}{V_D} = \cot^{-1}(L/D) = \cot \varepsilon_A$$

Fig 1.56   Influence of some geometry factors on land yacht stability and handling characteristics (Ref 1.63).

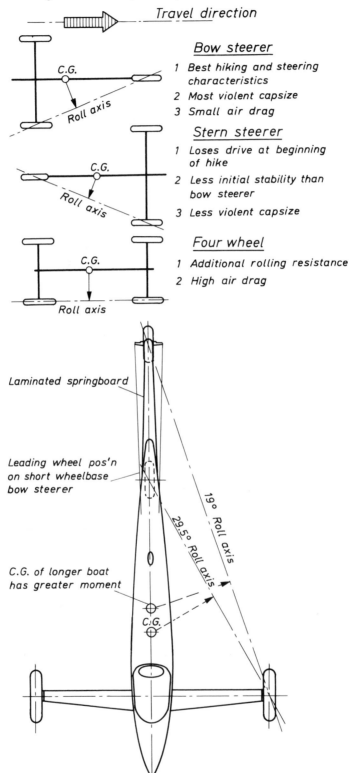

Travel direction

Bow steerer

1 Best hiking and steering characteristics
2 Most violent capsize
3 Small air drag

Stern steerer

1 Loses drive at beginning of hike
2 Less initial stability than bow steerer
3 Less violent capsize

Four wheel

1 Additional rolling resistance
2 High air drag

Laminated springboard

Leading wheel pos'n on short wheelbase bow steerer

C.G. of longer boat has greater moment

19° Roll axis

29.5° Roll axis

C.G.

Roll axis

Fig 1.57   Mini ice yacht, DN class, converted for land yachting purposes
(for use on ice the springboard is removed).

Total weight $W = 300–350$ lb
Sail area $S_A =$ about 70 sq ft

This, probably the most popular ice boat, is capable of reaching
a speed V_{Smax} just above $3 \times$ wind velocity, V_T. Much higher
performance figures have been claimed but dependable measurements are hard to obtain.

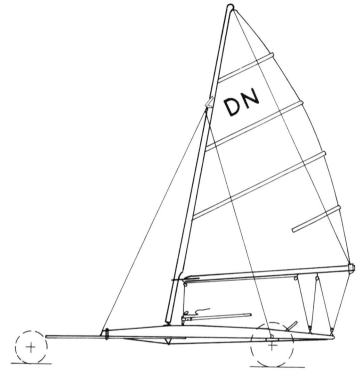

For a top high performance glider, equipped with a wing of laminar section, the L/D
ratio can be of the order of 50, i.e. the gliding angle ε_A is about 1°! In practice, the ice
boat of DN class, shown in Photo 1.23 and Fig 1.57, can reach a V_S of about $3 \times V_T$,
i.e. its overall L/D ratio is about 3. According to Ref 1.64 under ideal conditions, say
a steady 12 mph breeze and with ice made smooth and slippery by the sun, a well
tuned, modern, high performance skeeter can approach 60 mph i.e. its velocity
$V_S = 5 \times V_T$, the relevant L/D ratio being about 5. Much higher figures have been
claimed–146 mph is the speed record of the fastest E Skeeter, while speeds of
100 mph (87 knots) are supposed to be common. The L/D max ratio of the DN rig,
given in Ref 1.65, is between 8 and 9, rather exaggerated figures in the light of
available wind tunnel evidence, recorded in the course of investigation of the
aerodynamic characteristics of a Finn-type rig that has similar proportions to the
DN rig. Dependable measurements concerning ice boat performance are hard to
obtain, so it is difficult to establish their speed credentials.

Fig 1.58 Polar curves showing the performance characteristics of *Tornado* catamaran and two idealized ice boats developing overall *L/D* ratio in the order of 3 and 4. Performance is expressed in attainable V_S/V_T ratios.

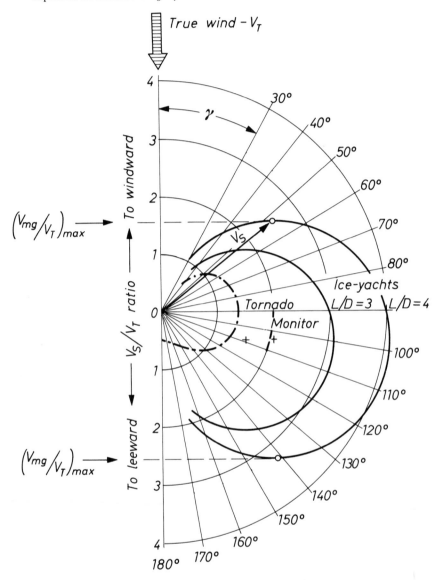

However, the fact remains that the speed potential of ice boats is spectacular when compared with the performance of any other type of sailing craft. Figure 1.58 demonstrates this point qualitatively. The three full polar curves represent the performance characteristics of the *Tornado* catamaran and two average ice boats, developing overall *L/D* ratios of the order of 3 and 4. There is also plotted a short performance line, representing the claimed speed record of the fastest sailing hydrofoil, *Monitor*, on the most favourable reaching course. The performances are expressed in terms of attainable V_S/V_T ratios on courses γ, ranging from close-hauled to running. Differences are indeed conspicuous.

The optimum performance figures and appropriate optimum course for tacking to windward or leeward, and for reaching at the maximum speed, can be seen immediately from polar diagrams. For ice boats the performance figures can be calculated from the following equations (Ref 1.1b):

To windward

$$(V_{mg}/V_T)\,\text{max} = \frac{1}{2}\left(\frac{1}{\sin \varepsilon_A} - 1\right)$$

at optimum
$$\gamma = 45 + \varepsilon_A/2 \qquad\qquad \text{Eq 1.12}$$

Reaching

$$(V_S/V_T)\,\text{max} = \frac{1}{\sin \varepsilon_A}$$

at optimum
$$\gamma = 90 + \varepsilon_A \qquad\qquad \text{Eq 1.13}$$

To leeward

$$(V_{mg}/V_T)\,\text{max} = \frac{1}{2}\left(\frac{1}{\sin \varepsilon_A} + 1\right)$$

at optimum
$$\gamma = 135° + \varepsilon_A/2 \qquad\qquad \text{Eq 1.14}$$

The equivalence of β and ε_A angles and their influence on the attainable V_m/V_T ratio in close-hauled conditions is depicted in Fig 1.59.

Let us verify Eqs 1.12, 1.13 and 1.14 by calculating performance figures of an arbitrary ice boat that develops the overall L/D ratio of about 4.0, and then comparing results with the relevant polar diagram given in Fig 1.58.

Applying Eq 1.8 we find from Table 1.4A that:

$$\varepsilon_A = \cot^{-1}\left(\frac{L}{D}\right) = \cot^{-1}(4) = 14°$$

From Table 1.9 given below we see that:

$$\sin 14° = 0.242$$

Hence, the expected $V_{mg}/V_T/\text{max}$ ratio when sailing to windward (Eq 1.12) should be:

$$(V_m/V_T)\,\text{max} = \frac{1}{2}\left(\frac{1}{0.242} - 1\right) = \frac{1}{2}(4.13 - 1) = 1.57$$

at the optimum
$$\gamma = 45 + \frac{14}{2} = 45 + 7 = 52°$$

Fig 1.59   Ice boat: geometry of sailing to windward. The equivalence of ε_A and β angles is apparent. Their influence on the attainable V_{mg}/V_T ratio is shown in the graph above.

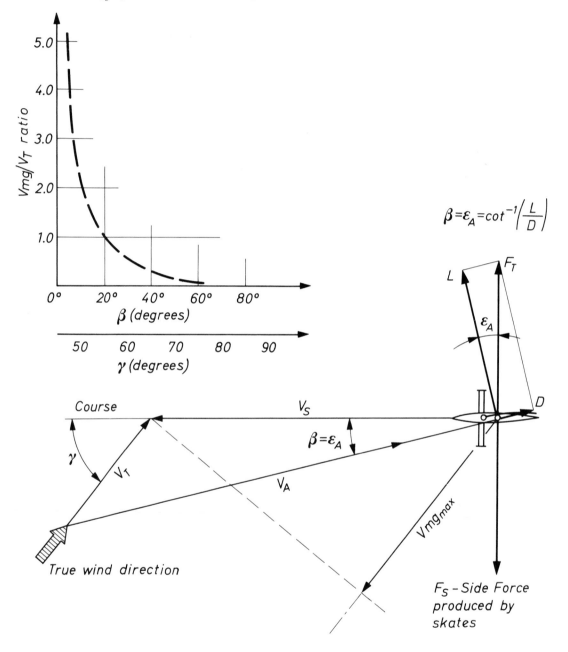

The maximum speed V_S or (V_S/V_T) max ratio should be (Eq 1.13)

$$(V_S/V_T)\,\text{max} = \frac{1}{0.242} = 4.13$$

at the optimum $\qquad\qquad \gamma = 90 + 14 = 104°$

and the (V_{mg}/V_T) max to leeward (Eq 1.14)

$$(V_{mg}/V_T)\,\text{max} = \frac{1}{2}\left(\frac{1}{0.242} + 1\right) = \frac{1}{2}(4.13 + 1) = 2.57$$

at the optimum $\qquad\qquad \gamma = 135 + 14/2 = 142°$

These figures are in good agreement with the polar diagram marked $L/D = 4$ in Fig 1.58. Equations 1.12, 1.13 and 1.14, as well as polar diagrams in Fig 1.58, are applicable on two conditions: first, the wind strength is sufficiently low, so that the limit of stability has not been reached; secondly, runner friction is negligible. In such a case, the derivation of the performance polar diagram, suggested by Barkla in Ref 1.1b, is relatively simple. It is shown in Fig 1.60, which illustrates the performance characteristics of an ice boat that can achieve β angle $= 30°$, equivalent to an overall L/D ratio of about 1.7 only. These rather poor performance characteristics, well below the average, were adopted for the sake of clarity of the drawing.

Vector O A represents the true wind velocity V_T. Since the apparent course β for an ice boat is constant and equal to the minimum aerodynamic drag angle $\varepsilon_{A\,min}$, hence the locus of point B (vertex of the velocity triangle) is a circle with the true wind vector O A, as a chord. Vector O B_3 equal to V_S max is the diameter of the circle that can be found from the relation: Diameter $= V_S = V_T/\sin \beta$.

With a little patience, readers with inquiring minds might themselves decode the method of plotting velocity triangles. By scaling the vectors and measuring the angles of Fig 1.60, the verification of Eqs 1.12, 1.13 and 1.14 can be performed, if desired.

A comparison of the drawing in Fig 1.60 with that in Fig 1.58 might be of some help when preparing performance polar diagrams for some other ice boats with different performance characteristics, i.e. different attainable ε_H angles.

Sailing boats

The conspicuous difference in performances of the fastest 'soft water' craft such as *Tornado* and *Monitor*, and average 'hard water' boats represented in Fig 1.58, can be regarded, in a way, as a measure of the price for contact of the hull (or just its appendages) with liquid water. In other words, supporting a fast sailing craft in a more or less upright position, by buoyancy or dynamic water forces, is a costly endeavour. Figure 1.61 illustrates this point in yet a different way. There are plotted the two hypothetical velocity triangles for *Icarus* and *Crossbow*, based on their published official speed record data in 1972. Assuming that both craft were sailed on

TABLE 1.9
Natural sines

	0′	6′	12′	18′	24′	30′	36′	42′	48′	54′
0°	.0000	0017	0035	0052	0070	0087	0105	0122	0140	0157
1	.0175	0192	0209	0227	0244	0262	0279	0297	0314	0332
2	.0349	0366	0384	0401	0419	0436	0454	0471	0488	0506
3	.0523	0541	0558	0576	0593	0610	0628	0645	0663	0680
4	.0698	0715	0732	0750	0767	0785	0802	0819	0837	0854
5	.0872	0889	0906	0924	0941	0958	0976	0993	1011	1028
6	.1045	1063	1080	1097	1115	1132	1149	1167	1184	1201
7	.1219	1236	1253	1271	1288	1305	1323	1340	1357	1374
8	.1392	1409	1426	1444	1461	1478	1495	1513	1530	1547
9	.1564	1582	1599	1616	1633	1650	1668	1685	1702	1719
10°	.1736	1754	1771	1788	1805	1822	1840	1857	1874	1891
11	.1908	1925	1942	1959	1977	1994	2011	2028	2045	2062
12	.2079	2096	2113	2130	2147	2164	2181	2198	2215	2233
13	.2250	2267	2284	2300	2317	2334	2351	2368	2385	2402
14	.2419	2436	2453	2470	2487	2504	2521	2538	2554	2571
15	.2588	2605	2622	2639	2656	2672	2689	2706	2723	2740
16	.2756	2773	2790	2807	2823	2840	2857	2874	2890	2907
17	.2924	2940	2957	2974	2990	3007	3024	3040	3057	3074
18	.3090	3107	3123	3140	3156	3173	3190	3206	3223	3239
19	.3256	3272	3289	3305	3322	3338	3355	3371	3387	3404
20°	.3420	3437	3453	3469	3486	3502	3518	3535	3551	3567
21	.3584	3600	3616	3633	3649	3665	3681	3697	3714	3730
22	.3746	3762	3778	3795	3811	3827	3843	3859	3875	3891
23	.3907	3923	3939	3955	3971	3987	4003	4019	4035	4051
24	.4067	4083	4099	4115	4131	4147	4163	4179	4195	4210
25	.4226	4242	4258	4274	4289	4305	4321	4337	4352	4368
26	.4384	4399	4415	4431	4446	4462	4478	4493	4509	4524
27	.4540	4555	4571	4586	4602	4617	4633	4648	4664	4679
28	.4695	4710	4726	4741	4756	4772	4787	4802	4818	4833
29	.4848	4863	4879	4894	4909	4924	4939	4955	4970	4985
30°	.5000	5015	5030	5045	5060	5075	5090	5105	5120	5135
31	.5150	5165	5180	5195	5210	5225	5240	5255	5270	5284
32	.5299	5314	5329	5344	5358	5373	5388	5402	5417	5432
33	.5446	5461	5476	5490	5505	5519	5534	5548	5563	5577
34	.5592	5606	5621	5635	5650	5664	5678	5693	5707	5721
35	.5736	5750	5764	5779	5793	5807	5821	5835	5850	5864
36	.5878	5892	5906	5920	5934	5948	5962	5976	5990	6004
37	.6018	6032	6046	6060	6074	6088	6101	6115	6129	6143
38	.6157	6170	6184	6198	6211	6225	6239	6252	6266	6280
39	.6293	6307	6320	6334	6347	6361	6374	6388	6401	6414
40°	.6428	6441	6455	6468	6481	6494	6508	6521	6534	6547

Fig 1.60   Velocity triangles for:

maximum V_{mg} upwind–(O B_1 A)
square reach–(O B_2 A)
maximum V_S on reach–(O B_3 A)
maximum V_{mg} downwind–(O B_4 A)

Diameter of the circle O $B_3 = V_{Smax} = \dfrac{V_T}{\sin \beta} = 2V_T$

β angle $= 30°$.

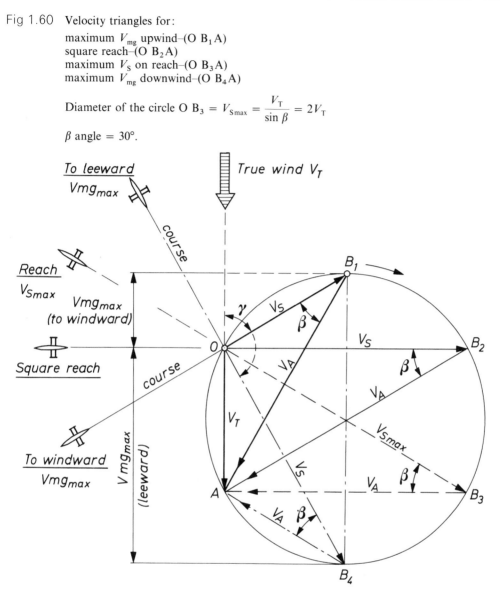

the most favourable sailing course γ (being slightly abaft the beam), one may find that their apparent courses β were about 46° and 39° respectively.

Existing wind tunnel evidence justifies a guess that the rigs of *Icarus* and *Crossbow* were capable of developing L/D ratio of 6 at least; this gives ε_A about 9.5°.

Since $\varepsilon_H = \beta - \varepsilon_A$, the hydrodynamic drag angles are approximately $\varepsilon_H = 46 - 9.5 = 36.5°$ for *Icarus* and $\varepsilon_H = 39 - 9.5 = 29.5°$ for *Crossbow*. These numbers indicate the possible room for performance improvements.

One may argue that, on a craft with a relatively poor hull but with good sails, further sail improvement will provide only a small result in improved overall β angle.

Fig 1.61   Hypothetical velocity triangles for the two speed record holders
            Icarus and *Crossbow*.

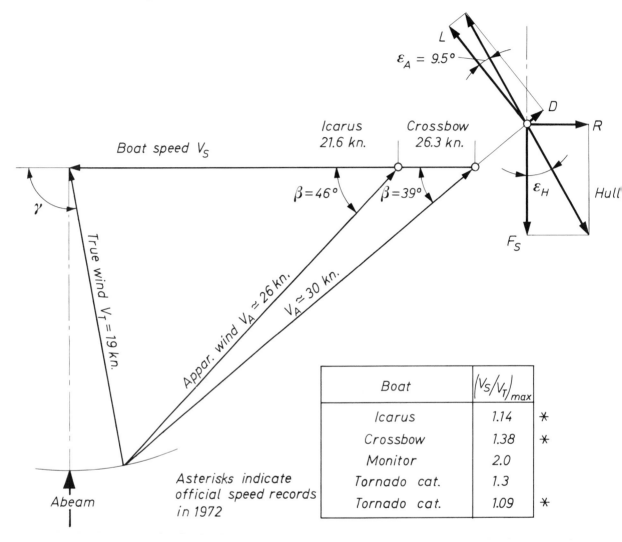

Boat	$(V_S/V_T)_{max}$	
Icarus	1.14	*
Crossbow	1.38	*
Monitor	2.0	
Tornado cat.	1.3	
Tornado cat.	1.09	*

Asterisks indicate
official speed records
in 1972

The same degree of hull improvement equipped with a good rig, however, may produce an outstanding performance breakthrough.

A numerical example may illustrate this idea more convincingly. Let us take the *Icarus* achievement as a basis. An increase of its rig efficiency by 50 per cent through reducing the aerodynamic drag angle ε_A from 9.5° to about 4.75° would reduce β angle to:

$$\frac{36.5 + 4.75}{36.5 + 9.5} \simeq 91 \text{ per cent of its initial value,}$$

i.e. the resulting total improvement would be about 9 per cent. It can be found in Table 1.4A that such an improvement would require a rig capable of developing L/D ratio of about 12 instead of 6. If, however, by some clever invention the hull drag angle ε_H could be reduced by 50 per cent, i.e. from 36.5 to 18.25, with rig efficiency remaining the same ($\varepsilon_A = 9.5°$), it would reduce β angle to:

$$\frac{18.25 + 9.5}{36.5 + 9.5} \simeq 60 \text{ per cent of its initial value,}$$

i.e. the resultant overall improvement would be about 40 per cent, quite a dramatic breakthrough towards higher speed. Viewed in the light of past achievement, improvements of this order of magnitude, in the case of sailing hydrofoils, do not seem beyond reason, particularly when attention is given to both components, the aerofoil and the hydrofoil, simultaneously.

Drag

Referring to ice boats again, it has been assumed in previous discussion that runners sliding on ice produce negligible friction, so small that it does not affect the overall L/D ratio, and therefore the craft's performance. Actually, the price paid for the sliding contact of runners with ice varies from one-quarter per cent to two per cent of the total weight of the craft, depending upon ice condition, shape of the runners, their cutting edge sharpness, etc.

Leonardo da Vinci (1452–1519) observed that: '...friction produces double the effort if the weight be doubled.' He was basically right, postulating that when one solid body slides on another, a force is needed to maintain the motion. To a fairly good approximation the friction drag D_f opposing the motion is nearly independent of the velocity and proportional to the weight of the sliding object and to the proportionality factor μ, the so-called friction coefficient, which is 'more or less constant' for a given pair of sliding surfaces, i.e. friction drag

$$D_f = \mu \times W \qquad\qquad \text{Eq 1.15}$$

where μ–coefficient of friction (for an ice boat $\mu = 0.0025 - 0.02$)
    W–weight of the sliding object.

It is believed that, depending on whether or not there is a relative motion between the sliding objects, one may get two different values of μ for the same pair of contacting materials, and the so-called 'stationary friction coefficient' seems to be greater. Friction coefficient may decrease substantially when the sliding object vibrates, or there are small bits of lubricant present. It can be just water in the case of contact between runners and ice; these variations of friction coefficient explain the erratic behaviour of ice boats observed in some conditions.

Apart from friction drag from the runners, an ice boat experiences a parasitic drag generated on all its structure elements, crew included, which produces no useful aerodynamic reaction. The drag of fuselage and crew, for example, is a *net aero-*

dynamic loss. In that respect the parasitic drag is unlike the drag of the sail, which can be regarded as a price that has to be paid for the beneficial part of the aerodynamic reaction–the lift and finally the driving force which makes the craft go 'against the wind'.

Since the apparent wind V_A strikes the ice boat's superstructure from practically the same very small angle β, whatever course sailed, the runners' friction drag D_f and parasite drag D_p can be added to the drag D associated with sail action. The overall lift/drag ratio $(L/D)_o$, including both additional drags D_f and D_p, can then be expressed:

$$(L/D)_o = \frac{L}{D + D_f + D_p}$$

i.e. that the overall $(L/D)_o$ ratio is smaller than that associated with the rig only. The example below gives a rough estimate of the relative influence of D_f and D_p on performance.

An ice boat sails on a square reach, as shown in Fig 1.55, and its characteristics are as follows:

overall $(L/D)_{o\,max} = 4.0$
sail coefficient C_L at $(L/D)_{o\,max} = 0.7$
sail area $S_A = 70$ sq ft
weight of the craft $W = 350$ lb
friction drag coefficient of runners $\mu = 0.02$ (2 per cent of the craft's weight)
parasite area (equivalent flat-plate area $A_p = 2.0$ sq ft)

When true wind $V_T = 10$ ft/sec (5.9 knots, i.e. a light breeze), and the boat reaches her ultimate speed $V_S = 4\ V_T$, the apparent wind $V_A = 4.1\ V_T$.

Hence the lift generated:

$$L = 0.00119 \times C_L \times S_A \times V_A^2$$
$$= 0.00119 \times 0.7 \times 70 \times 41^2$$
$$= 0.0583 \times 1681 = 98\ \text{lb}$$

Since $(L/D)_o = 4$, therefore the overall drag–

$$D_o = D + D_f + D_p = 98/4 = 24.5\ \text{lb}$$

The contribution of the runners' friction drag D_f towards the overall drag D_o, computed from Eq 1.15, is:

$$D_f = \mu \times W = 0.02 \times 350 = 7\ \text{lb}$$

which might be experienced on ice of moderate roughness and is independent of velocity. On ice smoothed by warm air or sun, the friction coefficient can be much smaller and the runners' friction drag can drop substantially to perhaps 2 lb, provided the shape and sharpness of the runners is correct.

The parasitic area A_p of the craft's structure, expressed as an equivalent flat-plate area, deserves some explanation before the magnitude of parasitic drag is estimated.

The purpose of streamlining of any object being affected by high velocity relative wind is to reduce drag or associated energy expenditure. This can be done by keeping the shape of the object as far away from the flat plate perpendicular to the wind direction as possible, since that is a form of very high resistance. Yet, the drag coefficient of such a plate has played a part in aeronautical calculations, in that the equivalent flat plate area is often used as an index of the parasitic drag offered by a given machine. Unless wind tunnel tests are made on an accurate model, parasitic drag can be estimated by adding the drag of each item expressed in terms of the area of a fictitious flat plate, perpendicular to the air flow, which has the same drag as the item. The symbol A_p is used here for equivalent flat-plate area in square feet and the parasitic drag can be calculated using the familiar formula:

$$D_p = 0.00119 \times C_D \times A_p \times V^2$$

The drag coefficient C_D for a flat plate averages approximately 1.2.

Assuming that the equivalent flat-plate area A_p of the parasitic area of the ice boat in question, including crew and some allowance for interference effects, is 2.0 sq ft (Photo 1.23), we may calculate the parasitic drag D_p at the boat's ultimate speed, when $V_A \doteq 41$ ft/sec

$$\begin{aligned} D_p &= 0.00119 \times C_D \times A_p \times V_A^2 \\ &= 0.00119 \times 1.2 \times 2.0 \times 41^2 \\ &= 0.0029 \times 1681 = 4.8 \text{ lb.} \end{aligned}$$

This drag will decrease or increase in proportion to the square of the velocity V_A, i.e. in the same manner as lift L.

Taken together, the friction drag of runners D_f, and parasitic drag D_p, contribute 11.8 lb, i.e. 48 per cent towards the overall drag of the craft at this particular wind speed. The remaining drag 12.7 lb is the drag associated with the sail itself. It means that the L/D ratio of the sail alone must be about:

$$\frac{L}{D} = \frac{98}{12.7} = 7.7$$

to produce the overall L/D ratio of about 4.0, when the two additional drags are added. Comparing these two (L/D) ratios, it becomes clear that the parasitic air drag D_p and runner friction drag D_f are rather costly components of the overall drag D_o, in terms of the wind energy they waste, and consequently in their adverse influence on the boat's performance. There is much room for improvement, particularly in fuselage design by streamlining its parasitic area. An ideal streamlined ice boat would then be one in which the total drag consists of the sum of the drag associated with an efficient sail (this drag cannot be avoided) and the drag due to skin friction over the fuselage and other essential parts only, without wake drag, due to separation and eddy formation. Poor aerodynamic shape prevents the attainment of streamline flow and gives rise to the shedding of continuous streams of eddies, which represent wasted wind energy. This requires a driving force in excess of that required to overcome the unavoidable aerodynamic drag, and such an excess is the measure of the aerodynamic inefficiency of a given design. Photographs 1.22 and 1.23 illustrate simple land and ice yachts that are certainly producing large components of parasitic drag.

Since the runner friction drag D_f is constant, i.e. independent of velocity, its relative contribution towards overall drag, and therefore overall $(L/D)_o$ ratio, varies depending on true wind speed V_T. It is higher at low wind velocity, and decreases gradually when V_T increases, and this explains why ice boats manifest certain sluggishness in light air conditions, but improve their performances in stronger winds.

The influence of the parasitic drag D_p and friction drag D_f on ice boat performance can be demonstrated graphically by means of the polar curve of coefficients C_L versus C_D. Such a polar diagram has special merit, particularly when used to illustrate and solve some problems concerned with performance. Figure 1.62 shows the polar curve of a hypothetical sail, discussed earlier, which has a maximum L/D ratio of about 7.7; its coordinates are C_L–O–C_D. If a straight line is drawn through the origin O, and tangent to the polar curve, the point of tangency will locate the maximum L/D for the sail alone. This point, together with associated drag angle ε_A minimum, and the lift coefficient C_L for which the latter is reached, are depicted in Fig 1.62. Since $L/D_{max} = 7.7$, therefore $\varepsilon_A =$ about 7.5° (which can be found in Table 1.4A).

Drag coefficient of the parasitic area A_p can be expressed as a function of the sail area S_A, i.e.

$$C_{Dp} = C_{D\,(\text{flat plate})} \times A_p/S_A = 1.2\, A_p/S_A.$$

The friction drag of runners can also be expressed in the form of a coefficient as a function of sail area S_A:

$$C_{Df} = \frac{D_f}{0.00119 \times S_A \times V_A^2}$$

Coefficients C_{Dp} and C_{Df} must be calculated for the same apparent wind velocity V_A.

Laying off both the parasitic and friction coefficients, $C_{Dp} + C_{Df}$ to the left of the origin O in Fig 1.62, gives a new origin O_1, for co-ordinates C_L–O_1–C_D of the complete ice boat. A line drawn through point O_1 and tangential to the polar curve locates the new, overall $(L/D)_o$ ratio of the entire ice boat, and the difference between attainable drag angles $\varepsilon_{A\,min}$ for the sail alone, and the overall drag angle ε_{Ao} for the complete craft, can be read off almost immediately from the graph. This new overall drag angle ε_{Ao} is 14°, which corresponds to $(L/D)_o$ about 4.0. It is noticeable that the optimum lift coefficient C_{Lo} is higher than the previous one, C_L. It is due to the fact that with the parasitic drag of the hull added, a greater lift coefficient is needed to achieve optimum performance. Subsequently the heeling force and heeling moment generated by the less aerodynamically clean ice boat will also be higher. The drag

Fig 1.62   Polar diagram for entire ice boat and for sail alone.

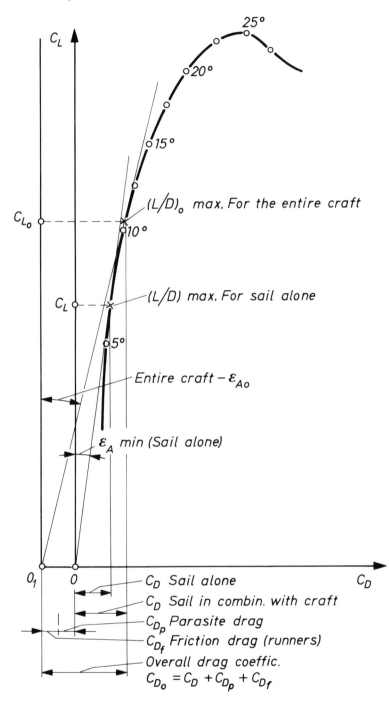

C_L

$25°$

$20°$

$15°$

$(L/D)_o$ max. For the entire craft

C_{L_o}

$10°$

(L/D) max. For sail alone

C_L

$5°$

Entire craft $-\varepsilon_{Ao}$

ε_A min (Sail alone)

0_1 0

C_D Sail alone

C_D Sail in combin. with craft

C_{D_p} Parasite drag

C_{D_f} Friction drag (runners)

Overall drag coeffic.

$C_{D_o} = C_D + C_{D_p} + C_{D_f}$

C_D

coefficient C_D of the sail itself set at a larger incidence angle, while operating in combination with the fuselage and runners, is seen to be greater too.

Since the runner friction drag coefficient C_{Df} decreases gradually when wind speed increases, the overall drag coefficient:

$$C_{Do} = C_D + C_{Dp} + C_{Df}$$

becomes progressively smaller. In consequence, origin O_1 of the whole craft polar diagram in Fig 1.62 migrates towards origin O which produces a desirable effect: a smaller overall drag angle ε_{Ao} and therefore improvement in boat performance.

Thus, while the boat accelerates or operates in variable wind conditions, neither the $(L/D)_o$ ratio, nor the lift and drag coefficients remain constant. At the beginning of the acceleration period and in light winds, the C_L coefficient is relatively large. When apparent wind increases the C_L coefficient decreases gradually and $(L/D)_o$ ratio approaches its maximum. Sooner or later, provided the true wind is sufficiently strong, the stability limit is reached and the heeling force approaches the maximum permissible value. Wind spilling becomes a necessary action which the helmsman is forced to adopt. The C_L coefficient then falls below that associated with $(L/D)_{o\,max}$ ratio, hence the boat's performance is bound to deteriorate. This effect of stability on ice boat behaviour is shown qualitatively in Fig 1.63A, adopted from Ref 1.66, and the characteristic downward bend of V_S versus V_T curves reflects it adequately. These curves give the variation of reaching speed V_S for two boats of different L/D ratio. As it is seen in true winds V_T up to about 10 knots, boat speed V_S increases almost in direct proportion to the V_T. Beyond a certain critical speed, as the sheet is eased progressively to maintain permissible heeling force, the relative speed V_S/V_T falls off pretty sharply.

In such conditions, both the simple formulae 1.12, 1.13 and 1.14, as well as performance polar diagrams presented in Figs 1.58 and 1.60, cease to be applicable.

From the attached sketch in Fig 1.63A, one may deduce that the permissible heeling force F_{Hperm}, roughly equal to lift L, can be calculated by multiplying the boat's total weight W by the ratio of the half-width B between runners, to the height H of the Centre of Effort of the rig, i.e.

$$F_{Hperm} \simeq L_{perm} \simeq W \times \frac{B}{H}$$

The limiting heeling force for ice boats is about $0.9–1.0 \times W$, while racing dinghies with trapezes for the crew rarely absorb a heeling force of more than about one-quarter of their total weight, i.e. $0.25 \times W$.

Figure 1.63B gives the qualitative comparison of ice boat and land boat speeds when reaching in varying wind strengths. Taken from Barkla's analysis (Ref 1.66), those curves were calculated on the assumption that both craft have the same rig. The only difference is due to the fact that land yachts suffer higher drag friction D_f than ice boats. It is estimated that the rolling resistance coefficient μ of land yachts running on modern tyred wheels is about 0.06 on tarmac and 0.10 on hard sand. In

Fig 1.63   A. The effect of stability and overall $(L/D)_o$ ratio on ice boat
performance.
B. Performance characteristics of ice-yacht and land yacht in
various true wind speeds V_T.

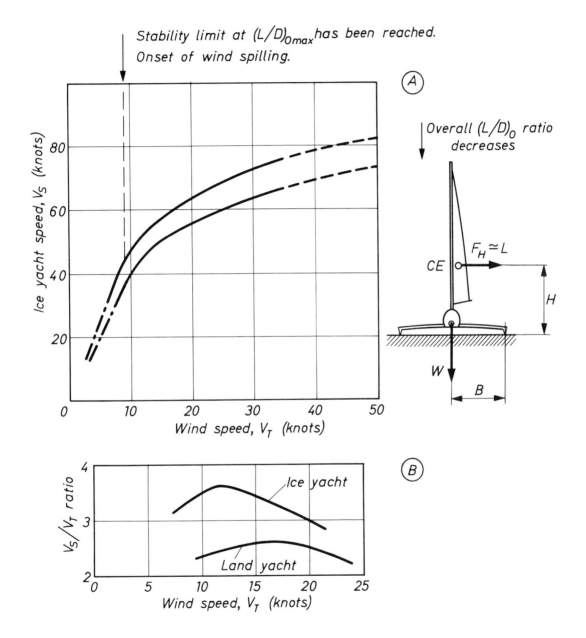

Photo 1.24   A, B and C–Some sail wing applications.

A. Sailing on Skates, as suggested in the *Illustrated London News* 31 January 1880.

B. I Hamilton's skate-sail developed about 90 years later (Ref AYRS Publ No 66).

C. Sail wing concept suggested in 1751, as taken from the book by Robert Pultock, *The Life and Adventures of Peter Wilkins, a Cornish Man* London 1751, J Robinson.

A

B

C

Boitard Fecit.
A Ganvrey Extended for Flight.

other words, the surface drag D_f varies from 6 to 10 per cent of the craft's total weight, depending on the supporting surface smoothness.

Photograph 1.24 demonstrates the simplest concept of sailing on skates in which the wind energy, with sails as a propulsive device, is exploited. The photographs are self-explanatory. The whole idea, suggested about 90 years ago, has been rediscovered quite recently.

Only some of the more important factors governing the behaviour of a variety of sailing craft have been discussed so far. There are many other variables or factors of secondary importance associated with trimming, tuning and adjusting of sails to meet particular demands. Besides, there are the boat's own antics which are frequently unpredictable. The great variations of weather and sea or unsteadiness of wind and water and their interface, make various interactions even more difficult to anticipate and estimate. Some of them will be discussed in following chapters, some have yet to be uncovered in the future.

Finally, the technically-minded sailor searching for perfection may easily miss just one variable. To expose it let us listen to the renowned Burgess (Ref 1.67), who was deeply involved in America's Cup challengers: 'I wish to point out that there are a great many variables which influence the speed of the yachts. Perhaps one of the most important and least regarded is what the skipper had for dinner the night before the race. It makes a great difference.'

References and notes to Part 1

1.1   a. *The behaviour of the sailing yacht* RINA paper, 1960.
      b. *Physics of sailing* (*Encyclopaedic Dictionary of Physics Vol 4*) H M Barkla, Pergamon Press, 1971.

1.2   *A survey of yacht research at Southampton University* T Tanner, JR Aer Soc, Vol 66, 1962.
      If the angle of heel is small, say of the order of 10°, the heel does not appreciably impair the potential efficiency of the sail and the hull, therefore it would seem reasonable for the sake of simplicity, to disregard the possible effects of:
      1. The vertical components of forces F_v and F_{vw}, (Fig 1.1A, B). They are in fact small, in comparison with the boat's displacement and in any case are opposed to one another.
      2. The trimming moments $M_{P(A)}$ and $M_{P(W)}$ (Fig 1.1A). The changes in trim of the hull can be minimized, particularly in dinghies, by shifting crew members longitudinally.
      3. The yawing moments $M_{Y(W)}$ and $M_{Y(L)}$, which may manifest themselves as weather-helm or lee-helm respectively. In the case of a well-balanced boat, little helm is required to keep a straight course. Carefully designed boats should have no excessive yawing moment, even when the heel angle is large.

1.3   In broad terms the feed-back is a process of influencing itself. Automatic control systems are based on this principle starting with James Watt's good old governor; if the steam engine goes too fast the regulator throttles the steam supply. In the sail-hull combination operating as a system, the sail can be regarded as the component which transforms wind energy into force. The hull can be regarded as the component responding to sail input in terms of variable hull speeds. This, in turn, affects the sail input by changing the apparent wind. The boat speed is, in a way, self regulating through a feed-back mechanism in which there exists a cause–effect relationship between the sail forces and hull speed.

1.4   *Aerodynamics* Vol 1 p 431 F W Lanchester, London, A Constable and Co 1907.

1.5   *Sailing Theory and Practice* C A Marchaj Adlard Coles Ltd and Dodd Mead & Co, 1964.

1.6   *Mechanics of sailing ships and yachts* K S M Davidson; a chapter in *Surveys in Mechanics* G K Batchelor, Cambridge, 1956.

1.7   *Designing for Speed to Windward* E Bruce, AYRS Publication No 61, 1967.

1.8   *The Characteristics of 78 related Aerofoil Sections from Tests in the Variable-density Wind tunnel* NACA Rep 460, E Jacobs, K Ward, R Pinkerton, 1933.

1.9   1. *America's Cup: Components of a Successful Defence* W F Ficker.
      2. *The Ancient Interface* 3rd AIAA Symposium on Sailing, California, 1971.
1.10 *Systematic Model Series in the Design of the Sailing Yacht Hull* Pierre de Saix Symposium–Yacht Architecture, HISWA 1971, Holland.
1.11 *The best course to windward* (*Yachts and Yachting*) H Barkla February 19, 1965.
1.12 Broadly defined, a heavy-displacement form of hull is a heavily loaded structure with a keel faired into the hull proper, in such a way that there is no sharp definition between the hull bottom and the keel proper. Deep and narrow yachts built at the end of the 19th century, so-called plank-on-edge yachts, basically slab-sided with no hollow at the garboards, are extreme examples of the displacement form. In contrast to displacement forms, the light-displacement or skimming forms have a distinct division between a basically flat-bottomed hull and the fin keel proper. An extreme example of this skimming form is the scow-type hull, capable by virtue of the lightness and flatness of the hull of developing, at high speed, a dynamic lift which reduces drastically the immersed volume of the hull (Ref 1.5). Displacement/length ratio $\Delta/(L/100)^3$, a criterion for boat lightness, defines load put on a given length L of the hull.
1.13 When talking about displacement/length ratio $\Delta/(L/100)^3$, a criterion for hull lightness and therefore a significant factor in the resistance characteristics of sailing craft, it is convenient to divide various designs into four broad categories, as shown in Fig 1.64:

Fig 1.64   Division of sailing craft of increasing length in terms of displacement/length ratio.

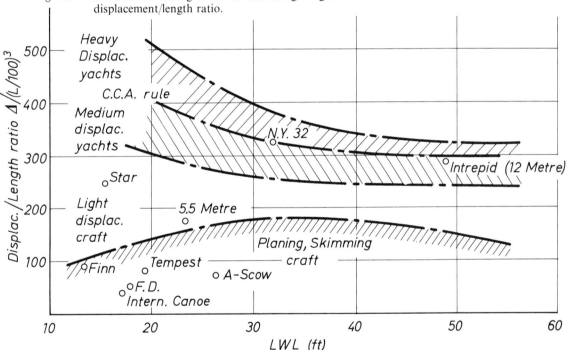

1. Heavy displacement yachts of the older, ocean-going type.
2. Medium-displacement cruiser-racers, built to the International Measurement Rule.
3. Modern light-displacement cruiser-racers, fin keel day sailers etc.
4. Planing and skimming craft of exceptionally low displacement/length ratio.
For further information consult *Sailing Yacht Design* R Henry and R Miller Cornell Maritime Press Inc, 1965.
1.14 The sail area/displacement (S_A/Δ) ratio is, beside the $\Delta/(L/100)^3$ ratio, the second important factor which decides whether or not a boat can plane. This ratio relates the potentially available power to weight.

As rightly pointed out by J Darby (AYRS *Dimensionless Ratios*) some writers err in dividing the sail area by the two-third power of the displacement i.e. $S_A/\Delta^{2/3}$. The argument against it is as follows: 'A ratio of driving force to inertia is required; the driving force is found by taking the product of the sail area S_A and a pressure P (standard atmospheric pressure does very well), and the inertia is the mass M. If the product $S_A \times P$ is in absolute units then comparing it with the weight $M \times g$ (where g = acceleration due to gravity) gives a dimensionless quotient; alternatively, $S_A P$ may be taken in gravitational units and the mass, to give the same result. If the pressure is left out, a dimensioned ratio results and the units have to be given, usually sq ft/ton. It then corresponds to the dimensioned speed/length and displacement/length ratios mentioned earlier.

The difference between the two can be illustrated by an example. Suppose a heavy crew brings the total weight of a Flying Dutchman to 700 lb for her 190 sq ft of sail, and a light single-hander in an International Canoe causes an all-up weight of 350 lb for 106 sq ft. The comparison is:

S_A/Δ	$S_A/\Delta^{2/3}$
FD 608 sq ft/ton	412 sq ft/ton$^{2/3}$
Canoe 678 sq ft/ton	363 sq ft/ton$^{2/3}$

The differences are in each case about 10 per cent but in opposite directions. One may argue that, in order to ensure that sail area/displacement ratio does not vary unduly with boat size, it is rather better that the $S_A/\Delta^{2/3}$ ratios are compared, to take some account of the scale effect. Such a method seems to obscure the mechanism of the two different phenomena.

1.15 *Plank-on-Edge* (*Yachts and Yachting*) W Radcliffe February 16, 1968.

1.16 *Britannia and her Contemporaries* B Heckstall-Smith, 1929.

1.17 a. *Yachting World Annual* 1972.

   b. *The History of American Sailing Ships* H I Chapelle

1.18 *The Tactical Implications of the Polar Curve of Yacht Performance* P V MacKinnon, Southampton University, Rep N.20.

1.19 *Ships* K S Davidson, 9th Intern Congress of Appl Mech–Brussels 1956.

1.20 The fullness or fineness of hull ends relative to midship sections has a considerable effect upon the wave pattern generated due to hull motion, in particular on the position and height of the bow and stern wave crests, and also on the dynamic lift produced on the hull bottom. This, in turn, affects to a large extent the wave-making resistance. In order to picture the fullness of the ends relative to the largest section of the hull, the so-called prismatic coefficient C_p is employed. It expresses the ratio of the volume of the immersed hull to the volume of a prism that has the cross-area of the greatest area of immersed section of the hull, and the same length as the LWL. The prismatic coefficient can be expressed as:

$$C_p = \frac{\text{Volume of the immersed hull (in cubic ft)}}{\text{Greatest section area} \times \text{LWL (in cubic ft)}}$$

The sketch in Fig 1.65 explains the terms used. It can be seen from it that the greatest section area of the hull is somewhere between stations 5 and 6, i.e. the position of maximum area section does not necessarily coincide with the maximum waterline beam.

Prismatic coefficients range from 0.50 to 0.70, depending on the relative boat's speed $V_S/\sqrt{L}$. Thus, conventional heavy displacement type craft are fine-ended, in comparison with fast planing full-ended boats reaching higher $V_S/\sqrt{L}$ ratios.

Since there is a distinct optimum prismatic C_p corresponding to minimum Resistance at any given $V/\sqrt{L}$, a proper C_p should be designed to suit the anticipated weather conditions. In general, boats designed for light weather conditions and operating at lower $V_S/\sqrt{L}$ ratios might have C_p in a range of 0.50–0.53; conversely, sailboats designed for heavy winds should have a higher C_p. This variation in C_p, which is contrary to what might perhaps be expected, was not generally recognized until Taylor's famous experimental data were published and applied initially to ships.

There is a controversy as to whether the fin keel proper should be included in the C_p calculation. It seems to be reasonable to consider that, on light displacement hulls resembling canoe bodies

Fig 1.65 Range of optimum prismatic coefficients for selected speed/length ratios.

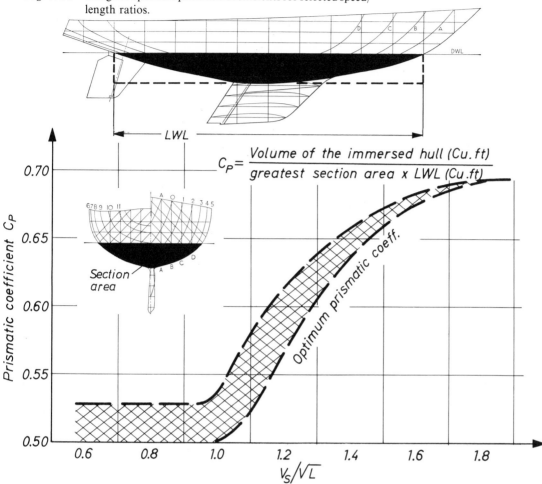

$$C_P = \frac{\text{Volume of the immersed hull (Cu. ft)}}{\text{greatest section area} \times \text{LWL (Cu. ft)}}$$

with clearly defined fin keels, only the hull proper should be taken for the prismatic coefficient, and in such a case C_p will be rather high.

Henry and Miller, in their book (Ref 1.13) which represents the recent state-of-the art of yacht design, advocate that in the case of cruising yachts, where there is less distinction between the hull and the keel appendage than in fin keel yachts, a different approach seems appropriate. A lean, sharp hull is a more efficient hydrofoil than a full, rounded one. Consequently there should be a relationship between prismatic coefficient C_p and the so-called lateral plane coefficient C_{Lp}, which is defined as the ratio of the projected area of the lateral plane A_{Lp} to the circumscribing rectangle LWL $\times$ D. Figure 1.66 is a plot of such a relationship between C_p and C_{Lp}.

One may expect that a yacht with low C_{Lp} coefficient should have a higher aspect ratio of her underwater part of the hull and possibly a lower wetted area; hence higher hydrodynamic efficiency than a shoal draught centreboard yacht might be expected. For this reason some of the new fin keel cruiser-racers are excellent performers in terms of V_{mg}. It does not, however, mean that they are necessarily seaworthy and of good course-keeping ability.

For further information the following references might be consulted:
Sailing Yacht Design D Phillips-Birt Adlard Coles Ltd, London, 1966, and International Marine Co, USA.
Skene's Elements of Yacht Design F S Kinney, Dodd, Mead & Co, New York, 1962.

Fig 1.66 Relationship between the prismatic and lateral plane coefficients.

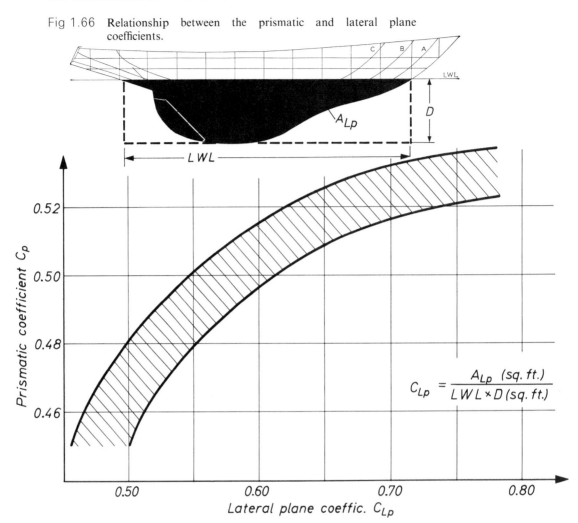

$$C_{Lp} = \frac{A_{Lp} \ (sq. \ ft.)}{LWL \times D \ (sq. \ ft.)}$$

1.21 *The Hydrofoil Boat; its History and Future Prospects* P Crewe, RINA Trans. Vol 100, p 338, 1958.
1.22 *Hydrodynamics in Ship Design* H E Saunders, SNAME New York, 1957.
1.23 *Safety, capsizing and self-rescue action* (*Flying Dutchman* Bulletin):
    1. Never trust the sea, it can change at any moment.
    2. If you sail to win, you will have to take the risk of capsizing.
    3. You must be able to right your boat, but realize it is far more difficult to right it in a rough sea.
    4. Do not take this risk when not enough rescue boats are around.
    5. When you train to make your boat fast, you should also train to capsize and right your boat in rough water.
    6. What types of buoyancy tanks are available:
       a. plastic foam-filled tanks–they are excellent but heavy;
       b. air tanks–they tend to leak–beware!
    7. Where should buoyancy be placed?
       a. it should give a maximum lift, therefore as low as possible;
       b. it should give stability when righting, therefore enough of it must be in the sides and fairly high up;

c. too much buoyancy in the sides will make the boat float high in the water, and it will blow away from you after a capsize.

8. Self rescue. *Sailing an FD to its ultimate speed you will capsize frequently.* Rescue boats are generally too far away. You must be able to right yourself…and quickly!

9. Always stay with your boat, hold on to your main or jibsheets.

   Some information about multihull capsizing can be found in AYRS Publication No 63 (1968).

1.24 *Comparative performance of the Flying Fish Hydrofoil and of the Tornado catamaran* Prof W S Bradfield (*The Ancient Interface*) Symposium on Sailing, California, Vol 10, 1971.

1.25 *The Development of a Hydrofoil Daysailer* Prof W S Bradfield, The American Institute of Aeronautics and Astronautics, April 1970.

1.26 *Predicted and measured performance of a daysailing catamaran* Prof W S Bradfield, SUYR No 25, 1968.

The simple dimensionless theory was obtained mainly as a result of some restrictions, as follows:

1. The motion is regarded as steady and rectilinear; water surface being 'flat', i.e. wave height is small as compared with the waterline length of the hull.

2. The rudder does not contribute to the side force.

3. Heel and pitch angles are assumed to be negligible.

4. Sail planform, camber and twist are assumed to be independent of wind force.

One of the further assumptions, that the residual resistance for light-displacement hull forms can be approximated by a linear function of $\Delta/(L/100)^3$, is well substantiated by Yeh (1965) *Series 64 Resistance Experim of High Speed Displac Forms*–Marine Technology, Vol 2 No 3, p 248. The theory was successful in predicting the performance of a daysailing catamaran.

1.27 *Wind Tunnel Tests of a 1/4 scale Dragon Rig* C A Marchaj, T Tanner, SUYR Rep 14, 1964.

1.28 *Yachting Monthly*, September 1972.

1.29 *My Big Boat and Me* J-Y Terlain (*Sail Magazine*), September 1972.

1.30 *The Shortest Distance Between Two Points* Allan Watts (*Yachting World*), June 1972.

1.31 *Crossbow* (*Yachting World*), October 1972.

1.32 *Onwards and Upwards* D Pelly (*Yachting World*), November 1972.

1.33 *Faith shall be easily shaken, hope quickly foiled* James Grogono (*Yachting World*), October 1972.

1.34 *Up, Up and Away* James Grogono (*Yachts and Yachting*), October 1969.

1.35 *Hydrofoil Sailing* A Alexander, J Grogono, D Nigg, Published by J Kalerghi, London 1972.

1.36 *Dogwatch* D Pelly (*Yachting World*), November 1972.

1.37 Since the upper part of a surface-piercing hydrofoil develops at certain angles of incidence and flow velocities a pressure (suction) below atmospheric pressure, it becomes susceptible to ventilation. Ventilation begins with air rushing down the surface of the foil, causing a more or less drastic loss of hydrodynamic lift. Atmospheric air communicates, usually through a spiral vortex, on the suction side of the foil. Ventilation, or air entrainment, can be controlled by the use of thin, chordwise fences which act as physical barriers to the passage of air. Such a series of fences is shown in Photo 1.25. The foils are designed in such a way that ventilation is tolerated down to the first fence below the surface. If a fence is submerged rapidly, air is carried with it and ventilation below the fence will take a considerable time to shut off. In practice, design against ventilation is a process of trial positioning and shaping of fences and can only be successful if based on greater understanding of the relevant phenomena. Ventilation may also occur on the suction side of dinghy rudders as well as spade rudders of heavy boats being driven hard. The presence of ventilation reduces rudder power drastically with subsequent tendency towards broaching. A small fence over the forward part of the suction chord, as shown in Fig 1.67, may prevent or at least delay the onset of the problem.

1.38 Dr R Cannon, while with the Baker Manufacturing Co, Wisconsin, USA, designed *Monitor*, a hydrofoil sailing boat 26 ft in length, shown in Photo 1.26. A speed of 30.4 knots was recorded in 1956 (35 knots has also been claimed). *Monitor*, sponsored by the US Navy, utilizes a system of two forward and one centre rear ladder-foil. Some control of roll is achieved through differential in-flight adjustment of foil angles. Rear foil trim adjustment is through a mechanical linkage to the mast, and steering is by rotation of the rear foil. Elevation of the hull above the water in flight is approximately 2.5 ft. The overall width is 21 ft and the boat carries 230 sq ft of sail.

Photo 1.25　On the left–a series of fences on the suction side of the hydrofoil. On the right–the fence near the top of the transom rudder. Such a horizontal fence prevents air entrainment (ventilation)–the main enemy of any surface-piercing foil.

Photo 1.26　*Monitor* designed by R Cannon: a speed of 30.4 knots was recorded in 1956; 35 knots was also claimed.

LOA–21 ft.
Sail Area 230 sq ft.

Fig 1.67   Rudder fence made of light alloy and glued on. Position about
3 in below the static waterline and parallel to it.

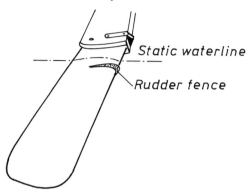

1.39 *Hydrofoil Ocean Voyager Williwaw* D Keiper, *The Ancient Interface*, 3rd AIAA Symposium on Sailing, California, 1971.

1.40 *The Hydrofoil Boat: its History and Future Prospects* P Crewe, RINA Transact Vol 100, p 329.

1.41 *More on Hydrofoils* Hugh Barkla (*Yachts and Yachting*), July 1958.

1.42 *HMCS Bras d'Or–An Open Ocean Hydrofoil Ship* M Eames, E Jones, Trans RINA, Vol 113, 1971.

1.43 *Hydrofoils: Optimum Lift-Off Speed for Sailboats* R M Baker Jr (*Science*), Vol 162, p 1273–1275, 1968.

1.44 *Preliminary Mathematical Analysis of a Rigid-Aerofoil, Hydrofoil-water Conveyance* R M Baker Jr, J S Douglas, AIAA Meeting *Quantizing the Ancient Interface*, 1970.

1.45 *Speed Week* Brian Cooper (*Yachting World*), November 1974.

1.46 *The Hydrofoil Sailing Vessel* C Hook (*Hovering Craft and Hydrofoil*), 1970.

1.47 *An appraisal of Hydrofoil Supported Craft* T Buermann, P Leehey, J Stilwell, Trans SNAME, New York, 1953.

1.48 *Research on Hydrofoil Craft* Prof S Schuster (*Hovering Craft and Hydrofoil*), December 1971.

1.49 *Hydrofoil Sailing* H Barkla (*Yachts and Yachting*), 1968.

1.50 *Sailing on hydrofoils* H Barkla, ANUSC Tech Paper No 3, December 1953.

1.51 *Beware–Low Flying Boats* R E Vincent (*Yachting World*), May 1972.

1.52 *High Speed Sailing* H Barkla, RINA Trans 1951.

1.53 *Analysis and Development of a Sailboat with Self-trimming Wing sail* G Fekete and B Newman, Techn Note 65, McGill University, Montreal, 1965.
Also by the same authors:
Development and Testing of a Sailboat with Self-trimming Wing sail Progress Report TN 71–3, 1971.

A theoretical analysis supported by wind tunnel tests has been made for a symmetrical sail with trimming sailplane. This investigation confirmed the static and dynamic stability of the wing sail, and gave lift and drag coefficients in good agreement with existing data for the same aerofoil section at comparable Reynolds numbers.

The above investigation formed the basis for the design of a small full-scale sailing craft having a rectangular wing sail mounted above a 16 ft canoe (Fig 1.50). The major modification to the initial project included the dynamic mass balancing of the complete wing sail.

1.54 *Wingsails* AYRS Publ No 14, 1951.

1.55 *Preliminary Analysis of the Self-tending Rigid Aerofoils for the Hydrofoil Water Conveyance* R Baker and R Gallington, 3rd AIAA Symposium on Sailing, California, 1971.

The above is a theoretical analysis of the stability performance of a rigid-aerofoil propulsion system for the hydrofoil sailcraft (discussed in Ref 1.44).

1.56 *Faster than the Wind* A B Bauer, 1st AIAA Symposium on Sailing, California, 1969.

1.57 *Windsurfing–a New Concept in Sailing* J R Drake, 1st AIAA Symposium on Sailing, California, 1969.

1.58 *The basic mechanics of sailing surface skimmers and their future prospects* Dr J Wolf (*Hovering Craft and Hydrofoil*), March 1972.

1.59 *Science and Civilisation in China* J Needham, W Ling, Vol 4, Cambridge, 1965.

1.60 *The Dynamics of Sailing on Land* D Rypinski, 3rd AIAA Symposium on Sailing, California, 1971.

1.61 To find out the relationship between boat speed V_S, true wind speed V_T, true course β, and apparent course γ, let us consider the velocity triangle and related angles as shown in Fig 1.68. To start with it is convenient to define:

$$V_S = A - B \qquad \text{Eq R.1}$$

since

$$\cot \beta = \frac{A}{C} \qquad \text{Eq R.2}$$

then

$$A = C \times \cot \beta \qquad \text{Eq R.3}$$

since

$$C = V_T \times \cos (90 - \gamma) = V_T \times \sin \gamma \qquad \text{Eq R.4}$$

substituting Eq R.4 into Eq R.3 yields:

$$A = V_T \times \sin \gamma \times \cot \beta \qquad \text{Eq R.5}$$

in turn

$$B = V_T \times \cos \gamma \qquad \text{Eq R.6}$$

substituting Eqs R.5 and R.6 into R.1 gives:

$$V_S = V_T \times \sin \gamma \times \cot \beta - V_T \times \cos \gamma$$
$$= V_T (\sin \gamma \times \cot \beta - \cos \gamma) \qquad \text{Eq R.7}$$

or

$$\frac{V_S}{V_T} = \sin \gamma \times \cot \beta - \cos \gamma \qquad \text{Eq R.8}$$

If we limit, at first, our attention to the simplest case, namely the beam reach condition, when:

$$\gamma = 90° \qquad \text{then} \qquad \sin 90° = 1.0 \quad \cos 90° = 0$$

therefore

$$\frac{V_S}{V_T} = \cot \beta \qquad \text{Eq R.9}$$

or

$$V_S = V_T \times \cot \beta \qquad \text{Eq R.10}$$

1.62 Some people are puzzled by the concept of apparent wind, and claim occasionally that it is impossible to obtain forward thrust from the apparent wind (read, for example, *Analysis of the apparent wind. It cannot provide thrust–in fact there is no such thing*, by W Stevenson (*Yachts and Yachting*), January 1966). The power of this 'unreal' wind is most conspicuous in the case of an ice boat, and this is perhaps responsible for a certain mystery surrounding ice boating and the heated emotions it may generate in its enthusiasts.

The answer, to those who are still doubtful of apparent wind, could perhaps be given using some arguments already expressed in discussion with W Stevenson. Of course one may say that

Fig 1.68

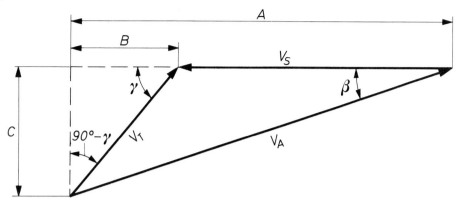

the apparent wind does not exist. It is a concept introduced only to simplify the study of the forces acting on a sail or any other aerofoil; by postulating an apparent wind we can consider a sail at rest with air flowing over it. Thus, for example, we can use a wind tunnel to simulate actual sailing conditions, in a similar manner to the aircraft designer who can study the forces acting on a stationary aeroplane model operating in an airstream generated in the wind tunnel. Those who have sailed ice boats know from experience that in marginal wind conditions, when the true wind is not strong enough to get the ice boat going, an eager crew must accelerate the craft by pushing it as fast as he can on a reaching course until the resultant, or apparent, wind becomes sufficiently strong and the boat 'ignites'.

The concept of apparent wind is certainly a vague idea, especially when first encountered. As in the case of 'force concept', or gravity concept and all other concepts which form the basis of any science, it requires time to become accustomed to the new idea in order to integrate it either with previous knowledge, or what is perhaps more important, to associate it with personal experience. In consequence the apparent wind is as real as true wind, or to put it in other words, the thing that makes apparent wind real is that it is a very useful concept. In fact the apparent wind V_A is a result of the vector addition of the true wind V_T and the reverse of boat speed V_S. As such it is frequently used through this whole book as it was previously used in *Sailing Theory and Practice*.

1.63 *How to Race Land Yachts* P Milne (*Yachts and Yachting*), November 1964.
1.64 *Ice boat and Catamaran Efficiencies* G Ellis, AYRS Publication No 66A.
1.65 *Speed on Ice* A Scantlebury (*Yachts and Yachting*), April 21, 1972.
1.66 *Faster ever Faster* H Barkla (*Yachts and Yachting*), February 25, 1972.
1.67 *The America's Cup Defenders* C P Burgess, Trans SNAME Vol 43 (1935).

PART 2

Basic principles of aero-hydrodynamics: aerofoil and hydrofoil action

'I can't believe that,' said Alice.

'Can't you?' the Queen said, in a pitying tone. 'Try again: draw a long breath and shut your eyes.'

Alice laughed: 'There is no use trying,' she said, 'one can't believe impossible things.'

'I dare say you haven't had much practice,' said the Queen. 'When I was younger, I always did it for half an hour a day. Why, sometimes I've believed as much as six impossible things before breakfast.'

Alice through the Looking Glass

LEWIS CARROLL

A Elementary concepts and assumptions

(1) Air and water: analogies and differences

Considering the yacht as a sailing machine that comprises four main parts–hull, sails, keel or centreboard, and rudder–and reflecting on her essentials, one soon discovers that a kind of action which, for want of a better general term, may be called the fin or foil action, is one of the first importance. Perhaps as far as a racing yacht is concerned, this fin action heads even the ability necessary to keep the boat afloat. Sails as aerofoils are, in principle, fins spread to the wind and extracting energy from the atmosphere; the hull, with an appendage such as a deep fin keel or centreboard, manifests a fin-like action, opposing a lateral force derived from the sail; the rudder which controls direction is nothing but a movable fin (Ref 2.1).

Fins or foils (aerofoils and hydrofoils) can be defined as relatively thin and flat bodies that, while immersed in a fluid, may be subject to two kinds of forces, arising from the relative motion between them and the fluid, termed the drag D and lift L. As shown in Figs 1.2A and 1.10, the lift component may, within a certain range of incidence angles, be many times greater than the drag component, and this fact makes possible both the flight of an aeroplane and the sailing of a boat close to the wind.

The analogy or comparison between the sail action and the lifting wing is an attractive one and, provided it is not pushed too far, one may reasonably expect that the methods and ideas commonly used in the study of rigid aerofoils may with some reservation be fruitfully applied to the study of sails. The same applies to the action of a fin moving through water. Fluid mechanics draws no qualitative distinction between physical phenomena associated with aerodynamics or the hydrodynamics of fin or foil action.

The science of *hydrodynamics* differs from that of *aerodynamics*, in so far as foil action at low velocities is concerned, in only two rather minor respects. The first of these refers to the numerical values describing the physical properties–the density and viscosity of air and water, as presented in Tables 2.1 and 2.2. The density of sea water for example is about 835 times the density of air at sea level. The second difference arises from the fact that while the hull of a yacht, part of which penetrates the surface of separation between air and water, is moving across this interface it causes waves to be formed. An unavoidable outcome of this is that the hydrodynamic drag of the hull contains an additional factor called 'wave resistance'. However, as far as deeply immersed appendages of the hull are concerned, water behaves in a similar manner to air, and it became customary in aero- and hydrodynamics to regard air and water as belonging to the same general class of substances known as fluids.

TABLE 2.1
Properties of air at different temperatures and standard atmospheric pressure
(See Note 2.2–References and Notes at the end of Part 2)

Temperatures °F	°C	Density ρ_A slugs/ft^3	Specific weight γ_A lb/ft^3	Kinematic viscosity v_A ft^2/sec
32	0.0	0.00251	0.0806	1.40×10^{-4}*
50	10.0	0.00242	0.0778	1.50×10^{-4}
68	20.0	0.00234	0.0750	1.60×10^{-4}
104	40.0	0.00217	0.0697	1.83×10^{-4}
140	60.0	0.00205	0.0660	2.07×10^{-4}

* Read Notes A and D in Appendix at the end of the book.

At 'normal' temperature $t = 15°C = 59°F$ and 'normal' atmospheric pressure (at sea level) corresponding to 29.9 inches = 760 mm of mercury equivalent to standard atmosphere = 14.7 lb/in^2, the mass density of air is $\rho_A = (\gamma_A/g) = 0.00238$ slugs/ft^3, where g = acceleration due to gravity = 32.2 ft/sec^2. The 'standard' dynamic pressure, called sometimes 'impact pressure', at sea level, can be expressed as,

$$q(\text{lb/ft}^2) = \frac{\rho_A \times V^2}{2} = (V \text{ ft/sec})^2/840 = 0.00119 \, (V \text{ ft/sec})^2$$

$$= (V \text{ knots})^2/295 = 0.00339 \, (V \text{ knots})^2$$

The kinematic viscosity of air v_A under normal, sea level conditions needed for the computation of Reynolds Number

$$v_A = \frac{1.57}{10^4} \text{ ft}^2/\text{sec} = 1.57 \times 10^{-4} \text{ ft}^2/\text{sec}$$

TABLE 2.2
Properties of fresh water at atmospheric pressure and standard gravity

Temperature °F	°C	Density ρ_w slugs/ft^3	Specific weight γ_w lb/ft^3	Kinematic viscosity v_w ft^2/sec	Vapour pressure P_v(psia)
32	0	1.940	62.42	1.93×10^{-5}	0.0885
40	4.4	1.940	62.42	1.66×10^{-5}	0.122
50	10.0	1.940	62.41	1.41×10^{-5}	0.178
60	15.6	1.938	62.37	1.21×10^{-5}	0.258
68	21.1	1.937	62.31	1.09×10^{-5}	0.339
80	26.7	1.934	62.22	0.930×10^{-5}	0.507
90	32.2	1.931	62.11	0.826×10^{-5}	0.698
100	37.8	1.927	62.00	0.739×10^{-5}	0.949
212	100.0	1.860	59.83	0.319×10^{-5}	14.70

$$1 \text{ standard atmosphere} = 14.70 \text{ lb/in}^2$$
$$= 2116.2 \text{ lb/ft}^2$$
$$\text{psia} = \text{pounds per square inch}$$

Specific weight of salt water $\gamma_w = 64.0$ lb/ft^3. For practical applications, the density of water $\rho_w = (\gamma_w/g)$ may be regarded as constant (independent of temperature and pressure).

$$\text{For fresh water } \rho_w = 1.94 \text{ slugs/ft}^3$$
$$\text{For salt water } \rho_w = 1.99 \text{ slugs/ft}^3$$

Corresponding to these densities, the dynamic pressure can be expressed:

$$q(\text{lb/ft}^2) = \frac{\rho_w V^2}{2} = 0.97 \, (V \text{ ft/sec})^2 \left.\right\}$$
$$= 2.78 \, (V \text{ knots})^2 \left.\right\} \text{ in fresh water}$$

or

$$q(\text{lb/ft}^2) = 0.995 \, (V \text{ ft/sec})^2 \left.\right\}$$
$$= 2.85 \, (V \text{ knots})^2 \left.\right\} \text{ in salt water}$$

The kinematic viscosity of water v_w needed for Reynolds Number computation at the 'normal' temperature of 15°C is $v_w = 1.23 \times 10^{-5}$ ft^2/sec. As shown in Table 2.2, water viscosity decreases appreciably as the temperature increases.

Vapour pressure–p_v given in Table 2.2 is the pressure at which water boils. The vapour pressure p_v is the equilibrium pressure which escaping liquid molecules will exert above any free surface; its magnitude increases with temperature. For boiling to occur, the equilibrium must be upset either by raising the temperature to cause the vapour pressure to equal or exceed the pressure applied at the free surface, or by lowering the pressure at the free surface until it is equal to or less than the vapour pressure.

Vapour pressure is of some importance in the case of highly loaded hydrofoils. A local drop of pressure on the suction side of the hydrofoil may, at some speed, be close to the vapour pressure p_v, at which the water begins to boil: this is called cavitation. Cavitation causes increased drag and loss of lift. An interesting point is that O Reynolds had been postulating cavitation on a theoretical basis long before cavitation was observed in practice (Destroyer HMS *Daring* trials in 1893).

(2) Ideal and real fluids, two-dimensional flow

When developing concepts of how lift and drag come into existence, it has been found that certain simplifying assumptions, like idealizations of reality, may make a complex problem easier to grasp and/or simple enough to be amenable to mathematical treatment.

The first assumption is that the foil may, if desired, be subjected to the flow of a so-called 'ideal' or 'perfect' fluid, i.e. a homogeneous fluid having no viscosity. This hypothetical fluid is a liquid that flows or slips over solid bodies without friction and is incompressible. Incompressibility is used here in the sense that fluid density ρ is not affected by pressure variation. This is true in the case of water and can also be accepted for airflow at low velocities. For example, an error in assuming air incompressibility when estimating drag would be about half a per cent at a relative speed of 87 knots and less than two per cent at 175 knots. It appears perfectly permissible to neglect the compressibility of air at the wind speeds experienced by sails.

By relative speed it is meant that the forces exerted, for instance, by the air on the sail do not depend on the absolute velocity of either air or sail, but only on the resulting velocity between them. So, if the relative velocity is the same, it is immaterial, as far as the physical phenomena are concerned, whether the foil moves' in a stationary fluid or whether a large bulk of fluid moves uniformly past a stationary foil. Thus at the outset we are concerned with relative velocities, and most experienced sailors will have realized that even in a flat calm it is possible to sail across the tide if the stream is running fast enough (Ref 2.3). For the sake of easier presentation and sometimes easier investigation (wind tunnel testing) it is more convenient to think of the foil as at rest with the fluid moving past it.

As distinct from an ideal fluid, real fluids such as air or water possess certain characteristics amongst which the two responsible for drag generation, namely adhesion and viscosity, cannot be ignored. Water, for example, is composed of submicroscopic chains of minute molecular structures of H_2O, schematic in nature, as shown in Fig 2.1, which may as well be called molecules. The tendency of water molecules to hold together or to cling to other matter is one of the most characteristic properties of water, called *adhesion*. Adhesion due to intermolecular attraction increases with closeness of contact.

'...for explaining how this may be,' says Newton in his *Opticks* (1703),

'some have invented hooked Atoms...I had rather infer from their Cohesion that their Particles attract one another by some Force which in immediate

Fig 2.1   Water is composed of minute sub-microscopic chains–minute
molecular structures–of H_2O, schematic in nature, which may as
well be called molecules. The tendency of water and other fluids'
molecules to hold themselves together due to intermolecular
forces, or to cling to other matter, is one of the most characteristic
properties of water.

'...for explaining how this may be'–says Newton in his *Opticks*–
'some have invented hooked Atoms...'

Contact is exceeding strong, and reaches not far from the Particles with any
sensible Effect...There are therefore Agents in Nature able to make the
Particles of Bodies stick together by very strong Attractions. And it is the
Business of experimental Philosophy to find them out.'

Viscosity is a molecular resistance, which fluid particles manifest against
displacement in relation to each other, and with respect to the surface of submerged
bodies moving through the fluids. This type of resistance presents itself in the form of
frictional or skin friction drag. Viscosity can therefore be regarded as a measure of
the ease with which a fluid will flow. The effect of viscosity can easily be appreciated
from drawing a knife on edge (or incidence = 0°) through heavy liquids such as
honey or treacle. Air which can be regarded as a thin 'dry water' possesses similar
characteristics although, at first sight, air viscosity may appear to be negligibly
small.

The second hypothetical assumption that simplifies the basic theorem concerning
both drag and lift generated by the foil is a concept of two-dimensional, uniplanar
flow round it. Such a concept can be approximated with sufficient accuracy in wind
tunnel experiments by an untwisted foil of finite span, situated between end plates or
wind tunnel walls, as shown in Fig 2.2 and Photo 2.1. Most students find it less

Fig 2.2 The flow around the foil between the sufficiently large end plates can be regarded as two-dimensional.

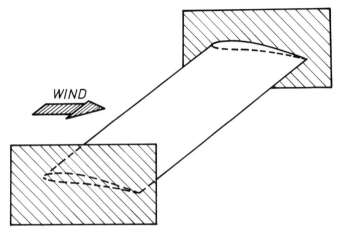

difficult to think of the two-dimensional flow pattern around such a foil than of the flow round the wing of infinite aspect ratio which does not exist in reality. The purpose of the end plates is to prevent the development of air flow around the tips of the foil, and the establishment of the complicated three-dimensional flows that normally persist far downstream behind the wing which are known as tip vortices. In the presence of tip plates the flow around the foil is about the best practical two-dimensional flow, i.e. exactly alike in all planes, perpendicular to the span. The study of the two-dimensional flow furnishes the foundation for the theory of the action of real foils of finite span.

(3) Potential frictionless flow patterns; interrelation between velocity, pressure and force

Air and water forces have their origin partly in the shape of an obstacle deforming the free motion of the fluid, and partly in the flow condition at its surface, both causes generally contributing at the same time. We are already familiar with a sail's ability to generate two kinds of forces: lift and drag. Paraphrasing Munk (Ref 2.4), it may be said that these air force components are of different dignity. The lift associated mainly with shape of the foil, which predetermines the flow pattern, is of higher dignity and may be called noble. The drag component caused by surface viscosity effects is of less dignity and must be considered as base. Drag is common, and it is a fact of general experience that a body in motion through a fluid always generates a resultant force which in most cases is just a resistance to motion and is frequently very dearly paid for. No doubt, lift is a much more positive kind of force in comparison with drag. To generate the noble lift in the most efficient way requires a special class of bodies, purposely and intelligently designed and operated. Strangely enough, we shall see later how friction resistance (which has been regarded

Photo 2.1 Model installation in the wind tunnel to measure section characteristics of NACA aerofoil (National Advisory Committee for Aeronautics). The infinitely long foil and uniplanar two-dimensional flow round it may be secured with good approximation if a foil having identical sections along the span is placed between flat walls in the wind tunnel, the walls running the full height of the airstream. The foil must go right to the walls, i.e. there can be no gap through which a substantial amount of air might escape.

by theoreticians as a villain, introducing enormous complications into the mathematician's dream land of perfect fluids and potential flows controlled by neat and tidy equations) must be employed in order to produce that noble lift. Without viscosity and friction, lift could not possibly exist in our real world–nobility appears to be inseparable from baseness.

Although the mechanism of lift generation is not too difficult to picture and explain, it seems that it is much easier to develop a mental image of its nature by considering step by step some simple flow patterns round the circular cylinder leading to lift generation. This should help in establishing the important relationships between the shape of the foil and the induced velocities, pressures and finally the resulting forces. The circular cylinder appears to be the right kind of shape to investigate because, apart from being an instructive example, it is familiar to

Fig 2.3   Ideal frictionless flow pattern around
circular cylinder.
Streamlines in so-called potential flow.
Theoretical pattern.
Streamlines recorded by camera. Very
low velocity of the flow. Boundary layer
adheres to the surface throughout. No
separation.
Velocities at various stations round
the cylinder.
$V = 2V_o \sin \Theta$.
Pressure distribution round the cylinder
in a perfect fluid shown in terms of
pressure coefficients $\pm C_p$.

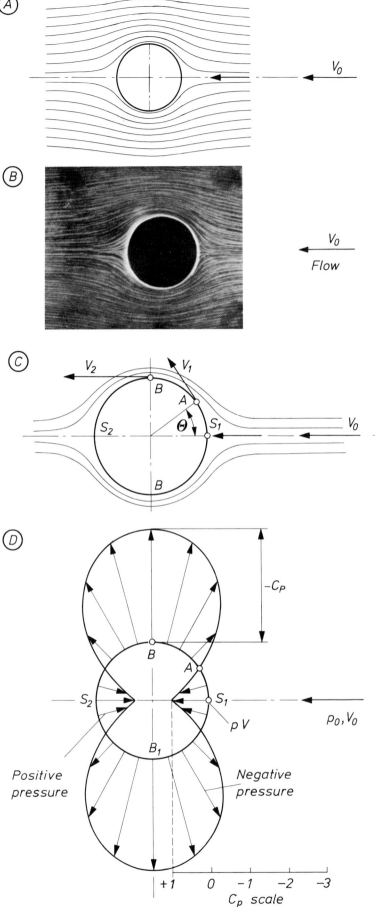

yachtsmen as a device supporting sails.

In Figs 2.3A and B are shown pictures of a two-dimensional, frictionless flow round a stationary cylinder immersed in a moving fluid. Such a flow pattern displayed by means of streamlines cannot be seen in normal conditions, and this produces certain problems for the student of aero- and hydrodynamics. Certainly, if only we could see those streamlines round the sail, hull and its appendages, many sailing mysteries would become easier to understand. Fortunately the streamlines can, in simple cases, be recorded in laboratory conditions; for instance by short exposure photographs of aluminium or oil particles suspended in water. After some experience it is possible to develop some kind of ability to 'see' movement of invisible air and water; and the ability to do this is made easier by employing, for example, wool streamers and observing their behaviour (Photo 3.26 in Part 3).

A stream of fluid (air or water) may be conceived as consisting of a number of particles moving in the same direction. The path of any particle can be distinguished and called a *streamline*. A streamline can also be defined as a line which runs in the direction of the velocity vector; in other words, the local velocity is everywhere tangent to the streamline.

In steady flow, where the velocity does not change, the shapes of the streamlines remain unchanged from one instant to the next.

In unsteady flow, the streamlines are continually changing their shape as the velocity of the flow varies.

In Fig 2.3A are displayed streamlines of the so-called *potential flow*, assuming that the non-viscous frictionless fluid closes in immediately behind the cylinder and therefore the separation which must actually take place in real viscous fluid does not occur. Such a theoretical flow pattern, or *potential flow* round the cylinder, can to good approximation be observed in reality for a short while immediately after the flow starts and viscosity effects have had no time to blur the flow picture. The Photograph in Fig 2.3B displays a low velocity flow at the beginning of motion, strikingly similar to the theoretical ideal case shown just above it. It can be seen that the boundary layer adheres to the back surface of the cylinder and there is no separation.

(a) *Bernoulli's equation*

The whole pattern of the potential flow, as shown in Fig 2.3A, is symmetrical about the horizontal axis passing through the cylinder centre and parallel to the remote undisturbed flow velocity V_0; the spacing of the streamlines indicates the magnitude of the velocity–the closer the spacing, the higher is the speed of the flow in this region; where streamlines are widely separated the fluid moves slowly. At a sufficiently great distance from the cylinder the streamlines are straight parallel lines with equal spacing between them, and this indicates undisturbed uniform flow. The picture of flow pattern given by streamlines is therefore not only a chart of flow direction but also a map of the so-called 'velocity field'.

Theoretically, the velocity V at any point on the surface of the cylinder is given by:

$$V = 2V_o \sin \theta \qquad\qquad \text{Eq 2.1}$$

where V_o is the velocity of undisturbed free stream well ahead of the cylinder or foil.

θ is the angle measured from the stagnation point S in which the flow is brought to rest or $V_o = 0$.

From it we may calculate the velocity distribution at various stations S, A, B around the cylinder as determined in Fig 2.3C. Hence:

at $\theta = 0°$, $\quad V = 0$ $\qquad\qquad$ since $\sin 0° = 0.0$
at $\theta = 30°$, $\quad V = 2V_o \times 0.5 = V_o$ $\qquad$,, $\sin 30° = 0.5$
at $\theta = 60°$, $\quad V = 2V_o \times 0.866 = 1.73\ V_o$ $\quad$,, $\sin 60° = 0.866$
at $\theta = 90°$, $\quad V = 2V_o \times 1.0 = 2\ V_o$ $\qquad$,, $\sin 90° = 1.0$

Following the above procedure it is an easy matter to find the complete picture of the velocity field round the cylinder surface. If the velocity distribution V is known, the pressure distribution can be found by invoking Bernoulli's theorem. It states that, along a given streamline, the sum of static pressure p (pressure head) and dynamic pressure $q = (\rho V^2/2)$ (velocity head) is constant, or in other words equals the total head. The changes in the flow shape and velocity, as we follow the fluid particle along its streamline, are accompanied by corresponding changes in pressure. An intimate knowledge of this pressure variation is of the utmost practical importance in studies of the flow round hulls and foils. If we represent the local static pressure at a point S_1 on the cylinder by p, and the velocity at the same point by V, as shown in Fig 2.3D, and further, as all the fluid originates from an undisturbed region where the static pressure is p_o and velocity V_o, then:

$$p + \frac{1}{2}\rho V^2 = p_o + \frac{1}{2}\rho V_o = \text{constant} \qquad\qquad \text{Eq 2.2}$$

where ρ is the fluid density. Or written in descriptive manner:

$$\text{Pressure head} + \text{Velocity head} = \text{Total head (constant)} \qquad \text{Eq 2.2A}$$

Bernoulli's theorem can be regarded as an expression of the law of energy conservation. One may interpret it as a mutual exchange between potential energy, represented by the static pressure p, and kinetic energy, represented by the dynamic pressure $q = \rho(V^2/2)$. The dynamic pressure q, representing the *kinetic energy* of the body is, so to speak, the amount of work it is capable of doing by virtue of its motion. The term potential energy is used for the work the body can do by virtue of its configuration. For example, a compressed spring is said to possess potential energy, like air under pressure.

At the stagnation point S_1 on the cylinder where the fluid is brought to rest or stagnates (Fig 2.3D), the fluid velocity V, as calculated earlier, is zero and the stagnation pressure p can be computed by applying Eq 2.2 along the streamline $O–S_1$ as follows:

Total pressure at   Total pressure in undisturbed stream
stagnation point S_1 some distance ahead of point S_1

$$p + \frac{1}{2}\rho \times 0 = p_o + \frac{1}{2}\rho V_o^2$$

$$p + 0 = p_o + \frac{1}{2}\rho V_o^2$$

$$p = p_o + \frac{1}{2}\rho V_o^2 = p_o + q_o \qquad \text{Eq 2.3}$$

The stagnation pressure p is therefore higher than the ambient pressure p_o by the amount of dynamic pressure q_o. The dynamic pressure term q, occurs frequently in our studies dealing with forces acting on foils moving in air and water. Table 2.3 and Fig 2.4 give values of q for various wind velocities. Table 2.4 gives corresponding values of speed expressed in ft/sec, m/sec, knots and Beaufort scale.

Fig 2.4   Dynamic pressure q against wind speed V.

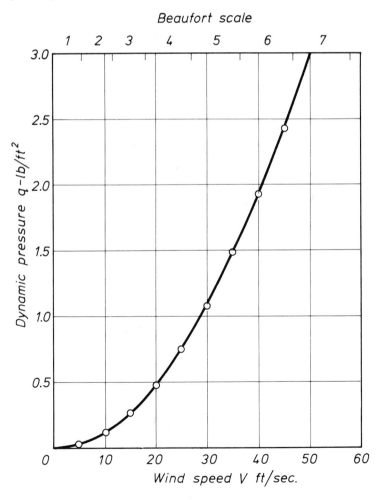

TABLE 2.3
Values of dynamic pressure q, for various wind velocities V (ft/sec)

V ft/sec	$q = \dfrac{\rho_A \times V^2}{2}$ lb/ft^2
0	0.0
5	0.02972
10	0.1189
15	0.2675
20	0.4756
25	0.7431
30	1.0701
35	1.4565
40	1.902
45	2.408
50	2.975
55	3.597
60	4.280
65	5.024
70	5.826
75	6.688
80	7.610

where $q = (\rho_A \times V^2/2) = 0.00119 \times V^2$ lb/ft^2 was calculated at 'standard atmospheric pressure' (sea level), i.e.

$$p_o = 2116.2 \ (\text{lb/ft}^2)$$
$$= 14.7 \ (\text{lb/in}^2)$$

Multiplying the above values of q by 835, i.e. the average ratio of water density ρ_W to air density ρ_A, the relevant dynamic pressure q for water flow is obtained.

Referring to Figs 2.3C, D and applying Eq 2.2, we may find that the local change in pressure $(p - p_o)$ is related to the local change in velocity by the expression

$$p - p_o = \frac{1}{2}\rho V_o^2 - \frac{1}{2}\rho V^2$$

$$= \frac{1}{2}\rho V_o^2 - V^2 \times \frac{V_o^2}{V_o^2}$$

$$= \frac{1}{2}\rho\, V_o^2\left(\frac{V_o^2 - V^2}{V_o^2}\right)$$

$$= \frac{1}{2}\rho\, V_o^2\left(1 - \frac{V^2}{V_o^2}\right)$$

since

$$q_o = \frac{1}{2}\rho\, V_o^2$$

TABLE 2.4
Corresponding values of speed, in four different units

ft/sec	m/sec	knots	Beaufort scale	ft/sec	m/sec	knots	Beaufort scale
1.69	0.52	1		81.07	24.71	48	
3.38	1.03	2	1	82.76	25.22	49	
5.07	1.55	3		84.45	25.74	50	
				86.14	26.25	51	
6.76	2.06	4		87.83	26.77	52	10
8.44	2.57	5	2	89.52	27.28	53	
10.13	3.09	6		91.21	27.80	54	
				92.90	28.31	55	
11.82	3.60	7					
13.51	4.12	8	3	94.58	28.83	56	
15.20	4.63	9		96.27	29.34	57	
16.89	5.15	10		97.96	29.86	58	
				99.65	30.37	59	
18.58	5.66	11		101.34	30.89	60	
20.27	6.18	12		103.03	31.40	61	11
21.96	6.69	13		104.72	31.92	62	
23.65	7.21	14	4	106.41	32.43	63	
25.34	7.72	15		108.10	32.95	64	
27.02	8.24	16		109.78	33.46	65	
28.71	8.75	17		111.47	33.97	66	
30.40	9.27	18		113.16	34.49	67	
32.09	9.78	19	5	114.85	35.00	68	
33.78	10.30	20		116.54	35.52	69	
35.47	10.81	21		118.23	36.03	70	
				119.92	36.55	71	
37.16	11.33	22		121.61	37.06	72	
38.85	11.84	23		123.30	37.58	73	
40.54	12.36	24		124.99	38.10	74	
42.22	12.87	25	6	126.68	38.61	75	
43.91	13.38	26		128.36	39.12	76	
45.60	13.90	27		130.05	39.64	77	
				131.74	40.15	78	
47.29	14.41	28		133.43	40.67	79	
48.98	14.93	29		135.12	41.18	80	
50.67	15.44	30		136.81	41.70	81	
52.36	15.96	31	7	138.50	42.21	82	12
54.05	16.47	32		140.19	42.73	83	
55.74	16.99	33		141.88	43.24	84	
				143.56	43.75	85	
57.43	17.50	34		145.25	44.27	86	
59.12	18.02	35		146.94	44.79	87	
60.80	18.53	36		148.63	45.30	88	
62.49	19.05	37	8	150.32	45.82	89	
64.18	19.56	38		152.01	46.33	90	
65.87	20.08	39		153.70	46.85	91	
67.56	20.59	40		155.39	47.36	92	
				157.08	47.88	93	
69.25	21.11	41		158.77	48.39	94	
70.94	21.62	42		160.46	48.91	95	
72.63	22.14	43		162.14	49.42	96	
74.32	22.65	44	9	163.83	49.93	97	
76.00	23.16	45		165.52	50.45	98	
77.69	23.68	46		167.21	50.96	99	
79.38	24.19	47		168.90	51.48	100	

therefore

$$p - p_0 = q_0 \left(1 - \frac{V^2}{V_0^2} \right) \qquad \text{Eq 2.4}$$

Equation 2.4 can be put in non-dimensional form dividing both sides of the equation by q_0,

$$\frac{p - p_0}{q_0} = 1 - \left(\frac{V}{V_0} \right)^2 = C_p \qquad \text{Eq 2.5}$$

or

$$C_p = \frac{\Delta p}{q_0} = 1 - \left(\frac{V}{V_0} \right)^2 \qquad \text{Eq 2.5A}$$

where C_p is a dimensionless pressure coefficient, in fact the ratio of the two quantities: static pressure and dynamic pressure having the same dimension–pressure lb/sq ft.

and $\Delta_p = p - p_0$ is differential pressure, i.e. a difference between the local pressure p and the ambient pressure p_0 at a given point along the surface of the cylinder or foil.

Thus, a decrease in the local velocity, giving V less than V_0, leads to a local increase in the value p, so that $(p - p_0)$ is positive and conversely, an increase in V, or acceleration to a value greater than V_0, leads to a decrease in the local pressure p, so that $(p - p_0)$ is negative.

This is the consequence of Bernoulli's equation which often causes difficulty at first sight, for it is rather instinctive to associate high pressure with high velocity and vice versa. A little reflection however shows that Bernoulli's theorem locates the region of higher pressure in places where the free motion of fluid is retarded. Since pressure may be regarded as a form of energy and Bernoulli's equation indicates that a balance is maintained between energy arising from the motion and that from the pressure in all parts of the stream; it becomes rather obvious and inevitable law that what has been lost in one form of energy must be recovered in another form (Ref 2.5). In other words, in the world of fluid mechanics it is difficult to have something for nothing.

In an ideal fluid, where no energy is dissipated into friction and subsequently heat, the energy conversion between the two forms–pressure energy on the one hand and kinetic energy on the other–involves velocity and pressure changes only.

If friction is present, as in the motion of real fluids, and part of the kinetic energy has been lost, i.e. has been dissipated into heat, the total head as given by the descriptive equation 2.2A (Pressure head + Velocity head = Total head) cannot be recovered or maintained in the course of energy conversion as expressed by this theorem. However, if losses due to friction and heat are small, as is the case when fluid flows slowly and has small viscosity, the Bernoulli principle can be applied as a reasonable approximation which gives good insight into the mechanics of foil action.

In the case of an ideal fluid devoid of friction, Eq 2.5A makes it possible to determine the pressure changes round the cylinder if the velocity changes have been already calculated. These calculations were in fact performed earlier for some stations shown in Fig 2.3C, namely S_1, A and B, applying Eq 2.1. For example, at the stagnation point S_1, the local velocity $V = 0$. Substituting this value into Eq 2.5A we obtain

$$C_p = \frac{p - p_o}{q_o} = \frac{\Delta_p}{q_o} = 1 - \left(\frac{0}{V_o}\right)^2 = 1$$

or

$$\frac{\Delta_p}{q_o} = 1$$

therefore

$$\Delta_p = q_o$$

And this means that the differential pressure Δ_p in the stagnation point is positive and higher than the ambient pressure p_o by the amount q_o, as already indicated when deriving Eq 2.3.

It should perhaps be stressed that at the stagnation point the pressure coefficient C_p can never be greater than 1. The dynamic pressure q_o can be regarded as the maximum kinetic energy available from the airstream that can possibly be converted into the static pressure exerted at the point where air is brought to rest, i.e. stagnates.

If the ambient pressure p_o is 2116 lb/sq ft, i.e. standard atmospheric pressure, and the cylinder is subjected to a wind of velocity 40 ft/sec, then at the stagnation point S_1, an increase in pressure $\Delta p = q_o$ over the ambient pressure p_o can be found in Table 2.3 or Fig 2.4; it is 1.9 lb/sq ft. In point B on the cylinder the local velocity V would be twice as big as V_o, i.e.

$$\frac{V}{V_o} = 2$$

Substituting this ratio into Eq 2.5A yields:

$$C_p = \frac{\Delta_p}{q_o} = 1 - (2)^2 = -3$$

and this means that at the point B there is a decrease in pressure below the ambient pressure p_o; and this decrease or negative differential pressure equals $\Delta p = -3q_o$.

If, as before, the dynamic pressure $q_o = 1.9$ lb/sq ft (corresponding to a wind velocity of 40 ft/sec) then the negative pressure Δp, or suction, at point B on the cylinder would be

$$\Delta p = C_p \times q_o = -3 \times 1.9 = 5.27 \, \text{lb/ft}^2$$

Repeating the procedure as presented above for a number of points round the cylinder, it is relatively easy to plot the pressure distribution; this is depicted in Fig 2.3D. Arrows pointing inward indicate a positive pressure coefficient $+ C_p$, or a positive differential pressure Δ_p at a given point relative to the ambient pressure p_o. Arrows pointing away from the cylinder surface indicate a negative pressure coefficient $- C_p$, or negative differential pressure $- \Delta p$. In yet other words, a *positive* differential pressure Δp means a pressure rise above the surrounding or ambient pressure; in our case the atmospheric pressure p_o, and vice versa a negative differential pressure $- \Delta p$ means a pressure drop below the ambient static pressure p_o.

(b) *Friction effect on flow pattern*

Since the pressures shown in Fig 2.3D are, in the case of a perfect fluid flow, symmetrically distributed relative to the vertical and horizontal cylinder axes $B–B_1$ and $S_1–S_2$, no net force acts on the cylinder. If we add (integrate) all the pressure forces round the surface these forces cancel each other, hence the resultant force will be zero. The cylinder would, in frictionless fluid, experience no resistance whatever. This result, completely contrary to common sense and experience, baffled for many years the cleverest mathematicians, amongst them the famous philosopher of the Age of Enlightenment, d'Alembert, who openly confessed in *Opuscules mathématiques* (1768):

> '...I do not see then, I admit, how one can explain the resistance of fluids by theory in a satisfactory manner. It seems to me that this theory, dealt and studied with the most profound attention gives, at least in most cases, resistance absolutely zero; a singular paradox that I leave geometricians to explain.'

This apparently strange state of affairs, known as d'Alembert's paradox, holding good for bodies of arbitrary shape, was studied not only by geometricians, but by many scientists for almost 150 years, without much success. Finally, at the beginning of the 20th century L Prandtl bridged the gap between the flow phenomenon which might be proved but not observed in reality and phenomena which could be observed but not proved. The subsequent rapid progress made since Prandtl, particularly in aerodynamics, was greatly facilitated by this concept, namely that the flow round a foil in a real fluid can be treated as consisting of the two distinct parts.

One part, that very close to the surface of the foil, is entirely affected by viscosity but its effects are limited to a thin layer immediately adjacent to the wetted surface. This restricted layer in which viscosity dominates was called by Prandtl the 'boundary layer'.

The second part consists of the flow outside this boundary layer, where effects of viscosity are negligible and therefore the flow may be regarded as that of an ideal frictionless flow. As such, it can be described to a high degree of accuracy, at least for some streamline shapes, by standard methods of classical mechanics of non-viscous fluids, which are in fact more or less pure geometry.

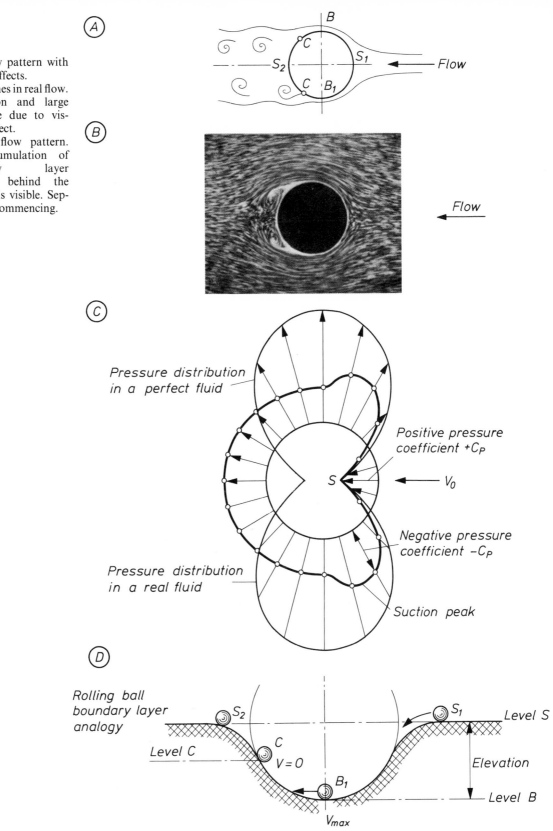

Fig 2.5 Real flow pattern with friction effects. Streamlines in real flow. Separation and large wake are due to viscosity effect. Viscous flow pattern. An accumulation of boundary layer material behind the cylinder is visible. Separation commencing.

A

B — Flow

B₁

S₂ — S₁

C

C

B — Flow

Pressure distribution in a perfect fluid

Positive pressure coefficient $+C_P$

S ← V_0

Negative pressure coefficient $-C_P$

Pressure distribution in a real fluid

Suction peak

D

Rolling ball boundary layer analogy

S_2 — Level S

Level C

C

$V = 0$

B_1

V_{max}

Elevation

Level B

S_1

Figure 2.5A demonstrates schematically the real flow pattern with friction effects. In contrast with an ideal frictionless flow, shown in Figs 2.3A and B, the ordinary fluids always exhibit a certain resistance, particularly to sudden alteration of flow pattern. The fluid particles, moving within the boundary layer adjacent to the wetted surface of the cylinder, suffer a certain retardation as they travel around the cylinder. Hence their velocity at point B is much less than it would be in the absence of viscous friction, and this implies that the kinetic energy of air particles is less than would be expected from purely theoretical consideration, i.e. $2V_o$. Part of the kinetic energy has been dissipated as heat. In order to travel along the path B–C the fluid particles would require the expenditure of the full amount of kinetic energy to reach the rear stagnation point S_2, but since this is not now available, the fluid stream close to the surface finds itself stopped in its track; the flow is unable to adhere to the cylinder surface. Having nowhere to go, the fluid particles pile up on each other somewhere below point C. The boundary layer thickens abruptly, as shown in Fig 2.5B, and the streamlines are forcibly pushed away from the cylinder contour. This rapid detachment of streamlines, beginning from point C, is called *separation*, which is followed by a more or less turbulent wake spreading downstream.

Shearing stresses due to viscosity, transmitted by the boundary layer, produce on the cylinder (or any other body of arbitrary shape) a force called *skin friction*. In turn, a certain modification to the flow pattern, which is a further consequence of boundary layer action, disrupts the pressure symmetry or equilibrium predicted by ideal fluid theory, as shown in Fig 2.5C, and thereby produces another kind of force, called pressure drag or *wake drag*. This pressure drag due to incomplete restoration of pressure, in particular over the rear side of the cylinder, can be estimated from Fig 2.5C by comparing the pressure distributions in the perfect and the real fluid. The major difference is seen to occur at the back of the cylinder where the pressure fails completely to rise to the stagnation positive value with $C_p = 1.0$. Actually, the pressure coefficient C_p in this region is negative, as indicated by arrows pointing away from the cylinder surface. It is self-evident that the resulting pressure drag, i.e. the net force caused by pressure drop in the direction of motion, could be minimized if the separation points were shifted as far rearward as possible. Thus the area of the cylinder affected by negative pressure (negative C_p) or suction would be reduced. The width of the wake behind the cylinder is, in a way, a measure of the drag magnitude–a larger wake means a bigger drag.

L Prandtl (Ref 2.6) compared the boundary layer action in the above conditions to that of the mechanical ball behaviour shown in Fig 2.5D. The ball may start rolling from the point S_1 moving down the track. During the descending period its potential energy corresponding to the elevation of its starting point S_1 (level S) relative to point B_1 (level B) is transformed into the kinetic energy measured by velocity V_{max} at point B_1. Ascending the slope towards point C, the kinetic energy of the ball is gradually transformed back into potential energy. The ball would reach the same elevation as it had at level S_1, provided that no energy were lost along the way. Since mechanical friction and other resistances cannot be avoided, the ball will

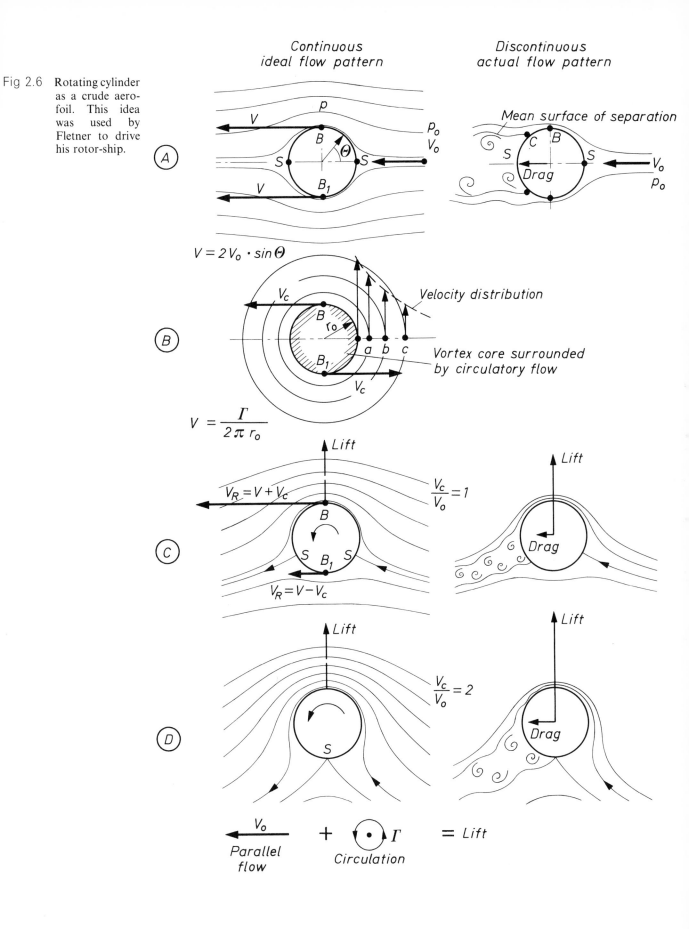

Fig 2.6 Rotating cylinder as a crude aerofoil. This idea was used by Fletner to drive his rotor-ship.

Continuous ideal flow pattern

Discontinuous actual flow pattern

Ⓐ

Mean surface of separation

$V = 2 V_0 \cdot \sin \Theta$

Ⓑ

Velocity distribution

$V = \dfrac{\Gamma}{2 \pi r_0}$

Vortex core surrounded by circulatory flow

Ⓒ

$V_R = V + V_c$

Lift

$\dfrac{V_c}{V_0} = 1$

Lift

Drag

$V_R = V - V_c$

Ⓓ

Lift

$\dfrac{V_c}{V_0} = 2$

Lift

Drag

V_0

Parallel flow

$+$

Γ

Circulation

$=$ Lift

not regain level S but somewhere in between points B and S_2, say at point C, the ball will stop after exhausting its available kinetic energy.

For a while we shall leave viscous and pressure drags. They will be discussed in more detail in following chapters. Now we confine our attention to lift generation.

(4) Circulation and Magnus effect

In developing the theory of lift it is convenient to introduce, as the next stage, another type of flow as depicted in Fig 2.6B; namely the steady motion of fluid in concentric circles round the cylinder. Such a flow can be initiated by a rotating cylinder submerged in a viscous fluid. The cylinder drags the fluid around with the help of viscosity, and the boundary layer is so thin as to be negligible. The cylinder itself may be regarded as a 'vortex core' surrounded by circulatory flow. Outside this core it is assumed that the fluid rotates in such a way that the velocity of circulation V_c is inversely proportional to the radius of streamlines, i.e. a distance r from the centre of the core. The spacing of streamlines shown in Fig 2.6B, and the velocity vectors becoming shorter in length, indicate a decrease in velocity with increasing radius. The relevant velocity profile or velocity distribution is also drawn. As in the case of rectilinear flow shown in Fig 2.3 the velocity of circulatory flow V_c is everywhere tangential to the streamline. Its magnitude is given by the equation

$$V_c = \frac{\text{Constant}}{r} \qquad\qquad \text{Eq 2.6}$$

where r = radius or distance of the streamline from the centre of rotation.

A similar kind of vortex motion can be observed in nature, in both air and water. If the vortex has considerable intensity and its core has a small diameter, one may expect that due to the high velocity of circulation near the centre, extremely high suctions (drastic drop in static pressure) can occur. It is a simple consequence of Bernoulli's theorem which states: 'Where the velocity is large the pressure is low.'

The tornado (Photo 2.2A), spinning dust whirls, liquid vortex over water drains, waterspouts and cyclones shown in Photo 2.2B are striking examples of the occurrence of vortex flows with just these properties, and all of them can be seen and/or felt. The drawing attached to Photo 2.2A demonstrates the relationship between the pressure p close to the tornado funnel and the ambient atmospheric pressure p_o, as can be measured far from the vortex centre. The broken, thin line indicates how quickly the pressure p drops towards the tornado centre. This pressure drop, together with high rotational velocity V_c may cause severe damage to buildings struck by the tornado along its path.

Another example of a similar whirling motion observed frequently in the atmosphere is shown in Photo 2.2B. It depicts the counter-clockwise rotation of a large mass of air in a low pressure system (in the northern hemisphere) as seen by a weather satellite. We will see later that aircraft in flight, and also sailing yachts on a windward course, trail behind them similar vortex flows which are produced at the cost of a continuous dissipation of energy.

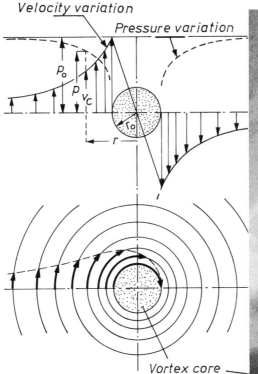

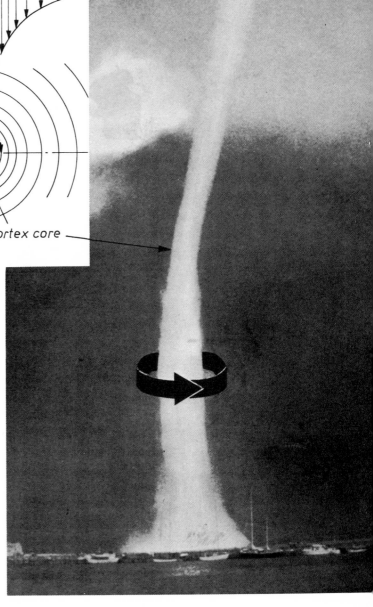

Photo 2.2A
Tornado.

Intense vortex system in which air may rotate with speed ranging from 150 to 450 ft/sec. The visible funnel (vortex core) consists of cloud droplets condensed due to expansional cooling resulting from markedly lower pressure in the vortex than in the surrounding atmosphere. Structural damage to buildings results in part from explosion when the atmospheric pressure outside is suddenly reduced and partly from force of the extremely strong wind. Damage from explosion may be reduced by venting or prior opening of windows to allow rapid equalization of pressure inside and outside the building.

Attached sketch closely approximates the relationship between velocity of circulation V_c (circumferential velocity) and pressure p in proximity of the vortex core. The variation of velocity V_c and pressure p with radius r is shown; p_o indicates the static (atmospheric) pressure well outside the immediate vortex action. As seen, the air particles closer to the vortex core pick up speed and this is associated with a more or less rapid drop in pressure.

You have seen a similar vortex system many times in the bathtub when draining water. One can find plenty of examples of concentrated rotation, i.e. vortices, in nature. Cloud patterns photographed by a satellite shown in Photo 2.2B indicate such a rotational effect in the earth's atmosphere on a grand scale. Photograph 2.2C illustrates rotating mass of matter on an astronomical scale.

Photo 2.2B   Counter-clockwise whirling of wind in the Northern
hemisphere–low pressure disturbance photographed by a
weather satellite. Since the central part of a cyclone is
characterized by strong winds, the navigators of sailing
vessels should avoid the dangerous semicircle in which a
vessel would tend to be carried by wind, and occasionally by
ocean current, into the path of a storm.

Photo 2.2C   The whirlpool galaxy NGC 3031 resembles the tip vortices
shed by any foil–be it sail, aeroplane wing or fin-keel–
generating lift and shown, for example, in Photo 2.27C.

Circulatory flow may be investigated further by introducing the concept of circulation designated by the symbol Γ and defined as a product of the tangential velocity of circulatory flow V_c and the path length (the circumference in the case of a rotating cylinder). If we consider one of the streamlines shown in Fig 2.6B as coinciding with the cylinder surface we may write that circulation

$$\Gamma = V_c \times 2\pi r_o \qquad \text{Eq 2.7}$$

hence

$$V_c = \frac{\Gamma}{2\pi r_o} \qquad \text{Eq 2.7A}$$

where $2\pi r_o$ is the circumference of the cylinder.

By substituting Eq 2.6 into Eq 2.7 we find that:

$$\Gamma = 2\pi \times \text{Constant} \qquad \text{Eq 2.8}$$

In other words, for the flow in question the circulation Γ has the same value for every closed path which encloses the cylinder just once; for example, paths a, b, c of Fig 2.6B have identical circulation to that around the cylinder itself.

The concept of circulation which has been introduced may be new to most readers and as such has not a readily understood physical and practical significance. Those who have difficulties in grasping this concept may perhaps find a consolation in the fact that it took man many years to evolve and understand the *concept of force* as a workable mental tool for explaining the varying interactions between objects in his environment. Today we use this idea almost unconsciously, conceiving forces as those *pushes* and *pulls* which tend to make bodies move or keep them at rest. An essential part of the hypothetical force concept is that forces are somehow in balance when the object under study is motionless, or when its motion is steady and when the forces are not in balance the object accelerates. One may debate a philosophical question whether or not forces *really exist*. From the standpoint of applied mechanics the fact remains that the force and circulation concepts, as well as many other concepts accepted in science and everyday life, allow us to predict events in the physical world and this somehow justifies their invention.

(a) *Rotating cylinder as a crude aerofoil*
A fortunate characteristic of the circulatory flow just described is that it may be combined with the parallel or rectilinear flow shown in Fig 2.6A (left side). It can be done by employing the principle of superposition, which allows the building up of complex flow patterns by the addition or superposition of two simpler flows. Such a composition of parallel flow and circulatory flow, depicted already in pure forms in Fig 2.6A and B respectively, is given in Fig 2.6C and D (left side).

In order to obtain the resultant flow pattern, the velocities V and V_c, fundamental quantities of both parallel and circulatory types of flow, must be added as vectors.

The problem is exactly the same as that of finding the resultant of the two forces by adding the components vectorially. For example, near point B in Fig 2.6C, the velocity $V = 2V_0$, resulting from parallel flow, is added to the circulation velocity $V_c = (\Gamma/2\pi r_0)$. The large resultant velocity V_R is therefore:

$$V_R = V + V_c = 2V_0 + \frac{\Gamma}{2\pi r_0} \qquad \text{Eq 2.9}$$

Whereas, near point B_1, the circulation velocity V_c, which is against the rectilinear flow, must be subtracted from the velocity V, thus giving a small resultant velocity:

$$V_R = V - V_c = 2V_0 - \frac{\Gamma}{2\pi r_0} \qquad \text{Eq 2.9A}$$

It is evident that the resulting tangential velocity V_R at any point round the cylinder differs from that of the parallel flow in Fig 6A, to an extent depending upon both the velocity of circulatory flow V_c, or circulation Γ, and the location of the point defined by the angle θ (Fig 2.6A–left side). This tangential velocity V_R is increased around the upper part of the spinning cylinder and decreased around the lower part.

One may expect therefore that, according to Bernoulli's theorem, the pressure on the underside of the cylinder as it is drawn in Fig 2.6D is larger than on the upper side and causes *lift* tending to push the cylinder upward. Thus a combination of parallel or rectilinear flow and circulation produces lift. This lift was not present in the simple parallel flow shown in Fig 2.3 because the pressure distribution was symmetrical. There is no such symmetry in the case of a rotating cylinder.

Figure 2.7 depicts the pressure distribution (in terms of pressure coefficients C_p) over a spinning cylinder in a perfect fluid. The arrows in the diagram indicate the intensity and the direction of the pressure forces on the cylinder surface from which they are drawn. The diagram clearly brings out the fact that the suction $(-C_p)$ over the upper surface of the cylinder, rather than the positive pressure $(+C_p)$ over the lower surface, is responsible for the major contribution towards the lift L, experienced by the cylinder.

The magnitude of the lift L per unit span b on the cylinder is given by:

$$\frac{L}{b} = \rho V_0 \Gamma \qquad \text{Eq 2.10}$$

or by substituting Eq 2.7 into Eq 2.10:

$$\frac{L}{b} = \rho V_0 (V_c 2\pi r_0) \qquad \text{Eq 2.10A}$$

Following the routine developed in aerodynamics we may find that the pressures on the cylinder are all proportional to the dynamic pressure $q = (\rho V_0^2/2)$ of the fluid stream ahead of the cylinder. The forces produced by these pressures are proportional to the size of the cylinder and hence proportional to the cylinder

Fig 2.7   Diagram of pressure distribution over the rotating cylinder in
perfect fluid.

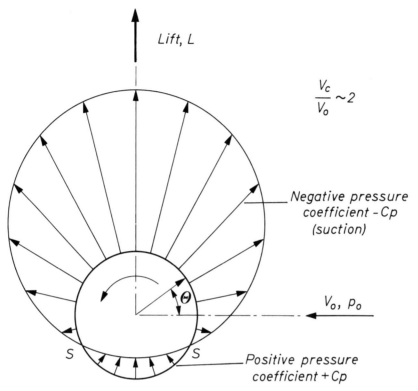

diameter $2r_o$. It is thus convenient to express the lift per unit span b by:

$$\frac{L}{b} = C_L \times \frac{\rho V_o^2}{2} \times 2r_o \qquad \text{Eq 2.10B}$$

where C_L is a factor of proportionality, or lift coefficient.
   Hence

$$L = C_L \times \frac{\rho V_o^2}{2} \times 2r_o \times b \qquad \text{Eq 2.10C}$$

The lift coefficient C_L for an ideal fluid can be expressed by:

$$C_L = \frac{L}{(\rho V_o^2/2) \times 2r_o \times b} \qquad \text{Eq 2.11}$$

Substituting Eq 2.10A into Eq 2.11 yields:

$$C_L = \frac{\rho V_o (V_c 2\pi r_o) b}{(\rho V_o^2/2)\, 2r_o b} = 2\pi \frac{V_c}{V_o} \qquad \text{Eq 2.12}$$

Equation 2.12 reveals that the magnitude of the lift coefficient C_L depends on the V_c/V_o ratio, i.e. a ratio between the peripheral or rotational speed of the cylinder $V_c = \omega r_o$ (where ω is the angular velocity of rotating cylinder in radians per sec) and the velocity V_o of the remote, undisturbed flow ahead of the cylinder.

The theoretical value of the lift coefficient C_L is much higher than is practically obtained by experiments in real fluid. This is primarily due to viscosity, which is responsible for the large wake shown in Fig 6C, D (right side) and therefore for the associated pressure drag. Since no indication of drag is given by theory based on an ideal fluid concept, the drag coefficient C_D can only be established empirically by measurement of the drag at a given value of the dynamic pressure $q = (\rho V_o^2/2)$ and projected area of the cylinder $2r_o b$. The real C_L coefficient may be obtained similarly by direct measurements of lift L.

Measurements of this kind, as well as of drag, have been made on cylinders rotating in air (Refs 2.7, 2.8) and the results are summarized in Fig 2.8, in which the measured lift and drag coefficients are plotted against the V_c/V_o ratio; V_c being the rotational speed of the cylinder and V_o the wind speed. There is also plotted the theoretical lift coefficient curve as calculated from Eq 2.12 for an ideal fluid. The evident discrepancy gives an idea of the cost one has to pay for the use of viscosity or viscous sheer in order to induce some degree of circulation round the cylinder so as to generate the noble lift, which otherwise could not be produced in an ideal fluid. It is evident that the fluid viscosity as an agent of base drag demands high payment for services rendered.

Considering the results for measured lift, we may notice that there is no lift for small rotational speeds, when the V_c/V_o ratio is less than about 0.5. Above it, lift increases in direct proportion to the rate of rotation V_c/V_o. The value of the lift coefficient that can be obtained depends upon flow conditions at the ends of the cylinder. Following the universal tendency to flow from high-pressure to low-pressure regions by every available path, the air escapes round the ends of a finite span cylinder. Such a leak of pressure obviously reduces the efficiency of the rotating cylinder as a lift-producing device.

Referring to Fig 2.8, we may find that by fitting end plates of diameter 1.7 times that of the cylinder (the aspect ratio AR of which was 4.7) the maximum lift coefficient C_L may be raised from just over 4 to the amazing value of about 9. With a cylinder of aspect ratio AR = 13.3 and without end plates, results are as good as those obtained with end plates on a shorter cylinder.

It may also be seen that a rotating cylinder is capable of producing a much greater lift than an aerofoil of the same projected area. However, this extra lift is dearly paid for, with drag several times greater than that of a good aerofoil. *A rotating cylinder may be regarded as a crude aerofoil, and the basic difference between aerofoil and cylinder as lifting devices is that the former may produce lift much more efficiently without mechanical movement of its parts.*

Fig 2.8   Lift generated by rotating cylinder in air at sea level may be expressed by:

$$L = 0.00119 \times C_L \times V_o^2 \times 2r_o \times b \quad \text{(see Eq 2.10C)}$$

where b–span or length of the cylinder

$2r_o \times b$–is equivalent to the foil area, S_A

Similarly the drag

$$D = 0.00119 \times C_D \times V_o^2 \times 2r_o \times b$$

Reid: Aspect ratio 13.3; Reynolds Number 3.9×10^4 to 1.16×10^5 no end plates.

Betz: Aspect ratio 4.7; Reynolds Number 5.2×10^4.

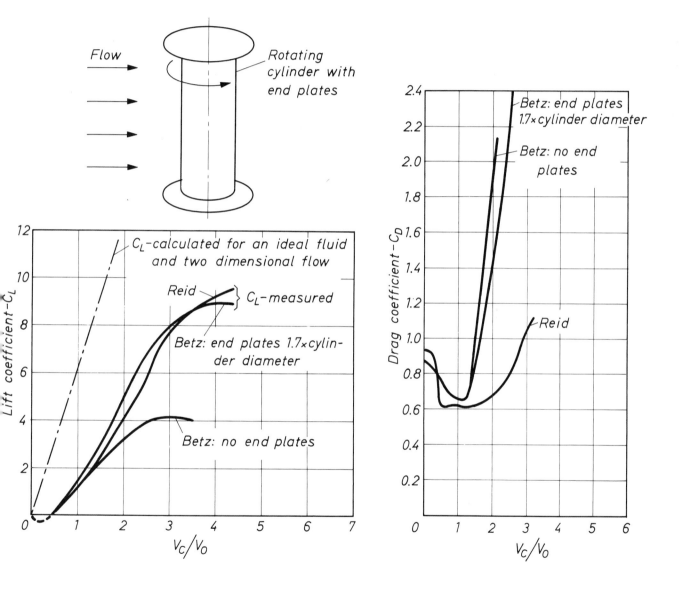

(b) *Fletner's rotorship*

The main interest of these discussions concerning rotating cylinders lies in the light they throw on the fact that the circulating flow about a cylinder will be found closely related to the flow round *a lifting aerofoil or sail*. Apart from that, the results of our study have definite and practical interest; *rotating cylinders have been used in place of sails*.

Figure 2.9A shows two coefficient polar diagrams, one for a cylinder with end plates and another one for a good gaff-mainsail. We can immediately recognize the remarkable superiority of a rotating cylinder as far as C_{Lmax} value is concerned, which is about 8 times greater than C_{Lmax} produced by the gaff sail. For this reason, the German engineer A Flettner developed in about 1925–26 the so-called 'rotor-ship'. The following is a brief account of the results given by the designer (Ref 2.8).

A schooner named *Buckau* was fitted with two rotating cylindrical towers as propulsive devices. These towers were built above the deck and were driven by an electric motor, the current being produced by a 45 hp diesel engine. The cylinder 9.1 ft in diameter and about 60 ft high could be rotated at various speeds up to 700 revolutions per minute; the direction of rotation was reversible.

The stability of the ship was greatly increased by the conversion. The weight of the two towers and driving plant was 7 tons, against a total weight of 35 tons of the former gaff rigging. The projected area of the towers was only about one-tenth of that occupied by the former rigging of the *Buckau* as a sailing schooner.

Since the rotor propulsive output is largely dependent on the ratio V_c/V_o (rotor peripheral speed/wind speed) and the peripheral speed V_c can be kept constant, the pressure on the rotating cylinders will rise only to a certain magnitude, even if the wind speed itself increases substantially. Hence the rotors can continue to function in very high winds. For this reason, *strong squalls have but slight effect on the ship, and pass almost unnoticed*. When, for example, the rotor is revolving at a peripheral speed $V_c = 24$ m/sec, in a wind of 8 m/sec, and a squall of 12 m/sec passes, the ratio of V_s/V_o which at first was 3 is automatically reduced to 2. In spite of the fact that the wind itself has grown much stronger, its effect on the rotor forces is automatically reduced. Thus practically any difference in pressure can hardly be noticed. Calculations based on laboratory experiments were amply confirmed by the first trial trips. It appears that a rotor offers some advantage as compared with the conventional rig of ships. One may control the wind pressure on the ship by rotating the cylinders at a suitable speed; the effect is instantaneous. Thus, the time- and manpower-consuming operation of shortening sail some time before the approach of a storm is unnecessary. Moreover, as shown in Fig 2.9B, the resistance of the stationary rotor is low, as compared with that of the old rigging with sails stowed.

The curves in Fig 2.9B display clearly the nature of the forces involved in Flettner's propulsion system, as compared with the resistance of the rigging of a sailing ship. Curve 'a' shows the wind pressure relationship for both towers with a peripheral speed of 24 m/sec. Curve 'b' shows the wind resistance of the con-ventional rigging, and curve 'c' that of the cylinder when stationary. It is plainly

Fig 2.9   Comparison of forces developed by rotating cylinder (rotor) and
gaff-type sail.

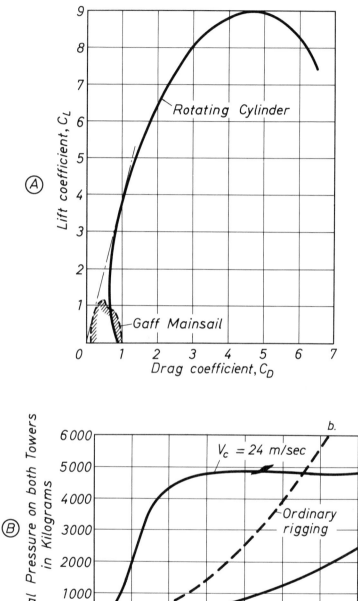

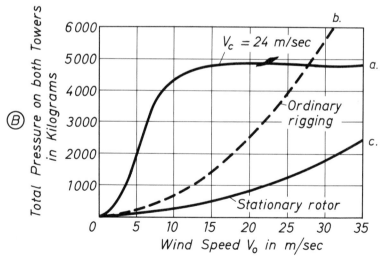

demonstrated how the forces on the rotating cylinder cease to increase beyond a wind speed of about 12 m/sec. This is a matter of great practical importance since no larger force is generated even in the highest winds than that which is determined by the peripheral controllable speed of the rotor. This advantage cannot perhaps be too strongly emphasized from the safety point of view.

The practical experiment with rotating cylinders used instead of sails was interesting and successful from a technical point of view. However, the ultimate failure of the invention was due to economic reasons. The anticipated application was intended for cheap freighters or fishing boats. The resulting expenses were too high and the supposed profit, in comparison with conventional, mechanical propulsion, became illusory, mainly due to the unpredictability always associated with wind, the necessary co-operator with both rotor and sail propulsion.

It is quite probable that the time will come when the Flettner idea will be reappraised and applied to special types of sailing craft. If oil fuel becomes uneconomic, the foreseeable alternatives are nuclear power or a return to sail- or wind-driven propulsive devices, at least for the transport of those commodities which do not command a high freight rate.

Figure 2.10 and the example given below should help those who might be interested in rotating cylinders as a propulsive device: Two rotors of span $b = 30$ ft and radius $r_o = 2.5$ ft, are used to propel a boat. Estimate the driving force generated by rotors when the apparent wind velocity $V_o = 40$ ft/sec, the angle of heading $\beta = 60°$ and the number of rotor revolutions $n = 5$ per sec. The velocity of circulation V_c induced by the rotating cylinder is:

$$V_c = \omega \times r_o = 2\pi \times r_o = 2\pi \times 5 \times 2.5 = 78.5 \text{ ft/sec}$$

Fig 2.10   Forces developed on rotor-vessel.

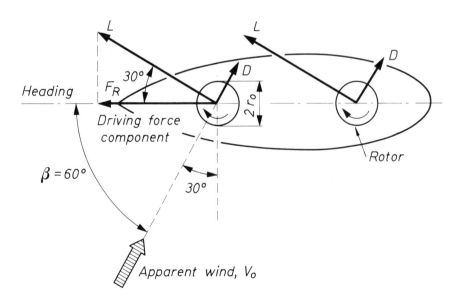

where ω angular velocity in radians per second $= 2\pi n$.

Hence

$$\frac{V_c}{V_o} = \frac{78.5}{40} = 1.97$$

From Fig 2.8 we find that at this ratio of V_c/V_o the C_L is about 5.0, and $C_D = 1.35$ (cylinder with end plates). Applying the equation given in Fig 2.8, we may calculate lift per rotating cylinder:

$$\begin{aligned}
L &= 0.00119 \times C_L \times V_o^2 \times b \times 2r_o \\
&= 0.00119 \times 5.0 \times 40^2 \times 30 \times 5.0 \\
&= 1430 \text{ lb}
\end{aligned}$$

Similarly drag:

$$\begin{aligned}
D &= 0.00119 \times C_D \times V_o^2 \times b \times 2r_o \\
&= 0.00119 \times 1.35 \times 40^2 \times 30 \times 5.0 \\
&= 386 \text{ lb.}
\end{aligned}$$

The total driving force F_R in the direction of motion, as presented in Fig 2.10, will be:

$$F_R = 2(L \cos 30° - D \sin 30°) = 2(1430 \times 0.866 - 386 \times 0.5) = 2090 \text{ lb.}$$

Generation of lift on a rotating cylinder lying crosswise in a stream of fluid is called the 'Magnus Effect', named after H G Magnus who published his discovery in 1853, under the title *The Drift of Shells*. Magnus carried out his experiments on the effect of the wind on projectile-shaped bodies, principally from the point of view of ballistics. A shell, rotated by the rifled gun-barrel, is affected by a side wind, which brings into play the above mentioned transverse force, perpendicular to the trajectory; rather annoying to the gunner since it causes an unpredictable vertical force on the shell to affect range. Lord Rayleigh has dealt with similar phenomena in tennis balls, in a short essay on *The Irregular Flight of a Tennis Ball*.

As a matter of fact, Rayleigh's study was undertaken to elucidate the swerving flight of a 'cut' tennis ball. The eccentricity in behaviour of a tennis ball accords with other practical experiences. For example, when a baseball pitcher throws a ball which follows a substantially normal trajectory for a certain distance and then breaks into a curve, it is of this aerodynamic phenomenon that he is taking a presumably unwitting advantage. The ball leaves his hand with a definite velocity and a definite rate of rotation. Both are progressively reduced by air resistance but the linear velocity falls off more rapidly than the rotational. The ratio between the two, originally below the critical value which is $V_c/V_o = 0.5$ as shown in Fig 2.8, accordingly rises in due course above it. A transverse force on the ball, corresponding to the lift on the cylinder, then develops and the ball is diverted from its path. If the pitcher's fingers had given it a more vigorous spin, the ratio of angular to linear velocity would have been above the critical value from the first; the deviation from the straight path would have been immediate and the resultant curve would have been of a variety known to the fields of play as 'round-house'. The same effect can be observed on the cricket field and a golf ball often shows similar characteristics, a slice appearing to start after a hundred yards or more of straight travel, and for the same reason (Ref 2.9).

The results obtained for the simple case of the circular cylinder, and the conclusions which may be derived from them, have a wide range of applicability. For example, when cylinders of any cross-section, including asymmetric foils, are subjected to a relative flow, the following statements may be proved to be correct:

    a. in the absence of circulation around the cylinder, lift cannot be generated: there are no differential pressures which might produce lift;
    b. if there is a circulation around the cylinder, no matter how it is achieved, then, as a result of differential velocities and pressures, lift is produced.

Consequently, any object, including the proverbial 'barn door' to which condition (b) is applicable, is potentially a lifting device. However, only certain shapes will produce a large lift/drag ratio and hence be effective enough to fulfil their desired functions.

The magnitude of lift L per unit span b perpendicular to the flow is given by Eq 2.10, applicable to two-dimensional flow only:

$$\frac{L}{b} = \rho \times V_0 \times \Gamma \quad \text{repeated} \qquad\qquad \text{Eq 2.10}$$

or

$$L = \rho \times V_0 \times \Gamma \times b$$

This result constitutes what is known as *Kutta–Joukowski theorem of lift*. It furnishes the foundation for the entire modern concept of fin or foil action.

(5) How lift is generated by a foil

The fundamental difference between the rotating cylinder and a foil as lift generating devices is in the manner in which the differential velocity and associated differential pressures are obtained on the upper and lower surfaces. The cylinder must be rotated to generate circulation and lift. The question arises, how is circulation created around a foil, where there is no mechanical device to initiate and support circulatory motion? The answer first suggested by Lanchester (Refs 2.5, 2.10) and developed by Kutta, Joukowski, Prandtl and others has been well substantiated by experimental evidence and photographs.

According to classical textbooks on aerodynamics (Refs 2.11, 2.12), it can be roughly discussed as follows. Consider the flow past a cambered, asymmetrical foil set at about zero angle of incidence as shown in Fig 2.11A; the flow has just started. Two fluid particles A and B above and below the stagnation point S_1 travel along

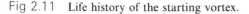 Fig 2.11   Life history of the starting vortex.

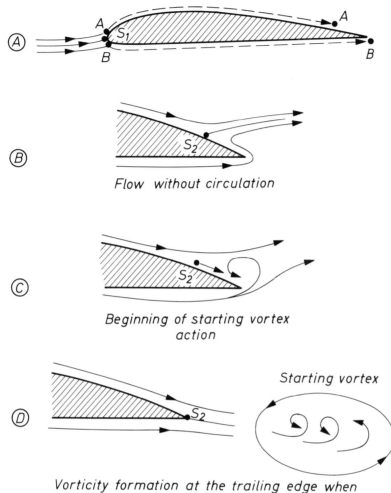

their respective surfaces at equal speeds and, since the upper surface is *longer*, A arrives at the trailing edge ahead of B. It then attempts to go around the sharp trailing edge as demonstrated on a larger scale in Fig 2.11B and Photo 2.3 in which the streamlines were made visible by the introduction of fine aluminium powder. This initial flow pattern is somewhat similar to that of the zero-circulation potential flow, depicted also in Fig 2.12A for a symmetrical foil set at a certain incidence angle. Without circulation present, the forward and rear points of zero velocity, or so-called *stagnation points*, occur at S_1 and S_2, which correspond to the $\Theta = 0°$ and $\Theta = 180°$ points on the cylinder in Fig 2.3C. Where these points actually occur on the foil depends on the angle of incidence α with respect to oncoming flow and foil section.

Photo 2.3   Streamlines round a foil at the very first moment after starting are similar to those presented in Fig 2.11B, flow without circulation. Rear stagnation point S_2 is situated on the upper surface (back) of the foil, at a certain distance from the trailing edge.

Camera is at rest with respect to the stationary foil.

It will be seen in Photo 2.3 that at the first moment of the motion the fluid has a tendency to *go around* the sharp trailing edge of the foil. It needs no mathematics to anticipate and demonstrate experimentally that no such flow of the viscous fluid, having to make an instantaneous turn around a sharp cusp of the foil, could be maintained for long. The fluid does not like this process, owing to the high velocity required at the sharp trailing edge and the large viscous and inertia forces brought into action. Consequently, the flow breaks away from the edge and the so-called *starting vortex* begins to operate between the trailing edge and the rear stagnation point S_2. The life history of this starting vortex is demonstrated in Fig 2.11 and Photo 2.4A, B, and C.

As the starting vortex rotates, a counter-rotation develops round the foil in the opposite direction to that of the starting vortex. This is caused by the viscosity forces involved in the process of transferring moment of momentum from the starting vortex, which one may imagine as a small spur-gear driving another bigger one; this mechanical analogy of circulation induced by the starting vortex is shown in Fig 2.13A. In fact the fluid viscosity and friction arising from it substitutes for the action of gear teeth and the induced counter-rotation of the mass of fluid around the foil appears as the circulation depicted schematically in Fig 2.12B.

The above analogy is closely related to the fundamental principle of mechanics, according to which a rotation, or in other words angular momentum (also called

Fig 2.12   Circulation developing round symmetrical foil section.

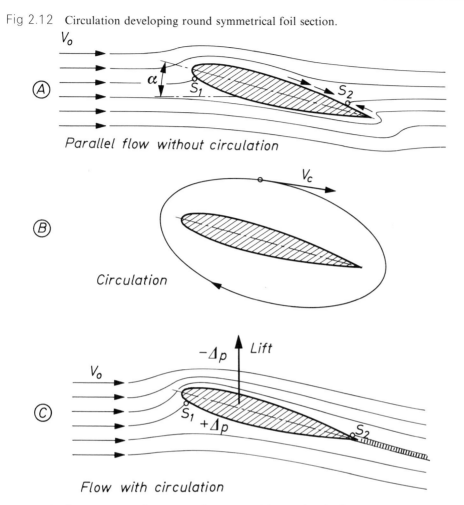

V_o

(A)

α

S_1

S_2

Parallel flow without circulation

(B)

V_c

Circulation

(C)

$-\Delta p$ Lift

V_o

S_1

$+\Delta p$

S_2

Flow with circulation

moment of momentum), cannot be created in a physical system without reaction. As a matter of fact, this principle is derived from the third law of motion presented by Sir Isaac Newton (1642–1727), which states: all forces arise from the mutual interaction of particles and in every such interaction the force exerted on the one particle by the second is equal and opposite to the force exerted by the second on the first; or as usually expressed–action and reaction are equal and opposite.

This idea is illustrated in Fig 2.13B, which may also serve as another mechanical analogy of starting vortex action, namely–if a man standing on a nearly frictionless platform tries to put into rotation a wheel, he will experience a reaction tending to rotate him in the opposite direction and finally, the product of $I_1 \times \omega_1$ should become equal to $I_2 \times \omega_2$,

where I_1 = inertia moment of the wheel
        I_2 = inertia moment of the man
and ω_1 and ω_2 are respective angular velocities.

Fig 2.13   Mechanical analogy of circulation induced by starting vortex.

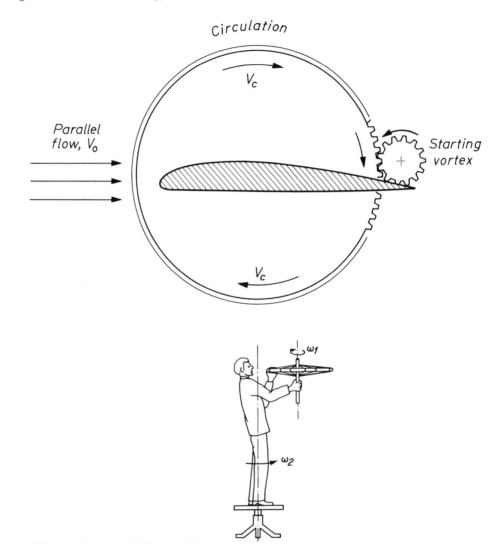

Of course, unavoidable friction at the rotating platform will complicate this relationship in a quantitative sense.

(a) *Kutta–Joukowski hypothesis*

It will be seen in Photo 2.4C that, when the starting vortex has fulfilled its function of initiating and developing circulation, it breaks away from the foil and passes downstream in the wake. This happens when the rear stagnation point S_2, distinguished in Fig 2.12C, has been brought close to the trailing edge, in which case there is no longer a velocity difference between the streamlines leaving the upper and lower

Photo 2.4   Three pictures which show successive stages of a starting
vortex may help the reader to grasp the mechanism of
circulation being set in motion.

A. The starting vortex in operation at the trailing edge.
B. Starting vortex leaving the foil.
C. Starting vortex further away.

Attention is drawn to the fact that the flow pattern will
appear different to two observers (or cameras)–to one who
follows the foil moving through the stationary fluid and to
another who is watching stationary foil immersed in moving
fluid. In Photo C foil moves forward and camera remains
stationary to still fluid. For comparison, Photo 2.5 depicts the
flow pattern when foil and camera are stationary and the fluid
flows.

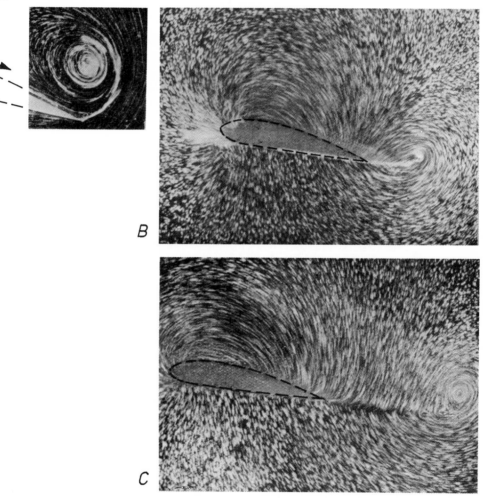

surfaces of the foil, and therefore there is no physical stimulus to maintain or support the starting vortex. The flow round the foil has then reached steady state with a fairly fixed magnitude of circulation and associated steady lifting force. The strength of the vorticity shed into the wake is, as a matter of fact, equal to that of circulation around the foil. To emphasize this salient point in lift theory, one may say that a physical role of the *starting vortex* is to shift the rear stagnation point towards the trailing edge so that the velocity of the flow leaving the upper surface at the trailing edge is equal to that of the flow leaving the lower surface. This assumption stipulated independently by Kutta and Joukowski, and since called the *Kutta–Joukowski condition*, is inseparable from Eq 2.10

$$\frac{L}{b} = \rho \times V_o \times \Gamma$$

which allows calculation of lift if the magnitude of circulation is known. On the other hand, the magnitude of circulation can only be determined if the two stream-lines marked A and B in Fig 2.11A, separated at the front stagnation point, rejoin smoothly the trailing edge as demonstrated in Figs 2.11D and 2.12C. If this condition is, for some reason, not satisfied Eq 2.10 cannot be applied and so the lift value cannot be predicted.

The existence of the starting vortex in the early stages of motion can be verified experimentally in a simple manner by dipping a flat plate (it can be a razor blade held half immersed in water) and moving it briskly in a direction inclined at a small angle to its surface (Ref 2.12).

In view of the opposite equality of the vortex strength in the wake and the circulation around the foil, it may be anticipated that the starting vortices must be shed whenever either incidence angle α of the foil relative to the flow direction or the flow velocity V_o changes. Any of these changes will inevitably lead to velocity differential at the trailing edge, which in turn will give a stimulus to starting vortex action. Thus the strength of circulation will be adjusted to the new conditions.

What has been just said is beautifully illustrated in Photo 2.4D taken from a German War Report No B44/I/41 by M Drescher (Institute for Unsteady Fluid Motion, Göttingen). It shows a starting vortex developing at the trailing edge after a sudden deflection of the flap from 0° to 15°. Picture *a*–cut out of a time retarding film, shows the starting vortex immediately after stopping the flap deflection; *b*–demonstrates the next instant; and finally *c*–the flow pattern approaching steady state circulation.

Some people have challenged the Kutta-Joukowski hypothesis; it has been argued for example that the foil, be it a wing or a sail, produces lift through the simple action of deflecting the wind. Accordingly, if the mass of air discharged from, say, a sail leech is thrown aside or bent sideways to windward, it produces a reaction on the sail, called lift. This concept, based again on Newton's laws, was discussed in *Sailing Theory and Practice* (Ref 1.5) and there is no point in prolonging the argument in this respect. What should be stressed is that the concept of a foil as *flow-deflector* does

Photo 2.4D   Single pictures cut out of a time-retarding film. Angle of incidence $\alpha = -5°$, flap deflection from 0° to 15°, Reynolds Number $R = 6 \times 10^5$.

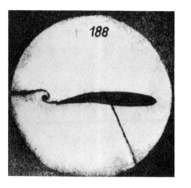 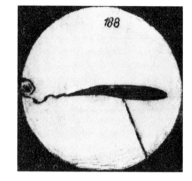 

Photo 2.4E   Hysteresis effect on the flow round a foil changing its angle of incidence rapidly. The flow actually recorded does not follow the flow pattern that would develop in stationary conditions at a given angle of incidence. In other words, the actual flow pattern in dynamic or unsteady condition depends on 'previous history' of the flow.

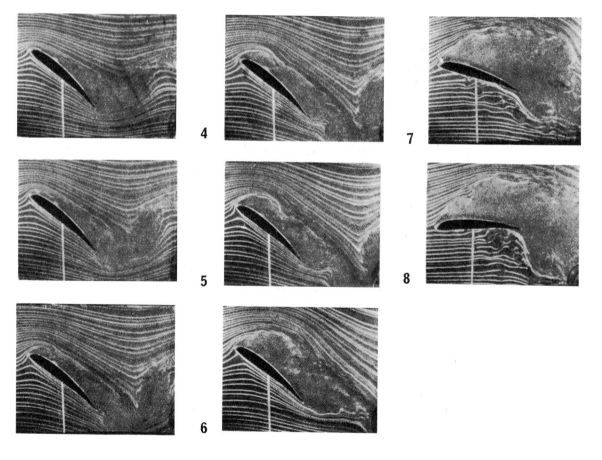

not invalidate the circulation theory of lift based on the Kutta–Joukowski hypothesis. This is simply another look at the same problem; a rather crude approximation to the mechanics of lift that the past generation of researchers, both scientists and flight enthusiasts were well aware of many years ago. Apparently they were not satisfied with such a theory, since a more sophisticated one was developed to cope with practical problems.

Although the interpretation or presentation of physical phenomena associated with the circulation concept of lift may differ in details, one fact cannot possibly be successfully challenged. Namely that, due to circulation, the air or water flowing over the upper (leeward) surface of the foil producing lift, as presented for instance in Fig 2.12C, does travel faster than that flowing over the lower (windward) surface. It is only on this condition that lift can be generated.

(b) *How quickly lift is developed*

Let us consider a practical problem–how quickly lift develops in terms of its magnitude when the angle of incidence of, say, a symmetrical foil shown in Fig 2.12C, is suddenly changed from 0, where there is no circulation, to α, at which circulation and lift must occur. One must realize that the circulation pattern around an aerofoil does not spring into existence without a certain time lag. A given mass of fluid must be accelerated against inertia forces and this takes time. Subsequently, both the circulation and lift normally associated with a given angle of incidence α at which the foil was set, do not reach their nominal values immediately, the full values being developed in the time taken to travel a certain number of chord lengths. This is shown in the graph of Fig 2.14A which presents the rate of growth of lift and circulation with time given in terms of number of chord lengths. It is seen that a half of the steady state value of lift L_o is reached almost immediately and about $0.9 L_o$ is attained after a lapse of time $t = (6c/V)$ during which the foil travels a distance equal to about 6 chords. This fact was first discussed by Wagner (Ref 2.13) who gave the theoretical reason for this peculiar behaviour of lift. In honour of the investigator this phenomenon is usually referred to as the *Wagner effect*. His findings, supplemented by other investigators to whom we shall refer later, may be considered to account for the *unsteady lift*, i.e. lift generated in unsteady conditions, to the same extent that the classical theory based on Kutta-Joukowski theorem accounts for lift generated in *steady motion*.

In the course of experimental studies on lift generated in unsteady conditions, i.e. when the angle of incidence changes more or less rapidly, it was found that actual lift build-up, experienced by foil RAF 30, given by the thin broken line in Fig 2.14A, has the nature but not the exact shape of the theoretical exponential curve presented by the thick broken line plotted in this figure. The solid line in Fig 2.14A represents the theoretical variation of circulation Γ.

It was found later that the actual lift build-up largely depends on how quickly the incidence angle is changed. An interesting example of measurement of transient variation of lift due to sudden deflection of flap is shown in Fig 2.14B which, in a way, supplements the pictures presented earlier in Photo 2.4D.

Fig 2.14 Growth of circulation and lift with time. This pattern of behaviour is applicable within the range of incidence angles below stall.

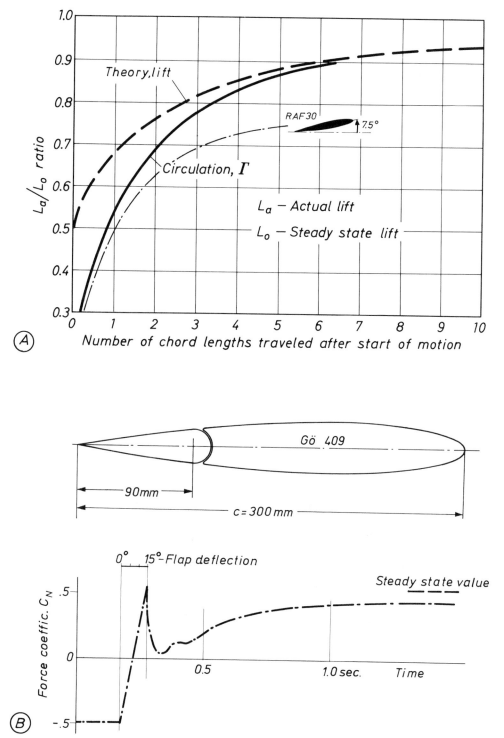

The tested symmetrical foil section is shown at the top of Fig 2.14B. A curve of the normal force coefficient C_N versus time is plotted below; C_N being the coefficient of the force normal to the foil chord and therefore almost equivalent to lift force. At the beginning of test the foil was set at an incidence $\alpha = -5°$ without flap deflection. The recorded C_N coefficient was about -0.5, and then the flap was deflected from $0°$ to $15°$. As reflected in the graph, lift jumped rapidly above the steady state value and then dropped again giving rise to periodic oscillation in C_N which gradually died out. Such a response demonstrates convincingly that the circulation about the foil cannot instantaneously assume the value which corresponds to steady state conditions determined by a given incidence angle.

Theory, as well as practical experiments depicted in Fig 2.14, indicate that circulation around a foil and associated lift never reach their steady state magnitude but merely approach it asymptotically as time goes on: thus the shedding of vortices never really ends, although they quickly become almost imperceptible. These small vortices are recognizable in Photo 2.4C as they stretch in the form of a thin vortex

Photo 2.5   Flow past an aerofoil at zero angle of incidence shown by smoke streamlines. Point of separation near the trailing edge of the profile. The presence of vortices (Karman vortex street) in the wake indicates that at zero angle of incidence there must exist a periodic circulation developing clockwise and anti-clockwise. This must give rise to alternating $\pm$ lift with average lift equals zero.

Evident instability of the flow and subsequently lift may, in some conditions, result in violent resonant vibration of the rudder, the latter depending on distribution of rudder shedding approaches the so-called 'natural frequency' of the rudder. The latter depending on distribution of rudder 'massiveness' and the amount and distribution of its stiffness.

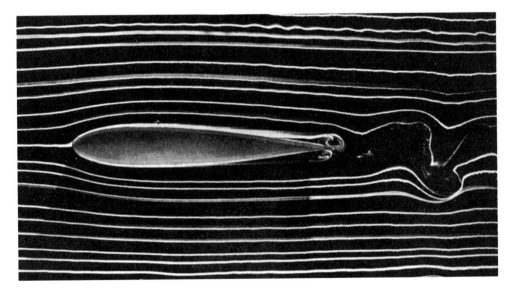

sheet between the cusp of the foil and the starting vortex seen at the right edge of Photo 2.4C. The effect of any viscosity in the fluid, however small, is to cause the two streams A and B, distinguished in Fig 2.11A, to slide downstream in the form of vortex sheets surrounding the foil, and the vortices of these sheets may be anticipated to act as the roller bearings between the surface of the foil and the mass of fluid outside it. Subsequently, the actual flow pattern takes the form depicted in Photo 2.5, where it is seen that the streams immediately adjacent to upper and lower surfaces of the foil do not reach the very trailing edge to join precisely at the cusp, as they would in the case of an ideal fluid. Instead, they leave the foil surface a short distance before the trailing edge to form a region of vortex motion which may develop into what is called a *Karman vortex street*. To maintain this system, vortices of opposite sign are shed alternately from the upper and lower surfaces of the aerofoil. The extent of this vortex street depends on the shape and incidence angle of the foil and, in the case of well designed foils operating at not too large incidence, the wake may be almost evanescent.

(c) *Practical implications*

Since at a small incidence the vortex wake is narrow and weak, the circulation round the aerofoil and the associated lift are sensibly constant. However, when the angle of incidence increases approaching stall angle, the oscillation in the magnitude of the circulation and lift may become an important fraction of the mean values. Fluctuations in the circulation can, in some conditions, make themselves felt in vibration of the foil and even set up a rather undesirable phenomenon of singing. Singing or rattling rudders and centreboards are good examples of those objectionable vibrations that may even cause damage due to material fatigue.

Several important deductions can be made from Fig 2.14:

a. It takes longer for circulation to establish itself around a foil of longer chord than a shorter one. This is because a larger mass of fluid must be set in motion. A spade-type rudder with short chord, i.e. of high aspect ratio, gives therefore a *quicker response* in generating side force than a flap-type rudder hung behind a long keel or hull, where the circulation path extends almost completely around the hull waterline. An unintentional experiment with a 6-Metre boat shown in Photo 2.6 provides an interesting insight into the significance of time lag in establishing circulation and its effect on directional stability. It will be seen that the hull depicted in Photo 2.6 incorporates both the so-called bustle and bulbous bow, the purpose of which is to cheat the sea into thinking that it is flowing past a slimmer or longer hull than it is actually, as measured by the rule. This new feature of yacht shape was loudly claimed as a revolutionary breakthrough towards faster hulls. There is a snag however. Bearing in mind that any change in side force generated by the hull requires a certain time in which a new flow pattern (circulation) is established, a long hull regarded as a hydrofoil is bound to respond slowly to rudder action. As reported in *Seahorse*

Photo 2.6   Modern 6-Metre boat incorporating bustle and bulbous bow.

of January 1976, '...her [6-Metre boat] worst feature is that she has tremendous directional stability: pull away the helm and it takes an age. Luff up to a sudden freeing puff, and she won't respond easily.' The above example demonstrates the fact that the probability of revolutionary breakthrough in racer performance is very slim indeed. It becomes almost routine that whenever the first model, as designed initially and incorporating *revolutionary* or *exploratory* features, is tank-tested, the impartial tank is reluctant to appreciate them. Cuttings and plasticine changes have to be made to the hull shape. Strangely enough, the more the model is run in the tank, the more *conventional* it becomes.

One may add that the great English yacht designer G L Watson tried the bulbous bow on a boat designed for himself in 1871, a small cutter by the name of *Peg Woffington*; apparently without success, since the invention was forgotten.

b. By taking into account the time lag in establishing circulation pattern, one can explain why the present breed of boat with high aspect ratio fin keel and separate rudder, modern 12-Metres included, is rather difficult to keep *in the groove*. In big waves with large components of water flow induced by the wave orbital flow, changing rapidly in velocity and direction, short-keeled boats respond quickly, yawing substantially and rapidly off a mean course. They require constant control which may be tiring in the long run.

c. Separation which occurs on the suction side of a foil, when the effective angle of incidence exceeds the stalling angle, is not washed away or dissipated instantly as the angle of incidence is reduced to that at which there is no separation. Fluid flow manifests a kind of *memory*, and the time lag in establishing the new flow pattern corresponding to the new conditions at any given instant, is known as *hysteresis*. The practical meaning of it is that, although the foil attitude relative to the flow may be altered instantaneously, the fluid *remembers* the initial flow pattern and it takes a certain time to achieve both a steady condition character of the flow round the foil and corresponding lift value.

Photo 2.4E demonstrates vividly this hysteresis effect on the flow pattern round a foil, the incidence angle of which was increased rapidly. Pictures 1–6 demonstrate how the full chord separation gradually develops. Subsequently, the foil incidence angle was reduced to a value at which separation does not occur in steady state conditions; however, as seen in Pictures 7–8, in spite of this reduction of incidence separation persists. This effect is of particular significance for sailing in light winds or drifting conditions. Once the attached flow round a sail has been destroyed, it takes a time to re-establish smooth flow with its associated high lift. Racing people are well aware of what it means in terms of boat speed reduction. Nevertheless, Chichester's experience is worth noting. In one of his Atlantic crossings he found that *Gypsy Moth* would ghost at about 1.5 knots under conditions of apparent absolute calm, but stopped dead as soon as he moved and rocked the yacht.

(6) A closer look into some foil characteristics

Figure 2.15A, based on velocity observations and measurements made by the late T Tanner (Ref 2.14) demonstrates that circulation around the symmetrical RAF 30 aerofoil section gives rise to a high velocity region, with the streamlines crowding together above the back of the aerofoil, and also a low velocity region, with a wider-spacing of the streamlines below the face. One may notice a general resemblance of the streamlines found in the real fluid (air) illustrated by broken lines to those theoretical streamlines represented by full lines calculated for the same lift coefficient.

The experimental values of the pressure coefficients C_p presented in Fig 2.15B, are plotted against x/c ratio, where x is the distance from the nose measured along the foil chord c; the theoretical pressure distribution is presented in a similar manner. It will be seen that the differences between the calculated and actually measured pressure distributions are small indeed, except for a short part of the tail where streamlines separate before trailing edge is reached. The agreement is much better than that observed in the case of theoretical and experimental pressure distributions round a rotating cylinder, as displayed in Fig 2.5C; it should be expected bearing in mind the absence of a large wake behind the foil and which exists in the case of a spinning cylinder.

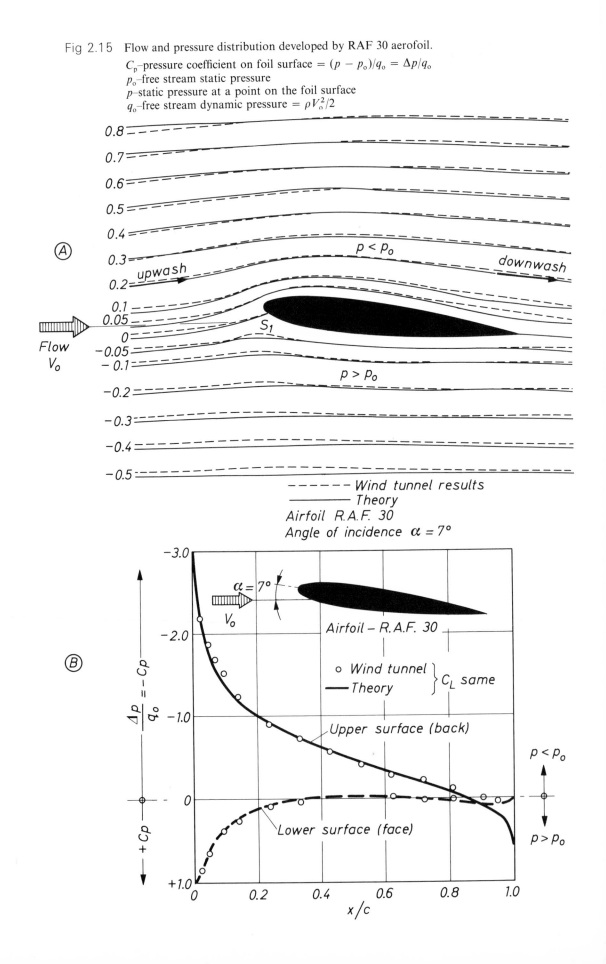

Fig 2.15   Flow and pressure distribution developed by RAF 30 aerofoil.

C_p–pressure coefficient on foil surface $= (p - p_o)/q_o = \Delta p/q_o$
p_o–free stream static pressure
p–static pressure at a point on the foil surface
q_o–free stream dynamic pressure $= \rho V_o^2/2$

One may find in Fig 2.15B that:

Firstly–the maximum positive pressure coefficient $+C_p = 1$ and the maximum negative pressure coefficient $-C_p = -3$; both occur near the nose,

Secondly–the negative pressure on the back of the foil makes an appreciably greater contribution to lift than the positive pressure on the face.

(a) *Methods of presenting pressure distribution*

A few words of explanation should be added about the forms of presenting pressure distribution. One way of displaying the pressures is that demonstrated in Fig 2.15B, another one is shown in Fig 2.16A. It is a relatively easy matter to determine these pressure changes round an aerofoil in a wind tunnel. For such a test the foil is equipped with a series of flush orifices of approximately $\frac{1}{16}$ in diameter each individually connected to a tube of a multiple manometer. For a given velocity V_o, the pressures p are read, then the atmospheric pressure p_o is subtracted and the remainder $p - p_o = \pm\Delta p$ is divided by the dynamic pressure $q = (\rho V^2/2)$ in order to find the already known non-dimensional pressure coefficient $C_p = \pm(\Delta_p/q)$.

Results of measurements can be presented in two different ways:

1. Coefficients $\pm C_p$, are plotted normal to the foil surface at the appropriate measurement stations. Such a presentation of pressure distribution given in Fig 2.16A depicts the pressures as they actually act–always perpendicular to the surface restraining the fluid.

2. Coefficients C_p can also be plotted normal to the chord at the appropriate stations, as shown in Figs 2.15B and 2.16B.

Several of the pressure readings are marked in Fig 2.16A and B by numbers 1, 2, 3, so that their relative position may be followed in the two different plots.

The variation of the pressure distribution with the angle of incidence α for a typical foil is presented in Fig 2.16C. It gives the answer to the often repeated question–which side lifts more, the upper or the lower surface of the foil. One may notice that with increasing incidence α the contribution of the upper surface to lift increases gradually until it finally contributes about 70 per cent of the total. It can also be seen in Fig 2.16, that the suction diagram for an angle of incidence of 20° when the foil is stalled differs considerably from the same diagram of 15°. This phenomenon is intimately connected with the fact that beginning from α about 15° lift starts to decrease with the angle of incidence. It is due to flow separation developing along the upper (suction) side of the foil (the start of the stall).

Some interesting information can be derived from pressure distribution, namely:

1. The location of the suction peak and its magnitude given by pressure coefficient C_p.
2. The load that the skin of the foil has to withstand and its distribution.
3. The centre of pressure location, i.e. position of the resulting force.
4. The relative magnitude of the resulting force which is proportional to the area of the pressure diagram (crossed zones in Fig 2.16C).

Fig 2.16 Pressure distribution diagrams for a foil at various incidence angles α.

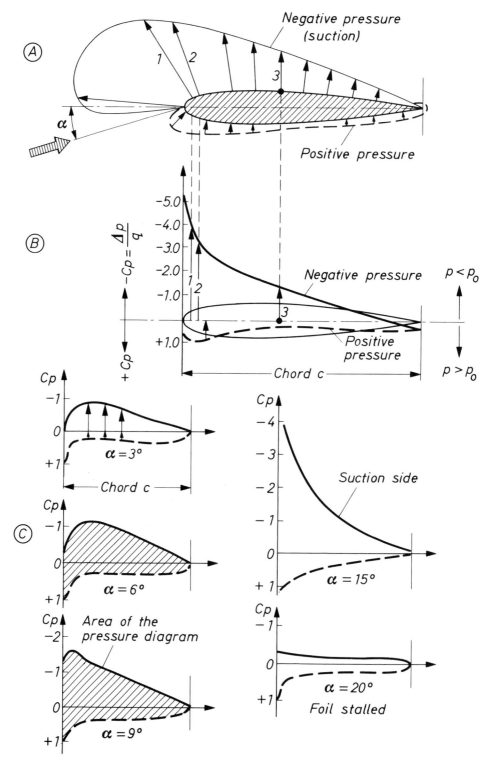

Referring again to Fig 2.15A it can be seen that the presence of a foil in the stream of fluid exerts a sort of *advanced influence* on the fluid, giving its motion an upward trend before it reaches the foil itself–*the upwash*. This upwash, increasing in magnitude as the flow approaches the foil, can be explained by the circulatory flow round the foil which necessarily induces an upward component in the flow pattern well before the streamlines reach the leading edge of the foil. For the same reason, some distance from the nose (about a quarter chord of the foil) the *downwash* builds up. These features of the flow pattern are marked in Fig 2.15A by the two appropriately labelled arrows.

Upwash and downwash must be taken into account when mounting instruments for measuring velocity and direction of flow. They should be placed well ahead of the leading edge: at least two chords of the foil, to give results reliable enough for serious tests. This applies to instruments measuring flow in the proximity of sails, as well as hull appendages.

(b) *Rate of lift growth with incidence* (*two-dimensional flow*)
Both theory and experiment lead to two conclusions:

Firstly–the magnitude of circulation Γ varies as, or is a function of, the flow velocity V_o, angle of incidence α and chord of the foil c. Accordingly we may write:

$$\Gamma = f(V_o \times c \times \alpha) \text{ ft}^2/\text{sec} \qquad \text{Eq 2.12}$$

where $f(...)$ defines 'is function of' or 'is proportional to' factors given inside brackets.

For an unstalled foil, the circulation Γ around it varies linearly with the angle of incidence α and if α is measured from zero-lift attitude which corresponds to $\Gamma = 0$, then

$$\Gamma \sim L \sim \alpha$$

where $\sim$ means 'is proportional to...'

As in the case of a circular cylinder (Eq 2.7), the circulation round the foil may be given by the product of the average velocity of circulation V_{cav}, and the path length which is approximately two lengths of the foil chord, i.e. $2c$, hence:

$$\Gamma = V_{cav} \times 2c \text{ ft}^2/\text{sec} \qquad \text{Eq 2.12A}$$

This equation explains in a different form why it takes longer for circulation to establish itself around a long path, measured by length $2c$, than a short one; a fact demonstrated experimentally by Wagner's and Drescher's tests presented in Fig 2.14.

Substituting expression 2.12 into Eq 2.10 we may find that lift L is proportional to some factors already familiar to us, namely:

$$L = f(\rho \times V_o^2 \times c \times b \times \alpha)$$
$$L = f(\rho \times V_o^2 \times S \times \alpha) \qquad \text{Eq 2.13}$$

where $\qquad\qquad \rho$ fluid density

$\qquad c \times b = S$ area of the foil

$\qquad\qquad \alpha$ angle of incidence

For reasons which will become apparent soon it is customary to express Eq 2.13 in a different way by introducing an empirical shape factor C_L, i.e. lift coefficient instead of incidence α, and dynamic pressure q which combines both ρ and V_o^2, hence,

$$\text{Lift} = C_L \times \text{Area} \times \text{Dynamic pressure}$$

$$L = C_L \times S \times q \qquad\qquad\qquad \text{Eq 2.13A}$$

Secondly–lift coefficient C_L for an unstalled foil at constant velocity V_o, is linearly dependent upon the angle of incidence α, and this relation can be expressed by the formula:

$$C_L = 2\pi\alpha = 6.28\alpha$$

if α is given in radians

$$1 \text{ radian} = 57.3°$$

$$\pi = 3.14$$

(see Appendix)

or

$$C_L = 0.11\alpha \qquad\qquad\qquad \text{Eq 2.14}$$

if α is given in degrees.

Which means that the lift coefficient C_L should theoretically increase by 0.11 when angle of incidence increases 1°. At 10° incidence the C_L would then be 1.1.

Equation 2.14, developed by classical theory of aerofoils, gives the value of lift coefficient for flat plate. To some extent the foil thickness, t, also affects the lift coefficient and for symmetrical foil sections these relations may approximately be expressed by:

$$C_L = 0.11\left(1 + \frac{t}{c}\right)\alpha \qquad\qquad\qquad \text{Eq 2.14A}$$

where α is in degrees and

t/c is thickness/chord ratio of the foil.

The reason is as follows: foil thickness means flow displacement and crowding of streamlines particularly in the region above and below the maximum thickness of the foil. As a consequence the average flow velocities along the sides of the foil sections are increased as compared with those past a thin plate. Hence the lift curve slope becomes slightly steeper for thick foils.

For asymmetrical sections the formula approximating the lift coefficients is:

$$C_L = 0.11(1 + t/c)(\alpha - \alpha_{L0}) \qquad \text{Eq 2.15}$$

where α_{L0} is the incidence angle corresponding to zero lift attitude. It is always zero for flat-plate or symmetrical foil sections and becomes negative and numerically greater with increasing camber.

Every theory seems to have certain limitations, being valid within its stipulated conditions, and the circulatory theory of lift is no exception. Its usefulness in predicting lift coefficient is restricted to a limited range of angles of incidence either positive or negative. In fact, Eqs 2.14 and 2.15 hold only to a certain point, i.e. the lift coefficient increases proportionally up to the angle of incidence at which, somewhere on the upper surface of the foil, separation occurs, as depicted in Photo 2.7. Separation can be defined as an abrupt departure of the streamlines from the contour of the foil that is supposed to be guiding them. Agreement between the theory based on the ideal fluid concept and experiments is tied up with the smallness of the wake behind the foil; and this agreement naturally ceases if, for some reason, the flow is no longer attached to the foil surface and the wake becomes excessive.

The resulting deviation from linearity of the lift coefficient C_L versus α, increasing with the angle of incidence until the maximum C_L value is obtained, is shown in Fig 2.17. It presents the experimental lift curves for three NACA symmetrical aerofoils (Ref 2.15).

Photo 2.7   Large angle of incidence. Point of separation close to leading edge. The separation can be defined as an abrupt departure of the streamlines from the contour of the foil that is supposed to be guiding them.

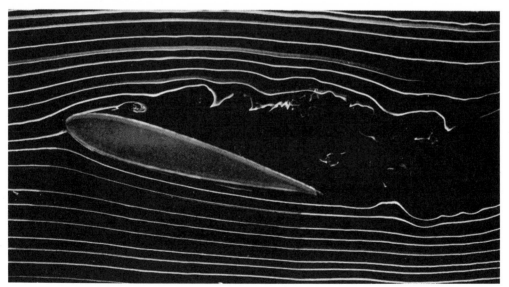

Fig 2.17 Aerodynamic characteristics of some NACA (National
Advisory Committee for Aeronautics) symmetrical sections
(two-dimensional flow).

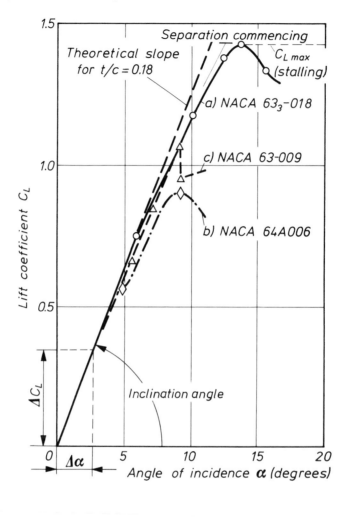

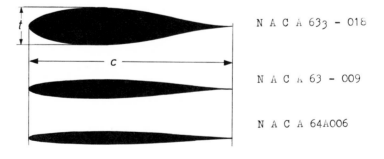

Let us take as an example the curve (a) for an 18 per cent thick aerofoil. Applying Eq 2.14A we might expect that the C_L value at incidence $\alpha = 10°$ should be:

$$C_L = 0.11(1 + 0.18) \times 10 = 0.11 \times 11.8 = 1.3$$

as indicated by the straight theoretical broken line.

In fact, the experimental value is about 1.2. On average, the experimental results give C_L values about 10 per cent lower than the theoretical ones found for the same angle of incidence and a more accurate empirical value of C_L is given by the formula:

$$C_L = 0.1(1 + t/c)\alpha \qquad\qquad \text{Eq 2.14B}$$

where α is in degrees.

This indicates that in general the *slope of the lift curve* is slightly less steep than that given by theory.

The term *slope* deserves perhaps some explanation. The slope of the lift versus incidence curve can be measured by an inclination angle Θ between the straight part of the curve and the *Angle of incidence* axis, as shown in Fig 2.17. In other words the slope is a measure of the rate of change of C_L coefficient with incidence angle α, i.e. it gives an idea how quickly lift rises per degree of incidence angle. This rate of change can be expressed as the ratio $\Delta C_L/\Delta\alpha$. Referring to Eq 2.14 it is evident that the ratio $(\Delta C_L/\Delta\alpha) = 0.11$.

(c) *Factors limiting lift growth*

As already mentioned, the lift generated by a foil is due to the pressure difference between the back and face surfaces of the foil. This differential pressure can only be most effectively maintained if the flow is attached to the geometrical contour of the foil. At small angles of incidence the streamlines have little difficulty in accommodating themselves to the foil surfaces, as demonstrated in Fig 2.15A. When the incidence angle is gradually increased, however, the streamlines may fail to maintain contact, especially on the back curvature of the foil, where they have to work their way against two kinds of resistance. *The first* is caused by viscosity and an unavoidable friction at the foil surface; *the second* is due to an unfavourable pressure gradient in the direction of the flow. The pressure gradient may in this case be defined as a rate of change of pressure intensity downstream, particularly important along the upper back surface of the foil (Note 2.16).

Referring to Fig 2.16, it can be seen that the intensity of suction increases rapidly as the incidence α increases. When incidence $\alpha = 3°$ the suction peak occurs about 25 per cent of the chord length from the leading edge. At incidence $\alpha = 6°$ the suction peak is situated about 20 per cent of the chord length from the leading edge and with increasing incidence it travels further and further towards the nose.

The pressure distribution, and the position of the minimum pressure in particular, have a large effect on the boundary layer (BL) flow, which in turn affects the flow outside the BL. Looking at Fig 2.18 which shows the largely expanded BL picture, we may notice that the local flow velocity V outside the BL changes in such a way

Fig 2.18   Picture of a typical turbulent wake due to separation on the curved surface. For the sake of clarity the BL is greatly exaggerated. Two consecutive photographs illustrate ensuing vortices downstream from separation point S.

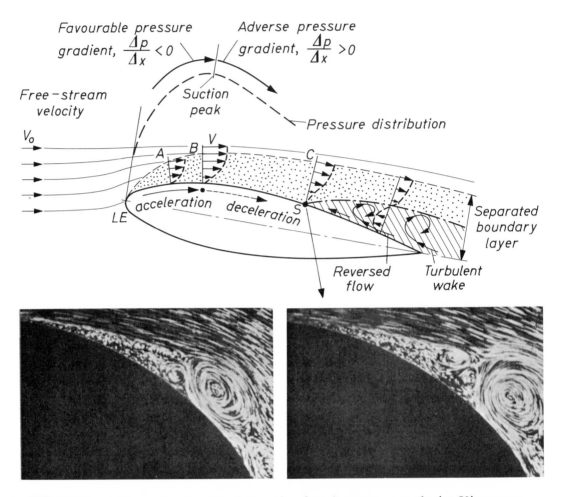

that the local pressure p along the foil section first decreases, as velocity V increases. The suction peak is reached somewhere at point B, where the flow velocity V is at its maximum. Downstream from the point B the local pressure gradually increases, approaching ambient pressure p_0 somewhere close to the trailing edge of the foil. A glance at Fig 2.16B should help in making this point clear; at the trailing edge the pressure coefficient C_p is close to zero.

Proceeding downstream from the leading edge (LE) up to point B, the fluid particles are accelerated; the favourable pressure gradient tends to accelerate the flow. The pressure is lower (more intensive suction) at point B than at LE, which is favourable to the flow because fluid particles move easily from a region of higher pressure to that of a lower one. As soon as the minimum pressure point B (suction peak) is passed, however, flow conditions become quite different. From this point the pressure gradient, becoming adverse, opposes the velocity, so that the fluid particles are decelerated instead of being accelerated. Since part of the fluid kinetic energy has been lost, due to the retarding action of viscosity, the fluid particles have not sufficient energy to make headway against the rising pressure or so-called adverse pressure gradient.

Consequently, this unfavourable pressure gradient will slow the flow down and bring fluid particles to rest before reaching the trailing edge. The increase of pressure becomes in a way an insurmountable barrier and the flow finds itself stopped in its tracks. When fluid particles fail to progress along the surface they accumulate, and this accumulation thereby produces separation of the main flow. As indicated in Fig 2.18, the separation ensues from point S and immediately downstream from this point a region of dead air or dead water appears–the fluid being driven backward in a turbulent manner. With increasing incidence the reverse flow progressively covers a larger and larger part of the back surface of the foil and the separation point S goes further and further towards the LE, as illustrated in Photo 2.7. The magnitude of circulation is being considerably reduced in comparison with that which the perfect fluid theory of lift prescribes, and as a consequence, the intensity of negative pressure developed on the back surface of the foil is also reduced.

From what has already been said one may conclude that:

a. Separation is less likely to occur if the boundary layer flows in the region of decreasing pressure, i.e. when flow is affected by favourable pressure gradient.

b. If the rate of pressure rise, i.e. unfavourable pressure gradient, becomes too high the boundary layer particles may slow down to a dead halt and separation takes place; this phenomenon is called a stall. With a streamline foil having a long tapered tail, the rate of pressure rise or adverse pressure gradient may, within a certain range of incidence angles, be so moderate that the boundary layer gets nearly all the way to the tail without separation (Fig 2.15A). The flow round the foil is then virtually the same as if there were no viscosity at all, and the pressure distribution on the surface of the foil is very much the same as in non-viscous flow. Consequently wake drag is negligible and only a very small drag arises almost entirely due to skin friction.

Therefore, one may rightly say that a body is streamlined if there is no boundary layer separation. With bluff bodies, on the other hand, the positive pressure gradient or the pressure rise is great enough for separation to occur well ahead of the rear of the body, with the resulting formation of a large wake. Such bodies always have a high drag, of which skin friction is only a small part. The foil which at low angle of incidence falls into the streamline category may fall into the bluff category when the incidence angle is large enough to encourage separation on the back surface near the leading edge.

Since, as shown in Fig 2.16C, the negative pressure on the back of the foil contributes to lift to a much greater extent than the positive pressure on the face of

Fig 2.19   The effect of trailing edge angle and trailing edge cutting-off on
            lift characteristics of an 18 per cent thick foil.

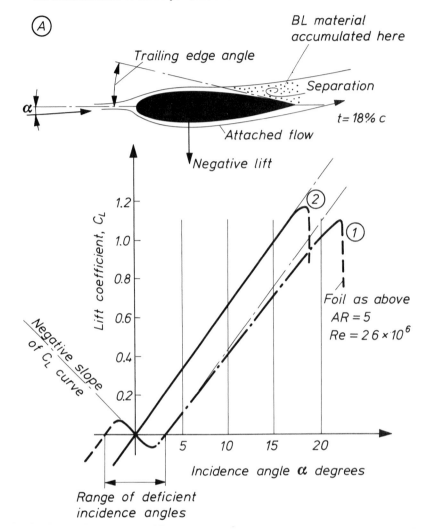

the foil, lift is bound to decrease as separated flow develops. In fact, the local pressure p in the dead air region, where the reverse flow occurs, is close to atmospheric pressure p_0. Thus, there exists for every foil a certain critical angle of incidence, the so-called stalling angle, beyond which lift no longer increases but, on the contrary, starts to decrease as the angle of incidence increases. This stalling behaviour of a foil, so strikingly different from that anticipated for an ideal fluid, is well demonstrated by the three experimental curves plotted in Fig 2.17. It can also be seen from this plot that the measured C_L curves lie much closer to the theoretical curve than that for the rotating cylinder presented in Fig 2.8. Evidently, this is due to a negligible wake in the case of a streamline foil, as compared with the rotating cylinder. Whereas the lift

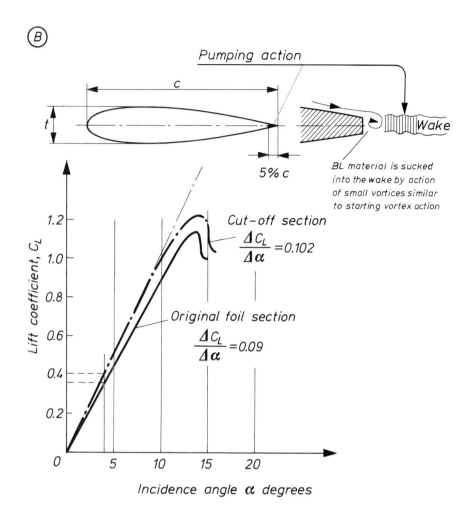

of the rotating cylinder is much higher than that of foils, drag is much more pronounced. As a result, the efficiency of the streamline foil as a lifting device, expressed in terms of L/D ratio, is far superior to that of the rotating cylinder.

It can be seen in Fig 2.17 that the separation and subsequent stalling angle of three different foils depend on their thickness. The thicker foil produces higher $C_{L max}$ and stalls at a higher angle of incidence.

It should not be assumed that the occurrence of separation defines the maximum lift coefficient. As a matter of fact, separation close to the trailing edge, depicted for example in Photo 2.5, has a relatively negligible effect on circulation and C_L value. Only with such a *full chord separation* reaching almost the leading edge, as demonstrated in Photo 2.7, does an abrupt change in circulation occur and it then coincides with a drastic drop in lift.

Even the best shaped foils are subject to separation and stall when the angle of incidence is increased beyond a certain critical value. Separation always starts below the angle corresponding to $C_{L\max}$ and, depending on its character and extension, the lift decreases, sometimes gradually, sometimes abruptly, as shown in Fig 2.17, curves a and c.

The stalling characteristics of the foil, its $C_{L\max}$ value, drag and L/D ratio, which are of great practical significance, depend on a number of factors. The most important of them are:

1. Section of the foil, its thickness and position of maximum thickness.
2. Reynolds Number (scale effect).
3. Aspect ratio and plan form.
4. Quality of the foil surface roughness, flexibility, etc.

(d) *Peculiar behaviour of some thick foils*

As an example, let us expose the influence of factor 1 on lift and show that not every foil which looks streamlined is equally efficient as a lift-producing device. Figure 2.19A illustrates this point. The curve marked 1 represents lift coefficient variation with incidence for the foil of aspect ratio AR = 5 and section as drawn above the lift curves. Location of maximum thickness is 40 per cent of the chord c behind the leading edge and the thickness ratio $t/c = 18$ per cent. For comparison, curve 2 gives the lift characteristics of a similar foil section of the same thickness ratio but with a different location of maximum thickness, which is 30 per cent of the chord c behind the leading edge, and hence the section has a more slender afterbody, i.e. smaller trailing edge angle as defined in Fig 2.19A.

When set at small angle of incidence, of the order of 1–2°, foil 1 produces a peculiar asymmetric flow pattern with partly separated flow and heavy boundary layer along the back of the foil. Since the effective curvature of the flow along the face is more pronounced than that along the upper side of the foil, higher average velocities are obtained along the lower side, where flow is attached. As a consequence, in the range of incidence angles $\pm 3.0°$ the foil generates negative lift, i.e. in the opposite direction to that in which it is expected. This is distinguished in Fig 2.19A by the negative lift-curve slope. Beyond this range of deficient angles of incidence the positive lift-curve slope is almost the same as that for the more slender type of section 2.

Such an undesirable effect of fullness of the afterbody on lift may be observed in the case of modern hulls incorporating so-called bustle and separate rudder hung on the end of the bustle. Since flow round the underwater part of the hull affects both the side force generated by the rudder as well as that generated by the hull itself it may happen that, due to flow separation at the blunt bustle or hull afterbody, the rudder working in the wake becomes deficient. It was reported that on the 12-Metre *Valiant*,

Photo 2.8   *Mariner*'s unusual configuration of afterbody was probably aimed, apart from the anticipated measurement benefit, at delaying flow separation and making the hull more efficient as a hydrofoil. The embryo transom stern may also advantageously affect, at some speeds, the wave pattern, and therefore the wave resistance of the hull.

at small rudder angles, almost no turning moment could be produced; a moment in the wrong direction had even been observed both in the towing tank and in practice. If the helmsman of *Valiant* gave a rudder angle smaller than about 10°, the boat would turn towards the opposite direction. This type of directional instability of a yacht is simply a consequence of negative lift-curve slope of the hull appendages combination similar to that shown in Fig 2.19A.

It has been found that cut-off trailing edges have an effect upon section characteristics precisely opposite to that characteristic just discussed of a foil with large trailing edge angle. (Hoerner–Base Drag and Thick Trailing edges–*Journal of Aeronautical Science*, Vol XVII, 1949 and also *Fluid Dynamic Drag*.) Due to the pumping action of negative pressure, originating behind the flat cut-off trailing edge, the boundary layer does not accumulate in the manner shown in the sketch incorporated in Fig 2.19A. Instead, the BL material is pumped away into the narrow wake. To be of benefit the edge should be sharp cut to encourage vigorous vortex generation. The mechanism of suction behind the trailing edge is similar to the self-bailer action which sailing men are well familiarized with. In this way, undesirable features of negative lift-curve slope, shown in Fig 2.19A in the range of small angles of incidence, can be cured.

Provided that the trailing edge cutting is not excessive, up to 10 per cent of section chord c, negligible drag penalty is incurred and, as seen in Fig 2.19B, the lift-curve slope is increased from $(\Delta C_L/\Delta \alpha) = 0.09$ (original foil section) to $(\Delta C_L/\Delta \alpha) = 0.102$ (cut-off section with 5 per cent c cutting). In practical terms it means that at a given angle of leeway a fin keel with cut-off section may produce about 10 per cent more lift than original section.

Britton Chance's *Mariner*, with its unusual configuration of hull afterbody, shown in Photo 2.8, was most probably aimed, apart from expected measurement benefit, at delaying separation and making the hull appendages configuration hydrodynamically more efficient and stable directionally. Although *Mariner* failed as a full-scale experiment in the 1974 12-Metre trials, it does not necessarily mean that her chopped afterbody is the only factor to blame. The whole concept is sound and if properly developed may prove to be a contribution to the art of yacht designing.

Now we shall look more closely into viscosity phenomena and their effects on drag which is another important component of the resulting aero- or hydrodynamic forces.

B Drag–viscosity phenomena

The problem of the reduction of resistance experienced by a body moving through a fluid has intrigued physicists, engineers and mathematicians for many years. Since, in general, full-scale experiments are time-consuming and expensive, early resort was made to the use of models in which changes in shape could be introduced easily and cheaply and the results measured with reasonable accuracy. No doubt men experimented with models very early in history, and one of the first recorded attempts in performing model tests can be attributed to Leonardo da Vinci who measured the resistance of models and drew the wake pattern they created (Photo 2.9).

The resistance of drag consists of different components that are not easily separable but which interfere with one another. They are very difficult to handle on purely mathematical grounds, therefore most of to-day's knowledge is necessarily based upon experiments. Thus, the solution of practical problems is still largely empirical.

As already noted in the previous chapter, a body subjected to a relative flow experiences at least two kinds of resistance, called *skin friction* and *pressure drag*. Both of these components of resistance are viscous in origin (Ref 2.17).

(1) The boundary layer, pressure drag and skin friction

Experience with bodies exposed to the flow of a real, viscous fluid has led to the differentiation between what are commonly known as streamline forms and other less fairly shaped, so-called blunt bodies, which offer accordingly a much higher resistance to the motion. The essential characteristic of a streamline form is that the

Photo 2.9   The flow of water and the origin of resistance seem to be
            fascinating problems.
               Wake and drag studies by Leonardo da Vinci.
               Drawing No 12579 at Windsor, England.

streamlines close behind the body almost without the formation of the turbulent
wake or dead water space.

   In the case of a perfect fluid this closing in of the streamlines would occur,
whatever the form of the obstructing body, so that one may say that in a perfect, non-
viscous, fluid all forms are streamline forms and no wake would be found behind
them. Such an ideal flow pattern round a totally immersed body in a perfect fluid is
depicted in Fig 2.20–the upper part of the drawing. It is seen that the flow is
symmetrical and the pressure distribution that might be measured over the body
contour A B C is also symmetrical. Since, by definition, a perfect non-viscous fluid is
incapable of transmitting any shearing force, which otherwise would occur due to
viscosity, it seems a legitimate inference actually made some time ago by d'Alembert,
that the body as shown in the upper part of Fig 2.20 would not experience any
resistance while moving through an ideal fluid.

   In the real world no perfect fluid exists and resistance to motion is always present,
its magnitude depending on the fluid properties (such as viscosity and density), on
the size and form of the body immersed, on the relative velocity between the fluid and
the body and also on some other factors which will be discussed later.

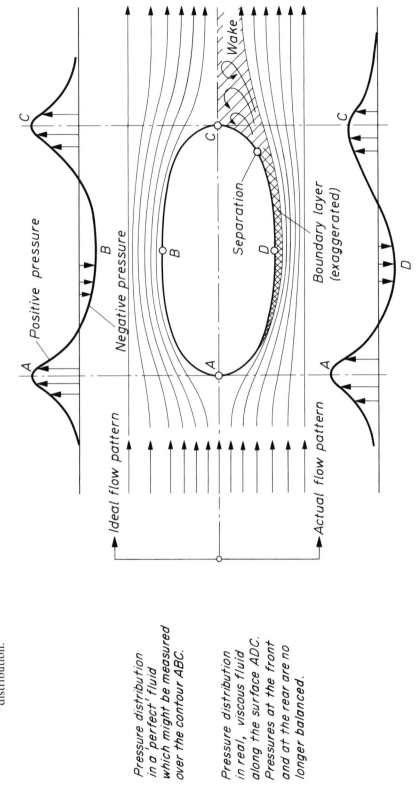

Fig 2.20 Ideal and actual flow pattern and associated pressure distribution.

Positive pressure

Negative pressure

A

B

C

Ideal flow pattern

Wake

Separation

Boundary layer (exaggerated)

A

B

C

D

Actual flow pattern

A

C

D

Pressure distribution in a 'perfect' fluid which might be measured over the contour ABC.

Pressure distribution in real, viscous fluid along the surface ADC. Pressures at the front and at the rear are no longer balanced.

Due to the retarding action of viscous forces, the streamlines do not follow the contour of the body back to its rear end. Instead, they separate from the surface somewhere as shown in the lower part of Fig 2.20, thus leaving downstream an eddying region of wake. As a result, the pressures over the rear part of the body cannot reach the same magnitude as those in front, and because they do not balance each other *pressure drag* occurs–the body is drawn downstream. This kind of drag, one may say, is caused by the fluid's inability to slow down without losing energy. On account of the obvious correlation between the size of the wake and the pressure loss at the rear part of the body, pressure drag is sometimes called wake drag (or eddy-making drag), larger wake indicating higher pressure drag.

Another kind of resistance to motion, skin friction drag, is the result of the shearing stresses in the fluid and is transmitted by the boundary layer as it passes over the surface of the body.

The total drag experienced by the body subjected to the flow as depicted in the lower part of the drawing in Fig 2.20 is simply a summation of frictional drag and pressure or wake drag. The first is the component of drag that proceeds from conditions inside the boundary layer, the second is the component produced by a breakdown of streamline flow due to the boundary layer action. Outside the boundary layer the actual flow can be regarded as the frictionless flow of a theoretical, perfect fluid, which is not subject to friction at the body surface. In fact, the flow pattern of a real fluid such as air or water round a streamline form, when no excessive wake exists, does not differ much from that of an ideal flow pattern as displayed in Fig 2.21. Any difference between the flow patterns is entirely due to the presence of the boundary layer that envelops the body.

Most readers probably have already observed the 'friction belt', or boundary layer, adjacent to the hull of a sailing craft due to adhesion and viscosity of water. For those who have not noticed this phenomenon yet, let us quote a splendid description by J Scott Russell–one of the first to be found in technical literature (Ref 2.18):

> 'I have also watched the action of this phenomenon, and the manner in which it takes effect appears to me to be this: The whole skin of the ship is covered with a thin layer of water, which adheres to it firmly and travels with it; to this first film a second is attached, which moves with it but which has to drag along with itself a resisting third film, which sticks to it; a fourth, fifth and sixth film, all in the same manner hang on to one another, until at last we reach a film which stands still. I have also watched the manner in which this phenomenon appears to grow and spreads from stem to stern.'

(a) *Laminar and turbulent boundary layer*

This idea of films or laminae of water which have successively every variety of speed, accepted later by scientists who developed the Theory of the Boundary Layer in the

Fig 2.21   Actual and ideal flow pattern over a streamline form.

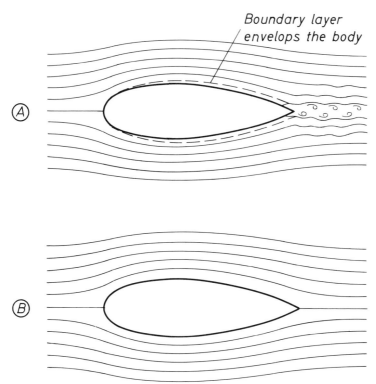

20th century, is shown in Fig 2.22 as if seen through a magnifying glass. It depicts a flow within the so-called laminar boundary layer and the picture presented would be the same no matter whether we consider the flow of water past the hull or of airflow past an aerofoil.

The boundary layer may be defined as a region of retarded flow, from full speed at the outer edge of the boundary layer to zero at the wetted surface. Friction drag is developed within this very thin belt of fluid immediately adjacent to the body surface through the successive water films, and is finally communicated to the body as friction forces acting parallel to the surface and opposing the motion.

From Fig 2.23 it will be seen that the fluid velocity increases gradually with the distance from the surface, i.e. it exhibits a certain velocity gradient. Sketch A in Fig 2.23 demonstrates the so-called velocity profile within the laminar boundary layer, from which we may calculate velocity gradient, dividing ΔV by Δy, $\Delta V/\Delta y$ (see Note 2.16). This ratio defines the change of velocity ΔV with distance Δy, measured from the surface. The name 'laminar' was derived from the early idea that in such flow the fluid could be imagined as a series of thin plates, or laminae, sliding one over the other. The fluid particles or molecules move smoothly along the path-lines of smooth curvature without intermixing. The transmission of momentum (the product of fluid mass and relative velocity between fluid particles) from the faster

Fig 2.22   A schematic representation of stages of the water flow around
the hull–laminar flow, transition, turbulent flow, separation
point and breakaway. The actual position of the transition point
depends on Reynolds Number, i.e. boat speed. Roughness of the
hull surface, particularly in the region of the bow, also matters.
Transition to turbulence will be hastened by random disturbances in the free stream ahead of the bow or the leading edge of
the foil.

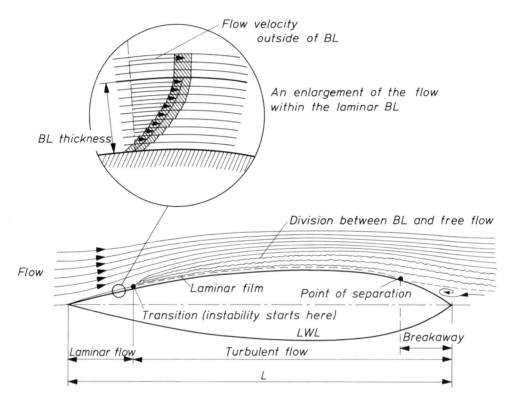

moving bands to the slower ones is carried out by the viscosity action. When laminar
flow transists into turbulent flow, shown in Fig 2.23B, this viscosity activity
continues as it did in laminar flow, but in addition there develops an exchange of
momentum between fluid particles moving from one band or stratum to the other
due to collisions. Since these particles have different average velocities, they transfer
different amounts of momentum into and out of the faster moving and slower
moving strata. The particles that move from the upper, faster stratum into the lower,
slower one have greater momentum in the direction of flow than the particles that
move from the lower stratum into the upper one. Thus, there is more momentum
transferred into the lower strata than transferred out of them.

Fig 2.23   Velocity profiles within the laminar, A, and the turbulent,
B, boundary layer.

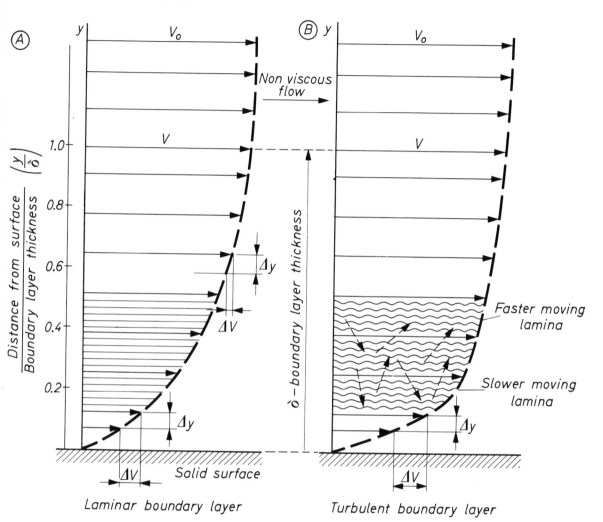

Laminar boundary layer   Turbulent boundary layer

Water and air, like every other fluid, manifest certain friction whenever there
exists a velocity gradient across the flow. The degree of viscous friction, different for
every fluid, is given by the appropriate coefficient of viscosity μ (Ref 2.20). Referring
to Fig 2.23 the coefficient of viscosity may be defined as the shearing stress or the
force required to move one of the two layers of fluid, each being of unit area, the two
separated by unit distance Δy, and the relative velocity between them also being of
unit intensity. In other words, the friction force acting on one of two adjacent layers
of a moving viscous fluid sets up a *shear stress* between the layers, which depends on
the fluid viscosity and the rate of change of velocity of flow within the boundary
layer across the flow, i.e. $\mu(\Delta V/\Delta y)$.

This shearing stress τ can therefore be expressed as

$$\tau = \mu \frac{\Delta V}{\Delta y} \ (\text{lb/ft}^2) \qquad\qquad \text{Eq 2.16}$$

where μ = coefficient of viscosity $\left(\dfrac{\text{lb sec}}{\text{ft}^2}\right)$

$\dfrac{\Delta V}{\Delta y} = \dfrac{\text{change of velocity}}{\text{small distance}}$ i.e. (velocity gradient)

Friction intensity, as shown by Eq 2.16, depends upon the rate at which a layer of fluid slides over the neighbouring one, hence when there is no relative motion between adjacent layers $(\Delta V/\Delta y = 0)$, there are no shear stresses and, where the relative motion or $\Delta V/\Delta y$ is large, the shear stresses will also be significant. It can be seen in Fig 2.23 that the velocity gradient $\Delta V/\Delta y$ is much greater at the surface than at a certain distance from it. The shear stresses and subsequent skin friction are therefore confined to a thin stratum of fluid close to the solid surface. Further away from this thin layer, the shear stresses reduce almost to zero and outside the boundary layer we shall assume that the flow is indistinguishable from that of perfect fluid where $(\Delta V/\Delta y) = 0$.

The diagram B in Fig 2.23, showing the velocity profile across the turbulent boundary layer, reveals that the velocity gradient $\Delta V/\Delta y$ at the solid surface is greater than that for a laminar boundary layer. As already mentioned, this is because the average velocity of the flow near the surface is increased by an exchange of energy between the particles travelling closer to the free stream and the particles already retarded at the surface. The velocity profile for turbulent flow differs therefore from that of the laminar profile: it is fuller near the surface and flatter away from it. Obviously, greater velocity gradient in the turbulent boundary layer produces larger friction than that generated by laminar flow. The problem of reducing skin-friction drag is then one of maintaining a laminar boundary layer as long as possible.

In order to gain further insight into the physical meaning and practical consequences of laminar and turbulent flows, we shall consider first the simplest flow on a smooth flat plate set parallel to the remote velocity, V_0 and demonstrated schematically in Fig 2.24. The flow within the boundary layer immediately downstream from the leading edge of the plate is usually laminar. The thickness of the layer grows relatively slowly and corresponding frictional resistance is low. Due to viscous properties of fluid, laminar flow has certain self-stabilizing or damping characteristics; it restores itself when abruptly disturbed. However, as the laminar boundary layer grows thicker and thicker, it becomes unstable and sooner or later degenerates, first into intermittently turbulent flow, and finally wholly turbulent flow. The character of the boundary layer flow then changes radically. The fluid particles no longer flow smoothly, in parallel lines, but instead start to oscillate in a direction perpendicular to the general flow, which is still maintained in the layer.

Photo 2.10   The cigarette smoke in a very quiet room streams at first in a form of laminar flow. At some point the smoke column becomes unstable and ultimately breaks into turbulent flow diffusing in an irregular manner into the surrounding air.

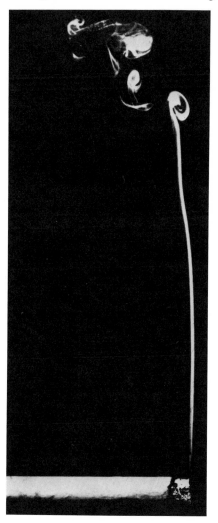

Photograph 2.10 shows a commonly known example, taken from every-day life, of transition from laminar to turbulent flow, while the cigarette smoke streams, undisturbed upwards in very quiet air. Transition from laminar to turbulent flow, which starts at a certain critical flow velocity, or critical Reynolds Number, can be attributed to the sudden appearance of small vortices inside the boundary layer, induced by unstable, minute boundary layer waves, which grow in amplitude as they travel downstream. They look like the breaking of 'white caps' on ocean waves and Photo 2.11 gives a good insight into boundary layer flow, as it becomes turbulent.

Photo 2.11 Smoke oozing from a hole in the upper surface of an aerofoil in a wind-tunnel gives a good insight into the boundary layer as it becomes turbulent.

Flow

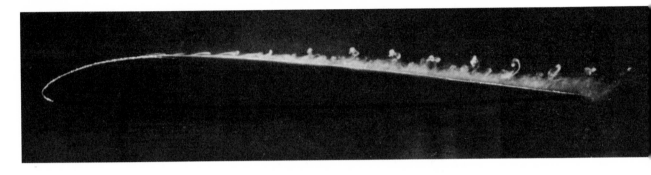

(b) *Calculation of skin friction*

The skin friction R_f of a flat plate, or surface with slight curvature, can be calculated according to the familiar formula:

$$R_f = C_f \times \frac{\rho V_0^2}{2} \times A \qquad \text{Eq 2.17}$$

where C_f = skin friction coefficient for the appropriate Reynolds Number
ρ = density of water = 1.99 slugs/ft³, air = 0.00238 slugs/ft³ (Tables 2.1 and 2.2)
V_0 = velocity of flow in ft/sec
A = wetted area in ft²

Therefore for salt water:

$$R_{fW} = 0.995 \times C_f \times V_0^2 \times A \qquad \text{Eq 2.17A}$$

Similarly for air at sea level:

$$R_{fA} = 0.00119 \times C_f \times V_0^2 \times A \qquad \text{Eq 2.17B}$$

As might be expected, the friction coefficient which enters into Eqs 2.17A, B, is not constant but is largely controlled by the character of the flow in the boundary layer. Reynolds' discovery and further contributions made by Reyleigh, Prandtl, Blasius and others, made it clear that the flow character depends upon the relative predominance of inertial and viscous forces, as represented by the value of Reynolds Number (Re) (discussed in Section C), the inertial forces favouring turbulent flow (higher Re), while the viscous forces promote laminar flow (lower Re). The three lines in Fig 2.24 represent the relationship between the skin friction coefficients C_f

Fig 2.24   Skin friction coefficient, C_f, for flow over the two sides of a *flat plate* with fully laminar boundary layer, with fully turbulent boundary layer and with a transition curve (Prandtl-Geber curve). Friction coefficients are given for the plan-projected area A, not the wetted area.

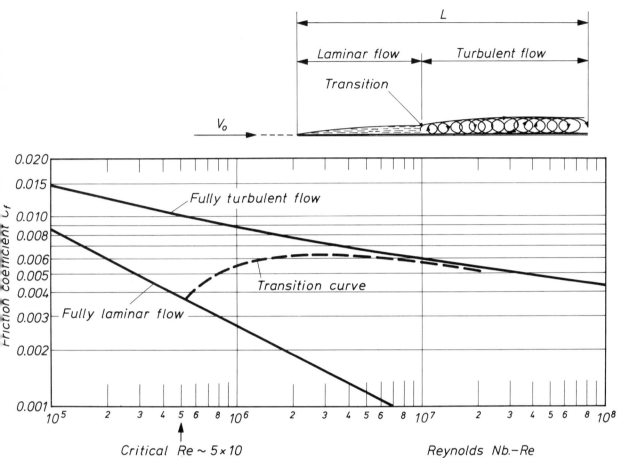

and Reynolds Number for the flow over the two sides of a flat plate with fully laminar BL, fully turbulent BL and a transition curve.

In order to appreciate the practical significance of transition, it is instructive to consider a situation in which the boundary layer is laminar on the forward part of the body, turbulent on the after part, with the dividing transition point between the two and shifting with every change of Reynolds Number. To give an example, we may find that at certain Re Number, in other words at a certain boat speed, the transition occurs somewhere along the hull, some distance from the bow, as shown in Fig 2.22. Experimental evidence enables us to assume tentatively that transition is likely to occur when the product of boat speed V_s (knots) and distance L (feet) is about five, i.e.

$$V_s(\text{knots}) \times L(\text{feet}) \simeq 5$$

which corresponds to the critical Reynolds Number Re_{cr} about 5×10^5 marked in Fig 2.24 (see Table 2.5) in Section C. Thus, if boat's speed $V_s = 2$ knots then distance L at which transition is expected to develop will be

$$L \simeq \frac{5}{V_s} \simeq \frac{5}{2} \simeq 2.5 \text{ ft}$$

If the speed of the boat increases, the transition point will gradually be shifted towards the bow.

Prediction of transition requires some empirical knowledge about factors such as surface roughness, pressure gradient, surface flexibility, etc., which may delay or promote flow change from laminar to turbulent. For a smooth, flat plank the transition occurs in the range of $\text{Re}_{cr} = 3$ to 5×10^5; for hull or foil when flow is affected by favourable pressure gradient, this range is shifted towards higher values of Re. If the critical value of Reynolds Number Re_{cr} (Fig 2.24) is assumed to be 5.0×10^5, then at Reynolds Number of 15.0×10^5 (1.5×10^6) the flat plate of length L would be expected to have laminar flow over the forward third of its length and turbulent flow over the remaining $\frac{2}{3}L$. The friction coefficient C_f would be the sum of one-third of the laminar flow coefficient for the relevant Reynolds Number of 5.0×10^5 and two-thirds of the turbulent flow value for $\text{Re} = 1.5 \times 10^6$.

The set of curves in Fig 2.25 show the variation of friction coefficient C_f of a smooth plate with Reynolds Number and position of mean transition point behind the leading edge. It can be inferred from it that transition is of some importance in estimating friction drag since this rather evasive phenomenon of transition is largely responsible for uncertainties while translating model experiments into full-scale prediction.

Referring to Eqs 2.17A, B and Fig 2.24, attention should be drawn to the fact that in estimating hull friction, its wetted area A designates the entire surface exposed to flow and, when such a surface is introduced into Eq 2.17, the value of the coefficient given in Fig 2.24 should be divided by 2. In the case of a flat plate or thin foil for which the plan-projected area A is usually given (so the exposed surface to the flow is double A), the coefficients presented in Fig 2.24 should be applied directly.

Depending on the shape of the foil and its attitude (incidence relative to the oncoming flow), the two components of resistance–skin friction and pressure drag–may change drastically and one of them may completely overshadow the other. For instance, Photo 2.12 pictures the flow round a flat plate at various angles of incidence. It is evident that when the plate is aligned with the flow direction, or set at small angles of incidence as in the photo at the top, when $\alpha = 9°$, drag is almost entirely the result of skin friction and the wake is negligible. Whereas at incidence $\alpha = 90°$ it is entirely pressure drag distinguished by a conspicuous wake.

As a matter of fact, a flat plate represents either of the extreme forms among the infinite variety of possible forms of a symmetrical body with regard to both types of

Fig 2.25 Variation of friction coefficient, C_f, of a smooth flat plate with Reynolds Number, Re, and position of mean transition point, TP, behind the leading edge.

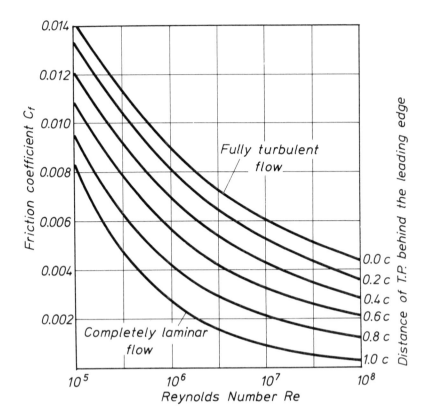

drag—frictional and pressure or wake drag. When the direction of motion is parallel or nearly parallel to the plate, the friction drag coefficient C_f is, on average, in the order of 0.004 to 0.008, as given in Fig 2.24. But when the direction of motion is perpendicular to the plate ($\alpha = 90°$), the drag coefficient (in fact pressure drag coefficient), C_D, is of the order of 1.9 for two-dimensional flow as given in Fig 2.26. This means that pressure drag can be 250–500 times greater than friction drag.

Friction drag and pressure drag together are frequently called *profile drag*, because they are determined to a large extent by the cross-section or profile of the body. Although friction drag seems to be small, it has far-reaching effects upon the character of flow round the body simply because, as has already been demonstrated, under certain conditions the flow, affected by friction in the immediate neighbourhood of a solid surface, separates from it. This is always accompanied by a more or less conspicuous formation of eddies in the wake and associated high pressure drag, a dominating component in bluff body drag.

Fig 2.26   Profile drag of flat plate at various angles of incidence (two-dimensional flow).

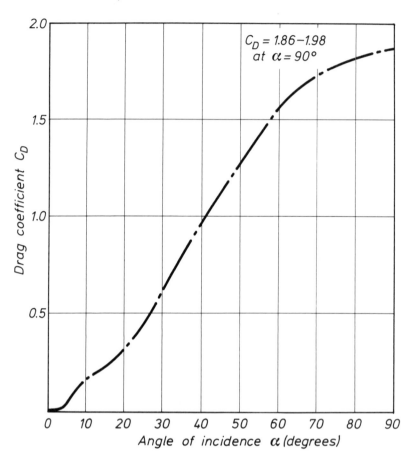

(2) Streamlining

If by some means, separation of flow is prevented, the wake or pressure drag could theoretically be reduced to friction drag only. One may rightly say therefore that flow separation can, in many circumstances, be regarded as an engineering problem of enormous practical consequence and every effort to delay or avoid separation can pay handsomely in terms of reduced drag, power required, or increased velocity. To reflect further on the drag generation mechanism let us consider an often met practical problem: how to fair most effectively a bluff cylindrical obstacle?

The fish and marine mammals owe their shape and their often spectacular, and sometimes almost incredible, speed performance to the inspiration of Nature (Ref 2.21). As far as shape is concerned, the lessons that may be learnt from fish can prove of value for any submerged bodies: foils or fins in particular. It is rather common to refer to the shape of a fish as a good streamline shape and it is not easy to

Photo 2.12   Flow patterns round flat plate at various incidence angles.

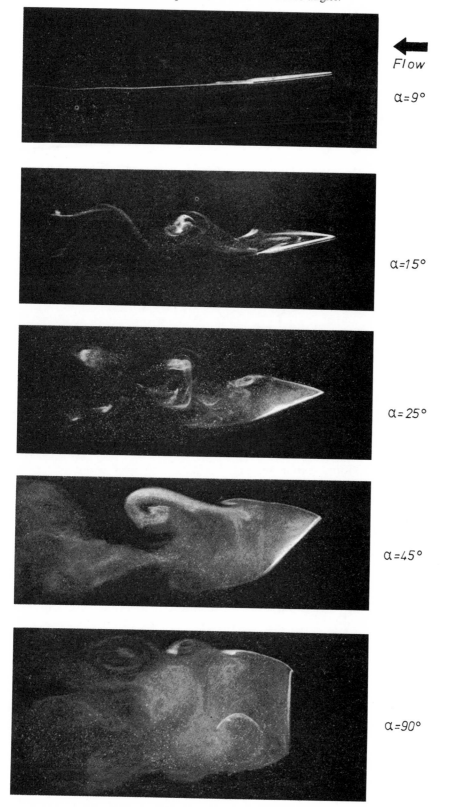

Flow

$\alpha = 9°$

$\alpha = 15°$

$\alpha = 25°$

$\alpha = 45°$

$\alpha = 90°$

Fig 2.27   Above–sketch of the cross-section of a trout, taken from Sir
George Cayley's Notebook (end of 18th century).
Below–for comparison, modern, low drag aerofoil sections
——— NACA section 63A016
- - - - LBN section 0016
oooo Trout

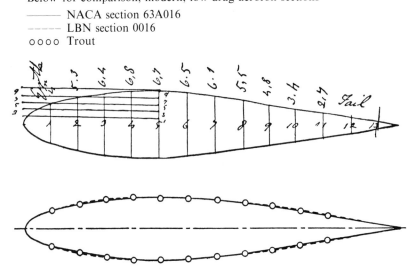

beat Nature in this respect. It explains perhaps why the descriptive expression 'cod's head and mackerel tail' is time-honoured by naval architects.

Figure 2.27 represents, for example, the cross-section of a trout, taken from Sir George Cayley's Note-Book (end of 18th century), being compared with modern low-drag aerofoil sections; the coincidence of shapes is striking. The common feature of these streamline bodies is that the afterbody curves to a finer taper than the forebody. Other characteristics are that the maximum thickness is well abaft the nose and that the widest portion of the tail tapers aft gently to a point. Such gradual tapering or streamlining is of advantage because of reduced wake drag by virtue of shifting the separation point close to the trailing edge, and Photo 2.13 displays it convincingly.

A word of warning seems to be appropriate here–the 'cod's head and mackerel tail' idea of streamlining, perfectly workable in the case of fully immersed bodies, fins, rudders and other appendages, should not be used as a guide while shaping hulls partly immersed in the two different fluids, water and air. Their interface imposes different requirements to cope effectively with wave making resistance.

The essential geometrical factors of streamline forms that determine drag are:

a. The thickness ratio t/c, i.e. the ratio of 'maximum' thickness t, to the length of the foil chord c.

b. The location of the point of maximum thickness t measured usually by the ratio of its distance from the nose to the length chord of the form (Partly exposed while discussing Fig 2.19).

c. The shape of the leading edge.

Photo 2.13   Effect of streamlining on the wake size.

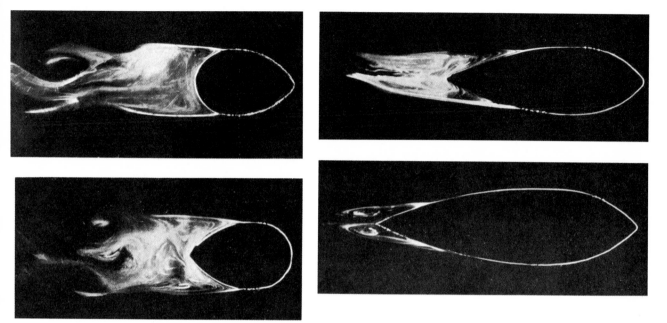

The dependence of drag on the thickness ratio t/c is best illustrated by comparative measurements on sections belonging to the same family, as shown in Fig 2.28 (Ref 2.7). The flow past each model of 7 ft span was two-dimensional and the experiments were performed at the Reynolds Number

$$\text{Re} = \frac{V_0 \times c}{v} = 4 \times 10^5.$$

One might expect that the minimum resistance of a circular rod or cylinder would be secured by fairing it by means of a long and slender form such as section 2 or 3 in Fig 2.28. However, the experiments clearly suggest that there is a certain optimum thickness ratio t/c which produces the lowest drag. More specifically, a fairing that will produce the lowest drag is the one for which the drag coefficient, obtained by dividing the drag per unit length by $(\rho V_0^2/2) \times t$, is a minimum. The profile drag coefficient for the sections depicted in Fig 2.28 is plotted against thickness chord ratio t/c. It consists of two parts, skin friction drag and wake drag. The section which has the lowest profile or total drag coefficient is one whose chord length c is about 4 times the thickness t. If a more slender section with longer chord is taken, the increase in skin-friction drag, due to greater wetted area, more than compensates for the slight drop in wake drag. Whereas, if a section with shorter chord is taken, the reduction in skin-friction drag is smaller than the rise in wake drag.

Expressing this in a different way, one may say that the greater wetted area over

Fig 2.28 Relationships between the wake (pressure) drag, friction and total drag of some streamline sections.

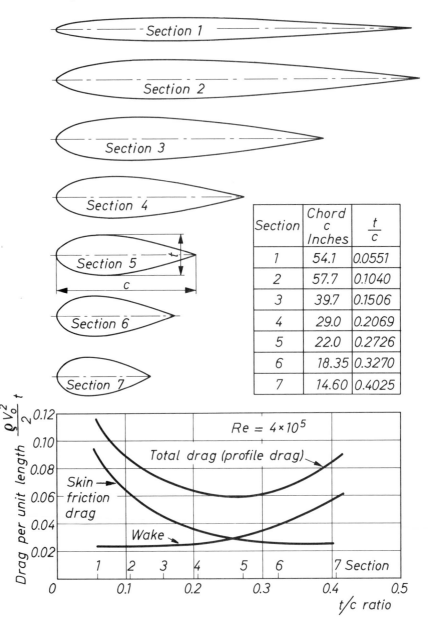

Section	Chord c Inches	$\frac{t}{c}$
1	54.1	0.0551
2	57.7	0.1040
3	39.7	0.1506
4	29.0	0.2069
5	22.0	0.2726
6	18.35	0.3270
7	14.60	0.4025

which the fluid flows accounts for the additional friction drag. The wake, or pressure drag, lessens because of the reduction of area of afterbody on which the pressure difference between fore- and afterbody acts. There must, therefore, be a point where the additional friction drag equals the reduction in pressure drag. So only up to this point will streamlining reduce the total drag.

Fig 2.29   Profile drag coefficients, C_D, for sections of various thickness
ratio t/c, as shown in Fig 2.28.

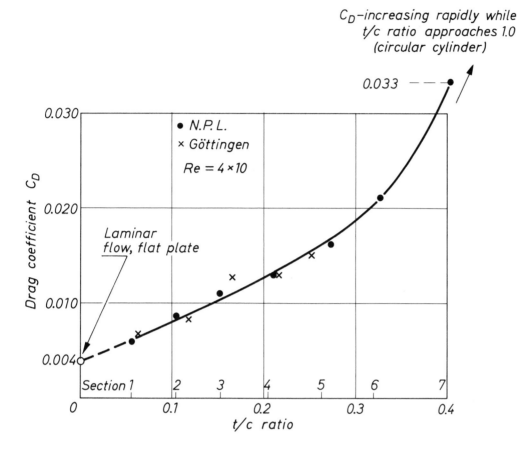

The curve, showing the change in drag coefficient C_D per unit area versus thickness chord ratio t/c, for the same family of sections as depicted in Fig 2.28, is given in Fig 2.29. The drag coefficient C_D increases with thickness ratio, at first slowly and then more rapidly. The curve must become much steeper at values of thickness ratio greater than those recorded. There is good reason to assume that when t/c ratio approaches 1.0 the drag coefficient C_D should differ only slightly from its value for a circular cylinder, which, for the same Reynolds Number, is 0.32 (Fig 2.35). So the rise in C_D over the range $0.4 < t/c < 1.0$ must be from 0.033 to about ten times this value; an enormous increase.

Figure 2.30 elucidates dramatically the advantage of streamlining. At the same wind velocity the drag on a round wire is almost ten times as big as the drag on a streamline section. The drag of a poor form, such as a wire, or circular rod, might be reduced by enclosing it within a fairing, as shown.

Alloy or plastic fairings, such as shown in Photo 2.14 and Fig 2.24 which can

Fig 2.30   The NACA section shown below and a wire which is ten times
thinner, i.e. its diameter $= \frac{1}{10}t$ have the same drag.
By fairing the circular rod or wire as indicated in the sketch
(*right*) it is possible to cut the drag by about $\frac{3}{4}$.

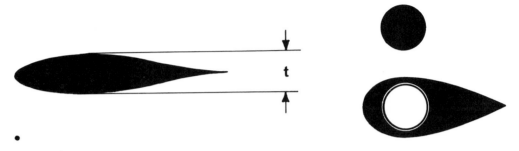

Photo 2.14   Some of the fairings (head-foils) tested. The centimetre scale
is shown on the upper edge of the sliderule. See also Photo
2.24.

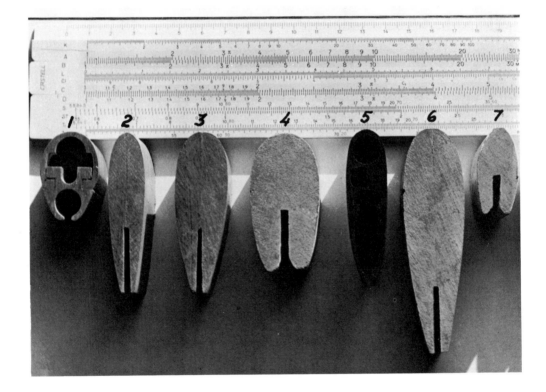

swivel round the headstay while supporting the jib or genoa are intended, amongst other functions, to reduce drag and provide a clean, aerodynamically effective leading edge. No doubt, they may fulfil this function if they are properly designed. Wind tunnel tests do not, however, confirm without reservations some of the claims that these fairings drastically improve aerodynamic efficiency of the headsail. The effect of a leading edge fairing on what might be called the 'entrance efficiency' of a thin aerofoil is discussed in chapter C.2 and also in Part 3, chapter D.3.

An important contribution to the air resistance of any yacht is made by the parasitic drag, due to the rigging and other items. In the complete structure of a yacht, there are various parts such as the mast, boom, spreaders, standing and running rigging, exposed members of the crew, etc., which take no part in generating driving force when sailing to windward. Since they produce drag only, we may distinguish this kind of resistance as 'parasitic drag'. Hull appendages, which do not contribute towards hydrodynamic lift, may also be grouped under the general heading of 'parasitic'.

For example, a 12-Metre yacht may have many hundred feet of rigging wire and, in the light of results just discussed, it is not difficult to anticipate that modifications to the rigging wire shape may bring a substantial total drag saving.

Interesting data in this respect are presented in Technical Memorandum No 140 Davidson Laboratory (Ref 2.22). The purpose of investigation was to compare the drag of five possible shapes of rigging wire for use on a 12-Metre yacht. Figure 2.31 presents results in the form of the drag per unit length of wire, as a function of the wind velocity in knots over a range of velocities from 15 to 45 knots, commonly

Fig 2.31   Rigging-wire drag per unit length against wind speed.

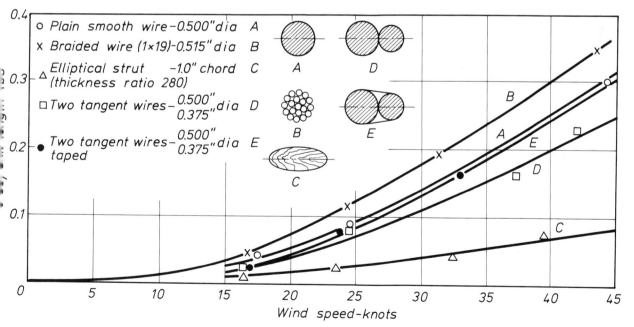

Fig 2.32 Dependence of rigging-wire drag on course sailed.

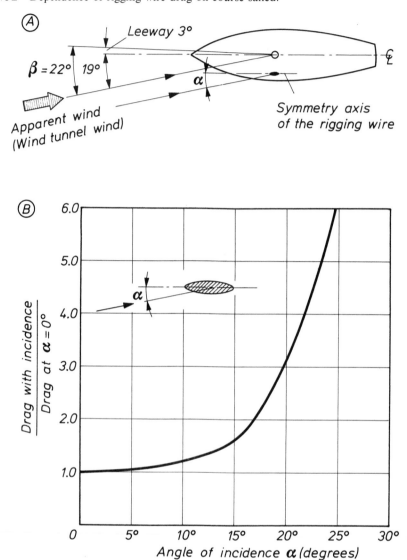

encountered when a yacht is sailing close-hauled ($\beta = 22°$).

The geometry of sailing attitude, simulated in the wind tunnel, is given in Fig 2.32A. All the rigging wires were mounted so as to span the test section of the tunnel completely. Two-dimensional flow around the wires was thus produced, so that data based on a unit length of wire would be obtainable without end effects. The rigging wires were of actual size, hence these tests were full-scale tests and not model tests. It means that scale-effect was not involved.

It can be seen that the form of rigging wire with the least resistance under the conditions tested is the eliptical strut C, which had a drag per unit length of only about one-fifth that of the braided wire over the velocity range tested. From the aerodynamic point of view strut C seems to be the most suitable form for rigging wire. Apart from lower drag, this section generates some lift which contributes towards driving force when a yacht is sailing close-hauled.

Circular wires, arranged in such a way that a smaller one is immediately behind a larger one, result in drag reduction below that of the larger wire alone. This can be attributed to the streamlining effect, produced by such a configuration. However, taping the wires results in a slight drag increase over this case, most likely due to the increased surface and roughness caused by the overlapping tape layers.

Drag values presented in Fig 2.31 naturally hold only for the case where the larger axis of the sections is at an angle of incidence $\alpha = 19°$, relative to the direction of wind flow, as shown in Fig 2.32A. In the case of oblique flow drag depends very much on the incidence angle; when this increases beyond a certain value drag may increase substantially. This is shown in Fig 2.32B.

The change in drag values up to an angle $\alpha = 5°$ is not significant, and then for greater values of α, drag increases rapidly (Ref 2.23). One must expect therefore that when the yacht bears away from a close-hauled attitude (β of order 25°), the advantage of using streamline rigging becomes less and less pronounced. The same applies to the shape of the mast, and there appears to be little to choose between a streamline (pear-shaped) section and a circular one, unless the mast can be rotated.

The general remarks made about the advantages of streamlining apply also to spreaders. Further savings in parasite drag are obviously possible by 'cleaning' the mast–using internal halyards where possible, by hoisting halyards up the mast on gantlines when not in use, by the use of internal tangs, etc. Some classes have changed from wood to light alloy masts without changing weight or dimensions. Hence, it may be possible to develop alternative staying arrangements which involve less rigging than is required with wood spars. It has been argued for instance by B Chance (Ref 2.24) that a large percentage of rigging is for the sole purpose of controlling forestay sag, and this can often be overdone.

It has been estimated that the parasite drag of the rig of a modern day-sailing keel-boat may contribute from 10 to 12 per cent towards the total drag. By carefully cleaning the rig, as demonstrated in Photo 2.15, the parasitic drag may be reduced substantially and the potential gain is well worth the effort, particularly when high performance is of primary importance. However, the desire for windage reduction must be tempered by consideration of other associated factors.

The spar-maker's art is largely based on his skill in finding the right balance between strength, stiffness (flexibility), weight and windage of the spar. Those factors are conflicting; a happy solution depends on how much emphasis is placed on a particular factor, for example, the weight aloft and strength, at the expense of windage, or vice versa. Rigging failures in yachts are nowadays more common than they used to be in the past. This may largely be attributed to the overwhelming desire for high performance that makes people willing to take the risk of accepting less generous safety factors and by driving their boats closer to their safety threshold.

Photo 2.15   In many of the top boats, attempts are made to have masts as free as possible of unnecessary windage. The lowest picture displays spreaders carefully faired into the FD mast. A good material to use here is a combination of micro-balloons and epoxy resin. Such a mixture has low density thereby allowing some weight to be saved aloft.

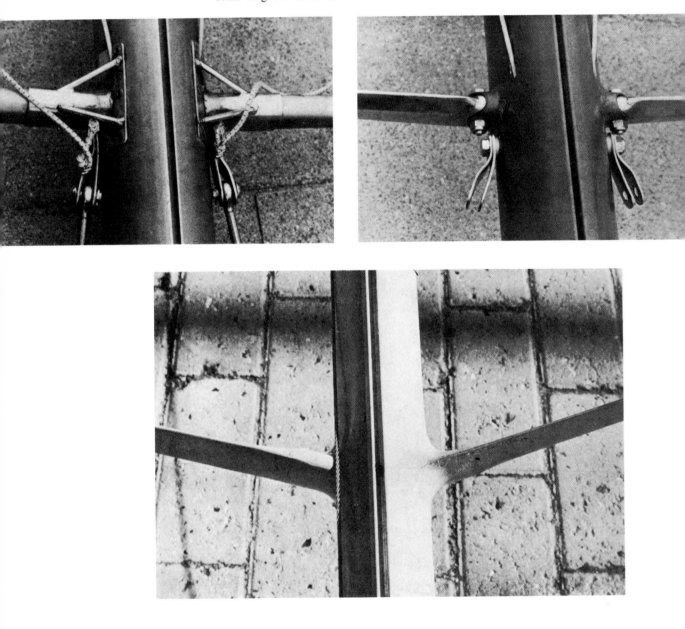

C Reynolds Number and scale effect

It has become customary to write the drag formula as follows:

$$D = C_D \rho \frac{V_0^2}{2} S$$

$$D = C_D q S \qquad\qquad \text{Eq 2.18}$$

The above equation (similar to Eq 2.13A for lift) reflects the so-called velocity squared law which is the very base of all aero-hydrodynamic computation. Its meaning is that the drag D, apart from being proportional to the velocity squared ($q = \rho[V_0^2/2]$) and the area of the obstacle S, is also proportional to the shape factor or drag coefficient C_D. This coefficient which can be established experimentally is equal to the quotient

$$C_D = \frac{\text{Drag force}}{\text{Dynamic pressure} \times \text{Area}} = \frac{D}{q \times S}$$

and can truly be defined only after an agreement has been reached about the choice of the reference area in the denominator; usually it is the plan or projected area.

The validity of the square law, as given by Eqs 2.13A and 2.18, holds only if the values of respective coefficients C_D or C_L are constant whatever the scale, the velocity or the fluid may be. One might suppose that the flow past a circular cylinder, for instance, would always follow the same pattern, hence the drag coefficient C_D would be fairly constant, but this is not the case. As a matter of fact, tests conducted under different conditions (different scale or velocities) do not always give the same coefficient. The velocity squared law is therefore not strictly true and can only be applied, like any physical law, within stipulated conditions. It might even be said that Eqs 2.18 and 2.13A are not 'laws' in a physical sense but empirical formulae or definitions of drag and lift. In order to determine the drag and lift coefficients for an arbitrary foil configuration, sails or yacht hulls in particular, one has to resort to experiments.

Immediate questions arise. Can the coefficients be expected to change with any simultaneous change of velocity or scale? Is there any law governing the change of coefficients? These are questions of the utmost practical significance since one may jump to the wrong conclusion while translating the model test results into full scale.

(1) What one can learn from golf ball behaviour

To illustrate the problem, let us consider the flow over a sphere such as a golf ball, which is probably the simplest (but very instructive) example of interaction between the character of boundary layer flow, the separation pattern, the wake size and finally the resulting drag (Ref 2.25).

Wind tunnel tests on the golf ball are summarized graphically in Fig 2.33, curve

Fig 2.33   In the range of velocities from V_1 to V_2 the roughened (dimpled) ball has less drag than the smooth one. Drag is related to the wake diameter indicated by arrows d.

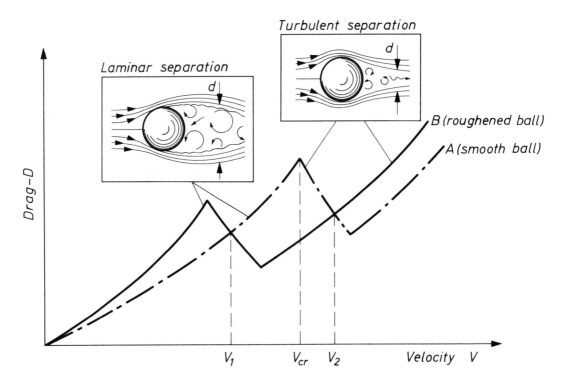

A, which shows the variation of measured drag D against wind velocity V. As the wind velocity increases, the measured drag also increases in proportion to speed squared law, as predicted by the general equation 2.18 but to a certain *critical velocity* V_{cr} only. Further increase of wind velocity causes drag to decrease and the measured drag is represented by the roughly descending section of curve A. By continually increasing the wind velocity we finally reach a condition at which drag, once again, goes up smoothly with velocity, but along a curve different from that for the first ascending section.

The peculiar behaviour of the measured drag suggests that two different flow patterns must exist, which dominate below and above the region of critical velocity V_{cr} and probably some intermediate pattern in the region of transition. One may ask, what are these two patterns and why should drag really be less at a higher velocity than at a lower one?

It has been observed in the course of experiments that, indeed, the physical reason for such a sudden decrease in drag is the existence of two fundamentally different patterns of flow, associated with distinct laminar and turbulent flows inside the boundary; these are shown above the curves in Fig 2.33.

It has already been mentioned that, in the presence of an unfavourable pressure gradient, the flow will not be able to remain attached to the body surface up to its rear end but, sooner or later, will separate. This is represented in Fig 2.34, which depicts the progressive development of the boundary layer leading to separation. If the flow velocity is sufficiently low, the laminar boundary layer can separate from the surface of the sphere (ball) before transition to a turbulent BL takes place. In such a case, the laminar separation point will be well upstream, on the surface, as shown in Fig 2.33. Since the flow behind separation creates a broad wake filled by vortices, the pressure drag will be correspondingly high.

Fig 2.34   Progressive development of the boundary layer in the presence
of an unfavourable pressure gradient strong enough to provoke
separation (exaggerated picture). Point of separation may travel
upstream or downstream depending on the character of flow
within the boundary layer.

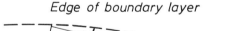

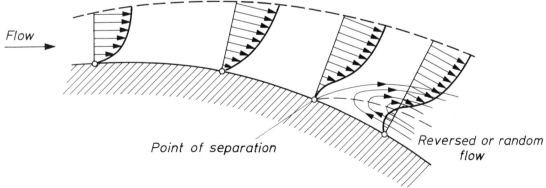

For a particular size of the sphere, there is a particular flow velocity referred to as the critical velocity V_{cr}, for which the point of transition and point of laminar separation are coincident. Any increase in flow velocity beyond this value of V_{cr} will result in the transition of the laminar layer to a turbulent one before separating.

The turbulent boundary layer can move up a steeper pressure gradient, without break-away occurring, than the laminar one. If in many practical cases the flow remains attached, and can support considerable adverse pressure gradients without separation, this is because the boundary layer flow is mostly turbulent. As such it has more capacity for mixing with and absorbing energy from the main-stream fluid than the laminar layer. It is therefore much more robust than the laminar layer, and hence much less likely to separate from the surface under the influence of adverse pressure gradient. It sticks to the surface better. However, although the turbulent boundary layer can support an adverse pressure gradient which is larger than that for a laminar boundary layer by a factor of 2.5, it is not a separation preventive.

When the laminar boundary layer becomes turbulent before separating from the surface, the separation point will shift farther back on the sphere and there will be an attendant decrease in the wake, as shown also in Fig 2.33. In general, when wakes are large, the pressure or wake drag is large too, and when wake is reduced for any reason, the pressure drag is also reduced. The wake drag is thus critically dependent upon the existence and position of separation, which in turn depends upon the shape of the object and character of the boundary layer flow as well. From what has been said one may conclude that even though the turbulent boundary layer has much higher skin friction than the laminar layer, the resulting total drag is lower.

Since the transition from laminar to turbulent flow can be beneficial in reducing wake drag, which is the dominating drag component of round bodies, one may foresee with a little reflection the effect of an artificial roughening of body surface. The minute roughness elements tend to act like turbulence generators by casting off eddies, which disturb the laminar boundary layer and induce earlier transition to turbulent flow. In consequence, the boundary layer may go turbulent before laminar separation occurs and subsequently the turbulent separation may take place at lower velocity. This is demonstrated in Fig 2.33 by the curve B which represents drag variation of a dimpled golf ball. It can be seen within certain velocity limits, from V_1 to V_2, the roughened ball has considerably less drag than the smooth one; at some speeds only one-fourth the drag of a smooth ball. It has been observed that with a swing which drives a dimpled ball about 230 yards in flight on the golf course, a smooth ball is driven only about 50 yards in flight. For this reason today's golf balls are dimpled; they were smooth in the early days of golf.

The transition to turbulent separation may also be artificially hastened by fixing a wire near or before the point of laminar separation, as demonstrated in Photo 2.16 (lower picture). One can notice a wire attached to the sphere upstream of the great circle, facing the oncoming stream. It is also possible to stimulate turbulence by means of sand-paper, studs, or other similar devices, attached to the surface, which may cast off small-scale vortices.

Photo 2.16   Picture of the flow past spheres.

A–with laminar separation (below the critical velocity).
B–with turbulent separation (above the critical velocity).

Shift in separation point can artificially be caused by induced turbulence due to the presence of the wire (case B). As a rule–the greater the turbulence of the on-coming fluid stream, the smaller the Reynolds Number for shift in separation point.

Flow

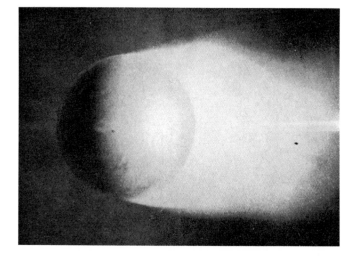

A

Flow

B

(2) Similarity law of Osborne Reynolds

Experiments carried out by Osborne Reynolds (end of 19th century) and by followers of his concept proved that the speed squared law expressed by Eq 2.18, is really based on the similarity of flow patterns. This means that *the drag coefficients C_D are expected to be equal when the flow patterns developed round objects of similar shape but of different scale are identical.*

This condition of flow similarity discovered by Reynolds might be stated as follows: if the flow velocity V multiplied by some linear dimension L of the body (the diameter or length) came to the same value, then the flow patterns as well as drag coefficients would be the same. This principle, which may be referred to as the $V \times L$ *similarity law*, is valid only if the fluid remains the same.

Thus, if one wishes to experiment on say, a one-tenth scale model in order to predict what forces will be developed on a full-size wing sail at 40 ft/sec wind, the test on the model should be performed at 400 ft/sec wind in order to keep the $V \times L$ product the same in both cases. Such a demand for high testing wind velocities sounds alarming. Happily, however, the rapid change in flow pattern in the case of sails occurs at a relatively low value of critical velocity above which coefficients remain practically unchanged. Obviously, for the sake of better reliability in simulating full size flow conditions it is desirable to have models as large as practically possible. Big models reduce the demand for large testing velocities, and certainly it is easier to manufacture them accurately.

It is rather evident that if the results of tests performed on foils in air (wind tunnel) are to be applied in water–while designing hull appendages for instance–then the similarity condition VL must somehow be amended by taking into consideration the differences in fluid density and viscosity.

To cut a long story short, O Reynolds found that if the quantity

$$\frac{V \times L}{v}$$ Eq 2.19

is kept constant, the flow pattern will be similar, and there will be no error due to scale effect. This quantity is known now as the Reynolds Number (Re).

In this way, the problem of how the coefficients change with the different parameters, such as velocity, size, type of fluid, etc., has been reduced to the problem of how they change with the only parameter–the Reynolds Number. And the answer is almost entirely left to experimental investigation.

The Reynolds Number, as given by equation 2.19, is non-dimensional and has the same numerical value whatever consistent system of units is used for V, L, and v.

In the old British system for instance:

V = velocity of flow *in ft/sec*
L = length of the body *in ft*/measured usually along the flow direction
v = is a term combining both the viscous and density properties of a fluid in

motion and is called the coefficient of kinematic viscosity–*in ft²/sec* (Tables 2.1 and 2.2)

Since at a normal temperature of 15°C (69°F) the kinematic viscosity of water is of the order of $v_w = 0.0000123$ or $1.23/10^5$ (ft²/sec), the Reynolds Number may be written:

$$\mathrm{Re} = \frac{V \times L}{1.23/10^5} = 81{,}300 \times V \times L \qquad \text{Eq 2.19A}$$

Similarly, for *air* at sea level, assuming $v_A = 1.57/10^4$ (ft²/sec)

$$\mathrm{Re} = \frac{V \times L}{1.57/10^4} = 6370 \times V \times L \qquad \text{Eq 2.19B}$$

Tables 2.5 and 2.6 give the Reynolds Numbers for various velocities of water and air flow calculated on the assumption that the characteristic length $L = 1$ ft. The corresponding values of speed in four different units, ft/sec, m/sec, knots and Beaufort scale are given in Table 2.4.

TABLE 2.5
Reynolds Number (Re) for various velocities of **water flow**

Velocity V (ft/sec)	V (knots)	Re
1	0.59	$0.81 \times 1 \times 10^5$
2	1.18	1.63
3	1.77	2.44
4	2.37	3.25
5	2.96	4.06
6	3.55	4.88
7	4.15	5.70
8	4.74	6.51
9	5.33	7.32
10	5.92	8.13
11	6.51	8.95
12	7.10	9.76
13	7.70	$1.06 \times 1 \times 10^6$
14	8.28	1.14
15	8.88	1.22
16	9.47	1.30
17	10.05	1.38
18	10.65	1.46
19	11.25	1.55
20	11.85	1.63

TABLE 2.6
Reynolds Numbers Re for various velocities of **air flow**

Velocity V (ft/sec)	V (knots)	Re
5	2.96	$3.19 \times 1 \times 10^4$
10	5.92	6.37
15	8.88	9.55
20	11.85	$1.27 \times 1 \times 10^5$
25	14.80	1.59
30	17.75	1.91
35	20.70	2.23
40	23.65	2.55
45	26.60	2.87
50	29.60	3.19
55	32.50	3.50
60	35.50	3.82

To summarize, if the Reynolds Number is the same, the flows are geometrically similar. With geometrically similar flows round the two bodies of different sizes, the corresponding streamlines will be geometrically similar, the relevant pressure distribution will also be similar, and hence the magnitudes of the forces will always have the same ratio to each other. Different combinations of model scale, velocity and fluid density give the same coefficients if the Reynolds Number is the same, i.e. smaller scale can be compensated by a larger velocity. If Reynolds Numbers are different, the coefficients may not be expected to be equal but they can be determined separately.

Models of sails, or other foils, may be tested in wind tunnels, or towing tanks, and the results of these tests may be used in the computation of the full-scale performance, provided that the relevant Reynolds Numbers are the same. Large inaccuracies might exist in drawing conclusions for full-size systems from model tests, unless due corrections or allowances were made to the Reynolds Numbers effects (sections C1 and 2).

The variation of the force coefficients with Reynolds Number depends primarily on the shape of the object; for instance, drag coefficient for rounded objects, such as circular or eliptical cylinders and even streamline foils are sensitive to Reynolds Number while drag coefficient for sharp-edged bodies, such as a flat plate set at an angle 90° to the flow direction, is not sensitive to scale effect at all.

Scale effects on drag, as well as on the lift, are of two principal kinds–the one due to variation of frictional coefficient, the other to change in general flow pattern, and specifically in the point of separation, on which the pressure or wake drag depends. Both are connected in their major and most spectacular manifestations with the character of the flow within the boundary layer. *The relative influences of those two components of drag are of opposite sign; the appearance of turbulent flow tending to increase friction may at the same time decrease wake drag component through a delay of separation.*

This mechanism of separation delaying is interesting from a practical point of view, since it may be employed in order to cut drag drastically.

(3) Ways of reducing drag

The separation and transition phenomena on a circular cylinder are similar to those observed for a sphere (ball); this is shown in Fig 2.35. The combination of the two

Fig 2.35   The total drag of a circular cylinder (two-dimensional flow) as a function of Reynolds Number.
It can be noticed that in a narrow range of Reynolds Numbers close to Re = 2×10^5 = 200,000 the drag coefficient can drop three-fold from about 1.2 to about 0.4.

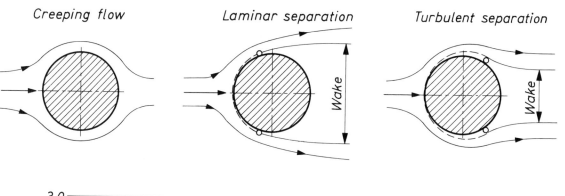

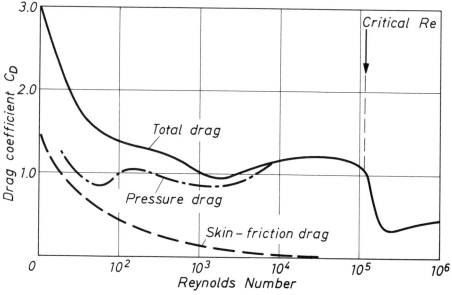

components of total drag, namely friction and pressure (wake) drag, produces a highly irregular drag coefficient curve against Reynolds Number.

A few explanatory words are perhaps needed in order to relate Figs 2.35 and 2.33. While discussing the results of tests on a ball, it has been stated that the drag of the ball depends only on the velocity. This is certainly true if the size of the ball (sphere) affected by the same fluid is kept unchanged. If we are going to compare the drag coefficients of cylinders or spheres of different diameters, and possibly moving in different fluids, it can only be done on the basis of Reynolds Numbers.

Referring back to Fig 2.35, it can be noticed that the relative contribution of the two drag components to the total drag changes radically with Reynolds Number. At very low Re (very low velocity), when so-called 'creeping flow' takes place, the drag coefficient results largely from viscous effects, i.e. friction. At high Re (high velocity) the inertia force prevails and the viscous force becomes progressively less and less important as the Reynolds Number increases. Finally, the total drag becomes primarily one of pressure or wake drag.

To recall T Karman's opinion (Ref 2.26), the Reynolds Number works in some cases almost like black magic. In a relatively narrow band of Reynolds Numbers, the drag of a cylinder, for instance, may suddenly decrease three-fold and the critical velocity at which the drag drops rapidly can, to a certain extent, be controlled by means of artificial roughness applied in the right place on the surface of the cylinder.

Considering the mast as a long cylinder, let us investigate the possibility of reducing parasite drag by changing deliberately the character of separation by means of turbulence-stimulating wires of very small diameter, placed along the mast at a position given by an angle $\pm 65°$, measured from the cylinder axis that is parallel to the flow direction. This is illustrated in Fig 2.36, which gives the drag coefficient curve for a bare cylinder of about 6 inches in diameter plotted against wind velocity V. There are also plotted the two other curves for the same cylinder, but with fine turbulence-generating wires of different diameter $d = 0.02$ and 0.005 in respectively. It is seen that the shape of the drag curve undergoes certain changes as the diameter of the wire is increased. Thus, the critical range of Reynolds Numbers, at which the drag coefficient suddenly drops from about 1.1 to 0.55, can be controlled by forced turbulence within the boundary layer, by means of small excrescences on the surface of the cylinder situated just before the suction peak occurs (compare the sketch in Fig 2.36 with Fig 2.5C).

Translating the result into more practical sailors' language–one should expect that in some wind conditions the drag of the mast-sail combination can be reduced by an artificial forcing of the turbulence inside the BL by applying turbulence stimulators.

An external halyard led down the side of the mast and positioned in the right place might fulfil the function of a turbulence stimulator. It has been reported that the 12-Metre *Courageous*, the America's Cup defender in 1974, used during races a mast which was: '...a subtle but significant variation from the conventional eliptical shape with a blunted forward face and two plastic strips of tiny triangles running up

Fig 2.36   The effect of turbulence stimulators on the drag of a circular
cylinder (mast).

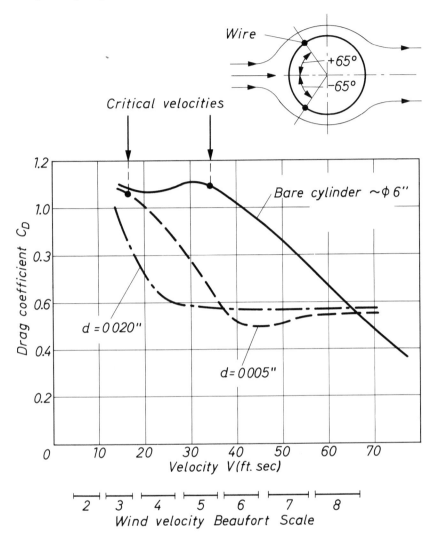

the front of the mast'. The *Courageous* mast was the result of an extensive series of
theoretical studies and sailing tests conducted by A Gentry, a research aerodyna-
micist. The mast section was designed to have its maximum effectiveness with the
plastic transition strips precisely positioned. The triangles moulded into the strips
generate tiny swirls of air that change the character of the flow close to the mast so as
to delay the lee-side flow separation. The optimum position of the strips was found
to be different on different mast shapes.

The drag coefficient curves for various sections as a function of Reynolds Number
are given in Fig 2.37. Limiting our attention to the circular cylinder it can be seen,

Fig 2.37 Drag coefficient, C_D, for various forms. When the span of a cylinder is finite a departure of the flow from the two-dimensional pattern occurs towards the ends. The effect of this change of flow on drag coefficient is shown by the two separate curves.

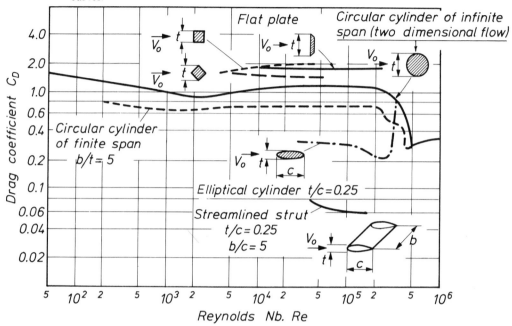

Photo 2.17 Surfaces of flow discontinuity behind the flat plates.

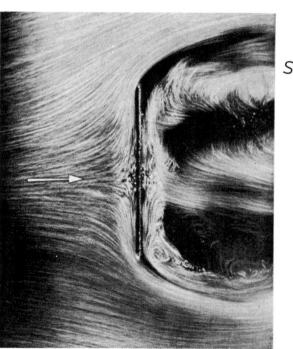
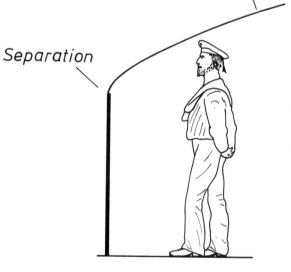

Fig 2.38   Action of vortex generators and their effect on lift.

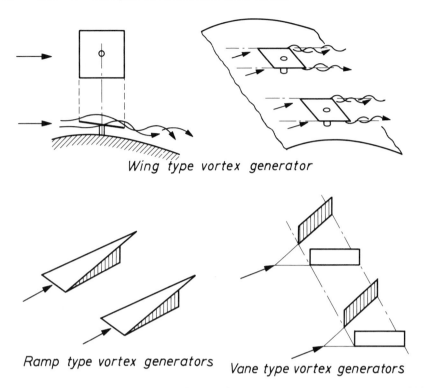

Wing type vortex generator

Ramp type vortex generators   Vane type vortex generators

A. Various basic configurations of vortex generators are possible

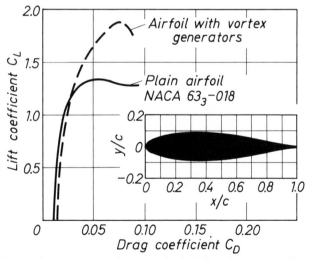

B. Aerodynamic characteristics of airfoil with and without vortex
generators

for instance, that when the cylinder is short ($b/t = 5$) and therefore a departure of the flow from the two-dimensional pattern occurs towards the ends, the drag coefficient is decreased substantially. For obstacles with very sharp edges, such as square cylinders or the flat plate shown in Fig 2.37, in which the point of separation is fixed regardless of the Reynolds Number, i.e. the flow pattern is always the same, the drag coefficient is nearly constant and independent of Re; Photo 2.17 illustrates this point. If fluid flows over a sharp edge it separates always at the edge. L Prandtl (Ref 2.26) explained this behaviour by adducing the special principle that the fluid seeks to avoid very large velocities while negotiating sharp edges, and forms surfaces of discontinuity instead.

Useful application of this principle is found in the screen employed on ships to protect the navigator or the watch from the rush of air without obstructing his field of vision; this is displayed in the sketch attached to Photo 2.17. The airstream is carried clear over the sailor's head and his eyes are protected by the surface of discontinuity. Windscreens of sports motor-cars work on the same principle.

Yet another way of generating small-scale vortices for improving the mixing between the 'tired' boundary layer and the 'strong' free stream outside of BL to make the BL less susceptible to early separation, is by the use of so-called vortex generators. In their simplest form, these consist of small vanes placed close to the surface, so that they behave like half wings shedding trailing vortices from their tips (Fig 2.38A). The fluid particles with high momentum (product of mass and velocity) are swept along helical paths towards the surface, mixing with and re-energizing the retarded particles at the surface. With the aid of this energy injection the boundary layer is able to keep moving without separation, so long as it is not confronted by too steep an adverse pressure gradient (Refs 2.27 and 2.28).

Unfortunately, vortex generators necessarily incur a drag penalty under flow condition when separation is not imminent, and their effectiveness depends critically on their position relative to the point where flow separation is liable to occur. Their effectiveness, advantages and cost in terms of additional drag, can best be demonstrated by the two curves obtained in the course of experiments on an NACA 63_3-018 aerofoil shown in Fig 2.38B. The vortex generators were attached along a spanwise line at 0.1 chord length from the leading edge. Their effect on lift is quite amazing–they raise the stalling incidence from 14° to 20°, with an increase in maximum lift from about 1.3 to 1.9.

It is known, for example, that the mast of one of the America's Cup challengers–*Gretel*–was equipped with thousands of tiny vortex generators.

Figure 2.39 presents, what H C Herreshoff, the author of the concept called jokingly 'the sailboat of the future' (Ref 2.29). The sketch shows several devices that provide boundary layer control and prevent separation over the leeward side of the sails; vortex generators are amongst them. There is also a small, high aspect ratio staysail, just in front of the mast, which acts very much like a leading edge slot in that it postpones the separation and stall.

Examining critically the location of vortex generators in close proximity to the

Fig 2.39   Sailing rig of the future?

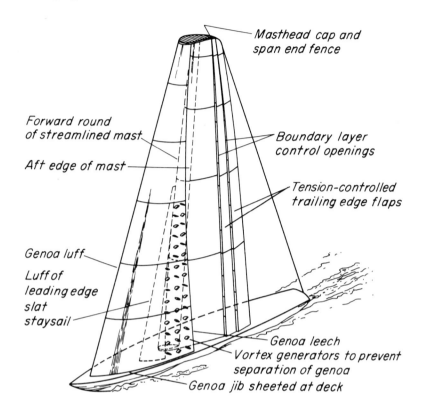

trailing edge of the genoa in Fig. 2.39, one may argue that they are distributed in the wrong place. The success or failure of this device in inhibiting the development of separation depends, as has already been mentioned, on their position relative to the region where flow separation is liable to occur. Certainly, they are useless if located downstream far away from the region of maximum sail camber, where separation has already developed.

It should be stressed that Herreshoff does not want us to consider his project too seriously as the rig of the future. This has been included here, however, as a constructive suggestion for those who are trying to refine their sailboats by applying existing experimental knowledge, and who enjoy bold approaches that might, in some cases, produce an optimum rig for a specific purpose.

(4) Variation of foil drag with Reynolds Number

Numerous tests have been carried out with the main objective of ascertaining the scale effect on drag and lift generated by foils. Let us first consider the Reynolds Number effect on drag. On a flat plate the skin friction coefficient C_f and the minimum profile drag coefficient C_{D0}, i.e. drag coefficient at 0 angle of incidence, are identical. On a thin symmetrical foil (or hydrofoil) there is still substantial equality of these. As the foil thickness increases, or as it loses its symmetry, acquiring a mean camber, the identity vanishes and the minimum drag coefficient C_{D0} increases well above that of skin friction C_f alone.

The values of minimum profile drag coefficients C_{D0} against Reynolds Number for some symmetrical NACA sections, are given in Fig 2.40 by means of broken curves. They are compared with those for a flat plate, represented by full line curves redrawn from Fig 2.24.

An interesting feature of the profile drag curve for the thinnest NACA section 0009, at Reynolds Numbers above the critical value (which is about 7.0×10^5), is that it follows closely the general trend of the transition curve for a flat plate. From what has been said earlier it is not surprising that the curves for thicker foil sections 0012, 0015, and 0018, show a clear tendency towards higher drag, C_{D0}, than that of the thin section 0009. The flow round thick sections is increasingly affected by the adverse pressure gradient developing behind the maximum thickness of the foil and which is almost non-existent in the case of a flat plate. As the curvature of the foil section increases with increasing thickness ratio t/c, an adverse pressure gradient of

Fig 2.40 Minimum profile drag, C_{D0} (friction drag + pressure drag) of symmetrical NACA aerofoils compared with flat plate drag characteristics (see Fig 2.24). Angle of incidence $\alpha = 0$.

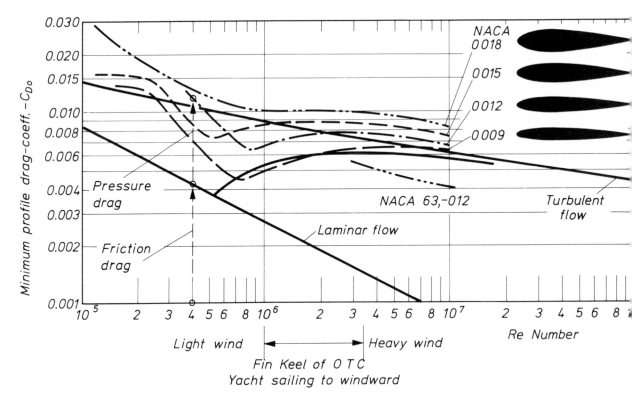

such a magnitude is produced that laminar flow can no longer be maintained, and transition to turbulent flow or even separation takes place. The transition is demonstrated in Fig 2.41–it shows the rate of growth of boundary layer thickness δ along the chord, c, established by experiments on a model of symmetrical section, depicted below the graph and set at zero incidence angle. The boundary layer thickens rapidly in the region about 0.3 c from the leading edge, where the maximum thickness of the section is observed. Rapid thickening of the boundary layer, a common feature of the flow in the presence of an unfavourable pressure gradient, is usually associated with transition from laminar to turbulent flow. The thickness of the boundary layer near the tail is in the order of 1.5–2 per cent of the chord.

Figure 2.42 illustrates schematically the three basic types of boundary layer flow. Type a has already been discussed: the laminar BL over the front portion of the section transits into the turbulent BL, and the flow is fully attached up to the rear end of the foil; the wake is relatively small.

Type b demonstrates so-called laminar separation of the BL, which may or may not develop into the turbulent BL. Such a laminar leading edge separation is associated with an abnormal increase of the profile drag coefficient as wake drag becomes very large.

Fig 2.41   Growth of the boundary layer thickness δ.

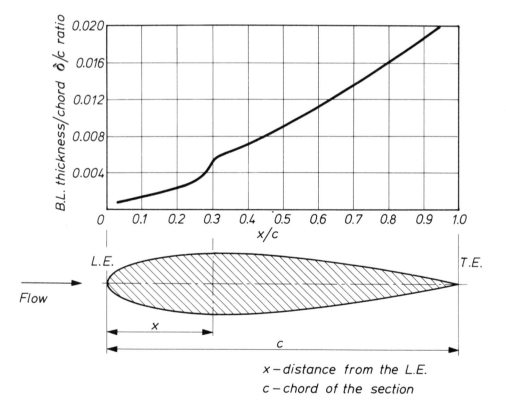

x – distance from the L.E.
c – chord of the section

Fig 2.42   Schematic representation of BL flow.
  a. Laminar flow over the front part of the section and turbulent
     over the remaining part. Flow fully attached.
  b. Laminar, leading edge separation, without reattachment.
  c. Rear or trailing edge separation.

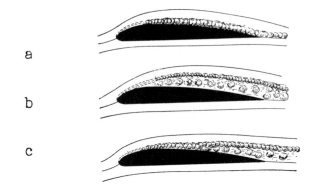

Laminar separation may at first be a local phenomenon, the flow subsequently changing to turbulent BL and re-attaching to the surface of the foil, by virtue of turbulent BL being the 'stronger' of the two and more able to survive an unfavourable pressure gradient. Accordingly, it may be expected that at 0° angle of incidence the separation point for the turbulent BL will be found closer to the trailing edge, the actual position of the separation point depending on the thickness of the section in question. This type of re-attached flow is marked c in Fig 2.42. If the Reynolds Number is sufficiently low, the separated laminar flow will not re-attach to the surface.

Bearing in mind those different types of flow, let us analyse Fig 2.40 further. The other striking feature of the four broken drag curves of sections 0009–0018, is their sudden departure from regularity at Reynolds Numbers below certain critical values that lie in the range between 5.0×10^5 and 8.0×10^5, depending on the foil thickness (Ref 2.30). For instance, from an examination of the drag curve of NACA 0012 aerofoil, it becomes evident that the rather substantial increase in drag observed below the critical $Re = 8.0 \times 10^5$ is most probably associated with the laminar separation type b in Fig 2.42. The resulting drag, three times greater than the friction drag component observed at the $Re = 4.0 \times 10^5$ on the flat plate, is certainly due to the occurrence of large wake drag, always associated with laminar separation. Such a separation, similar in principle to that observed on the circular cylinder and sphere shown in Figs 2.33 and 2.35, must occur as the Reynolds Number is reduced, even in the case of the excellently streamline section NACA 0012 set at zero angle of incidence. Considering the profile drag of this section at say $Re = 4.0 \times 10^5$, it can be found that it consists of friction drag about 0.004 and pressure or wake drag about 0.008, totalling 0.012. Those drag components are marked in Fig 2.40 by means of vertical arrows.

Considering the fin keel action of a Half Ton Cup-type yacht sailing to windward at speeds ranging from 2 to 5 knots, and assuming that the mean chord of the fin keel is about 5 ft, we may use Table 2.5 to show that the relevant Reynolds Numbers will be in the range 10^6–3.6×10^6. Those numbers are marked in Fig 2.40 below the horizontal Re Number scale. It can be deduced from this that laminar separation is not likely to occur in the specified case of the 5 ft wide fin keel. However, there are reasons to believe that laminar separation may take place, particularly in the case of narrow rudders in a very light wind (drifting conditions) or in strong winds when, perhaps after an unsuccessful tacking contest, the boat is losing her forward motion. It will be demonstrated that laminar separation greatly affects lift too.

The critical value of Re (at which transition from laminar to turbulent flow occur) depends to a greater or lesser extent on:

a. The roughness of the wetted surface.
b. The shape curvature of the wetted surface and associated pressure gradient along the surface.
c. The flexibility of the wetted surface, and chemicals used such as polymers, etc.
d. The angle of incidence.

If the Reynolds Number is measured in terms of the distance along the surface, then for values of Re less than about 10^5, the laminar boundary layer is stable and it is difficult to provoke transition. With increase of Re, however, the inherent stability of the flow within the laminar boundary layer decreases, the transition is more and more easily provoked; and with Re greater than 2×10^6, considerable care must be taken in keeping the surface smooth and the ambient turbulence of the oncoming stream small, if transition is not to be provoked. Usually, for a flat, smooth plate or plank, the transition to turbulent flow takes place at $\mathrm{Re} = 3.5 - 5.0 \times 10^5$.

The magnitude of the drag coefficient at sufficiently high Reynolds Number, at which the boundary layer flow is fully turbulent right from the leading edge, is determined primarily by skin friction. Tests reported in R and M 1838 (Reports and Memoranda Aeronautical Research Committee Great Britain) indicate that, when the flow is 'basically' attached to the foil surface, the pressure or wake drag varies linearly with foil thickness, t/c. It constitutes approximately the same proportion of the total or profile drag as its percentage thickness, i.e. a foil 10 per cent thick may be expected to have about 10 per cent pressure drag and about 90 per cent skin friction.

(a) *Roughness of the wetted surface*
Figure 2.22, shown earlier, displays schematically stages of the water flow around the hull. Although the laminar flow has turned turbulent, there remains a thin laminar film or a *laminar sub-layer* at the very surface of the hull. The thickness of this laminar sub-layer constitutes only a small fraction of the turbulent BL but its presence is of some practical significance; as long as the roughness of the surface is less than the thickness of the laminar film, it is buried in it. In such a case the roughness does not come into contact with the turbulent flow in the BL and therefore has

no effect upon it. So, the wetted surface behaves as though it were hydrodynamically smooth and no drag penalty is incurred through the rough surface. Roughness which protrudes through the laminar sub-layer gives rise to a resistance in excess of the smooth turbulent flow value. The thickness of the laminar film at a given distance from the leading edge decreases with an increase in velocity and hence the roughness of a height which does not protrude through it at low Re may do so at higher Re. The amount of so-called *admissible roughness* for given conditions, which causes no increase in drag as compared with the 'hydrodynamically smooth' surface, is of vital practical importance. It determines the man/hours necessary for polishing a given surface. A rough criterion for the size of the admissible roughness can be derived from published data (Ref 2.19) as follows:

$$k_a \leqslant 100 \frac{v}{V} \qquad \text{Eq 2.20}$$

where k_a = admissible roughness height
  v = kinetic viscosity
  V = velocity; all in consistent units

Equation 2.20 states that for a flat surface the admissible height, k_a, of the roughness element is independent of the length of the surface; it is determined solely by the velocity of motion and by the kinematic viscosity of the fluid.

For water-flow, admissible roughness k_a in thousandths of an inch, i.e. mils, is given by:

$$k_a = \frac{14}{V \text{ (ft/sec)}} = \frac{8}{V \text{ (knots)}} \text{ (mils)} \qquad \text{Eq 2.20A}$$

The above expression is obtained by substituting the kinematic viscosity of water

$$\left(v_w = \frac{1.23}{10^5} \text{ ft}^2/\text{sec} \right)$$

into Eq 2.20.

Similarly for air-flow, admissible roughness k_a is given by:

$$k_a \leqslant \frac{190}{V \text{ (ft/sec)}} = \frac{110}{V \text{ (knots)}} \text{ (mils)} \qquad \text{Eq 2.20B}$$

Thus, at the same speed, air-flow is able to tolerate an admissible roughness that is about 14 times greater than that in the case of water-flow.

Table 2.7 shows the 'admissible roughness' k_a determined by the approximate formula 2.20A for three hulls, say a model of 3 ft length and the two full-size hulls of lengths 20 ft and 50 ft sailing at speed/length ratios $V_s/\sqrt{L} = 1.0$ and 0.6. It follows that the admissible roughness for a model and its full-scale original at equal speed/length ratio $V_s/\sqrt{L}$ are different; for long hulls this may lead to a very small admissible roughness.

TABLE 2.7

L (ft)	3	20	50	
$V_s/\sqrt{L} = 1.0$	4.6	1.8	1.1	admissible
$V_s/\sqrt{L} = 0.6$	7.6	2.9	1.9	roughness
				k_a in mils

Whilst the approximate equation 2.20 is a good guide, it implies no variation of k_a along the hull. Since, however, the BL thickness is less near the leading edge or bow of the hull, the admissible value of k_a is smaller upstrean than towards the stern, or trailing edge in the case of an aerofoil.

The curves in Fig 2.43 take this fact into account (Ref 2.3). It will be noted that the

Fig 2.43 Approximate maximum admissible roughness, k_a, for the two different speed/length ratios, where x = distance from the leading edge.

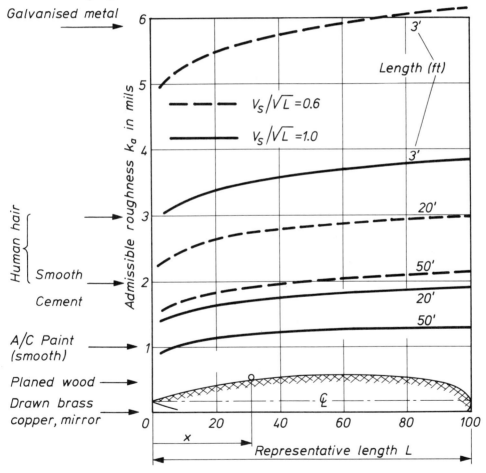

Fig 2.44   Drag characteristics of two NACA sections–the low drag section 63_1–012 and conventional section 0012. Two-dimensional flow.

In order to distinguish clearly between average lift and drag coefficients for a complete foil of a given planform, aspect ratio, etc., and so-called *section coefficients* corresponding to two-dimensional flow over a particular profile, the two sets of coefficients have been introduced in technical literature. They are designated by upper case and lower case symbols respectively, namely:

C_L, C_D–the average lift and drag coefficients for a foil of given planform and aspect ratio AR. In some NACA Reports, the C_L and C_D characteristics are published for standard rectangular planform of AR = 6. These coefficients, which might be called three-dimensional coefficients, are determined partly by the planform characteristics and partly by the cross-section of the foil.

c_l, c_d–the section lift and drag coefficients for two-dimensional flow, i.e. without end effect (induced drag).

Aerodynamic characteristics of the NACA sections given above depend entirely on the section shapes. They must be corrected to take into account the planform and aspect ratio effects which are discussed in following chapters.

degree of finish is more critical towards the bows and that reduction in speed, or size of the boat, increases the chance of obtaining a hydrodynamically smooth surface with a given finish or paint. To add practical meaning to the scale, 'admissible roughness k_a in mils', various finishes are quoted on the vertical axis, including the thickness of a human hair, which is about 2–3 mils (thousandths of an inch) in diameter.

Any rougher surface texture than indicated in Fig 2.43, apart from incurring an additional roughness resistance, should be considered as a possible source of transition from laminar to turbulent flow inside the BL. In general, small protuberances extending above the average surface level of an otherwise satisfactory surface are more likely to cause transition than are small depressions.

In the case of sailing yachts of 20–50 ft, admissible roughness lies between 1–3 mils (Fig 2.43). With reasonable care in finishing the hull surfaces there is no difficulty in meeting these demands. A hydrodynamically smooth hull can be obtained by rubbing down the dry varnish or antifouling paint with 400 grade wet carborundum abrasive paper. 'No roughness detectable to the fingertips should be allowed'–is a good criterion when assessing the quality of the final coat on a hull or foil.

It should perhaps be stressed that, from the hydrodynamic point of view, it does not make any difference whether the hull is painted with a graphite finish or any kind of plastic varnish, or whether the hull surface is matt or glossy, or has been treated with water-repellent silicons. Provided that the final coat does not release drag-reducing substances (polymers) into the boundary layer, the only factors that really matter, as far as skin friction is concerned, are the smoothness and shape or curvature of the wetted surface.

Figure 2.44 displays the effect of roughness on the drag characteristics of two

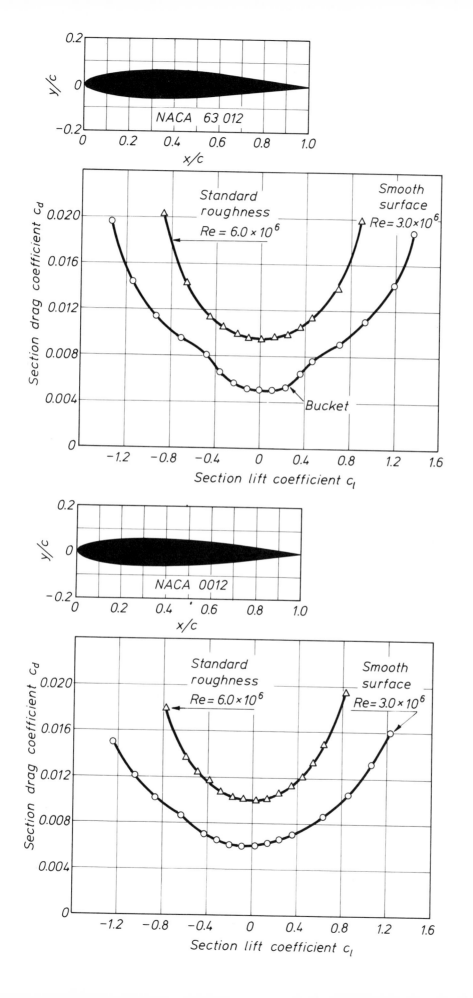

NACA 63 012

Standard roughness
$Re = 6.0 \times 10^6$

Smooth surface
$Re = 3.0 \times 10^6$

Bucket

Section drag coefficient c_d

Section lift coefficient c_l

NACA 0012

Standard roughness
$Re = 6.0 \times 10^6$

Smooth surface
$Re = 3.0 \times 10^6$

Section drag coefficient c_d

Section lift coefficient c_l

NACA symmetrical sections, 63–012 and 0012, one having smooth surfaces and another with 'standard roughness' applied at the leading edge (Refs 2.30, 2.31). The NACA models are usually finished by sanding in the chordwise direction with No 320 carborundum paper when an aerodynamically smooth surface is desired. The standard leading edge roughness selected by the NACA for 2 ft chord models consists of 0.011 inch (11 mils) carborundum grains applied at the leading edge, over a surface length of 8 per cent of chord measured from the leading edge on both upper and lower surfaces.

The so-called low-drag foil of the NACA 63 series shown in Fig 2.44 is characterized by a more rearward displacement of the maximum thickness of the profile to produce the desired pressure gradient over a longer portion of the foil. Its drag curve has a characteristic 'bucket' or region of reduced drag coefficient, c_d. This reduction in drag is due to the fact that laminar flow can be maintained over the larger part of the section before transition to turbulent flow takes place. An important feature of low-drag sections is the precision in design and manufacture which is essential for their success. The advantage of the bucket type drag curve is lost if the surface of the foils is not sufficiently smooth or if the foil is heavily loaded, i.e. working at lift coefficient, c_l, beyond the maximum width of the bucket. For the latter reason the use of a low-drag section for spade rudders frequently operating at high lift coefficient, would not be practical. There is also another reason against the low-drag section; achievement of a transition point well aft requires a relatively sharp leading edge and this, as we shall see in the following chapter, may lead to some reduction in the maximum lift available with resulting limitation in the steering power of the rudder.

Low-drag foils, as demonstrated by the broken curve marked NACA 63–012 in Fig 2.40, are effective at high Reynolds Numbers which, for most of the smaller sailing yachts, are beyond normal operating range. Finally, it is important to keep induced drag in mind when considering eventual advantages offered by low-drag foils. It will be seen that the total drag of appendages is overwhelmingly affected by the induced drag which is a function of aspect ratio. Relative to the variation of induced drag the profile drag reduction offered by NACA sections of 6 series is of secondary importance.

Low-drag sections of NACA 63 series with a relatively large bucket may, however, be used for fin keels. In the case of a low-drag fin keel, its success will depend upon

Photo 2.18　Picture of the transition from laminar to turbulent flow.

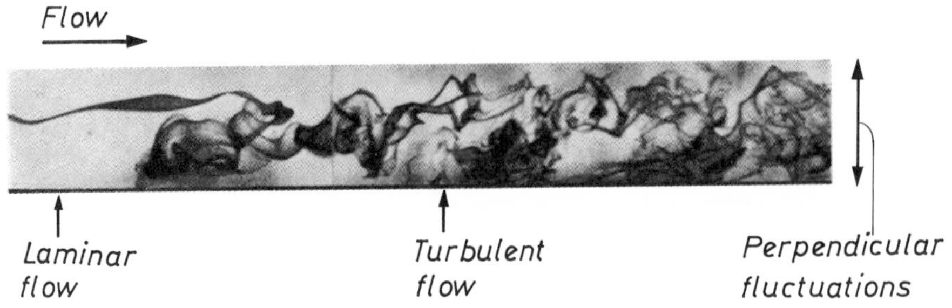

Flow

Laminar flow

Turbulent flow

Perpendicular fluctuations

the required standard of smoothness of the fin surface, which is quite stringent, and also on the degree of loading the fin is designed to sustain in anticipated sailing conditions, i.e. the value of C_L at expected angle of leeway. The lift coefficient should be sufficiently low to be located inside the bucket (Fig 2.44 and Fig 2.46 for comparison). This in turn will largely be determined by the wetted area of the fin keel and its aspect ratio.

(b) Shape (curvature) of the wetted surface and associated pressure gradient
From the preceding chapters two clear conclusions emerge:

Firstly–that a laminar boundary layer without separation gives the lowest drag
Secondly–boundary layer flow is very much affected by the pressure gradient; the
negative (favourable) pressure gradient preventing or postponing the transition
from laminar to turbulent flow.

Although the precise physical nature of the process of transition and its causes are not at present fully understood, nevertheless something is known about the underlying phenomena. The classical '...experimental investigations of the circumstances which determine whether the motion of water shall be direct or sinuous...' conducted by O Reynolds (1883) gave some initial insight into the subject.

By feeding into the water a thin thread of liquid dye the character of the flow can be made clearly visible. This is shown in Photo 2.18, which represents the result of an experiment similar to that first carried out by Reynolds. Laminar flow is distinguished by sharply defined boundaries between the dye and the stream of water; the fluid particles all along the stream are essentially in an axial motion. Sooner or later, with increasing distance or velocity (increasing Reynolds Number) a stage is reached when the fluid particles cease to move along straight lines and the regularity of axial motion breaks down. The flow pattern becomes subjected to increasingly irregular perpendicular fluctuations superimposed on axial motion. Gradually, the coloured thread becomes mixed with water and its sharp outline becomes blurred, the individual particles circulating back and forth through the thickness of the layer, instead of moving in orderly paths parallel to the surface to which it adheres. Photographs 2.10 and 2.18 give ample evidence of distinct differences between those two types of flow.

The transition from one type of flow to the other, commonly presumed to be abrupt, is not, however, sudden but is a super-position of the two boundary layers, a turbulent zone overlaying a laminar one, as shown in Fig 2.22 and also magnified in Photo 2.19 supplemented with a sketch. One may say that the laminar and the turbulent alternatives in a transition zone may coexist somehow, not at the same spot but in tandem: '...Each may have its zone of regency, the available territory parcelled out between the two. When that happens–and it is indeed the typical case–the parcelling is not fortuitous. It is controlled by definite laws' (Ref 2.9).

A number of investigators attempted to find out under what conditions small disturbances in the form of velocity fluctuation appear inside the boundary layer,

Photo 2.19 Picture of a single turbulent spot. Above–the transition and
turbulent flow resulting from the growth of turbulent spots
which travel downstream in a wedge-shape form.

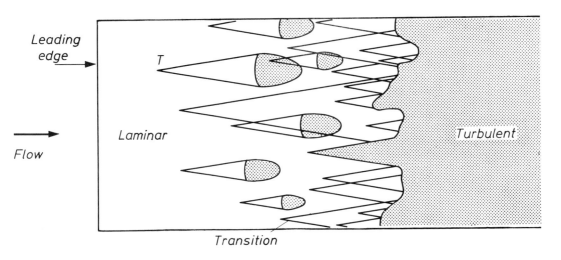

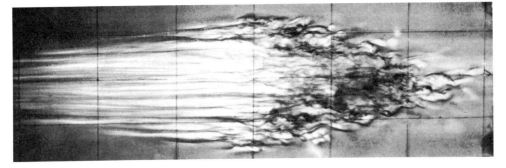

and why they increase or decrease with time. Schubauer and Skramstad (Ref 2.33) compared the boundary layer to the water surface. It is extremely rare to find conditions so *calm* that even tiny ripples do not exist. In practice then, some degree of unrest will always be present. The important question is whether little ripples will grow to sizeable storm waves, eventually producing transition to turbulent flow.

In the course of detailed measurements and studies on stability of laminar flow made by Schubauer and Skramstad (Refs 2.33 and 2.34), it was found that the laminar boundary layer, developing in the presence of favourable pressure gradient (see Fig 2.18), actually damps out disturbances if they are already present in the oncoming stream, or if disturbances are induced by surface imperfections. Conversely, in the presence of adverse pressure gradient the boundary layer is less efficient in damping action and the disturbances in form of velocity fluctuations reappear–magnified, giving way to transition and finally turbulent type of boundary

Photo 2.20 The model for tank testing is equipped with studs, i.e.
turbulence stimulators, in order to make the boundary layer
flow-turbulent. Studs are attached along the whole underwater part of the leading edge of the hull and keel. Some
people argue about the sense or nonsense of building a
smooth accurate model and then 'gluing rocks' on it.

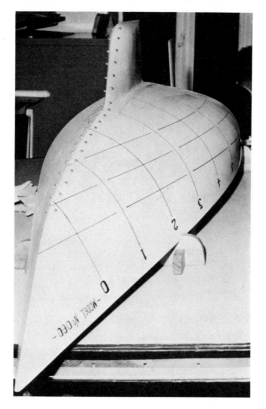

layer flow. The sequence of transition is such that gradually bursts of vibration of a
large amplitude occur and the singular points of transition, marked by letter T in the
sketch attached to Photo 2.19, become more and more frequent and of longer
duration. Periods of laminar and turbulent flow succeed each other in a random
sequence, with a clear tendency of progression of events leading up to fully turbulent
flow. It looks as if the laminar boundary layer, becoming thicker with increasing
distance from the leading edge, is losing its damping efficiency and becomes less
stable.

There are, of course, cases where the initial disturbances are so great due, for
example, to turbulence stimulators (Photo 2.20), that transition usually occurs at
once. The turbulence stimulators should not be mistaken for vortex generators,
which were discussed earlier. This problem is directly related to towing tank
experiments on small models. In order to make skin friction measurements
meaningful, it is usually necessary to have a clear picture of the state of the flow

Photo 2.21   The flow along the hull follows approximately the buttock lines and studs have no 'effect' on the BL flow which is laminar. However, on the keel proper the transition does not take place fully effectively at this relatively low speed.

throughout the boundary layer over both the model and the full-size hull. For reasons which are an inherent part of the performance prediction methods, it is desirable to use an assumption that the whole boundary layer flow is turbulent both on the model and the full-scale; this conveniently simplifies calculations and also covers existing ignorance and uncertainties. To satisfy this, stimulators are added along the whole length of the leading edge of the hull and keel, as shown in Photo 2.20. From recent examination tests on the effect of various turbulence stimulators it will be seen that studs or pins, very brutal obstructions to flow, may fail to produce transition if they are not properly distributed.

Some light was thrown onto this intricate problem by J van den Bosch and I Pinkster (Ref 2.35). Photo 2.21 illustrates a part of their tests performed on a yacht model of LWL 4.6 ft (1.4 m) equipped with a fin keel of NACA section 0010. The observation of the flow pattern inside the boundary layer was made possible by the injection of a potassium-permanganate solution; the distribution of the injection points is shown in the sketch attached to Photo 2.21. The laminar flow in Photo 2.21A, B and C can be recognized by long and distinct streaks, the turbulent flow by the rapid diffusion of dye.

It is seen in picture A that the prevailing boundary layer is laminar. The flow separates well behind the maximum thickness of the fin; in the region of the separation wake the dye accumulates due to the reversed flow. In picture B, flow is still laminar but, due to angle of leeway 3°, separation occurs closer to the leading edge of the fin. Picture C demonstrates the effect of studs on the character of the flow; on the keel proper transition takes place but the intended turbulent flow does not appear to be fully developed at this relatively low speed, the speed/length ratio, $V_s/\sqrt{L}$ being in the order of 0.3. It is noticeable that the flow along the hull follows approximately the buttock lines, not the waterlines (see Fig 2.49A) as might be expected, and for this reason the studs have no effect on the BL flow which is still laminar. In the course of experiments, two important facts were established:

Firstly–laminar flow is very persistent on the fore-part of models with V-sections and cut away fore-foot; the favourable pressure gradient in the flow direction tending to stabilize the laminar flow.
Secondly–comparative tests, with various turbulence-stimulating devices attached along the leading edge of the hull and keel proper, showed a considerable discrepancy in the measured values of the side force generated by the hull-keel combination.

As complained some time ago by A. Robb–Ref 2.36–for these scale effect reasons advertisements which appear with increasing frequency stating 'fully tank tested', may in some cases be almost fraudulent. Due to the notorious difficulties in

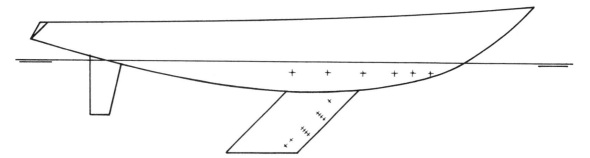

Model LWL – 1.4 m (4.6 ft) Points of injection.
Section of the keel proper – NACA 0010

A

V_s = .35 m/s
   = 1.15 ft/s
Leeway = 0°
No turbulence
stimulators

BL laminar. Flow separates well behind the maximum thickness of
the fin. The dye accumulates in this region because of reversed
flow in the separation wake.

B

V_s = .35 m/s
   = 1.15 ft/s
Leeway = 3°
No turbulence
stimulators

Flow laminar. Due to angle of leeway the separation occurs more
forward.

C

V_s = .30 m/s
   = 0.98 ft/s
Leeway = 0°
Studs along the
leading edge as a
turbulence

establishing a reliable picture of BL flow, both in the case of the model and its full-scale replica, all test results, below V_s/L of about 0.4, or perhaps even 0.5, should be suspected. Scale effect will be of the utmost importance in any attempt to find a quantitative correlation between the free or radio-controlled models and full-scale forms.

The stabilizing effect of favourable pressure gradient, tending to retard transition, has been exploited in the design of the so-called low drag aerofoils. In the NACA nomenclature, such foil sections begin with numbers 63..., 64..., 66... A way of providing the favourable pressure gradient is to move the minimum pressure point (suction peak) downstream by placing the maximum thickness of a section as far back as practicable. The general requirement for the maintenance of laminar flow would therefore be as follows. For a considerable distance back from the leading edge the local velocity V just outside the boundary layer, should increase continuously so that pressure will be falling in a similar manner. In such a case, the accelerating favourable pressure gradient will greatly facilitate laminar flow. Figure 2.45 shows the calculated velocity distribution expressed in terms of V/V_0 ratio, over three symmetrical sections set at zero angle of incidence, all 15 per cent thick but differing in their positions of maximum thickness (Ref 2.37). The maximum ordinate for the section A is located at 35 per cent of the chord from the leading edge, 45 per cent for B and 55 per cent for C.

As already mentioned, the low drag foils of NACA series 6 are distinguishable by a characteristic bucket or region of reduced drag, absent in the case of older NACA series 00 as displayed in Fig 2.44. If laminar flow with a short-chord keel is to be exploited then information is required on the maximum available width of the bucket or, in other words, the maximum available lift coefficient beyond which the low drag foil offers no advantage. This is entirely dependent on the geometrical features of the section, its thickness and distribution of thickness. By comparing the section characteristics of NACA 6 series given in Fig 2.44 section 63-012, Fig 2.46 section 63-009 and Fig 2.47 section 66-012, it can be inferred that, when thickness is reduced and other geometrical factors remain unchanged, the bucket becomes narrower (Figs 2.44 and 2.46); similarly, shifting the maximum thickness further aft brings the same effect (Figs 2.44 and 2.47).

The significance of the shape of the keel proper (or fin keel) with respect to its planform and cross-section was and still is rather underestimated. This is perhaps partly due to the fact that the underwater part of the hull is not as conspicuous as sails are (Photo 2.22) and partly because, once determined by the yacht designer, the keel or fin keel without tab leaves no opportunity for crew influence except to see that it is smooth.

The action of hull appendages is still a grey area to be investigated and its importance is certainly no less than that of the sails. Systematic tests are no doubt the best way to stimulate further progress towards better performance and more reliable craft.

Figure 2.48 demonstrates the generally accepted evolution of the underwater

Fig 2.45   Variation of velocity distribution with position of maximum
thickness for conventional section A and laminar or low-drag
sections B and C.

V–local velocity at any point along the foil surface.
V_o–undisturbed velocity ahead of the foil.

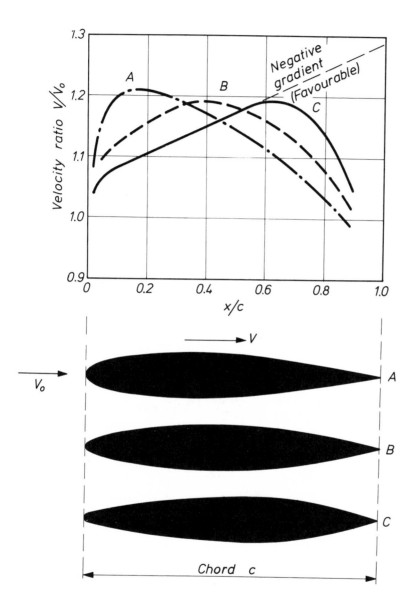

Fig 2.46   Drag characteristics of two NACA sections 63–009 and 0009 (two-dimensional flow).

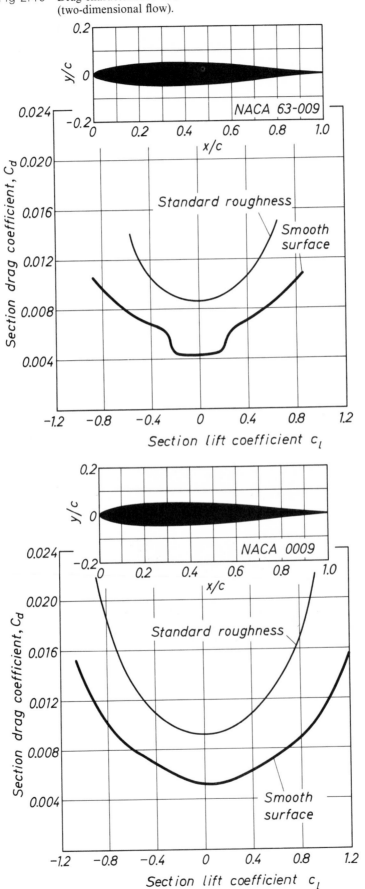

Fig 2.47   Aerodynamic characteristics of NACA 66$_1$–012 section. Two-
dimensional flow.

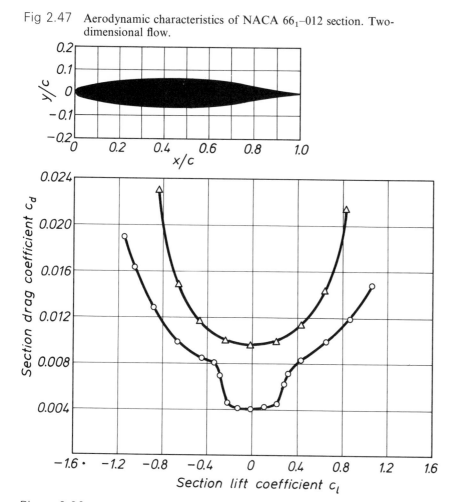

Photo 2.22   The significance of the shape of the keel proper was and still is
rather underestimated, perhaps because the underwater part
of the hull is not as conspicuous as sails are.

Fig 2.48   Evolution of the underwater planform of the English yacht
Calliope designed by Guy Thompson.

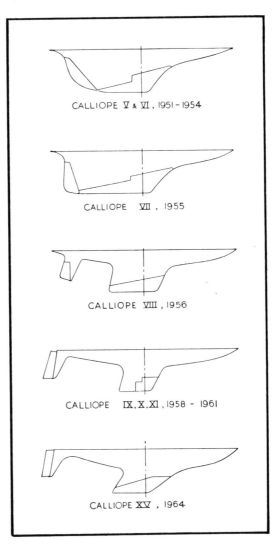

planform of the hull. The pictures present the development of *Calliope* designed by
Guy Thomson (Great Britain). One can clearly see the tendency to cut down, as
much as possible, the amount of lateral wetted area which in the past seemed to be so
important. As we shall see in the following chapters of Part 3 the theory of so-called
slender foils indicates that very little side force is developed by the area of fin behind
the maximum draft (Fig 2.49B) and almost the whole of the side force is produced by
the part of the keel immediately adjacent to the leading edge. So the justifiable
evolution of the fin would be towards the shark-fin which has been used by Olin
Stephens in *Clarionet*, as depicted in Fig 2.49A.

Fig 2.49

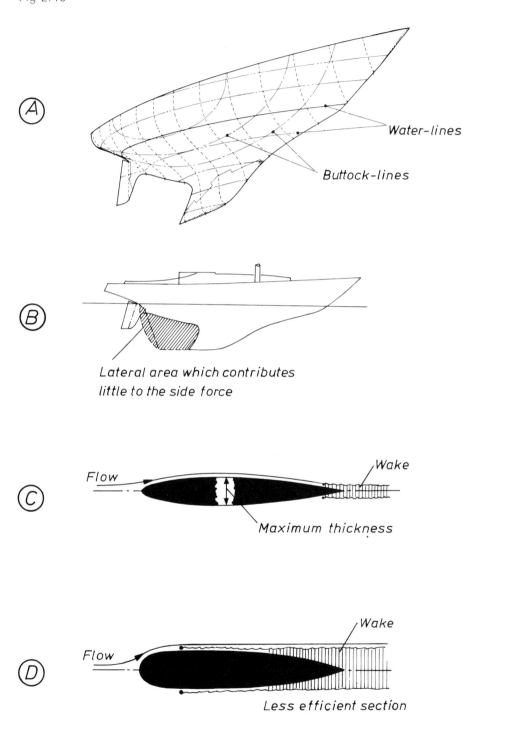

(A)

Water-lines

Buttock-lines

(B)

Lateral area which contributes
little to the side force

(C)

Flow

Wake

Maximum thickness

(D)

Flow

Wake

Less efficient section

This pruning of the wetted area can however be overdone and there is ample evidence that certain unpleasant drawbacks which have already been mentioned in Part 1 have to be accepted if better performance becomes of primary importance.

Further, possible progress in reducing friction drag might be achieved by proper distribution of the thickness of the underwater profiles of the fin. As already pointed out, laminar flow may be maintained to the point where the keel section reaches its maximum thickness. The gradual contraction of the flow along the profile, as shown in Fig 2.49C, may help to maintain the laminar boundary layer for a greater proportion of the total length of the fin and for higher values of Reynolds Number than in the case of the less efficient fin section shown in Fig 2.49D. Its blunt leading edge will favour the early onset of transition or even separation. Since, as shown in Fig 2.19, not every streamline section is equally efficient from a hydrodynamic point of view, the foil ordinates as given for instance in Ref 2.31 must be obeyed in the process of manufacturing. This knowledge could be used to advantage when designing or changing the shape of the fin. Experiments have revealed (in the Dragon Class) that the performance of a yacht can be improved by sharpening the blunt leading edge of the fin within the limits allowed by the class rules and shifting aft its maximum thickness.

(c) *Ways of reducing skin friction: polymers, flexibility of wetted surface*

During the 1967 annual meeting of the IYRU, the Permanent Committee had to cope with the rather difficult but interesting problem of the use of chemicals in yacht racing. The sailing fraternity has been intrigued by new possibilities of reducing frictional resistance, to increase speed by means of releasing or ejecting chemicals into the water immediately adjacent to the hull.

Two ideas which have attracted much attention among both scientists and racing helmsmen and which inspire a lot of mysterious speculation, are those of maintaining laminar flow over as large an area as possible of the wetted surface of the hull and of damping down turbulent flow or delaying transition.

In the past it has generally been recognized that friction is mainly a function of wetted area and its smoothness, but there are other factors which can offer the opportunity to reduce viscous drag (Ref 2.39). The immediate reaction of the IYRU towards chemicals was that such practices should be prohibited. Accordingly, a Sailing Instruction was produced: 'A yacht shall not eject or release from a coating or otherwise any substance (such as a polymer) the purpose of which is, or could be, to reduce the frictional resistance of the hull by altering the character of the flow of the water inside the boundary layer'–which will be applied in all yacht racing under these rules. However, the new idea of chemicals as drag reducers is open for the future, since there are feelings expressed also by some delegates to the Union that the IYRU should not be too restrictive on new and interesting developments of this kind unless undue costs are involved.

More recently, some experiments carried out in various laboratories have given rather startling results. Reduction of viscous friction by as much as 40 per cent in certain circumstances can be achieved by the addition of a few units of chemicals (certain polymers, Ref 2.40) per million units of water by weight.

In calm water friction drag accounts for a large part of the total resistance in the majority of water-borne vessels. In an average cargo ship, viscous drag amounts to 85–90 per cent of the total drag. In high speed surface craft, where the wave system is more conspicuous, skin friction is still above 50 per cent of the total resistance. In submarines and torpedoes running below the surface almost all resistance is viscous drag. The skin friction drag of a yacht is a substantial part of the total drag, varying from about a third at high speeds to almost the whole at low speed.

The problem of abnormal changes in the resistance quality is not a new one. For many years, researchers conducting ship-model tests in towing tanks had been puzzled by occasional and unexpected drops in the resistance of the so-called standard model. Such a model is towed from time to time, usually at regular intervals, to establish a sort of reference basis for the friction resistance of other models being investigated. As an example, a brass standard model of *Iris* (British despatch vessel–Admiralty Experimental Works at Haslar, England) was used to manifest considerable reduction in resistance of up to 14 per cent recorded in 1925 which could not be forecast or explained. These were called by the staff 'Iris storms' (Ref 2.41).

Scientists have recently found a reasonable explanation and proved beyond little doubt that the cause of such 'storms' is biological in origin. It has become known that solutions of certain long-chain molecules or polymers produced in Nature by some algae (*Anabaena-flos-aquae*, *Porphyridium cruentum*, etc.) can reduce friction considerably, depending on the concentration of their by-products. Tests of some polymers have shown no reduction in friction with molecular weight under 50,000. Molecular weight of at least 10^5 appears to be necessary to achieve a substantial decrease in friction drag, and the recorded differences in friction-reducing properties by both natural and synthetic polymers most probably reflect differences in their molecular weights and structure. Anyway, the requirements of an efficient drag-reducing additive are long molecules of high molecular weight with few branches and good solubility.

The mechanism responsible for the reduction of resistance to flow caused by even a very small amount of polymer diluted in water is not yet fully understood. However, investigators of this phenomenon have suggested that the relatively long and flexible macromolecules of polymers have a spring-like character and might damp transverse oscillations in the boundary layer that give rise to transition. The long molecules setting themselves along the lines of flow, resist transverse mixing of water in the boundary layer; in this way they may first delay transition, if the original flow is laminar, and eventually slow down the growth of turbulence. Another speculation is that the laminar sub-layer (Fig 2.22) thickens in the presence of polymer additives by physical absorption to provide a resilient wall layer. The explanation is probably to be sought in viscostatic effects.

Figures 2.50 and 2.51 represent the results of model towing tank experiments

Fig 2.50 The effect of polymer solution on resistance.

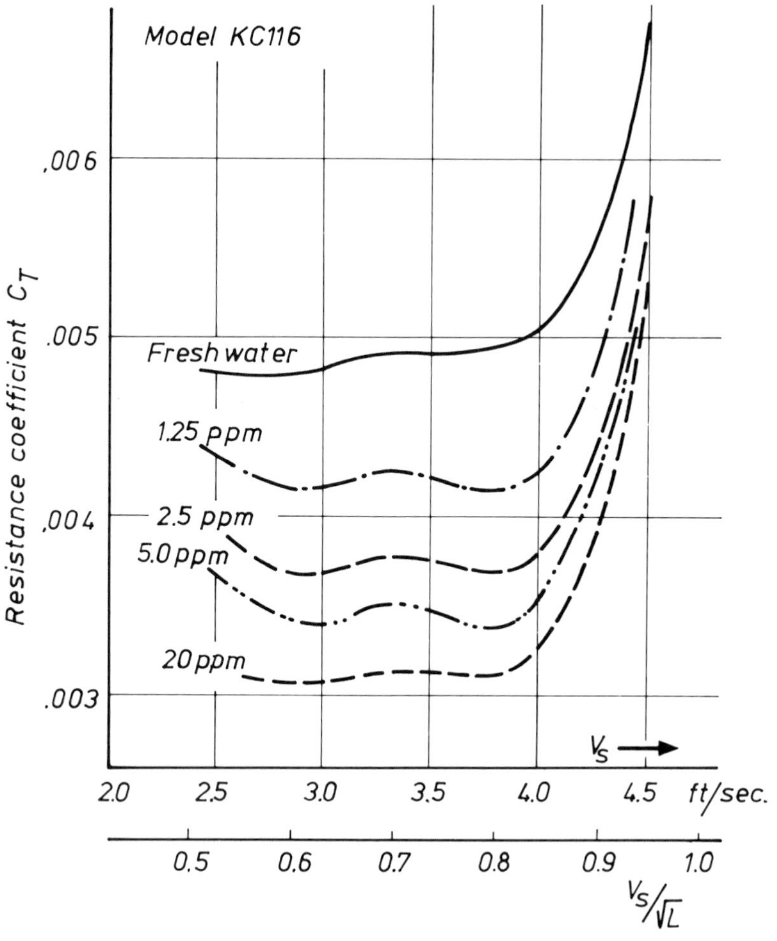

using dilute polymer solutions instead of water. The polymer used to vary the skin friction was Polyox WSR 301, a polyethylene oxide, having a molecular weight of 4 million. The experiments were carried out with a standard, eight ft model, KC116 (one of the British towing tank models). Figure 2.50 shows the variation of total resistance coefficient C_T (multiplied by 1000 on the scale) for a model which was towed at different speeds, first in fresh water and then in water containing polymer of various concentrations: 1.25 ppm (parts of polymer per million parts of water), 2.5 ppm, 5.0 ppm and 20 ppm (Ref 2.42).

Figure 2.51 gives the corresponding reduction in friction expressed as a percentage of resistance experienced by the model in fresh water. We can see that the addition of 20 ppm of polymer reduces friction between water and hull to about 60

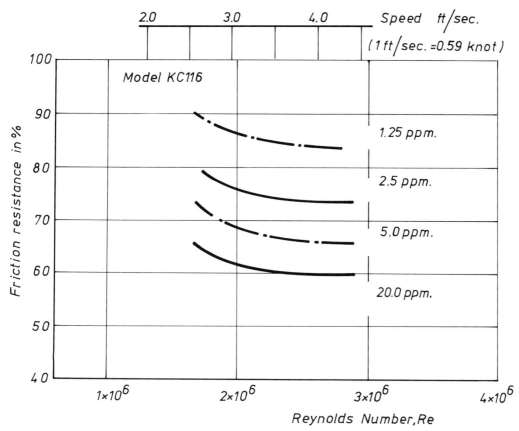

Fig 2.51 Reduction in friction resistance at various concentration of polymer expressed in ppm (parts of polymer per million parts of water).

per cent of the fresh water value–so the power required can be drastically cut. One can also notice that very few units of polymer are sufficient to affect skin friction enormously.

Discharging or ejecting of such friction-reducing chemicals over the bottom of the hull and leading edge of the keel proper through a series of small holes could radically change the performance of any yacht, particularly in ghosting conditions when skin friction is the most important component of total resistance. One may also say that such additives might radically alter the philosophy of yacht racing. Ejecting or releasing polymers along the wetted surface need not be continuous, it could be used as a new tactical weapon, for a limited period in crucial situations; sometimes a few seconds are just enough to win a race.

There are a number of problems to solve in how to eject most effectively the polymer solution. According to Milward (Ref 2.43) rough estimates of the quantities of polymer needed are given on the assumption that the test yacht is about Dragon or Soling size:

1. Weight of polymer required (W_p) to give an average concentration in the boundary layer of 20 ppm is:

$$W_p = 0.26 \, V_s \, \text{lb/min,}$$

where $V_s =$ knots.

2. If a 0.5 per cent solution of polymer is the most concentrated that can be readily handled, then the quantity of solution required is:

$$Q_s = 6.5 \, V_s \, \text{gal/min}$$

and the weight

$$W_s = 65 \, V_s \, \text{lb/min}$$

It is the quantity and weight of solution that are a problem.

Some recent work by Kowalski (Ref 2.44) however suggests that if the polymer solution is ejected in pulses of 1 sec duration in every 10 sec the same reduction in friction drag can be obtained. On this basis the quantities involved are greatly diminished becoming:

$$W_p = 0.026 \, V_s \, \text{lb/min}$$
$$Q_s = 0.65 \, V_s \, \text{gal/min}$$
$$W_s = 6.5 \, V_s \, \text{lb/min.}$$

This makes the ejection of polymer from a yacht a much more feasible proposition.

Perhaps it is interesting to add that scientists have speculated on the possibility that the mucous secretions (slime), which are also polymers, from the skin of fish enable them to move at increased speed for a given expenditure of energy. Secretion of slippery mucous may be gradual or abrupt and, what is perhaps most fascinating, irritability resulting from mechanical stimulation (it might be oscillations in the boundary layer) causes the mucous glands to increase their secretion.

Experiments with polymers are in the early stage of development, and its is quite possible that in the future someone will invent a paint containing polymers, similar in a way to ordinary antifouling paint, but from which the drag reducer could be slowly and continuously released. Since long-chain polymers ejected into the boundary layer near the bow do their friction-reducing work so effectively while sliding aft–'…it is not inconceivable that some ingenious inventor may find a way of tacking the long chain molecules by their ends to the bottom in such a way that they do their curious job of friction reduction while not getting washed away downstream' (Ref 2.42). And one wonders whether Nature did not get there first when she evolved hairy and scaly animals.

Effects somewhat similar to those associated with polymer soluble coatings were obtained by compliant coatings.

As we discussed earlier, transition from laminar to turbulent flow with its drastic increase in resistance is connected with the development of transverse oscillations within the boundary layer. Such an instability of flow is due to the fact that the inherent damping which normally comes from viscosity becomes, at certain Reynolds Numbers, insufficient to stop the build-up of boundary layer fluctuations.

Photo 2.23   Dolphins–mysterious swimming speed record holders. It has been claimed that a 40 knot burst speed with an 18 knot sustained speed were observed. Are they capable of maintaining laminar flow at very high Reynolds Number or is their 'muscle engine' exceptionally efficient? Both hypotheses are arguable.

Finally, the laminar flow turns turbulent. From this it follows that an artificial increase of inherent damping by mechanical or other means external to the boundary layer itself might lead to laminar flow stabilization.

Observation on the swimming of fish and sea animals like dolphins (Photo 2.23) suggests that they must experience an unusual amount of laminar flow to explain their performance at top speed. It has been observed (Ref 2.45) that the British bottle-nosed dolphin of length about 1.90 m (6.2 ft) is capable of reaching about 16 knots in a short burst of speed and can keep going at 12 knots for nearly a minute, so the Reynolds Number at the higher speed was in the order of 1.4×10^7. At such a large Re the prevailing flow should be turbulent and the friction so high that the dolphin muscle engine could not possibly give sufficient power output to reach the recorded speeds. According to Kramer (Ref 2.46) much higher speeds were observed–'...a 40 knot burst speed with an 18 knot sustained speed'. This is an extraordinary performance bearing in mind that it has been accomplished with the aid of a '...notoriously weak muscle motor' (two horse-power is about what

Fig 2.52   Stabilizing rubber-coating (artificial dolphin skin) devised by
M Kramer. Dimensions in mils (1 mil $= \frac{1}{1000}$ of an inch).

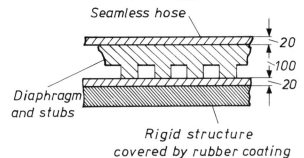

biologists estimate). In the case of seals and dolphins swimming at night among plankton (which becomes fluorescent whenever the water is disturbed, Ref 2.47), the former leave behind bright fluorescent wakes, while dolphin wakes are far less conspicuous. Such observations suggest that the dolphin somehow produces relatively little disturbance in its wake and probably suffers little pressure drag.

These findings have led to the suggestion that damping is one way in which the dolphin's skin functions. It would mean that the resilient dolphin skin acts as a widely spread damper to absorb oscillatory energy from the boundary layer and convert this energy into heat. In an attempt to simulate this suggestion, M Kramer developed a stabilizing coating of two rubber layers, separated by stubs, called 'Laminflo' of a total thickness of about 0.15 in, with the remaining space filled with damping fluid; it is shown in Fig. 2.52 (Ref. 2.48). Using this in tests with torpedoes, results of which are depicted in Fig 2.53, a maximum reduction of friction coefficient from 0.0026 to 0.0011 at a Reynolds Number of 1.5×10^7 (a reduction in drag of about 60 per cent) was attained (curve C). According to calculation, this drag reduction means that about 80 per cent of the total length of the model was fully laminar. In the same conditions the rigid, high-gloss reference model experienced almost fully turbulent flow.

The mechanism of damping explained by the inventor is as follows: as long as the boundary layer flow is laminar, the coating behaves like a rigid surface, provided the stiffness of the coating is sufficient to avoid surface flutter; the damping fluid within the coating remains at rest and the diaphragm surface is smooth. When the boundary layer tends to become wavy, pressure differences originate between the crests and troughs of the boundary layer waves. Thus pressure waves propagate along the coating and cause an oscillatory response of both the diaphragm, as well as the damping fluid inside the coating.

Inspired by Kramer's tests considerable interest has been generated in the development of drag-reducing compliant coatings (Refs 2.49, 2.50). Promising as the initial results obtained by Kramer were, unfortunately only a few experimenters since have been able to measure a reduction in skin drag using flabby skins. The observations indicate that there may exist a correlation between the turbulence

Fig 2.53   The drag coefficient of various models as a function of Reynolds
Number.

Curve A–rigid reference model.
Curves B, C and D–fully coated models with different stiffness of
coatings.

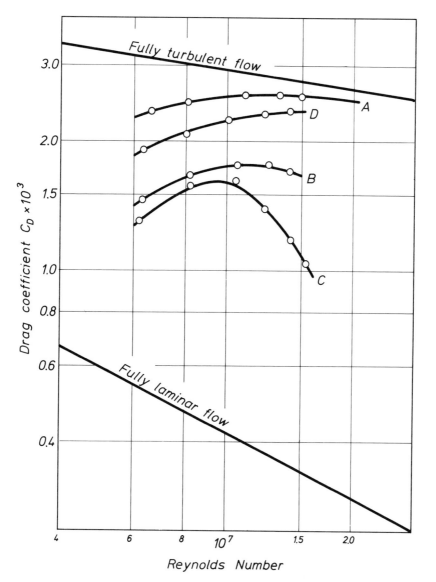

Fig 2.54   Variation of relative turbulence intensity.

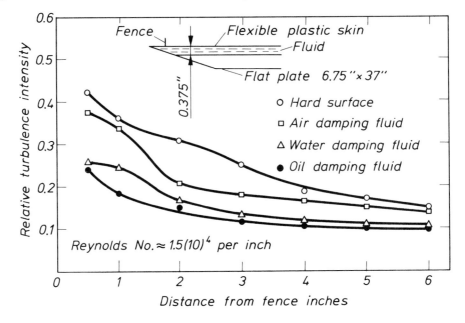

intensity within the boundary layer and skin friction reduction. There are good theoretical reasons supporting such an intuitive feeling. The test data given in Fig 2.54, taken from Ref 2.49, also supports this view. Exploratory tests were conducted in an air-flow to find out whether the compliant skin concept of drag reduction would work in aerodynamics. Instead of measuring skin friction coefficients, the turbulence intensities were recorded by using a hot wire anemometer near the plate covered by a flexible plastic skin (commercial name–*Clopay Frosty*) of about 2 mils (0.00225 in) thick. As shown in Fig 2.54, there was a $\frac{3}{8}$ in gap between the aluminium plate, which was 37 m long and 6.75 wide, and the flexible skin. This gap was filled with either air, water or automotive lubricating oil. The nominal wind velocity in the tunnel was 29 ft/sec. The turbulent intensity of flow was measured along the wake generated by a small fence as depicted in the sketch in Fig 2.54. From recorded tests it became apparent that the amount of turbulent damping depends on the viscosity of the damping fluid filling the gap between the hard surface and the skin.

The three coatings tested all showed less turbulence than the hard surface, and turbulent damping increased as the viscosity of the damping fluid increased. Investigators rightly conclude that this trend must certainly reverse itself if the damping fluid viscosities were to be increased to some large value. As viscosity approaches infinity it would seem reasonable to expect the coating to behave then as a hard surface.

It might be noted that such coatings are likely to be rather expensive if applied to ordinary yachts and it is not at all certain that they would be effective at the speed range within which yachts usually sail. They may, however, be useful in special cases, such as fast craft intended to beat speed records.

Some hopes of reducing skin friction may have been raised recently by the literature of one paint company. Their product as advertised:

'...is a revolutionary type of bottom treatment which increases the speed of sailing boats and motor boats...The dolphin has an ingenious way of reducing this friction. It simply uses water as a lubricant. In the outer surface of its skin it stores water, thereby creating a water-to-water boundary layer that reduces the friction...[advertised paint] works the same way thanks to this water-to-water effect' [sic!]

This concept of the dolphin's functioning was new to National Physical Laboratory (comments Paffett in Ref 2.42):

'Nevertheless, the claim made on behalf of the product appeared to be worthy of examination and amenable to quantitative assessment. Tests were therefore carried out with a plank-type test model, 1.14 long 0.76 m draught and 25.4 mm thick, with tapered entry and run and having turbulence studs at the fore end. The surface was of highly polished polyurethane varnish. Runs were carried out over a range of speeds. On completion the plank was raised, dried and coated with two coats of the product, which was allowed to dry overnight. The plank was then immersed for an hour, at the end of which the resistance experiments were repeated.'

Fig 2.55   Measured resistance results for a coated flat plate (Ref 2.42, J Paffett).

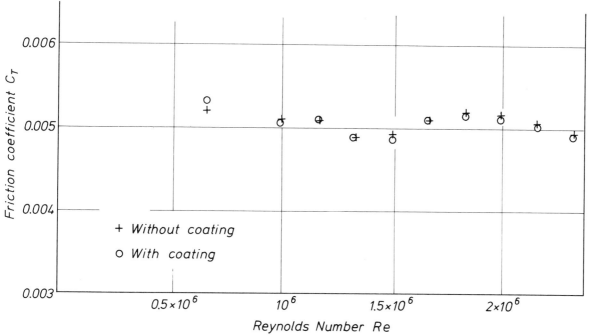

The results are plotted in Fig 2.55 from which readers may draw their own conclusions. There still seems to be plenty of scope for those friction-reducing inventions.

(5) Lift and Reynolds Number effect on foil lift efficiency

The maximum available lift coefficient, C_{Lmax}, is one of the most important aerodynamic characteristics that usually shows the largest Reynolds Number or scale effect. This determines, for instance, the maximum available lift or maximum hydrodynamic efficiency that the rudder operating at given Reynolds Number is capable of attaining. The observed failures of both rudders and fins while manoeuvring in rough weather conditions, when Reynolds Number changes rapidly from almost zero when a boat has lost way, to several millions when the boat has regained speed again, can be explained by just taking scale effect into account. As in the case of drag, the Reynolds Number effect on lift is associated with the shape of the foil section, particularly its leading edge, and also the roughness of its surface.

(a) *Effects of leading edge, camber and thickness on stall pattern*
While discussing low-drag sections (Fig 2.45) it has been mentioned that sharpening the nose prevents velocity and pressure peaks from occurring close to the leading edge. The effect of reducing the nose radius is aimed at extending the laminar flow over possibly the whole length of the section in order to maintain low drag. This effect is limited, however, to small incidence angles, and experiments suggest that sharpening the nose leads unfortunately to a reduction of the maximum lift coefficient; so that while a small radius is desirable for one reason its reduction should not be carried to extremes.

Figure 2.56 demonstrates this influence of the nose form on lift. The three foils tested and described as blunt, sharp and intermediate, are depicted. There is evidently a critical minimum value of the leading-edge radius, below which maximum lift suffers acutely (Ref 2.51).

The leading-edge shape that affects primarily the character of flow, and hence the concentration of pressure over the forward part of the section, determines the range of incidence angles in which separation and stall will occur. Any considerable sharpening of the nose results in an increase of the flow velocity at this edge of the section. This usually encourages leading-edge separation, followed by stall and consequent sudden loss of lift. With a suitable shape of nose and camber distribution over the front part of the section, such an eventuality may be avoided. It might be anticipated that when inertial forces are involved, the flow will have greater difficulty in negotiating a sharp leading edge than a more gentle one.

For similar reasons, a car driven through a gentle bend will not lose its road-holding capability as easily as turning round a sharp bend. In the latter case, the centrifugal force, which is an example of inertial forces, may throw the car outwards from the intended course; the sharper the bend the stronger the centrifugal force.

Fig 2.56   Lift curves of three sections with different shape of the leading
edge LE but the same thickness/chord ratio, 12 per cent.

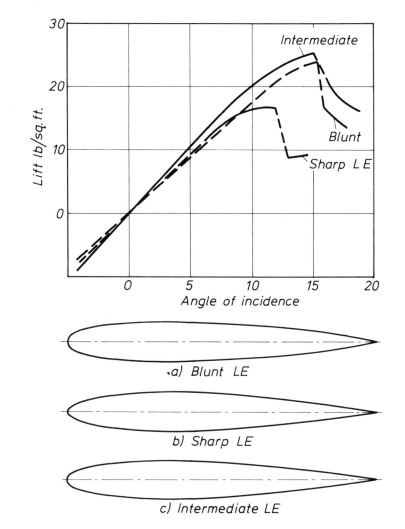

a) Blunt LE

b) Sharp LE

c) Intermediate LE

The reason for it can be found in any textbook on applied mechanics. Accordingly, an inertial force is produced in the mass whenever the mass, which can be a mass of air, is forced to change its motion, no matter what the velocity or direction.

A car of mass m entering with a speed v a road corner of radius r, may serve as an example to illustrate the inertial forces involved. The vehicle will be able to maintain a controlled motion in the circular path of the given radius, without skidding, only if the friction between the wheels and the road can resist the centrifugal force which is:

$$F = \frac{m \times v^2}{r}$$

One can infer from it that the smaller the radius of the corner r, the mass m and car speed v being constant, the higher the inertial force which will eventually cause a skid. In order to allow vehicles to drive safely at greater speeds round corners it is the usual practice to bank the corners, i.e. to lift the outer curve with respect to the inner curve. Otherwise, while approaching unbanked flat corners, the drivers are usually warned by the traffic signs what is the maximum speed limit with which the corner can be safely negotiated. Any attempt to corner faster may result in a skid.

The air particles cornering the leading edge of a foil are in a similar situation, with one further consequence arising from the so-called *principle of conservation of angular momentum*. Recalling the definition of this principle we write:

$$r \times m \times v = \text{constant}$$

from which it follows that, when the air particles of mass m, approaching the foil along a gentle curvature, are forced to make a sharp turn around the leading edge, from windward to leeward side of the foil, their speed v must increase as the radius r of their path decreases. In other words, the smaller the radius of path curvature, the higher the flow velocities and, in consequence of the Bernoulli Eq 2.3, the higher the suction becomes, as illustrated for example in Figs 2.16 and 2.76. Since the suction varies inversely as the radius of the flow curvature it implies that when r becomes zero (for an infinitely small radius of sharp leading edge) the flow velocity v should theoretically be infinite and this leads to infinitely high suction. In the actual state of affairs no infinitely large suction can appear, this is impossible in practice, and instead the flow separates at the sharp front edge. Nevertheless, on account of the high local flow velocities, the measured suction near the leading edge of foils is usually high.

The mathematical theory of the ideal fluid yields no information about the expectation of separation even in simple cases where intuition and common sense would predict separation almost with certainty, as shown in Photo 2.17.

Prediction of the angle of incidence α, at which the stall or maximum lift will occur, would, without experiment, be quite a difficult task. Numerous factors affect it. For instance, the stall (loss of lift due to flow separation) is related to the character and position of separation, which in turn depends upon the shape of the foil. The lift stall encountered by any foil when the angle of incidence is increased, originates in one of the two locations, or in both concurrently, and can be classified as follows (Refs 2.52, 2.53):

1. Trailing-edge stall–preceded by movement of the turbulent separation point S_r forward from the trailing edge with increasing angle of incidence. This is shown schematically in Fig 2.57A.
2. Leading-edge stall–an abrupt flow separation near the leading edge generally without subsequent re-attachment, as depicted in Photo 2.7 and Fig. 2.59A. This type of stall is due to the presence of a so-called short laminar separation bubble of very small chordwise extent (less than 0.01 c) near the leading edge.

Fig 2.57   Diagrams illustrating qualitative distributions of pressure round the thick and the thin foil with and without boundary layer effects. Bubble in sketch **B** is greatly exaggerated.

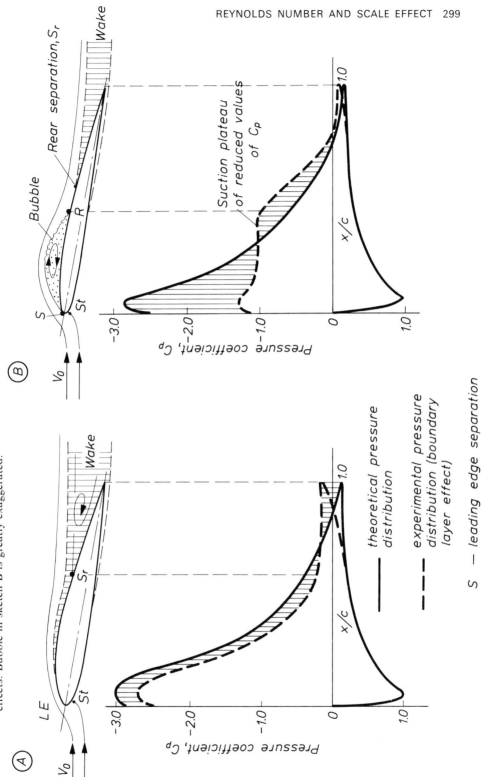

Fig 2.58   Lift curves of four symmetrical NACA aerofoil sections (Re = 5.8 × 10^6). Two-dimensional flow.

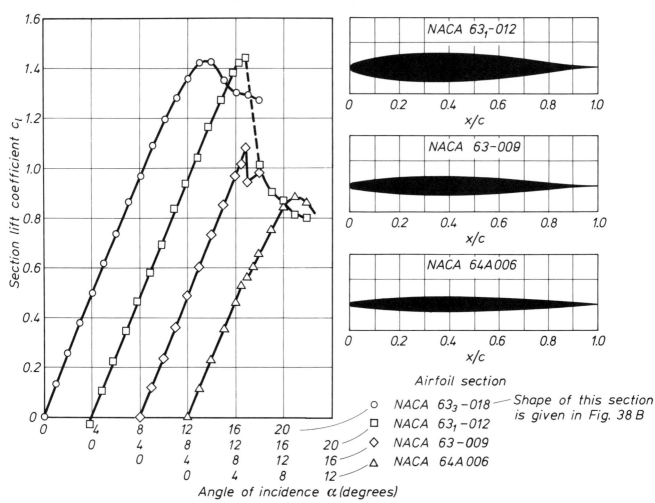

As incidence angle is increased, the bubble moves forward nearer the leading edge until it eventually bursts causing full-chord flow separation.

3. Thin-foil stall–preceded by flow separation at the leading edge in the form of a long bubble with re-attachment at a point R, shown schematically in Fig 2.59B. As incidence angle increases the re-attachment point R moves steadily rearward until it coincides with the trailing edge, at about which stage maximum lift is reached.

4. Combined trailing-edge and leading-edge stall, as depicted in Fig 2.57B, in which case the foil exhibits both types of separation.

Typical examples of the effects of those types of stall on the shape of lift curve and maximum lift are provided by the experimental data for the NACA 633-018, 631-

Fig 2.59   Flow round the cambered plate.
          A. Separated flow without reattachment (sharp separation)
          B. Separated flow with reattachment.

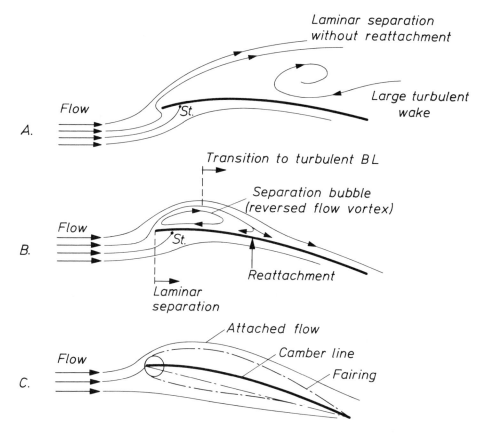

012 and 64A-006 aerofoil sections, shown in Fig 2.58. Ordinates and further details concerning these and other sections can be found in Ref 2.31. It is seen that for the thickest section 633-018, which experiences trailing-edge stall, the lift decrease is gradual and continuous; while sections 631-012 and 63-009 likewise, which undergo the leading-edge stall, show an abrupt discontinuity at the stall. The lift curve of the 64A–006 section, which experiences the so-called thin-foil stall, is characterized by a rounded peak, preceded by a slight discontinuity–a kink at an incidence of 5°. The lift characteristics of all four foils shown in Fig 2.58 are, as predicted by theory (see Eqs 2.14 and 2.14A), negligibly influenced by viscosity effects below the stall. This is reflected by the almost identical slope of all the curves representing the lift coefficient c_l, versus angle of incidence α, which are roughly parallel to each other up to the point when the stalling begins.

Since stalling is inseparably related to the behaviour of the boundary layer flow, the same factors which influence the boundary layer flow, i.e. Reynolds Number, leading-edge radius, pressure gradient, surface roughness, etc. also affect the stalling characteristics of the foil sections and therefore the maximum lift. A modification in any one of these factors may cause the stall of a given foil section to change from one type to another.

(b) *Flow pattern round thin, sail-like foils*

The flow round sharp edges of sails is of particular interest from the sailor's point of view. It deserves attention simply because the introduction of tell-tales made racing sailors acutely aware of their significance in detecting smooth air flow over sails, which in turn is so essential to sail performance.

Observation of tufts attached to a cambered plate of camber ratio 1/10, made by the author in the course of experiments in Southampton University wind tunnel, indicates that smooth flow at the leading edge occurs only at a unique angle of incidence which, according to Theodorsen definition could be called the ideal angle of attack (Ref 2.54). This angle was defined by Theodorsen, who first introduced this term in 1930, as that '…at which the flow enters the leading edge smoothly' and the lift at the very edge equals zero. At such an angle the wool-tufts on both sides of the cambered plate from leading to trailing edge flew straight aft. At any angle of incidence other than the ideal, a region of separation flow near the leading edge was observed and clearly indicated by fluttering yarns.

Subsequently, observations of flow were made on the same cambered plate with various, so-called, head-foils attached to the leading edge. The intention was to simulate the new headstay system, shown in Photos 2.14 and 2.24, which has become popular amongst serious racing yachtsmen. Experiments gave a similar picture of the flow, but with one small difference. The angle of incidence at which L/D ratio was maximum appeared to be slightly higher and separation did not occur so readily as before with changing incidence angle. In other words, the character of the flow at the leading edge was somehow less sensitive to angle of incidence variation. This in turn affected noticeably and advantageously the aerodynamic properties of the foil. The relevant data will be discussed in Part 3.

Attention is now drawn to Photo 2.25 which illustrates the flow on the leeward side of the cambered plate with one of the head-foils tested and shown earlier in Photo 2.14, namely Section 5. In principle, the picture of the flow is similar to that observed by some other investigators round the thin foils and represented schematically in Fig 2.57B. The peculiar behaviour of the flow, displayed in Photo 2.25A and the explanatory sketch, is almost certainly connected with an inability of the flow to remain closely attached to the foil surface, while passing from the stagnation point S_t situated on the windward side around the leading edge to the upper surface. Instead, flow separates from the leading edge as soon as the stagnation point S_t moves to the windward surface when the incidence angle increases. The separated flow passes above the surface of the foil and re-attaches further downstream in the manner described earlier as type 3 of the stall pattern. The exact mechanism of re-attachment is, however, obscure. All that can be said, for the present, is that for small incidence angles the flow re-attaches to the foil surface a

Photo 2.24   The new single and double-grooved headstay systems (some of them are shown here as examples) have become popular amongst serious racing yachtsmen. The rope luff of the headsail is hoisted within the foil groove which, swivelling freely, aligns itself with the apparent wind. See Photo 2.14.

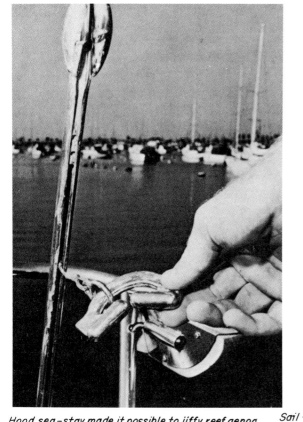

Hood sea-stay made it possible to jiffy reef genoa. Gismo in hand guides the luff of the sail between the rollers and into the groove.

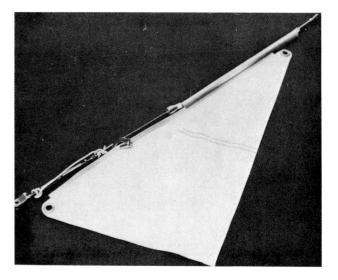

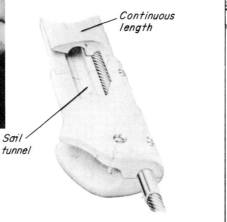

Continuous length

Sail tunnel

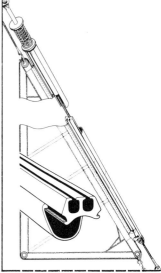

Photo 2.25   Flow over the lee-side of the rigid sail with head-foil attached
to the leading edge.

A. At small incidence angle α, flow re-attaches to sail surface
just behind the separation bubble in place marked R. Downstream from R tufts are lying flat along the surface.
B. Larger incidence. Fully separated flow indicated by
unstable tufts pointing towards the leading edge.

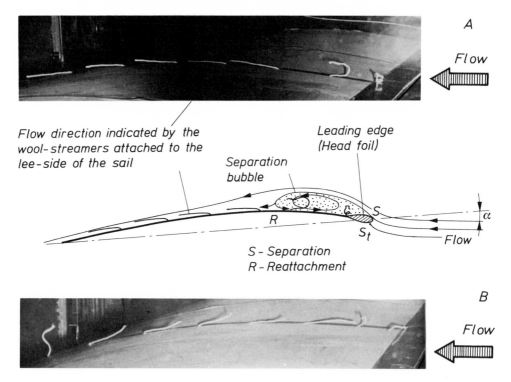

short distance behind the leading edge and then follows the foil surface up to the
trailing edge.

This is shown in Photo 2.25A, and the exaggerated sketch below the photograph
explains graphically the character of the flow. The region embraced by the separated
flow, between the point of separation S and re-attachment R and set into circulatory
motion, is commonly referred to as a 'long separation bubble'. It plays an important
part in determining the behaviour of the boundary layer on the foil surface and
consequently the pressure distribution. The size or extent of the bubble depends on
the incidence angle and it grows rapidly with increasing incidence until it extends
over the entire leeward side, at which stage its maximum thickness is in the order of 3
per cent of the chord length c. The full stall is then attained and any further increase
in angle of incidence gradually reduces lift. This is shown in Photo 2.25B and the tell-

tales indicate distinctly the full stall with reversed flow along the whole leeward side of the foil.

The presence of the long bubble behind the leading edge makes the pressure distribution radically different from that developed on a thick foil. This difference is distinguished in Fig 2.57, which illustrates the flow around the thick and thin foils and also the qualitative distributions of pressure for both foils including theoretical pressure distribution as a base for comparison (Refs 2.15, 2.53). It will be seen that the sharp suction peak near the leading edge of the thin foil is not realized; instead, a suction plateau of a reduced level extends over the region occupied by the bubble length. Within the extent of this suction plateau, the pressure gradient is about zero and this eases the flow, bringing the streamlines closer to the surface. There are indications that within this region the boundary layer becomes turbulent and this in turn facilitates the re-attachment.

Speculating about the role of the long bubble it appears that, by building up a kind of artificial thickness at the leading edge and over the front part of the foil, nature makes the flow round the sharp edge easier. Without a long bubble, which radically alleviates the difficulties associated with negotiating an unfavourable pressure gradient, the flow would separate and never re-attach to the surface of the foil. In other words, the long bubble may be regarded as an agent mitigating the severe consequence of sharp suction peak at the leading edge which would otherwise appear and might lead to sharp separation as shown in Fig 2.59A. However, the bubble has to be paid something for services rendered. The generally poorer aerodynamic qualities of very thin foil sections in terms of lift/drag ratio, except, as we will see later, at very low Reynolds Number, are due to the fact that drag is high. Plausible explanation may be given as follows: it has been mentioned that inside the bubble there is a flow set into a circulatory motion–a reverse flow vortex, indicated in Fig 2.59B. In three dimensions this vortex can be imagined as a rotating flat cylinder of fluid all along the lee side of the foil or part of it. Such a reversed flow vortex has a certain mass and is kept in rotational motion at the expense of the kinetic energy which can only be taken from the oncoming stream. The additional drag experienced by the sail-like foils is, in a way, a measure of energy lost. It can be assessed by comparing the data given in Figs 2.60 and 2.61, which represent some results obtained in the wind tunnel of Southampton University (Ref 2.55). Tests were made on a thin cambered plate, bent into the arc of a parabola of 15 per cent camber ratio, and on a typical modern low-drag aerofoil (NACA section 64A212) both with the same aspect ratio of 6 and at the Reynolds Number of about 2.5×10^5. The maximum lift coefficient C_L of the thick aerofoil is about 0.85 at an angle of incidence of 14° from the angle of zero lift; the maximum C_L for the cambered plate is 1.7 at an angle of incidence of 34° from the angle of zero lift and it develops an earlier peak in the C_L–α curve of 1.6 at an angle of 23°. Thus the cambered plate ultimately develops twice as much lift as this particular thick aerofoil, but at the expense of considerably increased drag, for the maximum value of lift/drag ratio of the plate is only 9.3 whereas for the aerofoil it nearly reaches 21, as seen in Fig 2.61.

Fig 2.60   Lift and drag coefficients for a thin cambered plate of 15 per cent camber ratio, f/c, and typical modern aerofoil (section NACA 64A212) both with an aspect ratio of 6 and at Re Number of about 250,000 (2.5×10^5). Rectangular planforms.

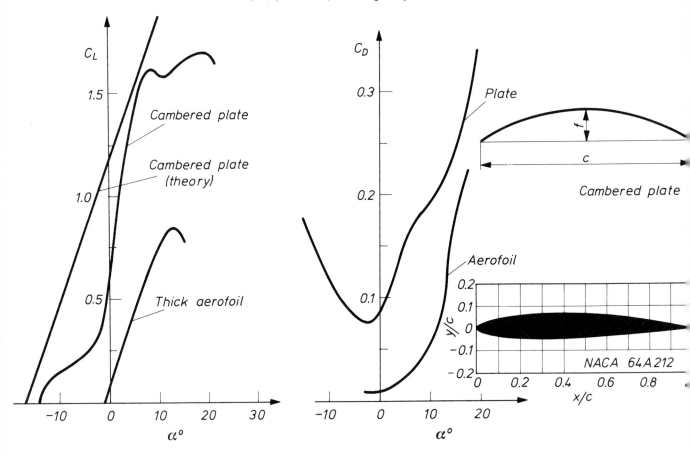

(c) *Thick versus thin foil controversy*

Thick aerofoils of greater camber can reach C_L values around 1.8 and still maintain high L/D ratio. They are therefore superior relative to thin foils. However, their superiority is lost at low Reynolds Numbers (low wind velocity) whilst the performance of thin foils is relatively unaffected by the Reynolds Number.

From what has already been said it may be inferred that, if a thin foil section, such as a sail, is to be employed successfully at various wind speeds and at different angles of incidence, it would at least be necessary to have its leading edge adjustable. The importance of the angle of incidence which gives maximum 'entrance efficiency' at the very leading edge has been pointed out while referring to Theodorsen's ideal angle of attack. Figures 2.59A and B clearly demonstrate the rule that the flow is ready to separate whenever velocity vectors are not tangent to the leading edge. This requirement for a smooth entry can only be satisfied by an adjustable curvature of

Fig 2.61   Lift/drag ratio of the two foils presented in Fig 2.60.

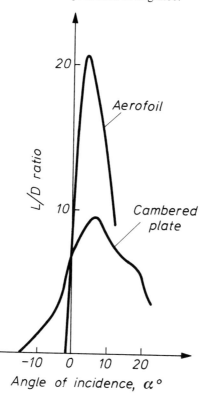

Fig 2.62   Flow in proximity of leading edge in two different conditions at
ideal angles of incidence.
a. Small angle of incidence–flat sail.
b. Larger angle of incidence more cambered sail.

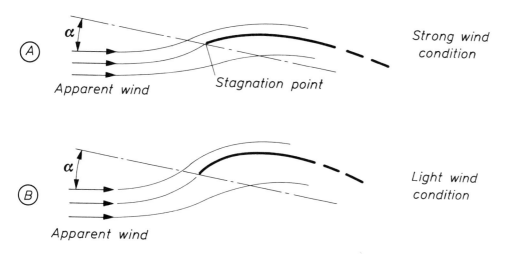

the foil at its leading edge. To make this point clear, let us refer to a headsail working in strong and light wind conditions, as presented in Fig 2.62A and B. Configuration A gives the leading edge curvature in a strong wind at which the boat usually points higher to the apparent wind and carries a flat sail to keep the heel angle small enough. Configuration B gives the curvature of the leading edge suitable for a light wind, in which the boat usually carries sails with larger camber. In such conditions, the optimum apparent wind angle is greater than that at stronger winds. These are just hints, and more information concerning sail setting with the aid of tell-tales is given in Part 3.

It is evident that, with a well-rounded nose to the foil section, the requirement of a smooth flow at the entrance edge is lessened. A thick nose is thus virtually equivalent to an adjustable leading edge. The thicker the section the greater is the possible change in the camber line of the foil and its incidence without provoking leading edge separation. For this reason, the curve of Lift/Drag ratio versus incidence angle for thick sections has a flattened peak. A thin section, on the other hand, does not lend itself to any such flexibility, hence its poorer characteristics in terms of L/D.

It is interesting to notice that the advantages associated with entrance efficiency were unconsciously obtained by rounding the 'front edge' of the early thin aerofoils developed at the beginning of the 20th century, well before theory could possibly justify further foil evolution that finally lead to 'thick' aerofoils for high speed work.

As depicted in Fig 2.59, sketch C, the thickness function may be considered as a fairing wrapped symmetrically around the so-called mean camber-line which serves as a skeleton of the thick foil for calculation and classification purposes. The radius of curvature at the leading edge or entrance is, as demonstrated earlier, an essential parameter controlling flow round the foil and hence lift and drag. The importance of the entrance efficiency has been discovered once again in the history of aero-dynamics–this time by sailors. Head-foils of various forms, displayed in Photos 2.14 and 2.24, combined with tell-tales or wool-streamers, are indicative of a new progress being made towards better understanding of how a sail really works and its requisites for high efficiency.

The process of stalling–whether it is abrupt or gentle, depends entirely on the nature of the boundary layer, and its behaviour in the presence of an adverse pressure gradient which builds up with increasing angle of incidence. This rather complex process has been compared to a contest between laminar separation near the nose (Photo 2.7) and turbulent separation near the trailing edge (Fig 2.57A), one or the other winning and thus determining the maximum lift (Ref 2.30). Which type actually wins is very much dependent on the Reynolds Number, or in other words on the ratio between viscous and inertial forces involved in fluid motion and which control the character of boundary layer flow. Figure 2.63 shows dramatically this effect on the symmetrical NACA 0015 section. It is seen that $C_{L\max}$ increases from about 0.85 at Re = 42,900 up to 1.55 at Re = 3,260,000. It can be expected therefore that within the range of Reynolds Number values encountered in full-scale sailing conditions the maximum lift coefficient $C_{L\max}$ for both fin keel and rudder

Fig 2.63   Reynolds Number effect on lift of NACA section 0015.

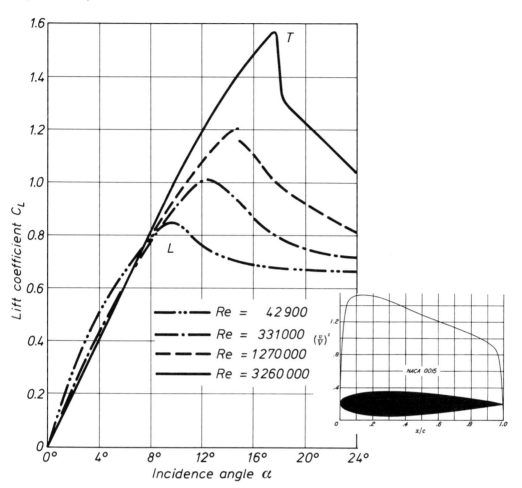

may change by a factor of about 2.0. Actually, the Reynolds Number for the fin keel varies from almost zero, when the boat begins to accelerate, to about 10 millions (10^7) in the case of large craft sailing fast. Of course, in steady-state sailing conditions the fin keel never operates at near stall angle of incidence but, while manoeuvring, accelerating and rolling, large angles of incidence may be reached and, from the standpoint of controllability and performance of the yacht, it is of interest to discover what maximum lift the hull appendages are capable of generating in unsteady motion. Reduced effectiveness of both the keel proper and rudder, caused by a sudden decrease of boat speed while manoeuvring, may well be augmented by a large drop in $C_{L\max}$, followed by a simultaneous increase of drag, also due to Reynolds Number effects. This is one of the factors which contribute to the often occurring deterioration of steering efficiency suffered in unsteady sailing conditions, which is particularly apparent in the case of a boat equipped with a high aspect ratio rudder of small area. The highest value of Reynolds Number in Fig 2.63 would correspond, for example, to a fin keel of 5 ft chord length and sailing velocity about 4.7 knots. Table 2.5 or Eq 2.19A can be used to calculate the relevant Reynolds Numbers for other foil dimensions and different sailing speeds.

Variations of maximum lift coefficients $C_{L max}$ with Reynolds Number, for some symmetrical NACA sections frequently used for fin keels and rudders, are given in Fig 2.64. The general tendency for maximum lift is first to increase slowly, then more rapidly, and finally to level off to a substantially constant value at a Reynolds Number well beyond that at which ordinary sailing yachts operate, i.e. Re greater than ten million. The reason is that in the low Reynolds Number range, distinguished in Figs 2.63 and 2.64 by the letter L, the laminar separation commencing from the leading edge takes place at comparatively low angle of incidence. Ultimately, at large Reynolds Number marked in the above Figures by the letter T, the robust, turbulent boundary layer is well established over most of the section length, thus the separation is delayed and shifted towards the trailing edge. In such a condition of substantially attached flow and small wake any further increase in Reynolds Number causes no appreciable change in $C_{L max}$. In the range of increasing Reynolds Number between L and T the transition point from laminar to turbulent boundary layer moves progressively from the trailing edge towards the leading edge. Since the tougher turbulent boundary layer adheres better to the contour of the section surface the separation wake becomes less extensive, shrinking in a manner similar to that shown for the circular cylinder (Fig 2.33). In consequence, the circulation increases and so does the $C_{L max}$. It appears that the highest maximum lift coefficient $C_{L max}$ is reached when the Reynolds Number corresponds to the occurrence of fully developed turbulent boundary layer, beginning very close to the leading edge where usually laminar separation takes place.

Of course, the particular variations in $C_{L max}$, as well as in C_D, are primarily dependent upon the foil section simply because the section geometry, its surface contour and its thickness at the leading edge, produce for every section a unique velocity and pressure distribution. For example, both C_L and C_D coefficients of a thick, asymmetrical section-60 shown in Fig 2.65 change with Reynolds Number in a rather astonishing manner bordering on sorcery. This becomes immediately apparent when its characteristics are compared with those of a thin, sail-like section-417 (Refs 2.56, 2.57). Reynolds Number effects may in some circumstances be of great practical importance and deserve some attention since in more recent years many people have become interested in improving the ultimate speed of sailing craft by using thick, rigid, or semi-rigid wing sails in the place of conventional thin, soft sails.

Sailing yachts operate in winds ranging from calms to gales, therefore the relevant Reynolds Numbers at which their highly tapered triangular sails work must necessarily vary over a large range. For a triangular sail of a 12-Metre size yacht the Reynolds Number varies from almost zero at the head to about 5 millions (5×10^6)

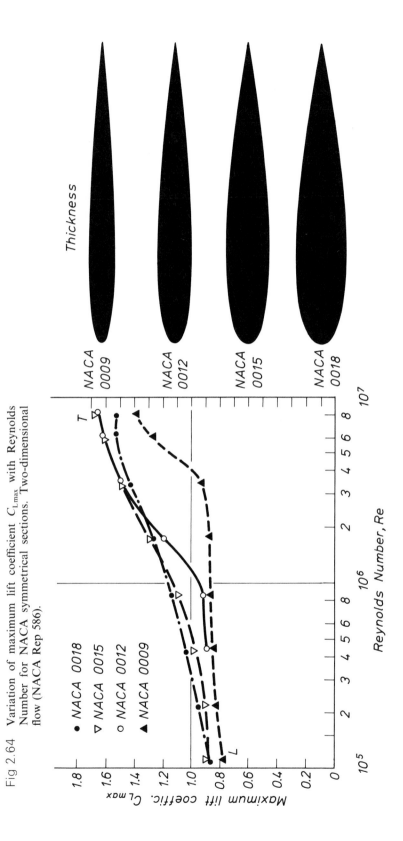

Fig 2.64 Variation of maximum lift coefficient $C_{L\,max}$ with Reynolds Number for NACA symmetrical sections. Two-dimensional flow (NACA Rep 586).

Thickness

NACA 0009
NACA 0012
NACA 0015
NACA 0018

● NACA 0018
▽ NACA 0015
○ NACA 0012
▲ NACA 0009

Maximum lift coeffic. $C_{L\,max}$

Reynolds Number, Re

Fig 2.65 Effect of Reynolds Number on aerodynamic characteristics of
thick (N 60) and thin (417a) sections.

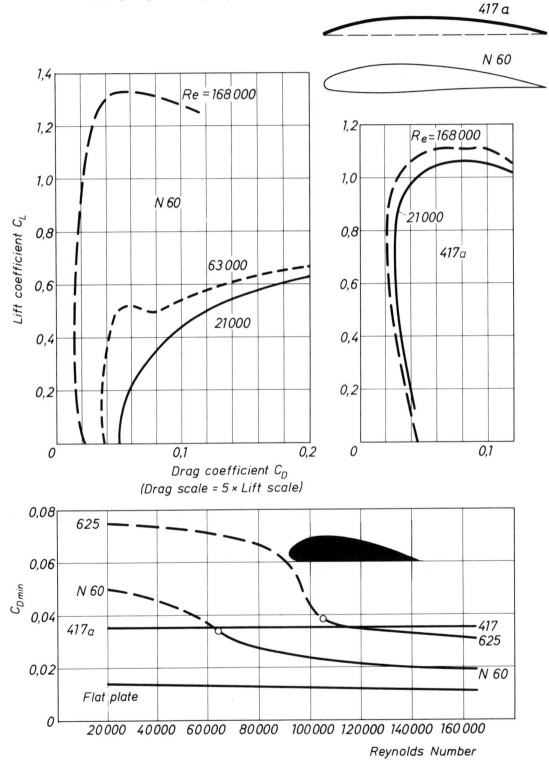

at the foot, assuming a wind velocity $V_A = 40$ ft/sec and sail chord about 20 ft. In the same wind the upper value of Reynolds Number for a One Ton Cup yacht would be about 2.5 millions (2.5×10^6). Table 2.6 and Eq 2.19B can be used to calculate the relevant Reynolds Numbers for other conditions.

Figure 2.65 gives good insight into the Reynolds Number effect on the aerodynamic characteristics of the two different asymmetrical sections at a low range of Reynolds Numbers. Two important conclusions can be derived from it:

Firstly–thick wing sails are nothing like as good as thin, soft sails in light weather conditions. The thick foil is very sensitive to Reynolds Number effect while the thin section displays remarkable lack of sensitivity in this respect (reasons are discussed in Part 3).

Secondly–comparison tests of the two different aerofoils belonging to different families can be most deceiving if testing is done at low Reynolds Numbers.

These differences between the thick and thin foil sections do not appear to be well known and appreciated. Readily available information on lift and drag coefficients of standard aerofoil sections refers to Reynolds Numbers which might be encountered only in rather strong weather conditions. Therefore, an assumption accepted by some enthusiasts, that the aerodynamic characteristics of rigid wing sails are better than that of ordinary thin conventional soft sails, is only partly true.

(d) *Rig of* **Lady Helmsman**

Disturbing inconsistencies, for instance between the full-scale rig performance of *Lady Helmsman* and that established in the wind tunnel while testing a 1/4 scale model of her wing sail shown in Photo 2.26, were plainly exposed in the discussion

Photo 2.26   The advances in spar and sail design, which led to *Lady Helmsman*'s supremacy in the International Catamaran Challenge Trophy, can largely be attributed to the wind-tunnel facilities at Southampton University, where the model of the rig was developed. The picture shows Austin Farrar with his ¼ scale model being tested. Observation of the flow, together with measurements of forces, helped to establish the optimum size and shape of streamlined mast relative to the sail proper.

Fig 2.66 Some results of tests on *Lady Helmsman* Una-rig ($\frac{1}{4}$ scale model, wind tunnel velocity 25 ft/sec).

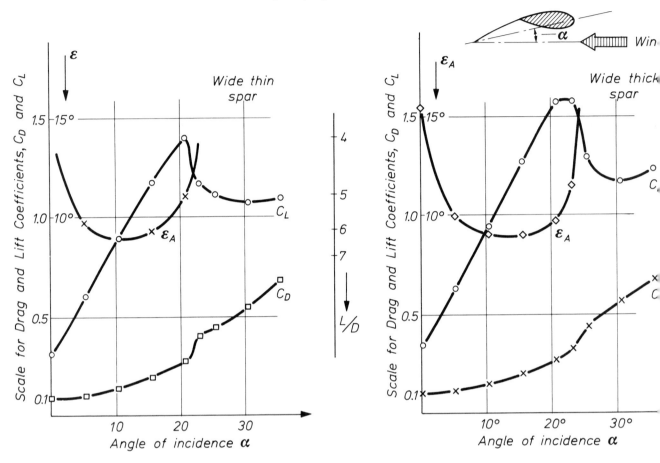

on the paper *Class C Racing Catamarans* published in RINA Transactions (Ref 2.58). The development of this Una wing sail for the British boat by Farrar was the result of several years' work aimed at exploiting any advantage which might be obtained by:

(a) precise control of the amount of twist,
(b) elimination of the mast interference.

Tests were inspired by the late General Parham who had already made exploratory experiments on wing-sails and curved spars in 1947.

According to the designer's own remarks: '...in the completed *Lady Helmsman* rig (which is a combination of thick rigid foil and sailcloth extension), the whole bore a strong resemblance to, and was based on, the Göttingen high lift glider section of 1928.' Some results of tests with the two different spars, performed by Farrar at Southampton University, are shown in Fig 2.66; finally, the wide-thick spar was selected as the better one. It is seen that the drag angle ε_A, the most vital factor controlling the performance of any fast sailing craft, is just below 9°, which corresponds to a *L/D* ratio of about 6.4. With her wing sail *Lady Helmsman* beat the American *Gamecock* in 1966 and also the Australian challenger *Quest II* in 1967. In a paper read at the RINA meeting the inventors claimed that '...these so-called wing sails have, in their most advanced form, shown an unquestionable superiority over the more conventional rigs with which the C-Class catamarans were equipped' (Ref 2.58).

This view was challenged by the late T. Tanner, as follows:

'...I believe I am right in saying that *Lady Helmsman* showed considerable superiority over other boats and I have therefore taken Fig 2.67 and tried to make a comparison with other rigs. If the results of wind tunnel tests on a 1/3 scale model of the X One Design rig (Ref 2.59) are plotted on this graph they show no appreciable differences. If then *Lady Helmsman* was so good, *wherein lies the reason for this superiority?*'

One more glance at Fig 2.67 will assist in this regard. The Lift/Drag ratio of *Lady Helmsman*'s rig, a factor which is so important in high speed sailing, is not any higher than that of the conventional sloop rig such as that of the X Class.

J. Fisk, the expert on the actual sailing of these craft, stated in discussion:

'...take all [Mr Tanner's] graphs and mathematics as being words of wisdom, because I do not understand these things well, but, when sailing against Una rigged and sloop rigged boats, the Una boats go much faster and closer to the wind. I do not know the reason but perhaps it can be proved in the wind tunnel some time.'

General Parham spoke rather bitterly: '...the graphs and figures pronounced by Mr Tanner seem to show that *Lady Helmsman* had a rig inferior in most respects to the normal rigs with which he compared her, and yet she won. Is it a question of sending Mr Tanner back to find some better figures or what? I do not know.' 'Admittedly,' commented A. Farrar, 'on Mr Tanner's figures a C-class catamaran would perform better with an X-boat rig, but I feel there must be some practical reason why it would not work. Certainly, any soft sail catamaran was out of date many years ago when fully battened sails were introduced, and then the ordinary sloop was beaten so frequently by the Una rig. There must be a practical reason for this and it is *the practical effect* that wins races.'

Who is right? Are wind tunnel tests reliable? What is this 'practical effect' that wins races and remains undetected by the wind tunnel?

The answer to these questions may be sought in Figs 2.65 and 2.69–this is the Reynolds Number effect that might be blamed for the discord in discussion concerning *Lady Helmsman*'s virtues. One may say therefore that all controversialists were, in one way or another, right in their opinions. Certainly, on the basis of the directly presented wind tunnel data, the wing-sail superiority could not

Fig 2.67 Comparison of characteristics of the two rigs tested in the same
wind tunnel at roughly the same wind velocity and the same
length of masts.

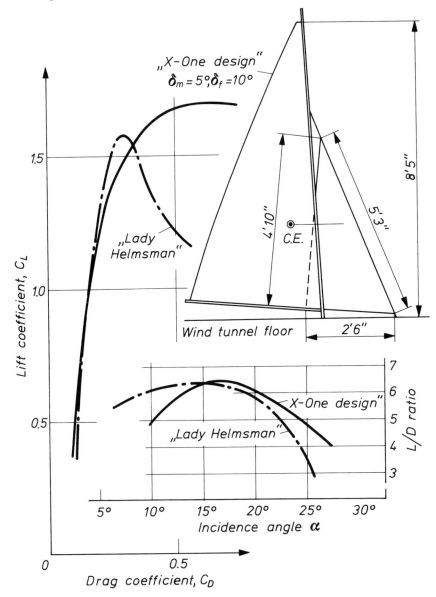

Fig 2.68   Comparison of *L/D* ratio of the rigid foil RAF 30 AR = 6.0
with some soft sail rigs tested in the Southampton University
wind tunnel.

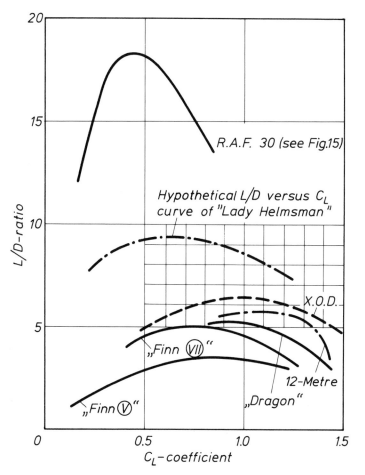

possibly be proved. However, bearing in mind the evidence of races and data
incorporated in Fig 2.65, it can be argued that, at relatively high Reynolds Numbers
actually attained in full-scale conditions, the *L/D* curve of the *Lady Helmsman* rig,
when plotted say against lift coefficient, should lie above the curves illustrating the
characteristics of soft sail rigs; Fig 2.68 may facilitate the reasoning along this line. It
demonstrates several curves of *L/D* ratio versus lift coefficient C_L obtained in wind-
tunnel tests on models of Finn, Dragon, 12-Metre and X.O.D rigs. There is also
plotted the *L/D* curve for the symmetrical section RAF 30 of aspect ratio 6.0, at the
Reynolds Number beyond the critical one. For comparison, as a pure conjecture,
there is drawn a hypothetical curve of *Lady Helmsman*'s rig which might reflect her
plausible characteristics at the Reynolds Number corresponding to full-scale sailing
conditions.

Fig 2.69 Dependence of lift coefficient C_L on Reynolds Number, for flat plate, cambered plate Gö 417a and thick aerofoil Gö 625 Schmitz F, *Zur Aerodynamik der kleinen Reynoldschen Zahlen*, Jahrbuch 1953 d WGL.

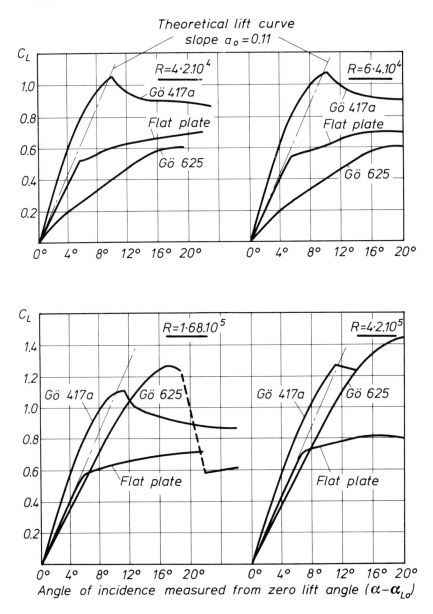

Angle of incidence measured from zero lift angle $(\alpha - \alpha_{Lo})$

Fig 2.70   Variation of lift with incidence for circular arc section of different camber ratio, f/c. Two-dimensional flow. Angle of zero lift α_{Lo} is almost independent of Reynolds Number, but its negative numerical value increases with camber.

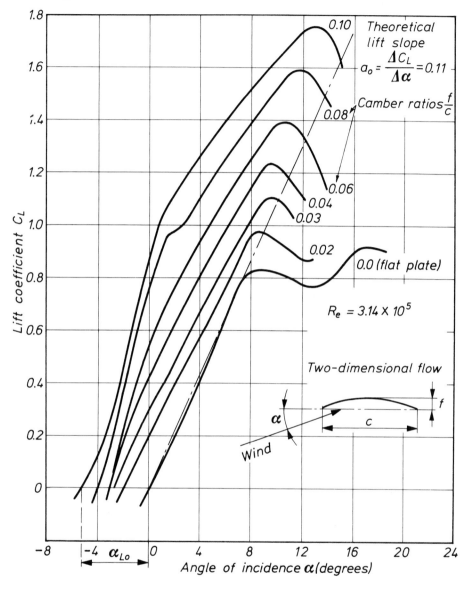

Fig 2.70A   Variation of drag with incidence for circular arc sections of
camber ratio, f/c, 0.0, 0.06 and 0.1 (Ref 2.60).

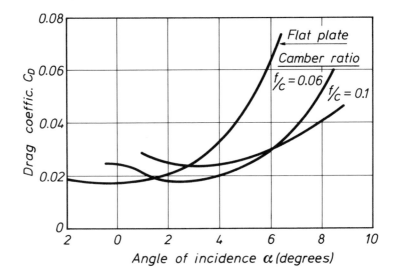

Since the Reynolds Number effect, as demonstrated in Figs 2.65 and 2.69, is indeed profound and rather unpredictable, it is difficult to guess how close the aerodynamic characteristics of her wing sail actually approached the known characteristics of rigid aerofoil sections, such as RAF 30, or others given in Ref 2.31. However, as far as L/D ratio is concerned, the full-scale wing sail performance of *Lady Helmsman* was certainly better than the measured characteristics in the wind tunnel. There is no other reason which might sensibly justify her racing record.

Unfortunately, very little correlation data between full-scale characteristics of aerofoil mast plus sailcloth extension and wind tunnel tests, at a sufficiently low range of Reynolds Number, are available to make any calculation or prediction reliable. This lack can be attributed to the sudden jump from low to high speed made in aeronautics since the Wright Brothers' flight. This way, low speed aerodynamics were somehow left behind, underdeveloped. The work done in the past on thin foil characteristics at low Reynolds Numbers has never been collected, except perhaps in F Schmitz's *Aerodynamik des Flugmodells*, published in Germany.

Figure 2.69 gives some information from German sources about the dependence of lift coefficient C_L on Reynolds Number for sections which are of particular interest, bearing in mind practical application in sailing, namely a cambered plate, a flat plate and a thick section (Refs 2.56, 2.57). Similarly as in Fig 2.65, to which Fig 2.69 is partly related, the same trend can be distinguished: the rate of change of lift coefficients C_L with incidence of thin and cambered plates is, within the incidence range up to the stall angle, little affected by Reynolds Number; while the rate of C_L change with α of thick foil is very much dependent on it. This can be assessed by comparing the actual slope of the C_L curve with the theoretical lift slope given by Eq 2.14 and plotted in Fig 2.69 in the form of a thin, broken line.

(e) *Thin foil properties*
Before analysing Fig 2.69 in some detail, let us recall Eq 2.14, written in the form:

$$C_L = 0.11\alpha \qquad \text{(Eq 2.14 repeated)}$$

where α = incidence angle given in degrees and

$0.11 = \dfrac{\Delta c_1}{\Delta \alpha}$ is the coefficient which gives the theoretical lift-curve slope for two-dimensional flow.

Putting $a_0 = 0.11$ into equation 2.14 it can be rewritten:

$$c_1 = a_0 \alpha \qquad \text{Eq 2.21}$$

where   subscript 0 at the letter a indicates two-dimensional flow.

The method of presentation of c_1 curves in Fig 2.69, which represent the two-dimensional lift characteristics, requires some explanation. Usually the c_1 co-efficients are plotted against the angle of incidence α, measured between the wind direction and chord of the foil c, as shown in Fig 2.70 in which there is drawn a series of c_1 curves for a family of circular-arc aerofoils of increasing camber ratio, beginning from the flat plate up to camber ratio $f/c = 0.1$ (Ref 2.60). It can be seen from it that the larger the camber the smaller the incidence angle α at which a given lift coefficient is produced. For progressively increasing camber the relevant lift-curves are bodily shifted somewhat to the left. This shift can be measured by the angle of incidence at which the lift coefficient is zero. For instance, for the foil of camber ratio $f/c = 0.1$ the *no-lift incidence*, denoted in Fig 2.70 by α_{L0}, is about $7°$. The incidence angle corresponding to $C_L = 0$ is always zero for a flat plate or symmetrical sections and becomes negative and numerically greater for asymmetrical sections with increasing camber ratio f/c.

To make the lift characteristics of various foil sections directly comparable it is convenient to draw the c_1 curves in such a way that the incidence angle is measured from the zero lift angle $\alpha - \alpha_{L0}$. This is equivalent to shifting the lift curves to the right until the *no-lift incidence* coincides with 0 of the horizontal axis. This is done in Fig 2.69, which illustrates the dependence of lift coefficients on Reynolds Number for three different sections. Taking the theoretical value for the lift-curve slope $a_0 = 0.11$ as a yardstick of foil efficiency in producing lift we may introduce an efficiency factor k by the relation:

$$a = ka_0 = k\,0.11 \qquad \text{Eq 2.22}$$

where $a = \dfrac{\Delta C_L}{\Delta \alpha}$ is the actual or measured slope of the C_L curve of the three-dimensional foil.

As demonstrated in Fig 2.69, for the flat plate the value of $k = 1.0$ is very nearly reached, but it becomes smaller as the Reynolds Number decreases; falling to $k = 0.84$ at Re $= 4.2 \times 10^4$. Thin cambered plates display in this respect anomalous characteristics. For example, the measurements on Go 417a section, for c_l less than 0.6, gave a high value of k about 1.4, and this is almost independent of Reynolds Number. More detailed measurements presented in Fig 2.70 indicate that the character of their lift-curve slope a_0 depends critically on the camber ratio. For instance, for the foil of 10 per cent camber the lift-curve has two approximately straight portions of different slopes. The slope of the lower portion displays a very high efficiency factor k in the order of 1.8 which is much higher than existing theory predicts. Unfortunately, the upper part of the lift-curve shows quite poor characteristics in this respect, the deterioration being more severe as the camber increases. The non-linearity of the lift-curves is noteworthy and the apparent departure from the theoretical slope can partly be explained by assuming that, within the range of incidence angles below the C_L maximum, flow over the thin section of camber higher than 5–6 per cent is never fully attached. It would mean that even at the so-called ideal angle of incidence, when the flow enters the leading edge smoothly, separation in some form almost certainly takes place. To make this point clear, it should perhaps be repeated that reasonably good agreement between lift theory, based on a non-viscous fluid concept and experiments on streamline foils of moderate thickness and camber, can be expected only in the case where the flow is attached to the foil surface everywhere and the fluid streamlines leave the trailing edge smoothly. This agreement no longer holds if, for some reason, the flow separates at either leading or trailing edge and part of the flow energy is somewhat dissipated.

Owing to the existence of separation bubble at the nose and/or separation at the trailing edge, a certain discrepancy between experiments and classical theory of lift must be expected. It is perhaps unfortunate that the leading edges of sails are usually either sharp (headsail), or blunt (mast-sail combination) because, for this reason, a relatively large penalty is incurred on them in terms of drag and, resulting from it, low lift/drag ratio as shown in Fig 2.61. However, it should be remembered that, as pointed out earlier, the sharp leading edges offer certain advantages in the range of low Reynolds Number in which streamline foils fail as efficient lift-producing devices. Referring again to Fig 2.69 it will be seen that the thick section Gö 625, like section N 60 in Fig 2.65, and any other thick conventional sections, have a poor value of efficiency factor k, at low Re up to about 10^5. The k value increases with Reynolds Number Re and at Re in the order of 10^6 the factor k, for foil sections of moderate thickness ratio t/c, becomes almost equal to 1.0.

Streamline sections of small thickness $t/c = 0.05$, such as the Joukowski profiles shown in Fig 2.71, produce similar lift-curves to that of the thin circular-arc sections of Fig 2.70. The lift-curves are non-linear and their slope generally decreases more or less rapidly with increasing coefficient. It appears that Joukowski foils of small thickness are not capable of generating such a high maximum lift as thin foils for the

Fig 2.71   Lift and Drag coefficents for Joukowski profiles (Ref 2.62).
Re = 4.2 × 10⁵. Two-dimensional flow.

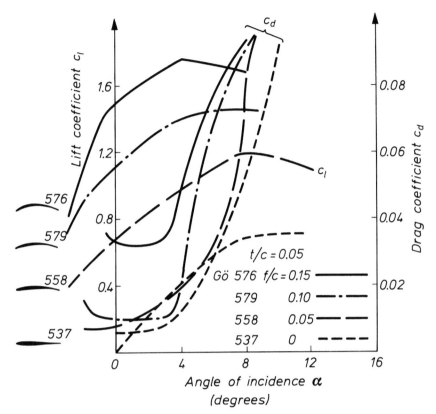

same camber ratio f/c as demonstrated in Fig 2.70; however, they have better drag characteristics in low range of incidence angles. Comparison of relevant drag data in Figs 2.70A and 2.71 reveals that the thin circular-arc sections produce about twice as much drag as Joukowski sections but only up to incidence angle of about 4°. Beyond this limit the drag coefficients of both types of foil become comparable. This implies that once separation takes place (most likely the same type of separation, defined earlier as *thin foil stall* in form of a long bubble at the leading edge), there is not much difference between those types of section in the range of Reynolds Numbers at which they were tested.

Our inquiries into the merits and demerits of various foil sections suggest that one should expect to gain something, in terms of lower drag, by introducing streamline thick foil in place of centreboards or rudders formed of thin, flat plate. But it does not imply, as one controversial journalist maintains, that '...most of the theorists will tell you that centreboards formed of thin flat plate will be infinitely less effective in resisting leeway than the thicker, aerodynamic form.' Such an overstatement, supported by evidence that '...in practice, the racing dinghies with flat plate type centreboards are very little different in performance to those with thicker aerodynamic forms,' certainly does not clarify the issue. The contribution of the centreboard or rudder profile drag to the total drag of a boat is relatively very small, nevertheless noticeable, particularly in the case of high performance boats sailed by experienced helmsmen. What really matters is not the almost negligible reduction in drag at low incidence angles, but better overall performance of streamline section as compared with, say, a flat plate section. The most important point to consider when designing an efficient rudder is to realize that its basic function is to provide a large lift force when needed; as when, for instance, a broaching condition threatens on a spinnaker reach. Bearing in mind this aim, it is not sufficient to have a streamline but thin section rudder. An analysis of the data in Fig 2.58 may give the answer to the question as to what should be the thickness ratio of, say, a spade rudder. It will be seen that the main disadvantage of thin sections with a thickness of 6 or 9 per cent of chord is that they stall at relatively small incidence angles and, what is more important, the available maximum lift coefficient is not much different from an ordinary flat plate. In these respects thicker sections, of thickness ratio 12 per cent or more, are better. Besides, it can be found in Ref 2.31 that, at incidence angles greater than about 5°, the thin section of 06 series produces much higher drag than any other section shown in Fig 2.58. Sections of thickness 9 and 12 per cent have comparable drags in a wide range of incidence angles. To conclude, there is no point in using streamline but thin sections for rudders or fin keels.

Pronounced influence of the section shape on lift and drag characteristics is illustrated in Fig 2.72 which represents the polar diagrams of three basically thin Göttingen sections investigated at the same Reynolds Number 4.2×10^5. The sections are different in leading edge shape and thickness distribution, and these are the factors that primarily affect the character of the flow over the front part of the foil, and hence, the pressure distribution and finally forces. It is a rather interesting feature of the lift curves presented that all the sections have an almost identical minimum drag angle ε_A given by the same tangent line and the vertical axis of the graph in Fig 2.72. Since the drag scale is expanded 10-fold as compared with the lift scale the angle of $\varepsilon_{A min}$ drawn has only a qualitative meaning.

Another interesting feature of those graphs is that the minimum drag angle ε_A equivalent to L/D max, occurs at different lift coefficient for each section. As already pointed out, the flow of air over a section with a sharp leading edge is likely to be smooth at one particular angle–the ideal angle of incidence–or close to it. When the leading edge is rounded on a comparatively large radius, as in the case of Gö 335 section, it is relatively easy for the oncoming current of fluid streamlines to divide at almost any point round the leading edge, without requiring any streamline of fluid, to turn a sharp corner that may lead to separation. The true leading edge in the aerodynamic sense, or stagnation point where the upper (leeward) and lower (windward) surface flows divorce, moves easily up and down the nose of the profile, as the angle of incidence changes. In the case of sharp leading edge, this is impossible,

Fig 2.72   Polar diagrams of three basically thin Göttingen profiles:
Gö 417a–cambered plate with nose and trailing edge rounded
Gö 335–rounded nose and sharp trailing edge
Gö 610–sharp nose and trailing edge (segment of circle).

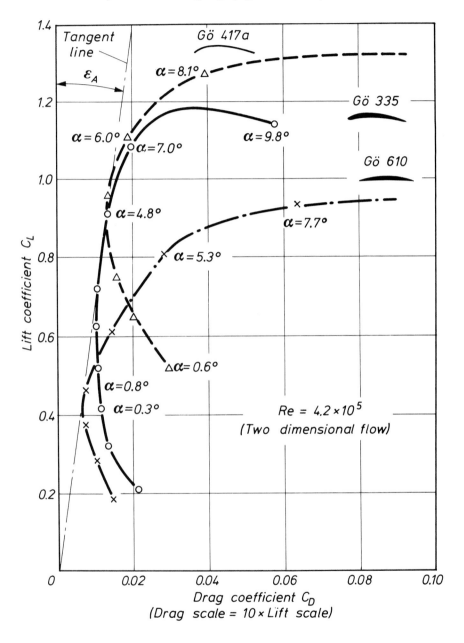

and the flow conditions change dramatically whenever the foil is set at an angle above or below the ideal angle of incidence. This inability of the flow to avoid separation, more acute the sharper the leading edge, is well reflected in Fig 2.72.

Let us look at the polar diagram of Gö 610 section which is a segment of a circle with a sharp leading edge. The maximum L/D ratio ($\varepsilon_{A\,min}$) occurs at an angle of incidence about 0.5°. The corresponding lift coefficient c_l is in the order of 0.45 only. It is noticeable that above and below this particular angle of incidence of 0.5°, which we may regard as an ideal angle, the polar diagram curve departs quite rapidly from the tangent line. This indicates that the flow conditions at the leading edge change dramatically and disadvantageously because of separation. A direct outcome of this is a rapid increase in drag.

For the section Gö 335 with its rounded nose, the lift coefficient at L/D max is about 0.8, almost twice as high as in the case of Gö 610 section. It occurs at incidence angle of about 3.0°, but, as seen in Fig 2.72, section Gö 335 is not as sensitive to the angle of incidence variation as section Gö 610. This is reflected in the graph by the rather flat polar diagram of section Gö 335, in the region where the tangent line touches the curve. In more practical language it would mean that this section is efficient over the much larger range of incidence angles than section Gö 610.

On account of the more distinct camber of the forward part of the section Gö 417A, its ideal angle of incidence is higher than that for the Gö 335 section, and as expected, the lift coefficient at L/D max is also higher–about 0.95. However section Gö 417A is almost as sensitive to angle of incidence variation as section Gö 610.

Different characteristics of the sections in question are partly attributed to the distinctly different leading edge flow conditions, and partly to camber effect. One may say, that no one section is better than the other. Much depends on what purpose the particular section serves, and there is no ideal section for all purposes.

It has been verified by experience that the lift coefficient C_L for conventional sailing craft should be as large as is compatible with stability requirements. From this standpoint, assuming equal L/D ratios, the conventional thin sail closely represented by section Gö 417A, appears to be superior to both sections Gö 335 and Gö 610, particularly the latter. Nevertheless, section Gö 610 has been successfully employed on hydrofoil supporting wings, as shown in Photo 1.16, Part 1. In such a case, if one wishes to avoid cavitation (see Note following Table 2.2–Properties of Fresh Water), special sections must be used, characterized by uniform, peakless pressure distribution on the suction side. It is found that conventional profiles with well rounded noses, whose maximum thickness lies further forward, are of little use for hydrofoils operating at high speed. This is because high peak of suction at the leading edge favours premature occurrence of cavitation. On the other hand, sections with uniform pressure distribution (pressure plateau), such as circular-segment profiles or the so-called ogival sections, and also low-drag NACA sections shown in Fig 2.45, whose maximum thickness occurs at half the chord, or even further aft, are much better in this respect.

(f) *Padded sails*

Some conclusions that may be derived from Figs 2.70, 2.71 and 2.72, together with certain implications resulting from Reynolds Number effects on C_L displayed in Fig 2.69, might well be of some value in guiding the future development of unconventional sails such as, for instance, the so-called padded sails or semi-rigid sails (Ref 2.61).

Rumours about the high efficiency of padded sails, and the possibilities of using them as a secret weapon, forced the IYRU to take immediate action by introducing in 1972 the rule which reads: 'In classes which require the sails to be made and measured in accordance with the IYRU sail measurements instructions padded sails are prohibited.' Of course, the IYRU ban does not exclude unconventional sails from classes where rules are more liberal, or there is no restriction at all.

The patent pending padded sail concept is illustrated in Fig. 2.73 and in the inventor's own words the semi-rigid sails have three special assets:

a. They appear less inclined to flog than ordinary sails, so they seem to set closer to the wind and go on working when other sails would have stalled.

b. The outside envelope of Terylene flattens the foam plastic at the edges (I had gambled on this, and it was one prediction that was satisfyingly achieved). As a result it is not necessary to taper the foam sheets at the edge, provided they are not too thick. This makes the fabrication of padding simple.

c. The foam on the model was 'good tempered'. It did not squeeze out, distort or behave in a awkward manner. It needed very little in the way of through stitching to hold it in place. This meant that we could take out one set of foam and put another in fairly quickly. The same should apply to full size sails, though there are obvious limitations as to what can be done on the foredeck or in the fo'c'sle of the Class III ocean racer going to windward in force 7. But ashore or in moderate weather it should be possible to swop the padding around quickly enough. Changing the stuffing of a dinghy sail should present few problems in any weather, doing the work ashore.

Fig 2.73   A section through a padded sail.

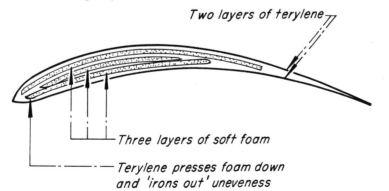

Wind tunnel tests described in Part 3 did not confirm, however, the inventor's expectations that '...a tubby cruiser setting a padded sail beat a slightly sleeker racing yacht to windward, just because the former was *properly* rigged in what...will be the 1972 style!'

A cross-section that looks beautifully streamlined to the layman's eye may display in the wind tunnel very poor characteristics in terms of lift and drag. Designing an efficient section which may fulfil predetermined functions imposed by conditions in which they will operate is not an easy job even for professionals. The so-called Joukowski profiles (from the Russian scientist who derived them mathematically at the beginning of the 20th century) depicted, for example, in Fig 2.71 and tested in Göttingen (Germany) during World War I, became obsolete soon after the speed range of flying machines changed. According to Mises (Ref 2.66) their leading edge portion is too massive in comparison with the thin tail and the maximum camber lies rather close to the middle of the chord, whereas a position within the forward third of the chord is preferred.

(g) *Significance of pressure distribution, mast effect*

Although the thin aerofoil theory as applied to sails is, as yet, far from satisfactory, certain features of the pressure distribution predicted on a purely theoretical basis (Refs 2.62, 2.63, 2.64) and measured on two-dimensional soft sails and cambered plates (Refs 2.64, 2.65) provide a partial explanation of the lift and drag characteristics, different in some respects from that displayed by thick foils. For instance, windward pressures on a cloth sail, as shown in Fig 2.74, are concentrated further towards the trailing edge than in the case of a well designed streamline thick section, demonstrated in Figs 2.15 and 2.16. Apart from that, the magnitude of the positive pressure near the trailing edge is somewhat greater than that observed on thick foils and this, as we shall see, tends to produce a large drag. A similar tendency is noticeable in Fig 2.75 which shows the pressures on the soft, full-scale sail, measured by Warner and Ober on the Marconi-rigged yacht *Papoose* (Ref 2.67). The striking feature of the pressure distribution on the windward side of the mainsail without battens (when part of the sail close to the leech is not flat, but curls to windward), is a concentration of positive pressure near the trailing edge.

An explanation should perhaps be given as to the value of the positive pressure coefficient C_p (at the stagnation point) recorded on the windward side of the sail in Warner and Ober's pressure plotting. It should never be greater than 1.0 for reasons already explained while discussing Eq 2.5A. Warner and Ober's recording of a positive pressure coefficient greater than 1.0 suggests some inaccuracies in measurements taken on *Papoose*. In fact, the experimenters admitted that they had met some difficulties in measuring the pressures on the full-scale sail. Their methods of testing, and the associated troubles, are given in Ref 2.67. In spite of the lack of accuracy however, their results of pressure measurement may serve in a qualitative sense.

Fig 2.74   Variation of pressure distribution with camber for $\alpha = 4.4°$.
Two-dimensional cloth sail. Re $= 3.2 \times 10^5$ (Ref 2.65).

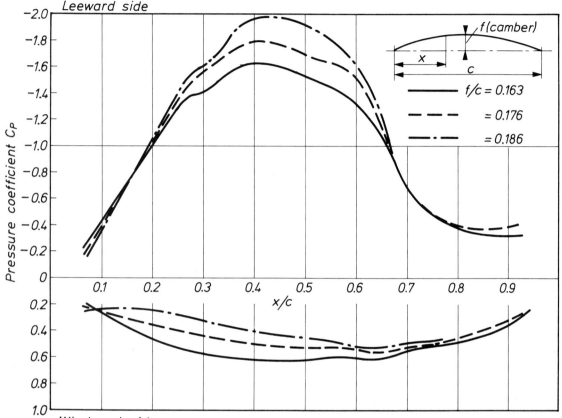

Let us refer now to the leeward pressure distribution, shown in Figs 2.74 and 2.75; the sails are characterized by having no large leading edge suction, so pronounced in the case of thick foils illustrated in Figs 2.15 and 2.16. Though the peak suction in Fig 2.74 is quite moderate, the overall shape of the pressure distribution curve is such as to develop large lift and unfortunately large drag too. The results in Fig 2.75 are noteworthy for the waviness of the pressure distribution found on the leeward side of the sail. Such a waviness is not recorded in Fig 2.74 because the furthest forward pressure tapping was at 0.05 c.

Warner and Ober concluded that the waviness recorded in their pressure distribution on the mainsail was due to mast interference. Their deduction may indirectly be confirmed by the picture of the flow round a pear-shaped mast section in the presence of a rigid sail, shown in Fig 2.80A, as observed in a wind tunnel. This reveals that just behind the mast there is a large, reversed flow vortex, similar to the separation bubble observed in the case of flow round the sharp leading edge of thin

Fig 2.75 Pressure distribution measured by Prof Warner on yacht *Papoose* (Figure taken from *Sailing Theory and Practice*).

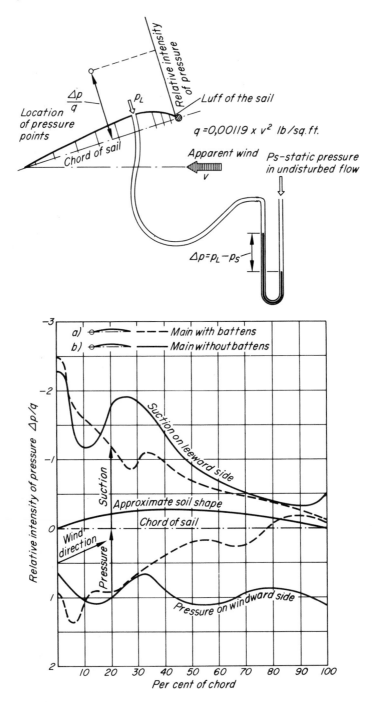

foils, displayed in Figs 2.57B and 2.59B. A plausible inference from it would be that, whenever and for whatever reasons a reversed flow vortex is formed at the leading edge of the foil, a wavy suction plateau of reduced values of pressure coefficient C_p, as illustrated in Fig 2.57B, should be expected.

In both cases, sail with and without mast, the reversed flow vortex with associated flatness in pressure distribution is evidence of viscosity effect causing losses of momentum in the airflow when getting around the leading edge. Referring to Thwaites' theory of sail (Ref 2.68), Boyd and Tanner (Refs 2.63, 2.64) demonstrated on a purely theoretical basis that waviness in pressure distribution, with characteristic flattening of the pressure peak, will occur on thin foils within a certain range of incidence angles and camber distribution. Thus, for instance, Fig 2.76 represents theoretical pressure distribution on the leeward side of a parabolic thin foil, of camber ratio $f/c = 0.188$, at two different angles of incidence $\alpha = 8.6°$ and $15.1°$ and three different positions of maximum camber X_m. Evidently, pressure distribution is very sensitive to variation in sail shape and position of maximum camber in particular. This fact is of acute practical importance, bearing in mind that the shape of a soft sail is never stable, and is therefore subject to large changes in aerodynamic characteristics over a period of time, due to:

a. Recoverable and irrecoverable stretch (as the resin filler breaks down and comes away from the fabric–Photo 3.11, Part 3).
b. Rearward movement of camber as the sail loading increases following wind or incidence angle increase.

Since the pressure distribution over the sail, and thus the driving force, is predominantly dependent on sail curvature, precise control over the sail shape when the yacht is sailing in variable conditions becomes imperative for a racing crew.

As an instance of this, *Lady Helmsman* was badly beaten in the World Championship in Bermuda early in 1967, explains Farrar (Ref 2.58) '...through using a sail which had stretched out of shape. The flow moved aft and was producing more drag than lift–in other words, more side force in proportion to driving force–than when in its prime.'

Certain features of the pressure distribution of Fig 2.76 (redrawn in Fig 2.77), which illustrate a yacht sailing to windward, are noteworthy and provide partial explanation of lift and drag characteristics. Thus, if leeward side pressures, as given by curve A, are concentrated far aft, they will tend to give large drag. The two pressure force vectors p_1 and p_2, acting normal to the sail curvature, illustrate the point. These vectors can be resolved into two components along and across wind direction, and we can see that the greatest contribution to the driving force comes from the forepart of the sail by virtue of both the magnitude and direction of the pressure force p_1. The high pressure p_2 near the leech gives no driving force component but mostly heeling force and drag. The same reasoning applies to pressures developed on the windward side.

One may anticipate that those distinctly different pressure distributions depicted

Fig 2.76   Theoretical pressure distribution on the leeward side of a parabolic thin foil of camber ratio $f/c = 0.188$ at two different angles of incidence (8.6°, 15.1°) and three different positions of maximum camber, x_m (Ref 2.64).

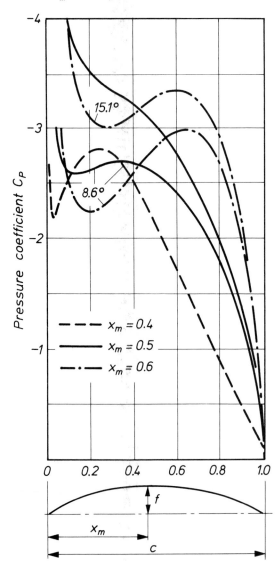

in Fig 2.77 by curves A and B will produce different sets of driving and heeling force components, one set being better than the other. The problem of simultaneously obtaining high lift and low drag is determined by the broad features of the distribution of pressure. In practice, it is important to know how specific alteration of the pressure distribution can be produced by suitable modification of the foil curvature.

Fig 2.77   Effect of pressure distribution on resulting sail forces.

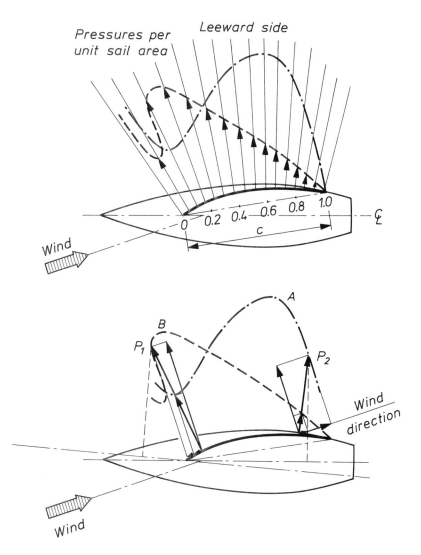

So far as the interrelation between the forces, pressures and shape of the foil section is concerned, a greater understanding is required of the pattern of flow, particularly close to the leading edge, and of the shape a sail adopts at various wind strengths. From both experimental and theoretical considerations it may be inferred that, by its very nature, the soft sail adjusts to some extent its shape so as to maintain attached flow at the leading edge (Ref 2.62). In most cases, however, crew intervention in modifying entrance efficiency aiming at possible minimum disturbance to the boundary layer, is the major factor. It should perhaps be emphasized that the many adjustments which a crew may deliberately introduce to make up what

Fig 2.78   Variation of lift coefficient C_L with angle of incidence for NACA
0012 section (two-dimensional flow). See also Fig 2.79.

a. Angle of incidence variation from 2° to 32° (in conventional
direction of flow).
b. Angle of incidence variation from 178° to 196° (in reversed
direction of flow).

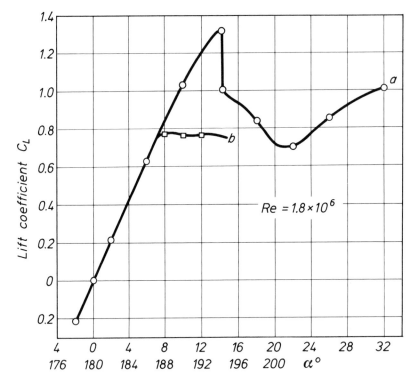

is known as tuning, is usually more important than having the best designed sail.
This way, the tuning problem is essentially reduced to the problem of how to achieve
the best pressure distribution in given conditions defined by wind strength, course
sailed and available stability.

   After what has been said about the effects of round and sharp leading edges on the
flow, one might guess when suddenly asked, that lift from a streamlined section such
as NACA 0012 for example, would be very poor if the attitude of section relative to
the flow were reversed by 180°, i.e. when trailing edge (TE) becomes the leading edge
(LE). Careful readers might not however be surprised to learn that the section in
question tested in the opposite direction (Ref 2.69) produces virtually the same lift
curve slope as in the case of conventional direction of flow, and Fig 2.78 demonstrates that lift increases in the same manner although only up to a certain point.

   As found in tests on thin foils, the sharp leading edge is no obstacle to the flow
going round and continuing, after a small local separation, along the suction side. A
separation bubble operating at the sharp leading edge acts as a mitigating agent

Fig 2.79 Pressure distribution round NACA 0012 symmetrical section at conventional direction of flow ($\alpha = 0°$ and $10°$) and reversed direction of flow ($\alpha = 180°$ and $170°$).

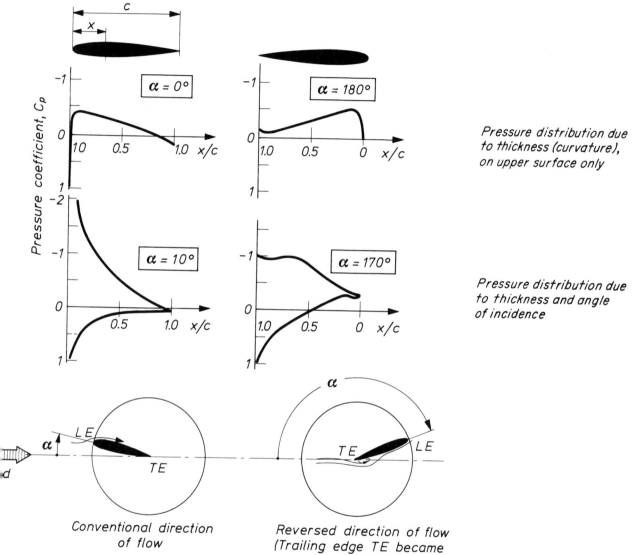

Pressure distribution due to thickness (curvature), on upper surface only

Pressure distribution due to thickness and angle of incidence

Conventional direction of flow

Reversed direction of flow (Trailing edge TE became leading edge LE)

Fig 2.80   Airflow over mast and sail. Masts are drawn to the same scale.

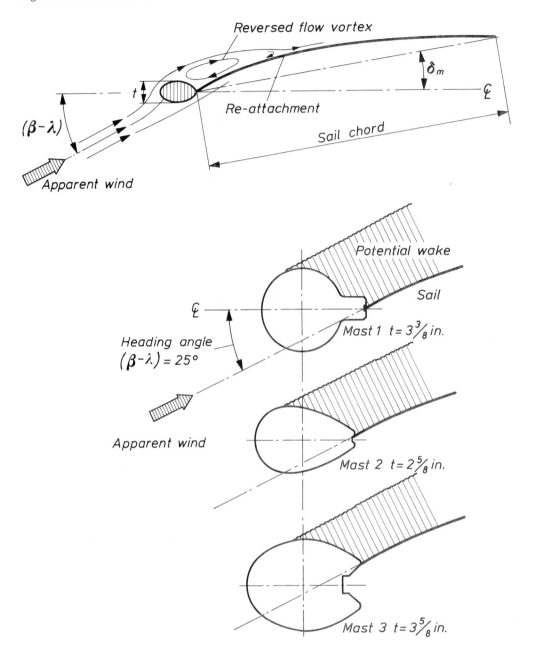

Fig 2.81   Airflow over mast and sail.

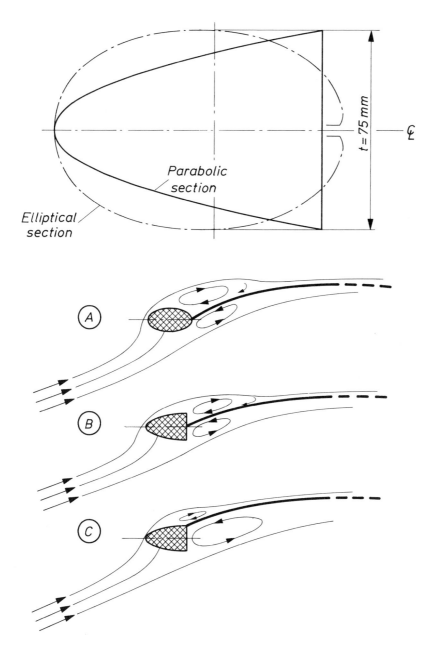

which, by means of gently displacing the oncoming streamlines, increases local velocity of the flow at the very edge of the foil thus helping to separation. Figure 2.79 depicts the pressure distribution recorded in the two opposite attitudes at identical incidence angles. As might be expected, in reversed attitude (at incidence $\alpha = 170°$) the pressure distribution along the suction side is clearly distinguished by the absence of a pressure peak and, although such a pressure distribution may produce high lift, the drag is usually also high. As seen in Fig 2.78, the maximum lift coefficient of a section with a sharp leading edge is heavily curtailed in comparison with that obtained in conventional attitude.

A mast at the leading edge of a sail determines the size of the separation bubble and it appears that the downstream extent of the reversed flow vortex right behind the mast is tangibly dependent on mast shape rather than its size. It is perhaps surprising to learn that mast 3 shown in Fig 2.80, the biggest one, was proved in the wind tunnel to be aerodynamically better than the two other mast sections–circular and pear-shaped. At heading angle $(\beta - \lambda) = 25°$, its potential wake shown by the crossed zone is less conspicuous than those of the others. Mast No 3 seems also to be good for structural reasons, that is, it may be the lightest consistent with stiffness and windage requirements.

Perhaps even more interesting results of investigation on efficiency of mast-sail combinations are presented in Fig 2.81 and Table 2.8. Each combination A, B and C was tested at the same selected angle of incidence $\alpha = 10°$, corresponding to heading angle $(\beta - \lambda) = 25°$, as shown in Fig 2.80A. Comparative tests were limited to measuring the aerodynamic characteristics of the models in the close-hauled condition only. In the combinations investigated, the mast was attached to the same rectangular model sail made of laminated plywood of chord 3 ft 4.5 in and of circular-arc camber $f/c = 0.1$. The tests were conducted in semi two-dimensional condition, i.e. the model sail nearly spanned the tunnel.

TABLE 2.8

	Configuration	Description	L/D
A	Elliptical section	Sail attached to centre line of the mast	6.7
B	Parabolic section	Sail attached to centre line of the mast	8.0
C	Parabolic section	Sail attached to leeward side of the mast	9.9

The relevant L/D ratios measured were practically independent of Reynolds Number in the range $0.5 - 1.0 \times 10^6$. The parabolic section in combination C with a sail sliding across the blunt rear part of the mast, demonstrates an improvement of the order of 40 per cent over the conventional elliptical section A.

It appears that:

Firstly–the efficiency of a mast-sail combination expressed in terms of L/D ratio depends almost entirely on the size of the wake covering the leeward side of the sail, just behind the mast.

Secondly–the shape of entry or, in other words, the very forward part of the mast, exerts a profound effect on the flow and resulting pressure distribution.

(h) *Roughness effect*

Whatever the basic section of a foil may be, there are certain secondary factors which superpose their effects on lift characteristics. One of the most obvious of possible disturbing influences is that of surface imperfection or roughness. A varnished plywood or GRP (glass reinforced plastic) rudder, for instance, can with sufficient care be brought to the smoothness of plate glass; or by improper finishing or through frequent damage to the leading edge during the sailing season, it may be left as rough as fine sandpaper. Such changes in surface texture may, apart from impairing drag, also affect lift characteristics. Figure 2.82 depicts the effect of surface roughness on the lift-curve of RAF 30 section (shown earlier in Fig 2.15) at four Reynolds Numbers (Ref 2.70). Tests were performed on two foils of:

1. Highly polished surface, obtained by the use of very fine grade abrasive and polished finally with rouge on a buffing wheel. Such a surface gave no detectable roughness to the touch;
2. Rough surface, obtained by using No 180 carborundum sprayed onto a coat of fresh varnish. Such carborundum grains average about 5 mils maximum dimension.

It can be seen that the value of maximum lift coefficient for the foil having a rough surface is little affected by the change of Reynolds Number, as compared with the large favourable increase in $C_{L\max}$ for the same foil with a polished surface. The foil showed approximately the same lift characteristics at the lowest Reynolds Number for both rough and smooth surfaces, but the differences between the lift characteristics gradually became pronounced as the value of Reynolds Number was increased.

As might be expected, some parts of a foil section are more sensitive to surface condition than others. The nose of a section and, in particular, the leading edge prove much more responsive than other parts farther back along the chord. The rear half of the section can in fact be deliberately roughened almost to the poppy-seed level, without noticeable adverse effect, but the slightest irregularity at the leading edge, as shown in the sketch attached to Fig 2.83, manifests itself immediately through decreased lift. As the roughness is moved away from the leading edge, the adverse effect becomes smaller. When the rough strip is directly over the point about which the leading-edge radius is taken (i.e. about 1.5 per cent of the chord) the adverse effect on lift almost entirely disappears. As a matter of fact, the nose, at a sufficiently large angle of incidence corresponding to maximum lift conditions,

Fig 2.82   Reynolds Number effect on the lift coefficient for smooth and
rough surfaces (Aerofoil RAF 30, see Fig 2.15).

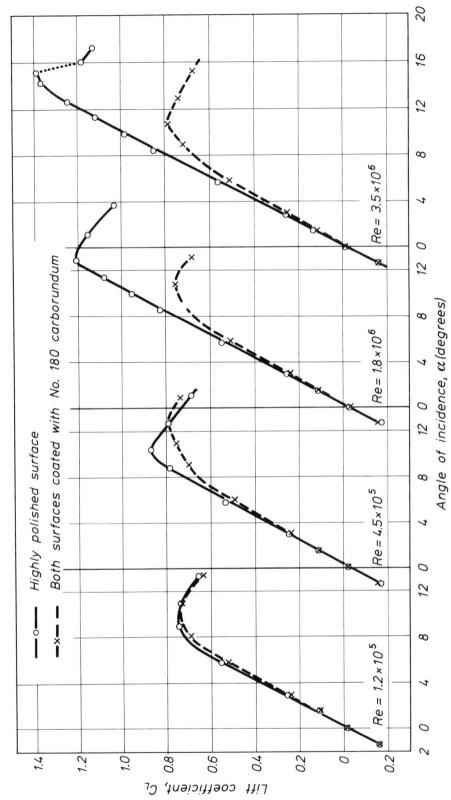

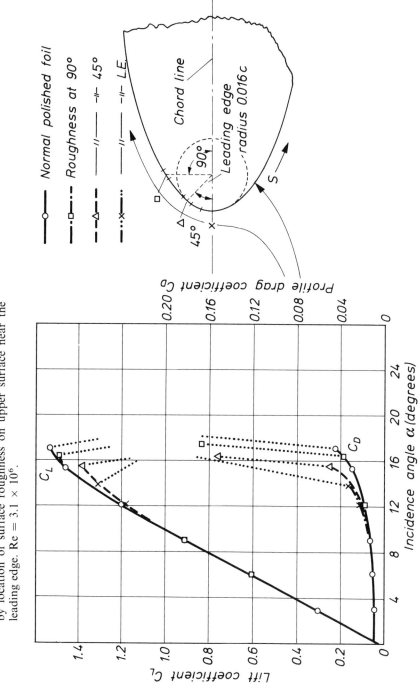

Fig 2.83 Aerodynamic characteristics of NACA 0012 section as affected by location of surface roughness on upper surface near the leading edge. Re = 3.1 × 10⁶.

Normal polished foil
Roughness at 90°
—"— 45°
—"— L.E.

Chord line
Leading edge radius 0.016c
90°
45°
S

Profile drag coefficient C_D
0.20
0.16
0.12
0.08
0.04
0

Incidence angle α (degrees)
4   8   12   16   20   24

Lift coefficient C_L
1.4
1.2
1.0
0.8
0.6
0.4
0.2
0

C_L
C_D

becomes actually a part of the suction side, while the stagnation point moves around the leading edge to take a position well below the extreme point where the chord-line intersects the nose. As seen, the greatest adverse effect of leading-edge roughness is on the value of $C_{L\,max}$ and, at low angles of incidence, the slope of the C_L curve is little affected.

In general conclusion it appears that the roughness of foil surface is always harmful, and the labour expended on perfecting the finish on the forward third of the chord of a rudder or fin keel is never likely to be wasted.

D Three-dimensional foils

Theories are nets;
Only he who casts will catch.

NOVALIS

So far we have been mainly concerned with foil action restricted to the two-dimensional flow condition, as illustrated in Fig 2.2 and Photo 2.1, in which all the streamlines of the flow lie in parallel planes perpendicular to the span of the foil. Accordingly, this supposition implies that there is no pressure variation along the span, no change in the streamline pattern and hence no change of lift and drag for every section along the span. The main object of investigation of two-dimensional flow was to obtain a relationship between the form of the foil section and the flow pattern, which in turn can serve as a means of determining lift and drag.

In practice, foils have a definite span and two or perhaps one free tip (as in the case of a spade rudder or fin keel attached to the bottom of the hull). No matter how the foils are mounted, vertically on the hull like sails or fin keels, or more or less horizontally like hydrofoils or wings, fluid in motion follows the universal inclination to flow from high pressure to low pressure regions by every available path. Examination of the flow pattern round any foil of finite span (Photo 2.27A, B, C) shows that at the foil tips the air or water tends to flow round the end from the underneath (or windward) surface where the pressure is higher than the ambient pressure, to the upper (or leeward) surface where the pressure is lower. The result of this is threefold:

1. The foil surface near the tip is much less efficient at producing lift.
2. This decrease in lift is accompanied by an increase in drag.
3. An additional disturbing air movement developing towards the tips modifies the direction of the oncoming flow near the foil, hence the effective angle of incidence along the foil span changes, as do lift and drag.

Photo 2.27A   At the sailhead the air tends to flow round from windward side, where the pressure is higher, to leeward side where pressure is lower.

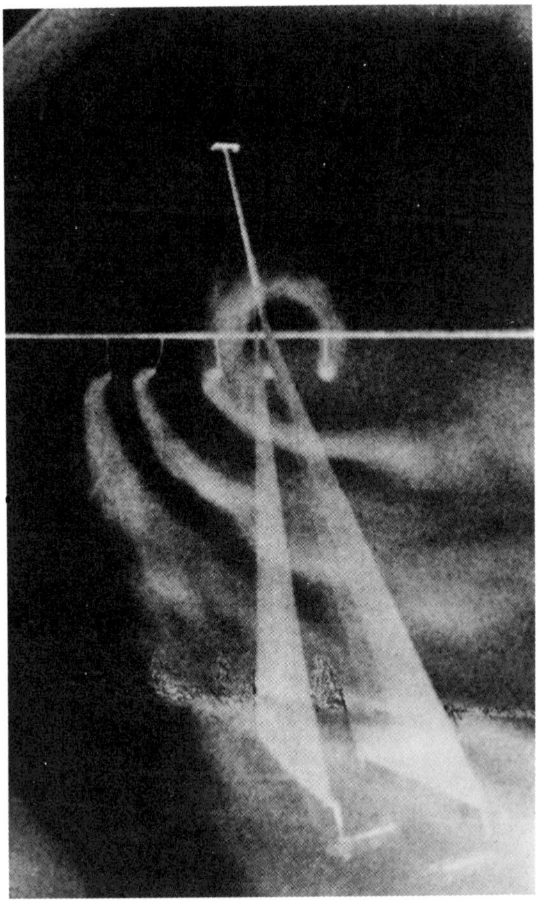

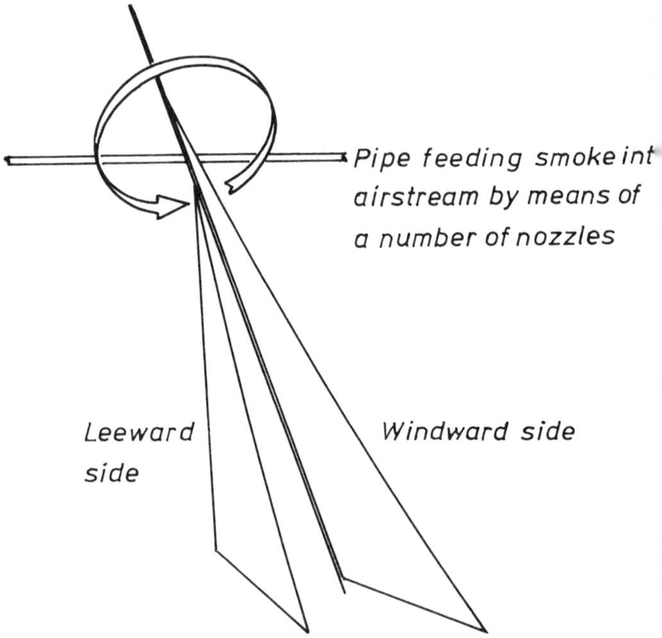

Pipe feeding smoke into airstream by means of a number of nozzles

Leeward side

Windward side

Photo 2.27B   Tip vortex developing at the bottom of the Dragon keel is similar, in principle, to that shown in Photo 2.27A and C.

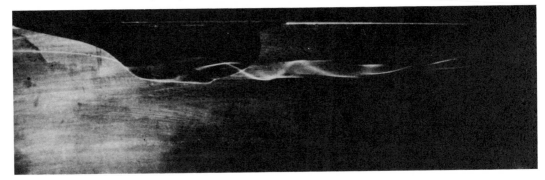

Photo 2.27C Picture of the tip vortex 16 chord lengths behind the foil tip. Cross on the photograph indicates the position of the trailing edge at the foil tip. Rotational flow within the vortex was made visible using small bubbles of hydrogen produced by electrolysing the water. The model foil was towed through water.

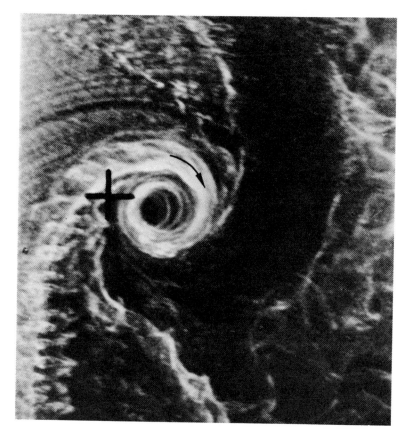

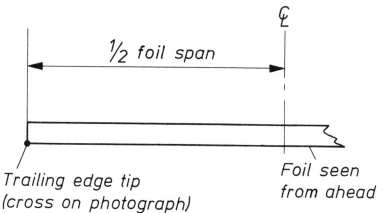

½ foil span

Trailing edge tip
(cross on photograph)

Foil seen
from ahead

All these effects are more or less detrimental, and their total result depends on the aspect ratio (AR) and planform of the foil; the greater the AR the smaller is the effect. The passage from the two- to three-dimensional flow round the foil is not simple. Admittedly, there is no easy shortcut, or quick by-passing to an understanding of the complicated causal interrelationship between the three effects just listed and the foil planform. However, if the art of sail cutting, trimming and tuning is to be developed on more rational premises, and not to be a hit-and-miss affair, one has to make some effort to acquire more knowledge about factors affecting sail performance. At the moment, as admitted by one sailmaker:

'Omar the Tentmaker, like every sailmaker, produces sails by a bewildering cross between the unfathomably abstruse reaches of mathematics and physics, the ancient and honorable experience of a long line of ancestors in the business, and perhaps just a dash of the occult thrown in. In any case what goes on in Omar's head when he designs your sail, or in his shop when he actually produces the sail, are both subjects about which you would like to know more.' (Ref 2.71)

From the aerodynamic standpoint sails analysed as lifting surfaces are complicated systems. One of the reasons is that the stress/strain relationship for sailcloth is non-linear, and deformation due to wind loading and other loadings, introduced by the crew by means of halyards, sheets, Cunningham holes, etc., cannot readily be determined. Hence it is difficult, if not impossible, to predict precisely what shape a particular sail will take in a given condition. Moreover, while racing, the shape of the whole sail configuration changes continually, as already mentioned, due to recoverable and irrecoverable stretch. One condition after another must succeed by virtue of feed-back, interaction and reciprocal causation to all geometry factors, such as camber, twist, chordwise and spanwise distribution of camber, sag of the forestay, mast bending, etc. Nothing is settled and any change in sail shape causes every performance factor to change to a greater or lesser extent. It is therefore vain to expect that there may be a final answer to all sailmakers' and sail users' problems. Many questions will remain open-ended, but there are certain theories and concepts developed in the course of progress made in aerodynamics which may help in gaining better insight into the question of how three-dimensional sails or hull appendages really work.

What follows is an attempt to present, step by step, some concepts and guiding principles of aerodynamics and hydrodynamics which, as working hypotheses, might help to uncover which of the many factors among the variables of a foil's geometry predominantly control its efficiency.

Without some guiding idea or theory one cannot even determine which features or factors to look for. Concepts or theories are frequently acquired by accident, through sailing experience but, as noted on many occasions in the field of experimentation, accident favours the prepared mind. As aptly remarked by Polany: 'To see a good problem is to see something hidden and yet accessible. This is done by integrating some raw experience into clues pointing to a possible gap in our knowledge. To undertake a problem is to commit oneself to the belief that you can fill in this gap and make thereby a new contact with reality' (Ref 2.72).

The fact that some foil characteristics, which are present in more complicated configurations, are also present in simple configurations, facilitates further discussion. Rigid lifting surfaces of fixed geometry are easier to analyse than flexible deformable surfaces such as soft sails. For this reason, when considering how three-dimensional flow develops if the end plates are removed, it is best to begin with a simple rigid foil. And then, by gradually introducing modifications to foil geometry such as taper, twist, sweep-back, etc., in an attempt to reproduce the actual foil or sail, examine how all those modifications affect foil characteristics in terms of lift and drag.

(1) Vortex system developed round a finite span foil

At the outset let us have one more look at the two-dimensional flow as shown in Fig 2.84A. If a rectangular foil is set at a certain angle of incidence relative to the wind, the circulation will develop and, according to Eq 2.12, its magnitude is proportional to wind velocity V_0, the chord of the foil c, and the incidence angle α, i.e.

$$\Gamma = f(V_0 \times c \times \alpha) \qquad \text{(Eq 2.12 repeated)}$$

where, as before, $f(...)$ means, is proportional to, or is a function of, the factors given between brackets.

Since chord c is uniform, it follows that circulation distribution at given V_0 and α will be uniform along the foil span. Hence one might rightly expect that the pressure distribution, lift as well as drag, should also be uniform, i.e. the same for every section of the foil along its span as shown by means of vectors in the sketch above the foil in Fig 2.84A. This is a consequence of Eq 2.10 $(L/b) = \rho V_0 \Gamma$ which relates lift per unit span L/b to circulation Γ.

The two-dimensional foil characteristics (lift and drag coefficients), measured in the wind tunnel as shown in Photo 2.1, are accordingly commonly called 'section characteristics'. They are usually given in technical literature by the lower case symbols c_l and c_d, in order to distinguish them from the complete three-dimensional foil characteristics of a specific planform, which are usually presented by using the upper-case symbols C_L and C_D (compare Figs 2.44 or 2.58 with 2.60 or 2.67).

The section characteristics c_l and c_d are intrinsically associated with the shape of the foil section profile as contrasted with three-dimensional foil characteristics C_L and C_D, which are profoundly affected by the foil planform. The detailed study of foils is greatly simplified by this concept of foil characteristics, because theory offers a method of estimating the properties of foils of arbitrary form provided the section characteristics are known (Ref 2.31).

When the end plates are removed, the flow will tend to spill over the free ends, as indicated in the right part of Fig 2.84B, i.e. from the side where positive pressure exists ($+C_p$) to the suction side distinguished by negative pressure $-C_p$. Such a flow

Fig 2.84   If circulation Γ is uniform along the span, the lift (loading) and pressure distribution will be the same for every section along the span.

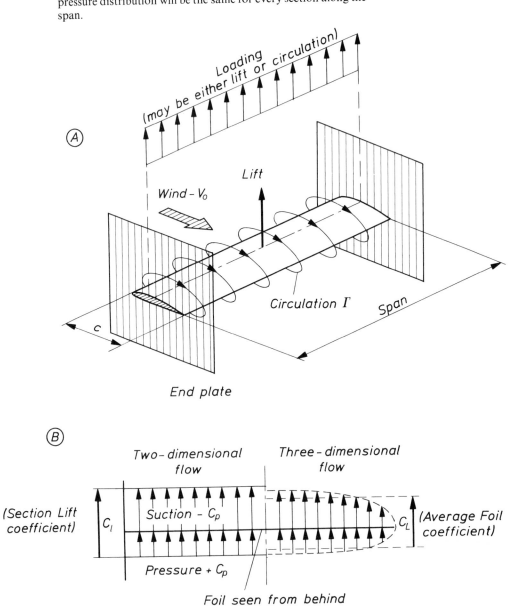

Fig 2.85 Total normal pressure can be obtained by adding numerically
+ Cp + $-Cp$ (see Fig 2.16 or 2.84B). LE–leading edge, CP–
centre of pressure.

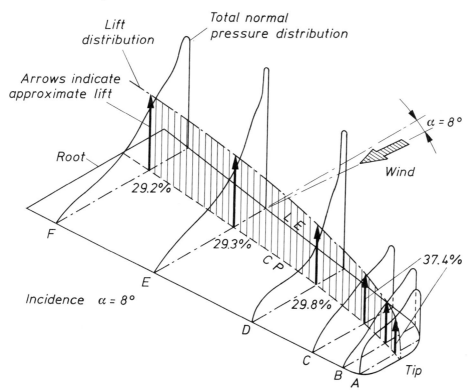

wipes out the pressure difference at the tips and reduces it over the entire span of the
foil. For this reason the characteristics of a foil of finite span when three-dimensional
flow takes place are worse than its section characteristics, and Fig 2.84B
demonstrates why the average lift coefficient C_L in three-dimensional flow is smaller
than that (c_1) for two-dimensional flow, assuming the same incidence angle. Figure
2.85 based on wind tunnel experiments may be of interest in that it shows
distribution of total pressure acting normal to the section chords A, B, C, E and F of
the foil. The arrows pointing upwards indicate approximate lift for a given section
and also lift distribution along the foil span. Reduction of lift close to the tip is quite
conspicuous.

Lanchester clearly understood this phenomenon when in 1897 he secured patent
No 3608 covering the use of end plates, called by him 'capping plates', at the wing
tips to minimize the pressure losses there–six years before the Wright brothers' flight
(Ref 2.11). In the patent specification Lanchester describes the capping plates'
action, to stimulate as far as possible the condition of a foil operating two-
dimensionally in order to minimize the dissipation of pressure. In his *Aerodynamics*
(Ref 2.10) he was the first to conceive that the important difference in the flow

pattern about two- and three-dimensional foils is traceable to the difference in spanwise lift distribution, which in turn is traceable to the disposition of circulation. This is implicitly incorporated in Eq 2.12 $\Gamma = f(V_0 c\alpha)$ with which we are already familiar, so we may deduce that, when length of the chord c of a foil decreases, the circulation decreases also. To account for the fact that the circulation must diminish to zero at the foil tips, Lanchester developed the concept illustrated in Fig 2.86, which is taken from his book.

Crudely judged, without reference for the time being to the exact structure, the tip vortices appear as a continuation of the circulation around the foil, trailing off downstream along the streamlines made up of the fluid particles twisting around the tips; with the form of the flow made visible they prove their reality as illustrated in Photo 2.27. These powerful tip vortices of strength equal to the circulation at the midspan of the foil affect the whole surrounding flow pattern and therefore the performance of the foil. And the loss of pressure at the foil tips or dissipation of energy to maintain the tip vortices is not the only disadvantage associated with their occurrence.

It is worth noting that, at the time when Lanchester evolved his concept of circulation and tip vortex effects, such evidence as depicted in Photo 2.27 did not exist and the whole idea is 'an outstanding example of a man of genius finding the correct solution to a baffling problem *without any experimental result to guide him*, a feat perhaps more appropriate to the world of ancient Greece than to our own' (Ref 2.5). This point, made by Sir Graham Sutton, deserves some attention in our scientifically orientated world in which the mission of science is seen as an attempt to introduce order among observables. Empiricism implies that knowledge can only grow by developing hypotheses that have meaning in terms of observations, experiments made or facts. Empiricists are therefore bound to insist that the only channel of cognition of physical phenomena is the sensory way, i.e. through human senses directly or through their magnified and sophisticated forms–instruments. Paradoxically, the empiricist Hume expressed the view that, '...it is impossible for us to think of anything, which we have not antecedently *felt*, either by our external or internal senses'. And he also set out to prove that pure empiricism is not a sufficient basis for science. Lanchester's concepts were not, in his time, related to any experimental evidence. This prompts a question of more general interest: are there any other means of cognition different from sense-perception?

By way of digression, this may perhaps lead us to the phenomenon of outstanding helmsmen who can tune and steer their boats in such an efficient manner that they give the impression of having some kind of theory to guide them. In most cases, however, when asked why they do such-and-such a thing in such-and-such conditions they are not able to answer in clear cut terms, which might indicate that there is no reasoned knowledge of the cause behind it. And it is rather hard to believe they are just those lucky ones in applying hit-or-miss, or trial-and-error routine. It appears that the common ingredient which distinguishes a good scientist, a good sailor, or a good artist is above all an intuitive feeling for nature.

Fig 2.86 Tip (trailing) vortex developing behind a foil of finite span
(according to Lanchester's own drawing).

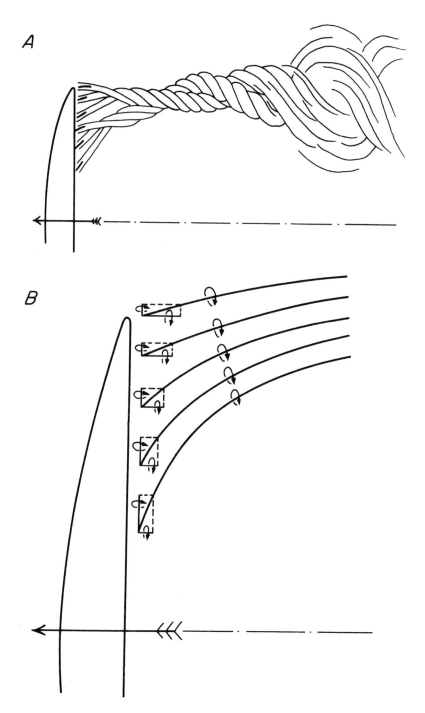

Lanchester's concepts, as depicted in Fig 2.86, may look perhaps trivial to students of modern aerodynamics. After all, there is nothing more simple than that which was discovered yesterday, but there is nothing more complicated than that which will be discovered tomorrow, as an eminent scholar wrote.

It has already been hinted that the consequences of flow modification due to tip vortices are more detrimental than they might at first appear. To be more specific, on account of the reduction of pressure at the foil tips to zero shown in Figs 2.84B and 2.85, followed by decrease in intensity of lift towards the tips, one could rightly expect that average lift value, for the same angle of incidence, must be smaller for low aspect ratio foils than for high ones. However, the observed influence of aspect ratio on lift is substantially greater than could be explained in this way. We must therefore investigate whether an explanation of this discrepancy can be found if the effect of tip vortices on the flow round the foil is taken into account.

As suggested by Prandtl (Ref 2.6), in order to obtain the simplest possible scheme, we shall assume that the lift or circulation is uniformly distributed over the foil span, as in two-dimensional flow, then the tip vortices will arise only at the ends and continue rearwards as free vortices. This is illustrated in Fig 2.87. As a further preliminary step, it is convenient to consider that the foil is replaced by a spanwise vortex similar in nature to the rotating Magnus cylinder (already mentioned in section A4), which as such is capable of developing circulation and lift. By reducing the diameter of the cylinder we may finally arrive at a *lifting line* or *lifting vortex* which carries circulation in the same manner as the foil. The centre of such an imaginary vortex or core of such a rotating fluid is located inside the foil and cannot be washed away by the fluid flow past the surface (Fig 2.88A). In the much magnified illustration of Fig 2.88B the vortex core is to be the foil itself. By means of such a vortex, one may even say mini-tornado, a part of the available energy of the wind is converted into the bound energy of the rotating mass of air attached to the foil. Since the lifting vortex is attached or 'bound' to the foil, it is also called a *bound vortex*, as distinct from the *free vortices* streaming from the free ends of the foil and no longer confined to it.

Lanchester, who developed this concept, was somehow inspired by the so-called 'Helmholtz theorem' published in 1858, which states that *a vortex once generated cannot terminate in the fluid; it must end at the wall* (as in the case of two-dimensional flow) *or form a closed loop*. This would mean that the vortex must be continuous, like a smoke ring. It may, however, have any shape; so Lanchester concluded that the bound vortex cannot end at the foil tips, but there must be some kind of continuation, drawn by him in Fig 2.86A, and this continuation must be in the form of a trailing or free vortex. The intensity of circulation of those free vortices must, according to Helmholtz, be the same as that of the bound vortex.

Since the Helmholtz theory requires that the entire circulation around a foil generating lift must take the form of a closed loop vortex, where is the missing link which might close the gap between the two free vortices? Theoretically, if air were devoid of viscosity, the tip vortices of, let us say, an aircraft flying hundreds of miles

Fig 2.87 Simplified, so-called horse-shoe vortex system formed behind a
finite span foil. The trailing vortices are solenoidal.

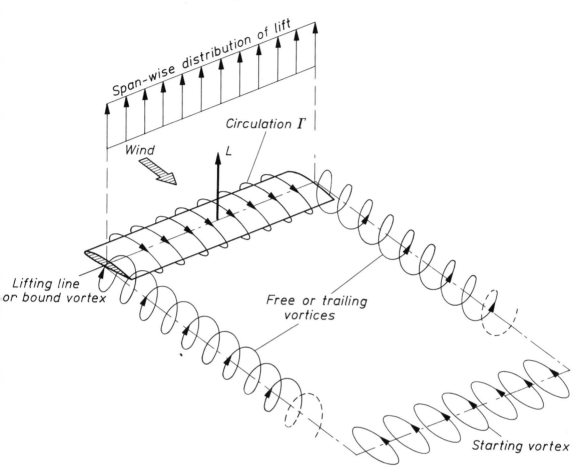

would terminate in the starting vortex shed at the trailing edge of the wing at the
airport where it took off (Fig 2.87 and Photo 2.4C). Such a vortex loop consisting of
a bound vortex, two trailing vortices and a starting vortex would, to use
Lanchester's own words, '…pervade the world for all time like a disembodied spirit.'

However, we are aware that in a real fluid such as air, the origin of circulation lies
in viscosity, i.e. vortices are formed as a result of viscosity and eventually can only
disappear by the action of viscosity and friction. It has been reported (Ref 2.73) that
trailing vortices persist behind heavy transport aircraft for about 10 km before they
gradually diffuse in the form of heat into the atmosphere. The up and down air
disturbances caused by trailing vortices behind the wing may reach velocities up to
± 4 m/sec (± 13 ft/sec); another aircraft flying into this wake may even be rolled
over. The so-called backwind and wind shadow areas (dirty wind) extending behind

Fig 2.88 Circulation streamlines determined by electrical method.

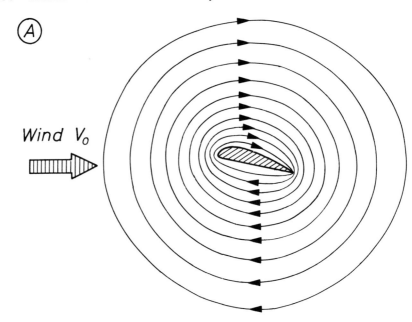

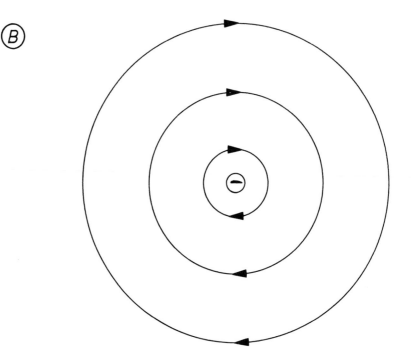

boats sailing to windward is another manifestation, more familiar to sailing people, of the trailing vortex action (Photo 2.27A).

(2) Mechanical and electromagnetic analogies

The preliminary account of the flow pattern round a foil given above is intended as an approximate sketch of the results derived during the development of a more complete theory. It may serve as a mental framework for correlating a more detailed investigation. The notions of *circulation* or *vortex* defining the hidden, invisible, phenomena are not particularly easy to grasp, otherwise they would appear more frequently in sailors' parlance; analogies and mechanical models may help in this respect. Even a scientist of Lord Kelvin's stature (who contributed towards the concept of vortex action) made the remark, 'I am only satisfied when I have designed a mechanical model of the object under examination. If I succeeded I have understood the phenomenon in question, otherwise I have not.'

Figure 2.89 represents a mechanical analogy of Lanchester's lifting-foil vortices: a

Fig 2.89   Mechanical analogy of Lanchester's lifting foil vortices (compare with Fig 2.87).

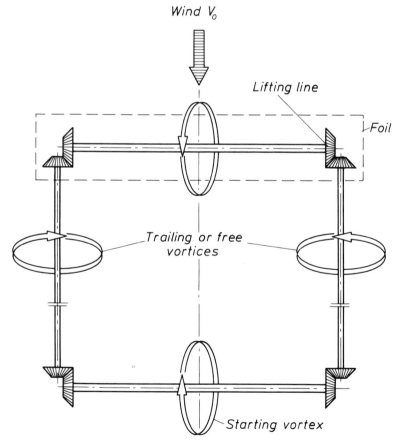

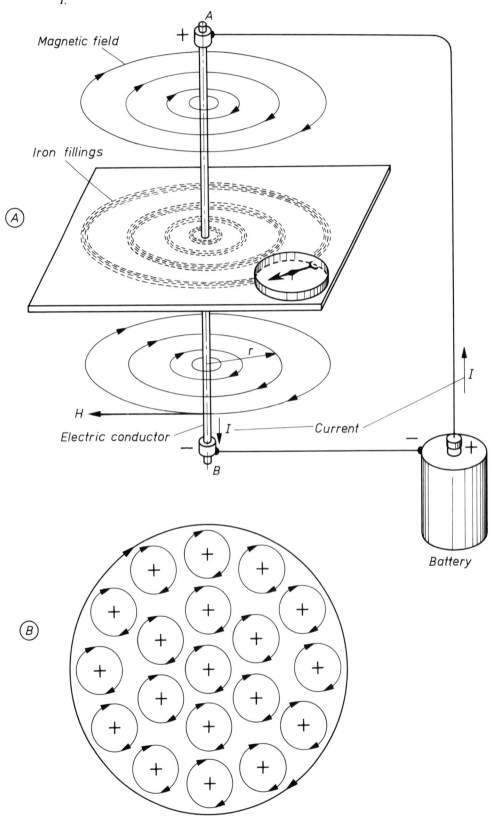

Fig 2.90 With the aid of iron filings or a small compass needle it is possible to show quite clearly the lines of magnetic field (magnetic force) around an electric conductor carrying current *I*.

Magnetic field

Iron fillings

(A)

H

Electric conductor

Current

I

Battery

(B)

Enlarged cross section of electric conductor A-B

cross shaft located inside the foil simulates the lifting line or bound vortex; it drives, by means of helical bevel gears, the two other shafts simulating free vortices rotating in opposite directions. Finally they are coupled by means of similar helical bevel gears to another shaft simulating the starting vortex.

Since most people today are more or less familiar with the concept of electricity and magnetism, and these phenomena are closely related to vortex motion and circulation, one may develop interesting analogies of great help to an understanding of some aspects of foil aerodynamics. For instance, the velocity field surrounding a rotating cylinder as shown in Fig 2.6B is much the same as the magnetic field around a wire carrying electric current.

There is an exact correspondence between the formulae concerning vortex motion and those concerning certain electromagnetic phenomena. In this analogy a vortex filament corresponds to an electric circuit, the strength of circulation to the electric current and the velocity of circulation to magnetic force (Refs 2.74, 2.75); this analogy is demonstrated in Fig 2.90A. According to the electromagnetic theory and experiments, the electric conductor AB connected to a battery is surrounded by a magnetic field, in which lines of flux or magnetic force H, assumed to be rotating in the direction shown, encircle the cable. These lines of force may be considered as spaced closely near the conductor and further and further apart at increasing distances from it, i.e. inversely proportional to the radius r from the wire. Thus magnetic field strength

$$H = \frac{I}{2\pi r} = \frac{\text{current}}{\text{circumference of the magnetic field circle}}$$

The whirling lines of magnetic force H which may be considered as one form of circulation, signifying a rotation about the axis of the conductor AB, can be detected with the aid of iron filings or a small compass needle (illustrated in Fig 2.90A).

As seen, the above equation is similar to Eq 2.7A

$$V_c = \frac{\Gamma}{2\pi r} \qquad \text{(Eq 2.7A repeated)}$$

defining velocity of circulation V_c.

The following can be proved both experimentally as well as theoretically to be correct. If the conductor AB in Fig 2.90B is composed of a number of small parallel wires bound together in a bundle (it can be of arbitrary cross section), the intensity of the magnetic field around the border of an anular area encompassing the small diameter wires in Fig 2.90B is equal to the sum of the magnetic fields around the elements making up the area; this constitutes the so-called 'Stokes theorem', applicable both in the case of current-carrying conductors and fluid circulation. The reader will find it helpful to keep these analogies in mind while following the development of trailing vortices behind lift-producing foils.

The current-carrying wire with its magnetic field shown in Fig 2.90A yields also a new physical phenomenon of great interest. This was first found by Faraday on

Fig 2.91 A current carrying conductor AB subjected to a magnetic field experiences magnetic lift *F*.

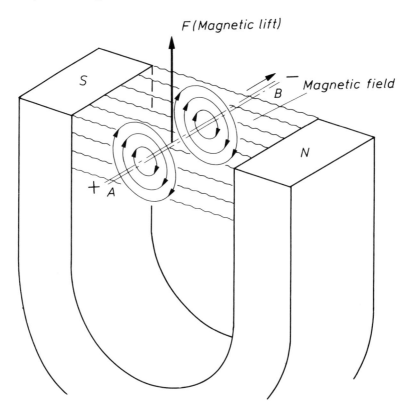

Christmas Day 1821 and can be simply demonstrated by placing the straight portion of an electric conductor into a magnetic field, in the gap between the poles of a permanent magnet, as in Fig 2.91. As soon as a battery is connected and the current in the wire has generated a rotating magnetic field round the wire, a force *F* springs into action tending to pull the wire AB out of the gap. This force acting in a direction perpendicular both to the current *I* and to the magnetic field generated by the permanent magnet can well be called magnetic lift, as being analogous to the aerodynamic lift shown earlier in Fig 2.6. Thus, as in the case of aerodynamic lift, the combination of circulation and parallel flow causes magnetic lift. The electro-magnetic force, like aerodynamic force, depends on the speed of motion (current) and the intensity of the magnetic field (circulation); this is the force that runs electric motors, moves pointers of various meters, is utilized in television tubes, etc.

(a) *Analogy for simple hydrofoil*
Electromagnetic analogy of foil action is presented in the diagram in Fig 2.92A adopted by kind permission from *Hydrodynamics of Ship Design* (Ref 2.75). A finite

Fig 2.92 Definition sketch for the electrical analogy on a hydrofoil.
Adapted from *Hydrodynamics in Ship Design* by H E Saunders,
copyrighted by The Society of Naval Architects and Marine
Engineers and included herein by permission of the aforementioned Society.

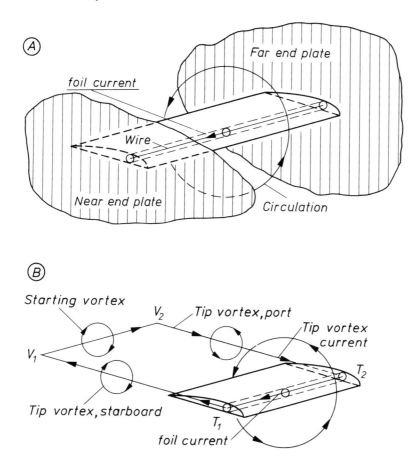

span is cut out of an infinitely long foil and fitted with imaginary conducting end
plates of infinite extent. It may be assumed that the lifting current enters the wire,
located inside the foil, through a thin conducting end plate attached to one end of the
foil normal to the wire; the current returns through the other end of the foil, likewise
normal to its axis (wire). To carry out the electrical analogy, the end plates are
connected to each other at a great distance astern, so that the lifting current flows
around a closed circuit. As soon as the lifting current flows in the direction indicated,
circulation combined with an oncoming stream will produce a lift force as described
earlier when discussing Fig 2.91:

'The lifting current concentrated in the imaginary wire over the span of the foil,
spreads at its junction with the near end plate and flows in a multitude of

minute paths to a great distance astern. There it flows across to the far end plate, as described presently. Then it flows back by many minute paths to the far end of the foil. There is no concentrated lifting current in either end plate' (Ref 2.75).

In sketch B of Fig 2.92, which is equivalent to Figs 2.87 and 2.89, both ends of the foil are devoid of end plates. The closed circuit necessary to carry the lifting current by the electromagnetic analogy is now considered to be formed by additional imaginary wires, which may well be called tip-vortex wires, forming a continuation of the imaginary wire through the foil. These wires trail away from the tips and are joined by another transverse wire at a point downstream where the flow started; that is, in the fluid which surrounded the foil when its forward motion began. This is already known as the region of the starting vortex. The manner in which this vortex is shed from the after edge of a foil starting from rest was depicted earlier in Fig 2.11 and Photo 2.4.

Following the description of Fig 2.92 given in Ref 2.75 and maintaining closely the original wording, in diagram B the foil portion of the lifting wire T_2T_1 is extended by the tip-vortex wires T_1V_1 and T_2V_2 back to the original starting position of the foil V_1V_2. The tip-vortex wires lengthen automatically as the foil moves ahead, away from the starting vortex. The starting-vortex wire V_1V_2 may pass between the tip-vortex wires in diagram B or between the infinitely large end plates in diagram A, extending from the tips back to the starting position of the foil. The vortex or lifting current passes around the complete circuit in the clockwise direction shown at B in Fig 2.92 with accompanying circulation around all four portions of the wire, indicated by the arrows. Corresponding to an electric current, the lifting current has the same strength in all parts of the circuit. The effect of a reduction of the foil span from infinitely two-dimensional flow to a finite length may thus be viewed as the effect of a bending of the bound vortex from the direction along the span back through 90° at the foil tips, i.e. downstream. Such a modification will, as we shall see later, affect the whole velocity field round the foil.

Circulation takes place around the foil as before, i.e. as shown in sketch A, and produces a corresponding lift. The circulation around the tip-vortex wires T_1V_1 and T_2V_2 causes the uniform lines of force above the rectangular area $T_1V_1V_2T_2$ to move down into it. Those lines of the uniform field beyond or outside the wires T_1V_1 and T_2V_2 move upward in an opposite direction. For a foil moving in a liquid this is understandable because the fluid underneath, where the pressure is greatest, is impelled by positive differential pressure $(+\Delta p)$ to escape around to the region above the foil where there is negative differential pressure $(-\Delta p)$ i.e. the pressure is lower. This it does by rolling around outboard of the tips from the under side to the top side, a phenomenon which has been explained earlier when referring to Figs 2.84 and 2.85.

(b) *Variation of circulation along the span*
In practice the circulation around a foil is never constant for the entire span. To take care of its spanwise variation the electrified foil is assumed to carry not a single wire,

Fig 2.93 Electrical analogy for different distribution of circulation. Adapted from *Hydrodynamics in Ship Design* by H E Saunders, copyrighted by The Society of Naval Architects and Marine Engineers and included herein by permission of the aforementioned Society.

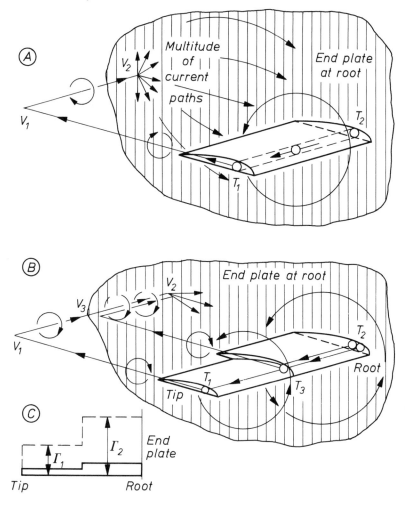

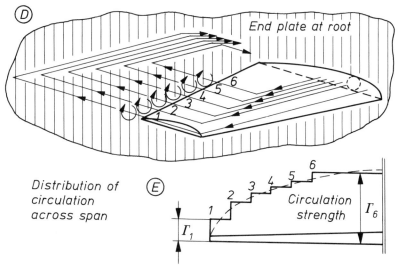

but a number of electric conductors as shown earlier in Fig 2.90B; some of them extend only across certain portions of the foil span. This concept is illustrated in Fig 2.93 by the cantilever foil attached at one end to a structure such as a hull which is relatively large compared with the foil section; the other end is free. Practical examples are a spade-type rudder hung closely under the wide, flat stern of a ship and the fin keel under the hull of a sailing yacht. In an extreme case, the entire underwater body of a ship represents a foil of very large chord length, relatively short span, and great thickness, cantilevered downward from an end plate of infinite extent, represented by the surface of the water on which the ship is floating.

Now consider first the cantilever foil at sketch A in Fig 2.93. The end plate extends from T_2 to V_2 and beyond in all directions; there is only one tip-vortex wire attached to the free end of the foil T_1; there is no vortex circulation in the region between T_2 and V_2, because the returning current is spread over an infinite area of the end plate in a multitude of current paths.

Consider next the single-stepped cantilever foil of diagram B in Fig 2.93. This has a small outer half, over which the circulation is only half of that in way of the larger inner half. The larger section T_2 to T_3 carries double the current that flows through the smaller section from T_3 to T_1. To represent this, the foil wire in the larger section $T_3 T_2$ is in the form of two conductors, each carrying equal currents. At its root T_2 it is attached to an end plate similar to that in sketch A. The lifting current from the longer wire $T_2 T_1$ comes off at T_1 and flows along the tip vortex wire $T_1 V_1$, then across the long starting-vortex wire $V_1 V_2$ to the end plate. The current from the shorter wire $T_2 T_3$ comes off at T_3, flows aft in the form of a trailing vortex, along the wire $T_3 V_3$, thence through the short starting-vortex wire $V_3 V_2$ to the end plate and back to the root of the foil.

Circulation takes place around both the tip-vortex wire $T_1 V_1$ and the trailing-vortex wire $T_3 V_3$; that around the outer tip-vortex wire $T_1 V_1$ is equal in strength to that around the single wire $T_1 T_3$ in the outer foil portion; the sum of the circulation around both $T_1 V_1$ and $T_3 V_3$ is equal to that around the inner or main portion of the foil, in way of the double wire from T_3 to T_2. Looking at the foil from ahead, as in diagram C of Fig 2.93 the circulation Γ about the inner or main portion is represented by the constant ordinate Γ_2; that about the outer portion as Γ_1, half as high as Γ_2.

Foils usually taper gradually in some fashion from root to tip; diagram D in Fig 2.93 is an example. Circulation then takes place about the foil and about the various tip- and trailing-vortex wires as though it were made up of many steps, each slightly greater than the one outboard of it. Diagram D assumes that there are six such steps, indicated in the distribution diagram at E. There are thus six wires at the root of the foil of which one terminates in the tip-vortex wire, and the other five in trailing-vortex wires.

Five hundred instead of five trailing-vortex wires may be imagined to produce smooth and reasonably fair spanwise distribution of circulation, indicated by the broken line in sketch E of Fig 2.93.

Analysing sketch E of Fig 2.93, a relationship can be established to describe the variation of circulation along the span and the strength of the trailing vortex, namely the strength of any individual trailing vortex leaving the foil at a given point is equal to the diminution in circulation at that section. For example, the strength of the trailing vortex released into the wake, say between sections 6 and 5, must be equal to the difference between the circulation strengths around sections 6 and 5. Thus if at sections 6 and 5 the circulation strengths are Γ_6 and Γ_5 respectively, the free or trailing vortex shed between those sections will have the strength equal to $(\Gamma_6 - \Gamma_5)$.

It can also be seen from diagram E that the more rapidly circulation decreases from a maximum value at the root to zero at the tips, the more intensive will be the vortices springing from the trailing edge and passing downstream. In other words, in general one may expect that the free vortices are strongest near the tip.

As compared with the simple vortex system depicted in Figs 2.87 and 2.92 diagram B, the actual vortex system generated by a foil is, because circulation is not constant across the span, more complicated. Instead of the two concentrated tip vortices shed from the foil ends there is a multiple system of small vortices or a sheet of free vortices streaming from the whole length of the trailing edge. Such a vortex sheet conceived by Lanchester and shown in his original drawing in Fig 2.86B, is unstable. As predicted by him, the filaments of the vortex sheet will evidently wind round one another like the strands of a rope (Fig 2.86A) into a pair of vortex tubes which extend downstream at a distance S, less than the span of the foil b apart.

For the same reason of mutual attraction two adjacent parallel cables in an electrical installation, carrying large direct current I flowing in the same direction, must be held apart by spacers to prevent them from drawing together and touching (Ref 2.75).

The trailing vortex system becomes of the type shown in Fig 2.94 and Photo 2.28. It consists of a bound vortex (lifting line) attached to the foil and, at a certain distance downstream, the two concentrated trailing or tip vortices, each of strength equal to the circulation Γ_c, around the central section of the foil. Thus, although the actual vortex system is considerably more complicated than that single horseshoe vortex of Fig. 2.87 suggested by Prandtl as a first approximation, the real picture approaches somewhat his simplification in which the trailing vortices, shed initially in the form of a sheet, roll up more or less rapidly behind each end of the foil into concentrated tip vortices. It has been found in the course of wind tunnel tests that the trailing vortex sheet rolls up 18 times more rapidly, in terms of chord lengths, behind a low aspect ratio triangular foil than behind a high aspect ratio rectangular foil (Ref. 2.76).

The origin of the trailing vortex system may be considered in yet another way as suggested by Glauert (Ref 2.12), and displayed in Fig 2.95, namely due to existence of a higher pressure $(+)$ beneath the foil than that above it $(-)$, a spanwise flow around the ends of the foil will take place. This motion superimposed on the parallel flow denoted by dotted and continuous streamlines ahead of the foil in Fig 2.95

Photo 2.28 (Compare with Fig 2.94.)

Picture of the trailing vortex sheet rolling up into concentrated tip vortices behind a low aspect ratio foil of AR = 2. Pictures were taken from behind the foil. The projections in the free-stream direction of the foil-tip positions are indicated in the photographs by the intersections of the vertical lines and the horizontal markers. Photographs of the wake at various stations behind the foil indicated by the ratio of d/c (where d is distance from the trailing edge and c is the length of the root chord of the foil) clearly show the rapidity with which the vortex sheet rolls up (Ref 2.76).

The trace of the vortex sheet was made visible by applying fine aluminium powder to the trailing edge of the foil.

Foil end ↓

Foil tip ←

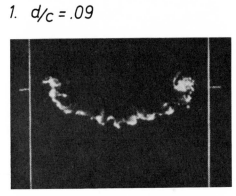
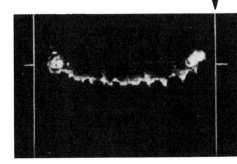
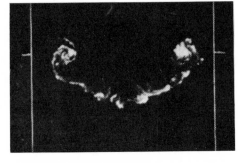
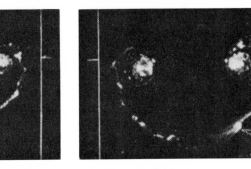

1. d/c = .09

2. d/c =.35

3. d/c =.60

4. d/c =.89

5. d/c =1.45

6. d/c =1.80

Fig 2.94   According to Prandtl the trailing vortices begin as a flat sheet at the trailing edge which rolls up into two vortex cores at a certain distance behind the foil. When developing the lifting line concept it has been assumed that each section along the foil span acts independently of its neighbouring sections, except for the induced downwash. Strict compliance with this assumption would require two-dimensional flow as presented in Fig. 2.84, that is, no variation of section shape, pressure, or lift coefficient along the foil span, and also that the lifting line or the foil leading edge is perpendicular to the flow direction. If departure from these requirements is not negligible, as in the case of so-called swept foils (discussed in following chapters) the foil properties, i.e. lift and drag, may change radically.

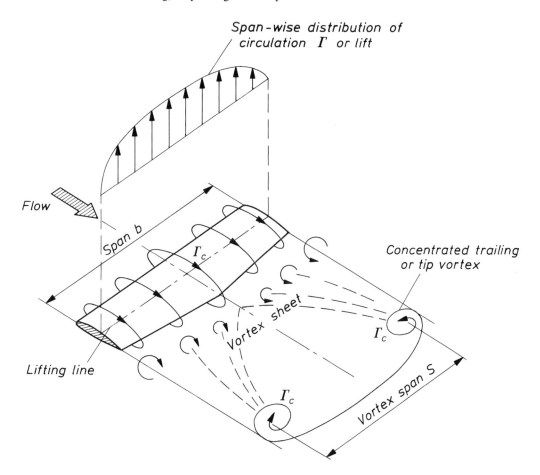

Fig 2.95   Origin of trailing vortex sheet.
           Note: Deflection of streamlines from straight lines when
      approaching the foil is greatly exaggerated for the sake of clarity.
      A. Foil seen from behind.
      B. Foil from bird's eye view.

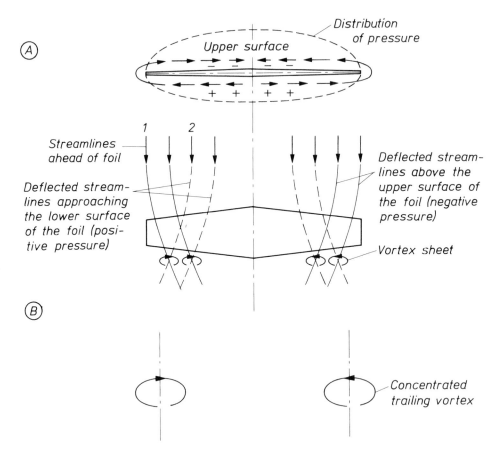

deflects the oncoming flow in such a way that the streamlines over the positive pressure side (broken lines) will be shifted outwards. Conversely, the streamlines approaching the suction side (solid lines) will be deflected inwards towards the foil centre. The flow immediately behind the trailing edge acquires therefore a swirling motion, most predominant at the tips, but with smaller swirls of vortices inboard which subsequently develop into a vortex sheet stretching along the whole trailing edge of the foil. The reason for this is rather evident: when the two streamlines 1 and 2 in Fig 2.95 from the upper and lower surfaces meet at the trailing edge, they are flowing at an angle to each other. This starts vortices rotating clockwise (viewed from behind the foil) from the left part of the foil and anticlockwise from the right part. The vortex sheet will ultimately roll up into concentrated trailing vortices at a

Fig 2.96 Induced velocities (downwash) due to the top vortex action.

A. Foil from bird's eye view.
B. Foil from behind. When the foil operates in a vertical altitude, as a sail or fin keel, the downwash could well be called sidewash.

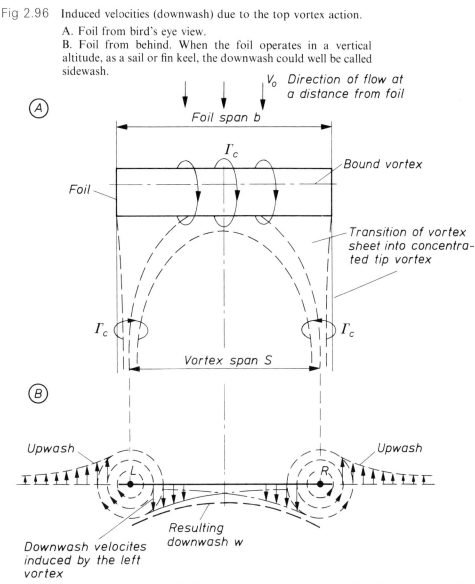

certain distance downstream in the manner already shown in Figs 2.86A and 2.94. Whatever the planform of the foil might be, the essential characteristics of the free vortex pattern and its formation are generally similar.

(3) Induced flow and associated induced drag

With the form of the flow made visible by various means, as shown in Photos 2.27 and 2.28, the tip vortices prove their reality. They give rise to a velocity field called the 'field of induced velocities'. This term, introduced to aerodynamics by Munk, was deliberately chosen to emphasize the similarity of the relationship between trailing vortices to that between electrical conductors and their magnetic field.

It is seen in Fig 2.96 that the tip vortices alter the direction of the streamlines approaching the foil from ahead and leaving its trailing edge. The manner in which they alter the flow at the foil is of great practical significance, since it will be shown that it may result in an enormous increase of so-called 'induced drag', the importance of which has been overlooked by sailing people.

As noted earlier, the trailing vortices developed by a three-dimensional foil alter the foil characteristics in a detrimental sense, as compared with two-dimensional flow. Each of the tip vortices L (left) and R (right) has associated with it a circulatory motion as depicted in the upstream view in Fig 2.96B. For the region between the vortex cores L–R the fluid particles leaving the trailing edge are given downward acceleration acquiring induced velocity w called downwash. In the vertical plane of the foil span, as much fluid must move up as move down. Accordingly there is an induced upwash outward and sideways from the foil tips, along the line of the extended span. This upwash is not constant but fades gradually with increasing distance from the vortex core. Migrating birds, flying in line abreast or in familiar 'V' formation are making use of this phenomenon. The ascending current produced by the wings of neighbouring birds reduces the expenditure of energy necessary to support flight. And birds do not consider it worth flying in the middle, i.e. behind the others where downwash prevails.

The downwash and upwash velocities w distributed over the span of the foil follow the law expressed earlier by Eq 2.7A, i.e.

$$V_c = w = \frac{\Gamma}{2\pi r}$$

where r is the distance from the vortex core.

Assuming that undisturbed flow velocity V_0 (Fig 2.96) and foil attitude (incidence angle) remain unchanged, and hence the circulation is established, one can find the downwash velocity w from the above expression. It is inversely proportional to the distance r from the vortex core. Since both vortices L and R contribute towards downwash, the resulting induced velocity w is obtained by adding the velocities associated with both trailing vortices. This downward flow observed behind the trailing edge of the foil between its tips must not be confused with downwash induced by the bound vortex, i.e. circulation round the foil, as shown earlier in Fig 2.15A. One difference is that the latter is accompanied by a corresponding upwash in front of the foil and the diagram of vertical velocities is symmetrical, as shown in Fig 2.97A which demonstrates the two-dimensional flow with imaginary end plates. In such a flow pattern imposed by the presence of the foil and called by Lanchester a supporting wave, the kinetic energy of the fluid stream is basically conserved. It means that the dynamic pressure of the flow $q = (\rho V_0^2/2)$ measured in front of the foil, after some partial and temporary transformation felt by the foil as a static pressure differential Δ_p across the foil surface, which manifests itself as lift, is almost

Fig 2.97   Induced velocities in foil vicinity.

A. Upward and downward velocity components due to bound vortex (circulation) action only. Two-dimensional flow.
B. Downwash induced by trailing vortices only (three-dimensional flow).

The downwash velocity w varies along the flow direction.

If downwash underneath the foil is, say, w its value some distance downstream increases to $2w$.

Note: for the sake of clarity the upward and downward velocity components are exaggerated as compared with V_0 vector.

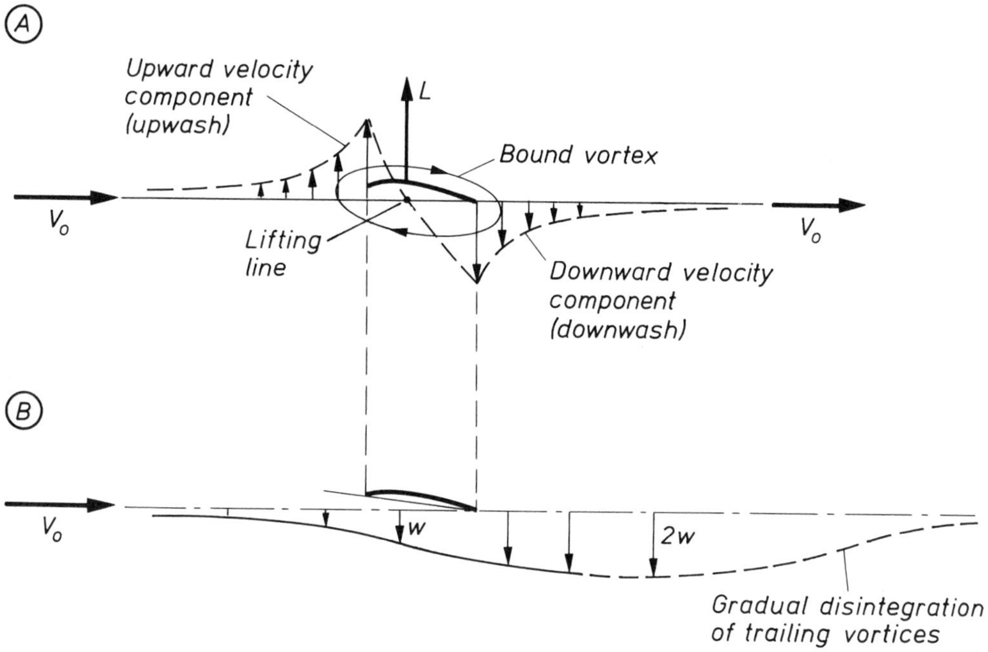

wholly recovered behind the trailing edge.

Since lift L is perpendicular to the relative motion velocity V_0, no work is done by the flow on the foil in order to generate lift. As a matter of fact the dynamic pressure would be wholly recovered only in the case where there are no losses due to friction and the flow is attached over the foil without separation, i.e. when air or water were truly devoid of viscosity. Thus Fig 2.97A can only be regarded as a plausible approximation of reality.

It is most important to understand that if the foil is of finite span and the flow is three-dimensional, an additional downward motion (downwash) due to the action of the trailing vortices displayed in Fig 2.97B must be superposed upon the components of induced velocities drawn in sketch A in Fig 2.97. The bound vortex is seen to contribute an upwash upstream and downwash downstream of the foil, while the tip vortices together contribute a downwash of increasing magnitude with increasing distance from the trailing edge. But to a certain point only, beyond which

Fig 2.98 Simplified perspective view of velocity field induced in the immediate neighbourhood of a rectangular untwisted foil.

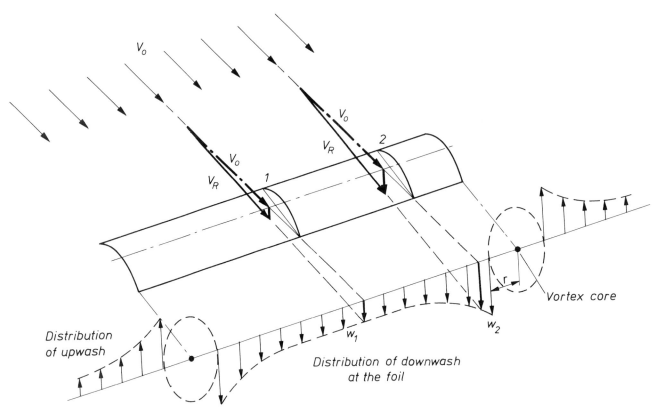

these vortices begin to disintegrate and finally their effects disappear. The bound vortex combined with the trailing vortices give the resulting downward current. Its effect may be visualized by considering that the foil with its circulation velocity field about its lifting line (Fig 2.97A) is moving through a region where the surrounding fluid is sinking or dropping slowly beneath it, by the magnitude of the induced velocity w along its path (Fig. 2.97B). This is the so-called 'climbing analogy' explanation of real foil action. With reference to the fluid stream in its vicinity, the foil is climbing continuously along a path having the slope w/V_0.

In contrast to the quasi-conservative system of the flow depicted in Fig 2.97A, the flow system given in Figs 2.97B and 2.98 may be called a 'dissipative system'. In such a flow, energy is not conserved even though the fluid be non-viscous. The generation of the trailing vortices requires a quantity of kinetic energy–or rather a continual removal of kinetic energy from the flow, which is not recovered by the foil. It is, in fact, lost to the foil by being left behind in the wake. In what follows it will be shown that this continual expenditure of flow energy appears to the foil as an induced drag which might also be called the trailing vortex drag.

It should perhaps be stressed that the influence of the trailing vortices is not

Fig 2.99 Streamlines about jib and mainsail showing local wind speeds and flow angles. Although the flow field created by the jib-mainsail combination is different in a quantitative sense from that generated by a single foil, it is in a qualitative sense similar. One has to go several boats' lengths upstream or downstream before a free stream velocity $V_0 = 10$ knots would be reached. The above Figure is included herein by kind permission of Arvel Gentry (Ref. 2.78). See also Fig 256, *Sailing Theory and Practice*.

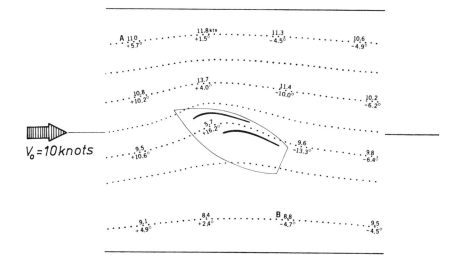

limited to the region just behind the foil but the downwash velocities are imported well ahead to the fluid stream approaching the foil. In general, the velocity fields created by the circulation around the foil, as well as the velocity fields caused by the tip vortices, are large compared with the size of the foil itself. This is seen in Fig 2.99 which represents the velocity field developed by a sailing boat in the close-hauled condition (Ref 2.78). The upwash as well as downwash may be perceptible several chord lengths in front and behind the foil respectively. Experienced helmsmen are quite well aware of this fact and exploit it tactically while racing.

(a) *Effective angle of incidence and induced drag*

Now, let us examine the effects of downwash on the forces developed by the rectangular foil of finite span, shown in Fig 2.98. If we add, as vectors, the downwash velocity w at the foil to the oncoming flow velocity V_0, we find that the resultant local velocity V_R is deflected downwards through an angle α_i whose tangent is w/V_0, i.e.

$$\alpha_i = \tan^{-1}\frac{w}{V_0} \quad \text{or in radians}$$

$$\alpha_i = \frac{w}{V_0}$$

This angle is usually called the induced angle.

Since the induced downwash w is relatively small when compared with the velocity V_0, its superposition on V_0 does not materially change the magnitude of the relative motion between the foil and the flow in its vicinity, i.e. $V_0 \simeq V_R$. However it changes the direction of the local flow, desginated V_R, 'felt' by the foil section, and this is of considerable importance. Its effect is twofold:

Firstly, the downwash reduces the effective angle of incidence and it is seen in Fig 2.100 that:

$$\alpha_{ef} = \alpha - \alpha_i = \alpha - \frac{w}{V_0} \qquad \text{Eq 2.23}$$

where α is the geometric angle of incidence measured between the direction of the flow at a distance from the foil and the foil chord.

Hence the lift generated at the effective incidence α_{ef} is smaller than would be expected from the geometric angle of incidence α. For a foil of infinite aspect ratio, i.e. in two-dimensional flow, the induced angle of incidence is zero. Therefore the effective angle of incidence α_{ef} is equal to the geometric incidence α so that

$$\alpha_{ef} = \alpha$$

For finite span foils therefore the geometric angle of incidence is the sum of α_{ef} and α_i, i.e.

$$\alpha = \alpha_{ef} + \alpha_i \qquad \text{Eq 2.23A}$$

Thus, in order to produce the same lift L, as is produced in the two-dimensional flow condition, the finite span foil must be set at the geometric angle of incidence α which is larger by α_i than the effective incidence α_{ef} for the foil of infinite aspect ratio.

The broken line section in Fig 2.100 gives the respective position of the foil operating at greater angle of incidence in order to compensate for the downwash effect.

Secondly, lift L_1, generated by a finite span foil is tilted backward through an angle equal to the induced angle α_i relative to lift L. Lift L_1 is substantially equal to L but there is now an additional drag component D_i, in the direction of flow velocity V_0, which is called 'induced drag', since its origin is to be found in the action of the induced flow by the trailing vortices. In a sense, the induced drag is part of the lift generated by the finite span foil. So long as the foil produces lift we must have induced drag. For this reason the induced drag is sometimes called drag due to lift. We can never eliminate it altogether no matter how cleverly the foils are designed. There are however certain planforms, notably so-called elliptical planforms, which produce less induced drag than the other forms.

It should perhaps be remembered that the forces of lift and drag are not normally referred to the local flow direction, but to axes perpendicular and parallel to the direction of undisturbed relative flow some distance from the foil outside the

Fig 2.100   Geometry of induced angle and induced drag at the immediate
vicinity of the foil.

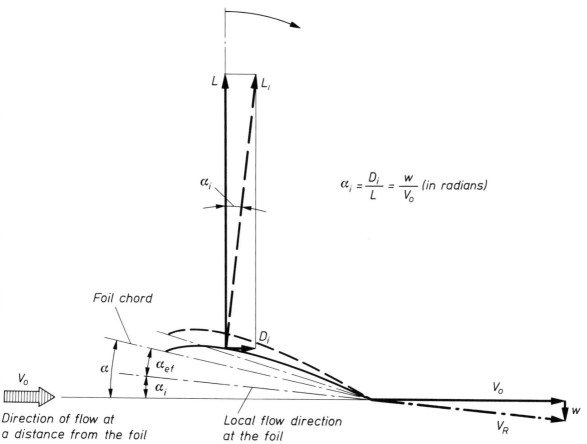

$$\alpha_i = \frac{D_i}{L} = \frac{w}{V_0} \ (in \ radians)$$

immediate zone of its influence and therefore clear of any induced flow deviations.

We may now look at the lift–drag relationship by applying Newton's second Law, frequently referred to as the action and reaction law. Figure 2.101, which gives an idealized version of the actual influence of the foil on the flow, fits reality nearly enough to suit our purpose. It is based on Prandtl's concept of the so-called 'swept area', a surrounding circular area A of diameter b equal to the foil span. The effect of the foil on the flow stream of velocity V_0 within this area is presumed to be constant while outside that area the flow is ignored. This fictitious concept is very convenient in developing the induced effects formulae.

Applying the action and reaction principle one may find that lift generated by a foil can only be provided by a downward acceleration to the fluid particles affected by the presence of the foil. In other words, the foil reaction L must be equal to the downwash momentum imparted to the mass of fluid m acted upon some distance downstream, i.e.

Fig 2.101   Perspective view of the swept area A assumed to be deflected by
the foil. See Fig 2.97B.

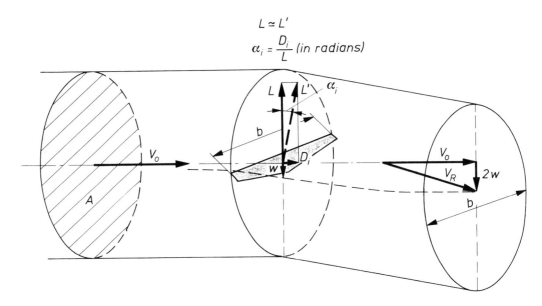

$$L = m \times 2w \qquad\qquad \text{Eq 2.24}$$

The mass of fluid flowing through the swept area A per sec is

$$m = \text{(mass density of the fluid)} \times \text{(volume per unit time)}$$
$$m = \rho \times (A \times V_0)$$

Substituting the above expression into Eq 2.24 yields $L = \rho A V_0 2w$.
Since

$$2w = V_0 2\alpha_i \quad \text{from Fig 2.101}$$

where α_i is given in radians

therefore

$$L = \rho A V_0^2 \alpha_i$$

The induced angle is then

$$\alpha_i = \frac{L}{2\rho A V_0^2} \qquad\qquad \text{Eq 2.25}$$

For the foil of span b the swept area A is

$$A = \frac{\pi b^2}{4} \quad \text{see Fig 2.101}$$

Lift expressed in the conventional way is given by

$$L = C_L \frac{\rho V_0^2}{2} \times S$$

where S is the foil area

Substituting expressions for A and L into Eq 2.25 gives

$$\alpha_i = \frac{C_L \times (\rho V_0^2/2) \times S}{2\rho(\pi b^2/4)V_0^2} = \frac{C_L S}{\pi b^2}$$

Since by definition the aspect ratio AR of a foil is given as

$$AR = \frac{b^2}{S}$$

then

$$\alpha_i \text{ in radians} = \frac{C_L}{\pi AR} \qquad\qquad \text{Eq 2.26}$$

where $\pi = 3.14$.

The corresponding equation in degrees is

$$\alpha_i = \frac{C_L}{\pi AR} \times 57.3 = 18.24 \frac{C_L}{AR} \qquad\qquad \text{Eq 2.26A}$$

And this is the induced angle of incidence at the immediate vicinity of the foil or, more precisely, at the lifting line which is located inside the foil (see Fig. 2.88).

Putting expression 2.26A for α_i into Eq 2.23A we obtain

$$\alpha = \alpha_{ef} + \alpha_i = \alpha_{ef} + 18.24 \frac{C_L}{AR} \qquad\qquad \text{Eq 2.27}$$

From the geometry of Fig 2.100 it is seen that

$$D_i = L \times \alpha_i$$

where α_i is in radians.

Or, in coefficient form

$$C_{Di} = C_L \times \alpha_i \qquad\qquad \text{. Eq 2.28}$$

Substituting Eq 2.26 into Eq 2.28 yields

$$C_{Di} = \frac{C_L^2}{\pi AR} \qquad\qquad \text{Eq 2.28A}$$

The average downwash angle a short distance behind the foil is seen from Fig 2.101 to be twice the value given in Eq 2.26A, i.e.

$$\alpha_i = 36.5 \frac{C_L}{AR} \qquad \text{Eq 2.29}$$

(b) *Elliptic planform, elliptic load distribution*

The relatively simple relations given by Eqs 2.26, 2.27 and 2.28 were obtained by Prandtl for the so-called elliptic lift distribution associated with elliptic planform. The curve representing such a distribution is the half of the ellipse shown in Fig 2.102. Lift L, has the maximum value L_0 in the median plane of the foil and drops gradually to zero at the foil tips.

Since, according to Eq 2.12, $\Gamma = f(V_0 c \alpha)$, i.e. circulation is proportional to the local chord c of the foil and lift in turn is proportional to circulation (see Eq 2.10), the elliptic distribution of lift can be obtained in the simplest way by taking a foil of elliptic planform having geometrically similar cross-sections with parallel chords over the whole span so that there is no twist. Such an untwisted foil may be called a planar foil and an elliptical lift distribution can be realized by making a foil consist of two semi-ellipses, as illustrated in Fig 2.121 sketches 4, 5 and 6 (Ref 2.79). It is perhaps worth noticing that nature widely employs these elliptical shapes. Many air- and water-borne creatures are equipped with foils of this particular planform or similar, for example bird wings or the dorsal fins of sharks and dolphins (Photo 2.23). The most important reason for the special emphasis which has been placed upon the elliptic distribution of lift is that of all forms of loading, this one leads to the smallest possible induced drag and uniform induced angle of incidence.

In general, the downwash w, which affects both the effective angle of incidence and induced drag, varies in magnitude along the foil span; how it actually varies depends largely on the planform of the foil. Figure 2.98, for example, demonstrates in a qualitative sense the distribution of downwash for a rectangular untwisted foil.

In an attempt to find out under what conditions the induced drag D_i will be a minimum for a given value of lift and given aspect ratio, Prandtl (Ref 2.79) demonstrated that D_i minimum occurs when downwash w is uniform along the foil span, as illustrated in Fig 2.102. Such a distribution of downwash is observed in the case of an untwisted foil of elliptical planform which, in a sense, can be regarded as an ideal planform. Aerodynamically, the merit of a foil can therefore be measured by the closeness with which the load distribution curves over the foil span approximates to the semi-elliptic form (Ref 2.12).

The mathematical processes involved in the development of Prandtl's concept are too complex to be summarized, but in Note 2.80 there is a much less rigorous explanation which involves only elementary mechanics. There is, however, a certain restriction involved in Prandtl's theory of the lifting line and is the *uniformity* of the undisturbed flow in which the foil operates. There are cases which do not satisfy this condition. For instance, a sail affected by the wind gradient, a case

Fig 2.102   If a foil is elliptically loaded the downwash is uniform.

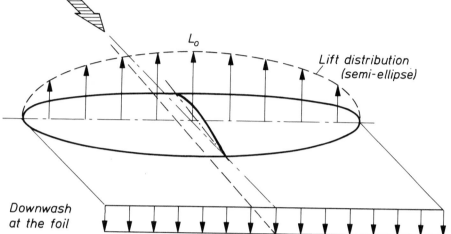

illustrated in sketch A of Fig 2.114, where the airstream speed has a maximum at the sail head and drops to zero at the water surface. In such circumstances, the minimum induced drag is given by the condition that the induced downwash angle α_i (not the downwash w) should be constant along the sail height. Figures 2.98, 2.100 and 2.102 can be of some help in clarifying this point. Non-uniform flow velocity along the foil span implies certain modifications to the ideal planform which may not be elliptical but not far from it if twist is deliberately used to achieve desirable distribution. This problem will be discussed in following chapters.

We shall interpret Eqs 2.27 and 2.28A in some detail

$$\alpha = \alpha_{ef} + \alpha_i = \alpha_{ef} + 18.24\frac{C_L}{AR} \qquad \text{Eq 2.27}$$

$$C_{Di} = \frac{C_L^2}{\pi AR} \qquad \text{Eq 2.28A}$$

It is seen that both the induced angle of incidence α_i and induced drag coefficient C_{Di} are tied up closely and almost exclusively with aspect ratio (AR). In order to determine the effect of a change of AR on foil lift characteristics it is convenient to fix a value of lift coefficient, say $C_L = 0.5$, and calculate the effect of AR variation on induced angle of incidence.

Figure 2.103 illustrates the effect of AR variation on the lift-curve slope of symmetrical foils of thickness ratio $t/c = 0.1$ and Table 2.9 gives the result of calculations of induced angle α_i and also geometric angle of incidence α based on Eq 2.27.

TABLE 2.9

AR	∞	6	4	2	Remarks
α_i	0°	1.5°	2.3°	4.6°	$\alpha_i = 18.24 \dfrac{0.5}{AR} = \dfrac{9.12}{AR}$
α for $C_L = 0.5$	4.6°	6.1°	6.9°	9.2°	

For an unstalled foil of infinite aspect ratio AR_∞, the circulation Γ around it varies almost linearly with the angle of incidence α, i.e.

$$\Gamma \sim L \sim \alpha$$

where $\sim$ means 'is proportional to'

and the lift coefficient can, according to Eq 2.14, be expressed as

$$C_L = 0.11\alpha$$

Such a theoretically ideal foil marked $AR = \infty$ in Fig 2.103 may serve as a yardstick. Relative to this foil all other foils of $AR = 6, 4,$ and 2 must be set at higher incidence

Fig 2.103   Theoretical C_L versus α curves for symmetrical foils of different aspect ratio. For reasons which are given in section 5 d and e the lift-curve slope for the foil of $AR = 2$ should be regarded as approximate. The experimental values of the C_L for foils of AR above 2 are fairly well given by the following formula:

$$C_L = \frac{0.11}{1 + 2/AR} \times \alpha$$

This is applicable to untwisted foils of elliptic planform.

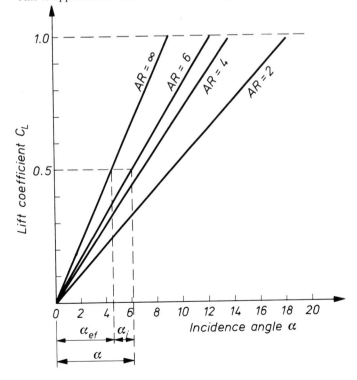

α by the amount of induced angle α_i in order to obtain the selected value of lift coefficient $C_L = 0.5$ which occurs when the effective angle of incidence α_{ef} is about 4.6°. As can be seen in Fig 2.103 the geometric angle of incidence of the foil of AR = 2 is twice as big as that for the foil of infinite AR. The incremental differences in α for other aspect ratios and lift can be read or deduced from Fig. 2.103. The influence of the profile or section of the foil upon the lift-curve slope of the two-dimensional foil is relatively small, so that it has little practical significance. Therefore, the slope of the lift curves for finite aspect ratio foils may be considered as depending almost entirely upon the aspect ratio, or more precisely, upon the spanwise distribution of lift if the foil is not of elliptical planform. For reasons which will become apparent in section 5d and e the lift-curve slope for AR = 2 in Fig 2.103 should be regarded as approximate.

For comparison, Fig 2.104 demonstrates the experimental C_L versus α curves for aerofoils of the same section Gö 389, but of different aspect ratios from AR = 1 to AR = 7 (Ref 2.66). The almost linear portions of these curves clearly have different slopes and the practical consequence of this is that, depending on AR, the foils of the same section characteristics will, at the same geometric angle of incidence α, develop different forces. For example, at $\alpha = 10°$ the foil of AR = 5 produces twice as much lift as the foil of AR = 1.

Fig 2.104   Experimental C_L versus α curves for aerofoils of the same section (Göttingen 389) but of different aspect ratios AR. The lift-curve slope becomes progressively less steep when the AR of the foil decreases.

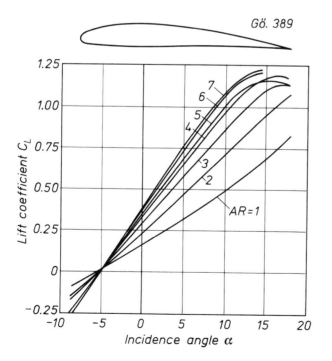

Turning now to the induced drag equation 2.28A, let us calculate and plot the drag for various lift coefficients and aspect ratios. Figure 2.105 illustrates graphically the results of the calculations which are presented partly in Table 2.10.

TABLE 2.10
Elliptic planform

C_L	Induced drag C_{Di} for aspect ratios				
	1	2	4	6	8
0.1	0.0032	0.0016	0.0008	0.0005	0.0004
0.2	0.0127	0.0064	0.0032	0.0021	0.0016
0.3	0.0286	0.0143	0.0071	0.0048	0.0036
0.4	0.0509	0.0254	0.0127	0.0085	0.0064
0.5	0.0795	0.0398	0.0199	0.0133	0.0100
0.6	0.1146	0.0573	0.0286	0.0191	0.0143
0.7	0.1560	0.0780	0.0390	0.0260	0.0195
0.8	0.2037	0.1018	0.0509	0.0340	0.0255
0.9	0.2578	0.1290	0.0645	0.0430	0.0322
1.0	0.3183	0.1591	0.0796	0.0530	0.0398
1.2	0.4584	0.2292	0.1146	0.0764	0.0573
1.4	0.6239	0.3120	0.1560	0.1040	0.0780

(c) *Sail plans*

The virtues of high aspect ratio as a promoter of aerodynamic efficiency are evident, but it is plain too that they are subject to a law of diminishing returns. The higher the aspect ratio to start with, the less the advantage that is to be derived, in terms of induced drag, from each succeeding increment of AR.

In yacht designing there has been a tendency over many years towards higher aspect ratio of sails, and no doubt the tall rigs have been found advantageous in unconventional fast sailing craft or conventional yachts sailing on triangular olympic courses when their windward performance counts most. In both cases the maximum available L/D ratio is of primary significance.

However, the aerodynamic efficiency of the rig cannot be expressed by means of a single number such as AR or L/D ratio. And although high AR and, associated with it, high L/D ratio are of some value when sailing upwind, the sail shape–its AR, planform, camber, etc.–must be guided by the particular needs or requirements imposed by the actual or prevalent course sailed, wind velocity and stability. It is obvious that from consideration of heeling moment, a low aspect ratio rig is preferred, but it will be demonstrated that even ignoring stability effects the low aspect ratio rig is in some conditions superior to the taller rig.

.Returning now to induced drag, a further question arises: how big is the contribution of induced drag towards the total drag. As indicated by the curves in

Fig 2.105 Variation of induced drag C_{Di} with aspect ratio AR. C_{Di} scale is
expanded 5-fold for the sake of clarity. The polar curve of
induced drag is the same for all foils of the same plan-form,
varying only with the AR.

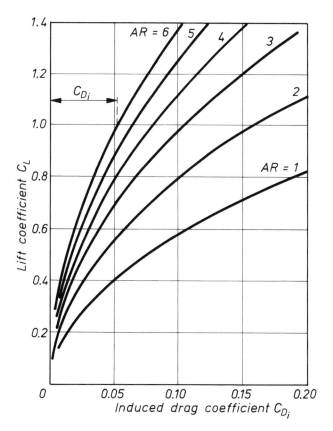

Fig. 2.105, which are parabolas as a matter of fact, the induced drag C_{Di} arises only
when the foil produces lift L. It was shown however in Section B.1 that no matter
whether a foil produces lift or not, there is always a drag resisting the relative
motion. This drag is usually called profile drag because it is determined to a large
extent by the cross-section or profile of the foil. It chould be recalled that profile
drag, which is normally measured under two-dimensional flow conditions and
referred to in technical literature as the section drag c_d, consists of two components,
friction drag and pressure drag. The relationship between these different kinds of
drag, C_{Di} and c_d, is shown diagrammatically in Fig 2.106.

A glance at the controlling factors listed in Fig 2.106 reveals that the induced drag
C_{Di} depends almost entirely on the geometry of the foil planform, as contrasted with
profile drag c_d, which is intrinsically affected by the shape of the foil section. One
may rightly say that the aerodynamic characteristics of a foil of finite span are

Fig 2.106 Analysis of the total drag components.

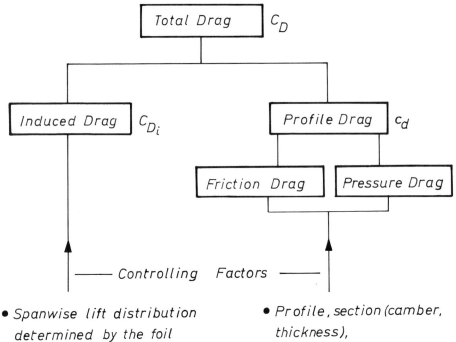

determined partly by its cross-section and partly by its planform. Hence, the total drag C_D of a foil of an elliptic planform can be written:

$$C_D = c_d + C_{Di} = c_d + \frac{C_L^2}{\pi AR} \qquad \text{Eq 2.30}$$

For any other foil planform, rectangular, triangular, tapered (planar or twisted), the total drag will always be higher.

A polar diagram in which C_L is plotted against C_D with angles of incidence marked on the polar curve is a very convenient way of demonstrating the mutual relationship between these drag components and lift. Comparisons between foils can be more readily made from polar diagrams than from separate curves for lift and drag plotted against incidence angle.

Figure 2.107 illustrates such a polar diagram for a rigid elliptical foil of AR = 4 and cross-section Clark Y. Apart from the polar diagram itself there is also drawn the induced drag parabola. A breakdown of the total drag given by Eq 2.30 is demonstrated at $C_L = 1.0$ to show the relative proportions of the induced and profile

Fig 2.107   Polar diagram of an elliptic foil of AR = 4 section Clark Y.
Induced drag is a function of lift alone and has nothing to do
with the angle of incidence except to modify it through the
introduction of an induced angle.

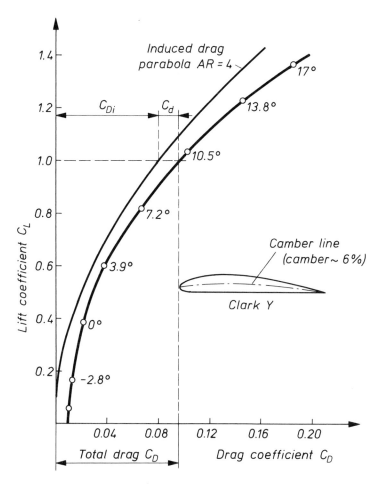

drags. The division of the total drag of a foil into profile and induced components
makes it possible to isolate and study the effects of the section shape and planform
upon the total drag and finally on foil performance.

It is evident that the drag components c_d and C_{Di} vary depending on lift, and total
drag is increasingly dominated by the induced drag component when lift or incidence
angle increases. At an angle of incidence of about 10°, when $C_L = 1.0$, the induced
drag C_{Di} is 5 times greater than the profile drag c_d. On the basis of presented theory of
induced drag, or in other words the theory of foil planform, it is possible to predict
the aerodynamic characteristics of any elliptic, planar foil for which the profile
characteristics (i.e. c_l and c_d), measured in two-dimensional conditions, are known.
This is certainly true in the case of rigid foils, such as shown in Fig 2.107, and there is

ample literature in which the profile characteristics can be found, for example Refs 2.30, 2.31 or 2.57.

Once the section characteristics, c_l and c_d, are known, it is an easy matter to introduce corrections for induced drag by applying Eq 2.30 and for induced angle of incidence by applying Eq 2.27.

Proof that induced drag and induced angle of incidence are not directly dependent upon foil profile is found in the confirmation of the transformation Equations 2.27 and 2.30 by numerous tests on foils or wings of all kinds. It should be pointed out with reference to Eqs 2.27 and 2.30 that they are also applicable, with sufficient degree of accuracy, to tapered untwisted foils of constant profile within the range of taper ratio:

$$\frac{c_t}{c_r} = \frac{\text{tip chord}}{\text{root chord}} = 0.3 - 0.5 \quad \text{as shown in Fig 2.108.}$$

The effect of such a linear taper upon the induced drag and induced angle of incidence of a foil of AR = 4–6 is negligible.

The values of C_{Di} and α_i for a moderately tapered foil are in the order of 2–5 per cent greater than the relevant values for elliptic foil. These differences in predicted characteristics may become even smaller if the foil tips are rounded. However, very severe tapers resulting in pointed tips lead to much less favourable characteristics than those for elliptic or moderate-taper foils. Triangular sails, as we shall see are thus notable examples of the worst planform from a purely aerodynamic point of view.

Fig 2.108   The aerodynamic characteristics of an untwisted tapered foil of
c_t/c_r ratio 0.4 are not much different from that of elliptic foil.
The attached sketch demonstrates how to draw an ellipse.

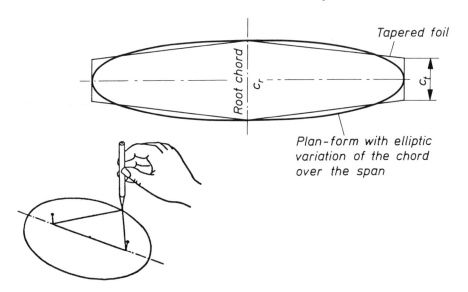

Tapered foil

Root chord

Plan-form with elliptic
variation of the chord
over the span

Once again reverting to the polar diagram plotted in Fig 2.107, it should be added that it represents the aerodynamic characteristics of an isolated foil or wing, which is not affected by ground or sea proximity, and does not experience any kind of interference effect.

If the foil is now attached to any structure, be it a monohull, catamaran or aircraft fuselage, there is an increment of drag to be added to the foil drag at all angles of incidence. This increase, which is termed parasitic drag, can be dealt with in the same way as already described when discussing Fig 1.62 in Part 1. Apart from parasitic drag which results in decreasing the L/D ratio of the complete machine, there is another factor which is advantageous and should not be overlooked. This is the interference effect due to the presence of any surface close to the foil ends, for example the deck or sea surface in the case of a sail, or the bottom of the hull in the case of appendages such as a rudder and fin keel. It will be shown that the manner in which the foil is attached to the hull, and in particular the distance between the boom and the deck or sea level, is of importance. Empirical efforts to exploit this are illustrated by Photos 2.29–2.31.

(d) *Triangular planform*

Figure 2.109, which is drawn in a similar manner and to the same scale of C_L and C_D as Fig 2.107, illustrates the aerodynamic properties of a Finn-type sail of AR = 3.1 made of Terylene, for comparison with those of the rigid planar Clark Y foil; measurements were made in the wind tunnel on a 2/5 scale model (Ref 2.82). The sketch attached to Fig 2.111 depicts the sail planform, vertical and horizontal camber distribution, and finally twist, measured relative to the boom. In the graph of Fig 2.109 is plotted the parabola of induced drag for aspect ratio 3.1 for elliptic planform, which may be used as a yardstick. One may rightly suspect that the soft triangular and twisted sail of non-uniform vertical distribution of camber might not experience elliptic loading. This is indeed the case. Thus, the direct determination of the actual induced drag of a soft sail from experimental measurements of the lift and drag forces which it produces is by no means as simple as it is in the case of a rigid wing of elliptic planform. However, to give a clearer picture of the real nature of sail aerodynamic performance than that demonstrated by the experimental polar diagram in Fig 2.109 (thick curve with angle of incidence values written along it), one must study the detached values of drag components in order to establish how they contribute towards the total drag. Once their relative contribution is known, one may attempt to find out the means of reducing those drag components.

In the case of an elliptic foil the total drag, as can be seen in Fig. 2.107, is simply the sum of induced drag and profile drag. Therefore, the profile drag coefficient c_d, at any value of the lift coefficient, can be measured off directly as the horizontal distance between the induced drag parabola and the polar of the foil at the appropriate value of C_L. Since a foil, such as the triangular, twisted sail characterized in Fig 2.109, does not experience elliptic loading there is an additional drag which is masked behind the measured total drag, which can be written

$$C_D = C_{Di(elliptic\ planform)} + c_d + \text{Additional Drag}$$

Photo 2.29   *Tackwind*–one of the World Sailing Speed Record entries.
The manner in which the sail is attached to the hull is of
some importance. *Tackwind's* solid sail rotates about the
masthead–this first asymmetrical wing sail boat can sail on
either tack. Despite the elliptic planform, her sail, operating
close to the sea, will not offer the full advantages associated
with elliptical loading (*Yachting World*–November 1974).

Photo 2.30   Aerodynamic efficiency of the rig, no matter whether the sail
is soft or rigid, will depend to a large extent on the gap
between the sail-foot and sea.

Photo 2.31   C-Class cat *Patient Lady III*.

C-Class is the ultimate development class where rules allow for almost unlimited variation in sail concept. The double-slotted wing sail of *Patient Lady III*, designed by D Hubbard, consists of three panels: a leading-edge panel which serves as the mast, a central panel and trailing edge panel. The two after panels are hinged from the leading panel so that it can be cambered to sail on either tack. Further details–see Ref 2.90.

Fig 2.109 Polar diagram of the Finn-type sail. Components of the total
drag C_D are distinguished (see Fig 2.111 in which the sail shape
is shown).

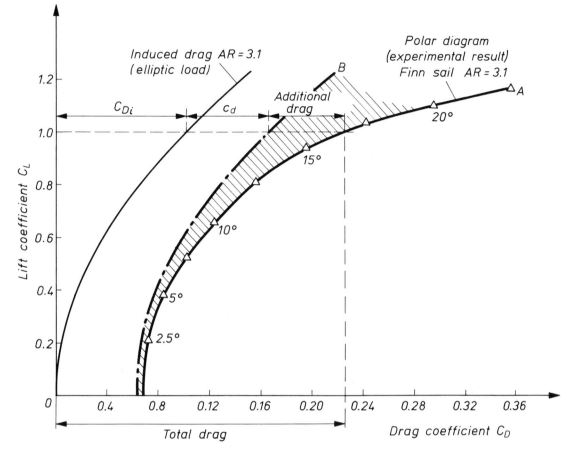

This additional drag might be regarded as a price paid for the departure from
elliptic loading. In order to assess the magnitude of additional drag we must separate
it from the remaining two kinds of drag C_{Di} and c_d. To do so one must know the value
of profile drag c_d. In what follows, the aim will be to assess the order of losses in foil
efficiency when it deviates from elliptic loading associated with elliptic planform.

Unlike the section characteristics c_l and c_d of streamline thick profiles, the two-
dimensional characteristics of a sail section, such as are represented by cambered
plates, particularly with the presence of a mast at the leading edge, are neither
numerous nor readily available. Nevertheless, from what has been published, one
may assess the sail profile drag c_d with a reasonable degree of accuracy. Figure 2.110,
which is based on tests made by Herreshoff (Ref 2.83), demonstrates the section
characteristics of two sail-like profiles having a camber of about 12 per cent. The
only difference between sections 1 and 2 is the position of the maximum camber in

Fig 2.110 Section characteristics c_l and c_d of two sail sections with masts.
Camber about 12 per cent.

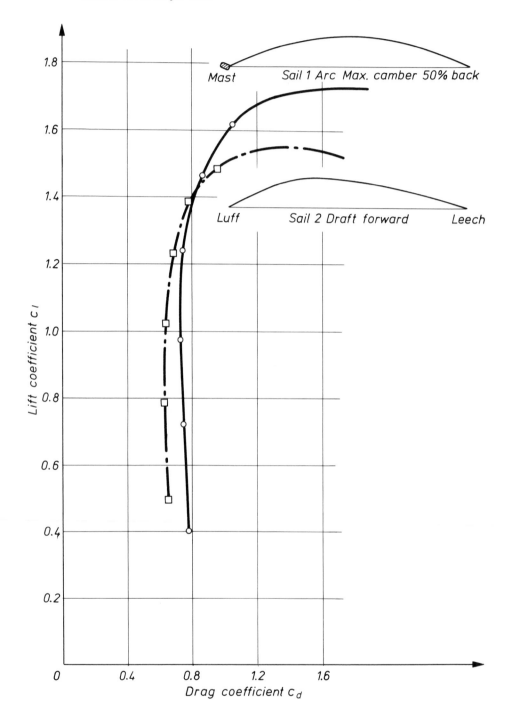

relation to the mast. It is seen that within the range of lift coefficients used in close-hauled conditions the c_d coefficient of section, which has the maximum camber $1/3$ chord back from the leading edge, varies little–from 0.062 to 0.064. The other sail of section 1, with the maximum camber halfway back from the mast, produces a higher drag coefficient for the same c_l. If we adopt the drag values of section 2 as roughly representative of the Finn sail given in the sketch of Fig 2.109 and add them to the induced drag parabola, shown as a thin continuous line in that diagram, we obtain the resulting polar curve B, shown as a thick broken line. It would represent the aerodynamic characteristics of the Finn-type sail if its load distribution were elliptic and therefore the induced drag would be minimum. Since such a loading was not achieved by this particular sail configuration and cannot truly be achieved by any soft, triangular sail of high taper ratio, a certain penalty must be paid in terms of an additional drag which is distinguished by the cross-hatched zone between polar A (experimental results) and polar B (calculated results).

Because there exists, according to Eqs 2.28A and 2.30, a linear relationship between the induced drag coefficient C_{Di} and lift coefficient C_L^2, followed by an approximately linear relationship between the $C_{Di} + c_d$ and C_L^2, another type of graph is frequently used in evaluating model test results. Figure 2.111 displays such a presentation where the scale for drag coefficient C_D is plotted along the vertical axis while that for C_L^2 is written along the horizontal axis. Since the variation of C_{Di} with C_L^2 is given by a straight line, the remaining components of the total drag, namely c_d and additional drag can readily be determined from the C_{Di} line upwards.

Now, if we compare the triangular sail characteristics of Fig 2.111 with the elliptical rigid foil characteristics shown in Fig 2.112, the differences in foil performance become immediately apparent. The sail inferiority is certainly due to much higher drag profile drag c_d but above all, to an enormous additional drag contribution towards its total drag. At $C_L = 1.0$ the Finn-type sail produces twice as much drag as the rigid elliptical foil. The differences in calculated C_{Di} values, resulting from unequal aspect ratios for both foils, are relatively small.

The question of what can be done in order to reduce sail drag can only be answered if we know what the factors are which determine the additional drag.

The causes of high profile drag c_d of a thin foil operating in the presence of a mast at its leading edge have already been discussed. The flow conditions behind the mast, which acts as a turbulence generator, are largely responsible for the poor aerodynamic qualities of mainsails. No substantial improvement can be expected in this field unless rotating streamline masts are employed, a solution which may not be practical for some non-aerodynamic reasons associated with structural integrity.

The main three factors which affect the magnitude of additional drag are:

a. Taper ratio
b. Twist
c. Vertical distribution of camber

If the planform of given aspect ratio AR with its most important parameter, the

Fig 2.111 Components of total drag C_D plotted against C_L^2. The 'Additional drag' represents the penalty paid for the departure of actual loading from elliptical loading.

This figure displays in different form the foil characteristics shown earlier in Fig 2.109. See Part 3–Tests on Finn sail.

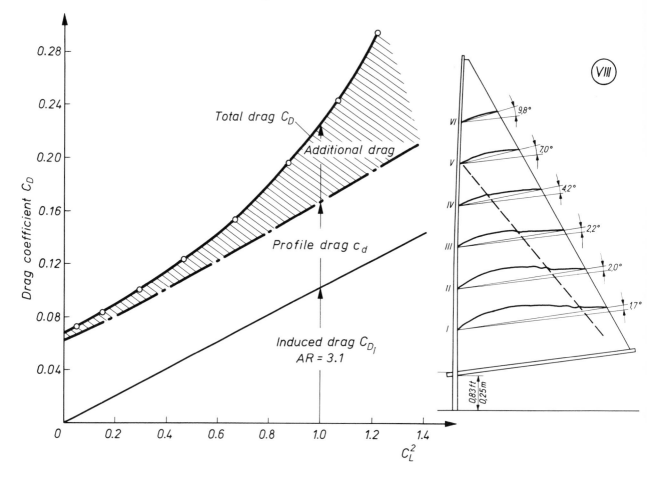

taper ratio, remains invariant, one may eliminate factor (a) and investigate the effects of (b) and (c). In the case of a soft sail it is almost impossible to isolate those two factors completely. The kicking strap or boom vang, which in principle is intended to control the sail twist, unfortunately affects the camber distribution too. One may argue that twist appears to deserve more attention than camber distribution. Most yachtsmen are probably aware of the apparent desirability of minimizing twist, but are possibly unaware that by so doing they may considerably affect the induced drag and also what might be called the vortex drag. This term must be explained. It has already been said that induced drag can be defined as drag due to lift or, in other words, drag associated with the production of a resultant lift

Fig 2.112   The total drag components plotted against C_L^2, of the foil Clark
Y shown earlier in Fig 2.107.

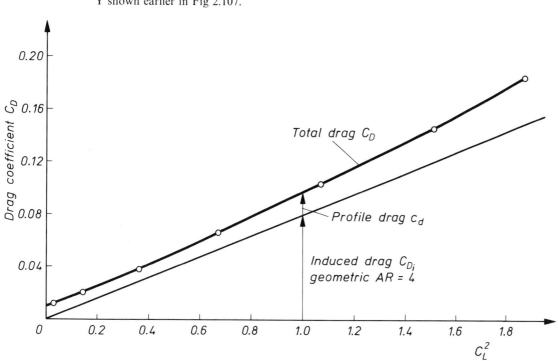

force. Such an auxiliary definition is perfectly correct in the case of a planar, i.e. untwisted, foil. However, as pointed out by T. Tanner (Ref 2.84), the above definition may not be sufficient in the case of twisted foils. If a sail were represented in an experiment by a rigid sheet-metal model with a certain amount of twist, depicted for example in the sketch attached to Fig 2.111, it would be incorrect to assume that the induced drag was the increase of drag associated with the generation of a resultant lift. With such a twisted model one part of it could produce positive lift whilst another part was producing equal negative lift. And in this case there would be appreciable induced or vortex drag despite the resultant lift being zero.

It is a matter of terminology and preference whether such a drag due to twist alone be called induced or vortex drag. Since their very nature is exactly the same, a large part of additional drag can be regarded as additional induced drag and added to the minimum induced drag which is represented in Fig 2.111 by the straight line. This additional induced drag is the price one has to pay for applying a triangular planform which is destined to failure in achieving the unique elliptic loading which leads to minimum induced drag.

Figure 2.113 illustrates how big this additional drag, mainly due to twist, can be in the case of the Finn-type sail described earlier when discussing Figs 2.109 and 2.111. The results of wind tunnel tests shown in Fig 2.113 were obtained by measuring

Fig 2.113 The total drag components plotted against C_L^2. The 'Additional drag' represents the penalty paid for the departure of actual loading from the elliptical loading.

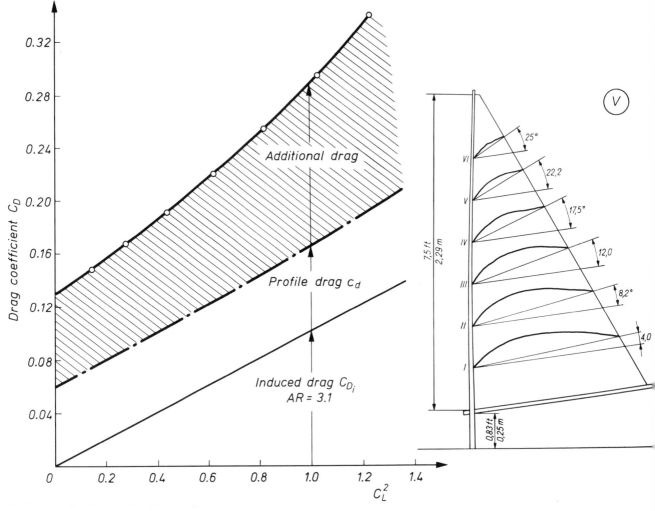

forces developed by the same model as before, the essential difference being the magnitude of twist. Instead of about 9.8° measured between section VI and the boom shown in Fig 2.111, the relevant twist was increased, by easing the kicking strap tension, to about 25°. The effect of such an increase in twist, in terms of additional drag, distinguished by the crossed zone in Fig 2.113, is devastating and noticeable over the whole range of lift coefficients. At the lift coefficient $C_L = 1.0$, which can be regarded as an average lift coefficient in close-hauled conditions, the combined additional and induced drag is more than 4 times as large as the profile drag which is a sum of friction and pressure drag. Twist in the order of 25° or even more is not uncommon and is frequently observed in practice.

According to Milgram and Tanner (Refs 2.83, 2.84), the fact that the total drag of sails is strongly dominated by its induced drag component has escaped a large number of yachtsmen who have gone to extraordinary lengths to reduce profile drag by experimenting with various mast sections where, in fact, the right vertical distribution of twist and camber, aiming at diminishing the induced drag effects, would have been more rewarding.

Apart from the induced drag hidden behind the additional drag in Figs 2.111 and 2.113 there are also other kinds of drag lumped into this category. Those drags, resulting from incorrect vertical distribution of camber and sail cloth porosity, have not been clearly distinguished as yet. Before their influence on sail efficiency is discussed in following chapters, we must look first at the sail operating in proximity of the sea, the sea being understood as a flat surface stretching horizontally beneath the sail.

(4) Mirror-image concept of the sail or foil: effective aspect ratio

'(7) We know *nothing* of how the vastly important wing tip vortices (so important in aviation) affect our sails. Here is an aircraft approaching:

Our sail is *one* of these wings. There is a small vortex up aloft and there must be a big one, greatly modified by the hull and the sea, low down at the foot and, in a sharply heeled keel yacht, what happens? Does the sea's surface act like the top wing of a bi-plane and give the same adverse effect on the lower wing, the sail?'

Gen H J PARHAM, *Yachts and Yachting* 1956

The uninformed but technically curious yachtsman frequently asks such questions as, 'is a sail a wing operating in a vertical position, i.e. is the hull of a yacht equivalent to the fuselage of an aeroplane, and the sail equivalent to one wing projecting from the side of the fuselage?' Or, 'is a sail equivalent to a pair of wings, but with an upper tip and a lower tip?' Other statements such as, '…in a boat like the Finn, where the sail is carried down close to the hull, it may be reasonable to consider the sail as equivalent to a single wing'; or, '…a wing is required solely to give lift; drag is the price you have to pay for this lift. The sail is not required to give lift but driving force so, if the problem is defined in such a way, one cannot see any connection whatsoever between a spinnaker and a wing.'

These questions and statements taken from yachting magazines sound contradictory and are confusing. As so frequently happens, these controversialists are, in one way or another, often right in their opinions of effects, although not often in

Fig 2.114   Mirror-image concept makes easier to grasp the distinction between the hidden essentials and delusive appearance. For the sake of simplicity the downwash and upwash velocities in sketch B are drawn qualitatively for real sail only. For comparison see Figs 2.96, 2.98 and Photo 2.27.

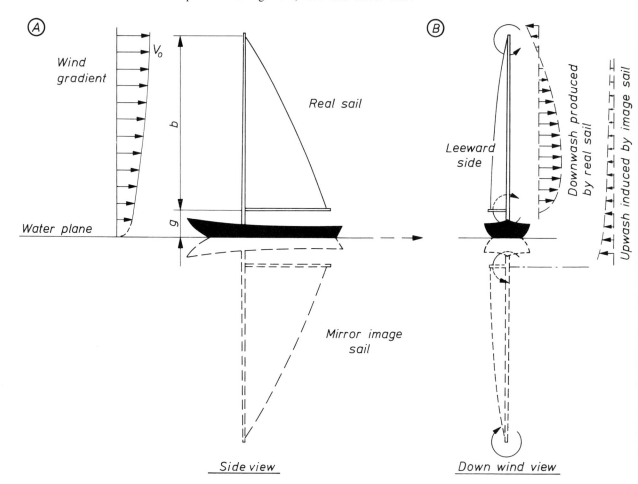

their opinions of cause. No doubt the proximity of the sea will have an effect on the flow round the sail, and therefore on loading distribution, and of course on the associated phenomena, induced drag and induced angle of incidence.

In order to obtain a picture of the influence of the water-plane on the flow round the sail, the best way is to introduce a concept used in physics. This is the so-called 'mirror-image method', by which the effect of sea presence can be obtained by replacing the water-plane by an inverted mirror-image of the boat as shown in Fig 2.114, and considering the new but easier problem of interacting of flow around the two sails, or rather two boats since the hull below the sail should not be ignored. This imaginary boat may be referred to as an image of the real one, i.e. its image in a mirror at the position of the water-plane. The problem is thus reduced to that of two boats in a stream of air devoid of water boundary. *The flow pattern and interactions resulting from a combination of a real and a mirror-image boat symmetrically placed on the other side of the boundary will be identical with the flow pattern on a real boat over a flat plane.* Critical readers may ask why one must construct imaginary worlds in order to explore certain features of the real world? The only justification is that, in this particular case, the understanding is facilitated by the fact that we already know how the flow pattern around a single foil looks. Figures 2.96, 2.97 and 2.98 illustrate it. But we know little or nothing about the effect of the water surface on a sail. Now, if we can remove the water-plane and replace it instead with the mirror-image sail, which develops exactly the same flow pattern as the one we are already familiar with, then such a problem-solving device as the mirror-image method is of some value. The following examples should explain it better than words.

From diagram 2.98 it is clear that the foil generating lift induces downwash between its tips and also an upwash outwards and sideways from the tips. It has already been mentioned that migrating birds instinctively apply the technique of flying in the region of upwash generated by the neighbouring birds' wings. It leads to the often observed side-by-side or V-stagger formation. Birds flying this way recover the part of the energy lost by their neighbours. Because of the reduction in drag due to smaller resulting downwash, less power is required to maintain the forward velocity and hence each bird flying in a V formation is losing less energy than when flying singly.

Two foils operating in a side-by-side condition, be it the sail illustrated in Fig 2.114, or any foil, keel, rudder, etc., experience exactly the same kind of interaction. An upwash is induced by each sail on the other so that the resultant downwash at each is less than would exist if the two sails were so far apart that their interaction was negligible. The arrows drawn in Fig 2.114B indicating downwash and upwash, as they affect the real sail only, are intended to give a rough idea of their relative significance and distribution in a qualitative sense. Thus, for example, the magnitude of the upwash due to the mirror-image sail on the real sail varies along its height (the spanwise direction) and has a maximum value at the sail foot, near the gap between it and the deck, and decreases continuously towards the top of the real sail.

Since, according to Eq 2.28, induced drag depends in turn on the amount of resulting downwash, i.e. on the induced angle of incidence α_i, one may rightly expect that the induced drag of the two interacting sails depicted in Fig 2.114 will be smaller than that of an isolated, non-interacting sail. One may further expect that the gap between the two interacting sails will be of some importance and experimental data presented in Fig 2.115 confirm it (Ref 2.85). It is thus seen that when the gap g is, say, 4 per cent of b ($g/b = 0.04$), the effective induced drag $D_{i(ef)}$ of the sail is about 20 per cent less than it would be when there is non-interaction due to the presence of the water-plane.

In the discussion which follows, the non-interaction condition should be understood to be when the sail or foil is operating in isolation, i.e. far away from the sea or

Fig 2.115   The effect of gap on induced drag of a split (mirror-image) foil.

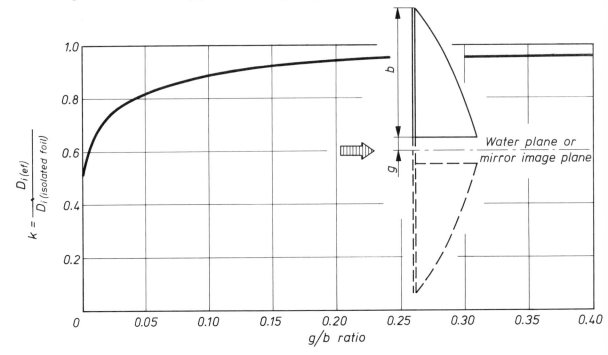

$$k = \frac{D_{i(ef)}}{D_{i(isolated\ foil)}}$$

g/b ratio

other foil with which it might interact, an unrealistic condition as far as sails and hull appendages are concerned and this means that interactions must always be taken into account in practical situations.

The graphical presentation given in Fig 2.115 can be expressed mathematically as

$$D_{i(ef)} = k \times D_{i(isolated\ foil)} \qquad \text{Eq 2.31}$$

This effect of the presence of the mirror-image sail on drag is equivalent to an increase in aspect ratio of the real sails. The divided foils of the mirror-image system shown in Fig 2.115 may subsequently be considered as a single foil which would be formed by pushing its halves together, and so closing the gap. If the span of each divided foil is b and the gap width g becomes zero, the comparable resulting foil has a span of $2b$, so the effective aspect ratio AR_{ef} is twice that of the actual geometric aspect ratio of AR of the real part of the mirror-image combination, i.e.

$$AR_{(ef)} = 2AR$$

and accordingly, for the reasons illustrated by Eq 2.28A and Fig 2.105, its induced drag will be reduced by half ($k = 0.5$), i.e.

$$D_{i(ef)} = 0.5 D_{i(isolated\ foil)} \qquad \text{Eq 2.31A}$$

It should be emphasized that expression 2.31A may hold only when the gap g between the end of the foil and the dividing plane, be it the water-plane or deck in the case of the sail, or the flat bottom of the hull to which the rudder is attached, is practically sealed. It is evident from Fig 2.115 that even a small longitudinal slot facilitating a leakage of the flow will influence adversely the induced drag.

As a matter of fact, the effective aspect ratio AR_{ef} may be either greater or smaller than the geometric aspect ratio AR. To make this point clear consider Eq 2.28A, which is applicable to elliptical loading only and can be written more generally as:

$$C_{Di} = \frac{C_L^2}{K\pi AR} \qquad \text{Eq 2.32}$$

The coefficient K in the denominator depends on the resulting effect of (a) advantageous interaction between two foils, and (b) penalties paid in the form of additional drag for the departure of the actual loading from the elliptic loading which produces a minimum induced drag for a given geometric aspect ratio AR, as shown by the hatched zones in Figs 2.111 and 2.113.

The effective aspect ratio AR_{ef} can therefore be defined as:

$$AR_{ef} = K \times AR \qquad \text{Eq 2.33}$$

and the coefficient K may well be called the sail efficiency factor. Hence Eq 2.32 can be rewritten as:

$$C_{Di} = \frac{C_L^2}{\pi AR_{ef}} \qquad \text{Eq 2.34}$$

If an ideal sail, such as might be represented by a wing sail of semi-elliptical form, similar to that shown in Photo 2.30 (catamaran US2 following Tornado Class Cat), had its foot somehow tight sealed to the water surface the sail efficiency factor K in Eq 2.33 might approach 2. Such a theoretically possible benefit is difficult to achieve in reality, for three reasons:

1. In customary working arrangements the clearance at the sail foot is large enough to prevent the adjacent hull structure from serving as an effective inner end plate. Figure 2.115 clearly indicates that even a small gap of only a few per cent of the sail height will destroy much of the potential advantage to be obtained from a reduction in induced drag.
2. Unavoidable disturbances in air flow caused by the hull may decrease the gains one might expect from sealing the sail foot to the deck.
3. In the case of a conventional, highly tapered and arbitrarily twisted triangular sail it is rarely possible to obtain the optimum vertical load distribution of the semi-elliptical form.

For all those reasons, the actual sail efficiency factor K of an ordinary soft triangular sail rarely reaches a K value of 1.0. Hence, according to Eq 2.33, one may expect that the effective aspect ratio AR_{ef} of traditional sails may even be lower than the geometric aspect ratio AR.

To appreciate the significance of the effective aspect ratio and related induced drag in the upwind sailing condition one must know how much the variation of aspect ratio will affect the magnitude of the driving force coefficient. To calculate this effect we shall employ an equation, which gives the driving coefficient C_R in terms of lift, drag, and course sailed β relative to the apparent wind V_A (see Fig 1.10 Part 1):

$$C_R = C_L \sin \beta - C_D \cos \beta$$

According to Eq 2.30:

$$C_D = c_d + C_{Di} = c_d + \frac{C_L^2}{\pi AR}$$

Fig 2.116   Influence of effective aspect ratio AR_{ef} on driving force coefficient C_R. On the basis of well-established aerodynamics theory appropriate to wings of high dihedral angle–which is roughly equivalent to the heel angle of the sail–one may derive a relatively simple formula which relates the effective AR_{ef} to the geometric AR:

$$AR_{ef} = AR(1 - 2\Theta^2/10,000)$$

From it one may calculate, for instance, that a rig of $AR = 3.0$ operating at heel angle $\Theta = 30°$ has an effective aspect ratio of

$$AR_{ef} = 3(1 - 2 \times 30^2/10,000) = 3 \times 0.82 = 2.46$$

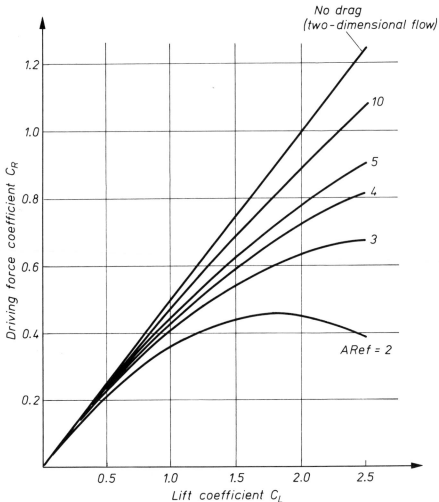

For a non-elliptically loaded sail the effective aspect ratio AR_{ef} must be introduced instead of the geometric AR, so we may rewrite the previously given expression for C_R in the following form:

$$C_R = C_L \sin \beta - \left(c_d + \frac{C_L^2}{\pi AR_{ef}} \right) \cos \beta \qquad \text{Eq 2.35}$$

To expose the significance of effective aspect ratio alone we may deliberately ignore the effect of profile drag c_d, as being constant for a given rig, and concentrate on the effect of the induced part of the total drag which depends so much on the crew's ability to tune the rig to given sailing conditions. Therefore Eq 2.35 above takes the form:

$$C_R \sim C_L \sin \beta - \frac{C_L^2}{\pi AR_{ef}} \cos \beta \qquad \text{Eq 2.35A}$$

The results of calculations are plotted graphically in Fig 2.116 for various values of AR_{ef}, namely 2, 3, 4, 5, 10 and finally for no induced drag at all, which represents a two-dimensional flow condition in an ideal fluid devoid of friction, i.e. $AR_{ef} = \infty$. It was assumed that $\beta = 30°$ is a reasonably representative course in sailing to windward. It is seen from the graph that the primary requirement for improving yacht performance in close-hauled sailing is to increase the effective aspect ratio of the sail; which is not the same as a simple increase in geometric aspect ratio. This is equivalent to reduction of the additional induced drag to the minimum possible by proper sail tuning, sail design, or both.

However sound and convincing the conclusions just derived concerning additional drag, interactions, effective aspect ratio, etc. may appear, and however well the theoreticians may agree among themselves, they cannot be given full confidence until they have been through wind tunnel tests at least, not to mention full-scale tests in racing.

The wind tunnel experiments on a Finn-type sail, shown earlier in Figs 2.111 and 2.113, clearly confirmed the expected gains from tuning, in that lesser additional drag resulted from smaller twist. To give another example relevant to the same rig, it can be seen from Fig. 3.20 in Part 3 that by reducing the distance between the foot of the sail and sea level from 10 in, equivalent to about 11 per cent of the sail height (Run VII), to 6 in (Run IX), i.e. by 40 per cent, with camber and twist remaining the same, the L/D ratio increased from 5.05 to 5.55, some 10 per cent. The lower induced drag, which can be estimated from Fig 2.115, is however insufficient to account entirely for this increase in L/D ratio, the other cause which must be looked for is therefore increased lift.

Some physical insight into this important foil-slit effect may be facilitated by remembering that the difference in pressure between the windward and leeward sides of the foil causes air or water to flow through any opening in the foil and that the kinetic energy thereby lost appears as drag. Apart from this, the presence of a gap or slit in the foil, or between the foil and a plane to which the foil is attached, will

Fig 2.117   The effects of the hull on the driving force of the Dragon rig (see Photo 3.30, Part 3).

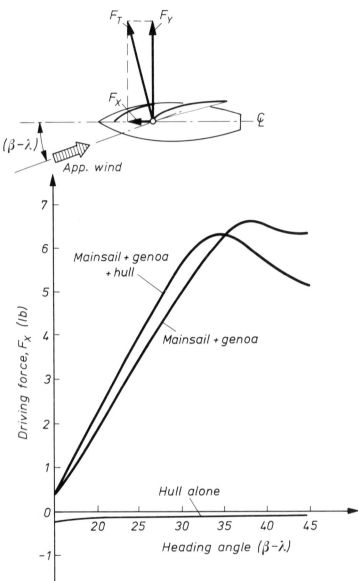

inevitably lead to unfavourable distortion in the lift distribution which, except when the gap is very small indeed, will manifest itself as reduced total lift (see Fig 2.119).

Some mistakes in this respect were made even in the course of the testing routine adopted by Sir Richard Fairey's Experimental Wind Tunnel for Research (Great Britain), operating about 40 years ago (shown in Fig 43 of *Sailing Theory and Practice*). The model sails attached to the complete hull, including its underwater part, were tested in conditions where the wind tunnel floor, simulating the sea, was

Fig 2.118   The effect of the hull on the heeling force of the Dragon rig.

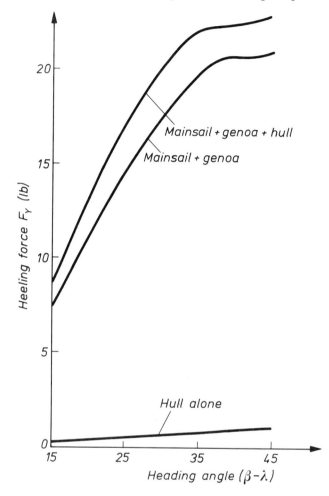

well below the keel! This mispresentation of real conditions certainly leads to wrong performance characteristics.

The following example may illustrate the above point further. As might be expected, the part of the hull without sails which projects above the water develops in close-hauled conditions drag and a certain amount of side force that together result in negative driving force. This was confirmed by wind tunnel tests on a 1/4 scale Dragon rig, the results for which are displayed in Figs 2.117 and 2.118 (Ref 2.86). However, rather surprisingly to some people, although the hull alone develops negative driving force, when included with the rig, the driving force of the whole model is actually increased. This contribution of the hull, by virtue of restricting and/or closing the gap between the sails and water-plane, is demonstrated in Fig 2.117, which presents the driving force F_x and heeling force F_Y components for a range of heading angles $(\beta - \lambda)$ (see Fig 3.14 Part 3).

Photo 2.32   It may sound strange, but in order to understand how a real
sail works one has to consider the effects caused by the
presence of its mirror-image companion.

Although the foregoing conclusion may sound strange at first, it is inescapable. In order to understand how a real sail works it is most essential to anticipate and take into consideration the interference effects due to the presence of its symmetrical ghost companion depicted in Photo 2.32. To express it in a different way, on account of the existence of the water-plane below the sail foot the only sensible approach to sail aerodynamics is to regard the visible sail as a part of the mirror-image system, as illustrated in Fig 2.114, and not as an isolated, real sail only set as if an empty space devoid of a boundary or with boundaries located in an unrealistic place. Interactions are tangible, physical phenomena and should not, as such, be underestimated. For all these reasons, the sail of elliptic planform shown in Photo 2.29, will not offer the full advantage which one may expect from such an ideal planform! Clearly, neither the existence of the water-plane nor its substitute, the mirror-image of the real sail, appear to be fully appreciated by the designer.

Figure 2.119 is intended to clarify the above point. For the sake of simplicity of presentation the spanwise distributions of local lift coefficient C_l and induced downwash velocity w are plotted for an uncomplicated, split rectangular foil of

Fig 2.119 Spanwise distribution of local lift coefficient C_l and induced downwash w for a split mirror-image rectangular foil of combined AR = 6. Compare the downwash curves with those in Figs 2.98 and 2.102.

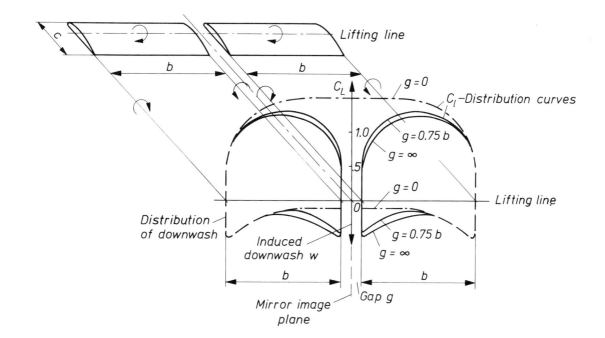

Fig 2.120 Explanatory sketches supplementing Photo 2.29.

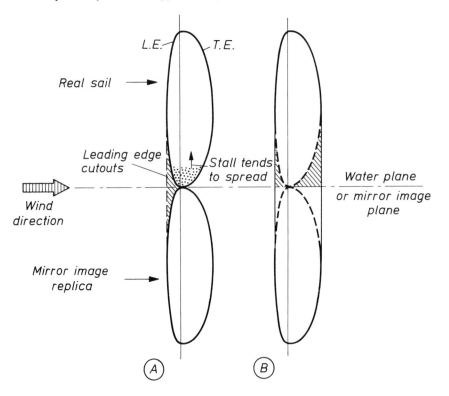

combined AR $= (2b/c) = 6$. As might be expected, the magnitudes of local lift coefficient C_1 and downwash w vary along the span and depend on gap g, between the inner tip and the mirror-image plane. Thus, when $g = 0$, i.e. when there is no gap between the two foils, the C_1 distribution and downwash distribution curves are given by the broken lines. The two other full lines represent the relevant distributions when the gap $g = 0.075\ b$ (7.5 per cent b) and when $g_1 = \infty$, which means that the foils are so far apart that there is no interaction between them. It is clear from the C_1 distribution curves that the effect of gap is to decrease the lift at most of the spanwise position. However, when the gap g is very small, below 0.7 per cent of the foil span b, a slight increase in lift will occur. The explanation given in Ref 2.85 is that '…the leakage flow, issuing as a jet, rolls up into a single discrete vortex a little inwards from the tip but away from the suction surface. This augments the suction pressure and hence the lift at the tip region near the slot.'

From the induced downwash curves it becomes evident that due to the presence of a gap larger than 0.7 per cent of b the two tip vortices are shed and they, of course, induce downwash in excess of that which would exist due to trailing vortex action when there is no gap. Such an additional downwash will result in higher induced drag.

Referring again to Photo 2.29, the visible elliptical wing sail should be considered as a part of the mirror-image system. When the tip of a real sail touches its image the geometric aspect ratio of the whole system is theoretically increased twofold as depicted in Fig 2.120A. Disregarding now the dividing water-plane and considering the flow round the two sails as if round one, we may examine the actual flow conditions in the region of the junction at midspan by taking into account certain hints derived in the course of testing single foils. It is, for example, known that even fairly large cut-outs in foil area do not have serious aerodynamic effects if they are located close to the trailing edge (Ref 2.87). However, any modification at or near the leading edge, such as that rather unfortunate junction distinguished by a hatched zone in Fig 2.120A, may cause premature separation, which not only affects the immediate part of the sail downstream, but often spreads spanwise. This is shown by the dotted zone in Fig 2.120A, so that a large portion of the sail may be stalled, with a consequent considerable increase in drag and decrease in lift. In other words, a foil loses significant effectiveness as a lifting surface even when a relatively small part has stalled.

The reason for possible premature separation is that, due to intensive local upwash induced by the mirror-image sail, the part of the real sail adjacent to the inner tip has a much higher effective incidence than the upper parts of the sail. And when the average lift generated by those upper parts of the sail is such that stalling conditions have not been reached yet, the junction part of the sail may have already suffered full stall with consequent decrease in lift and increase in drag. This effect of higher effective angle of incidence is augmented by another one, associated with decreasing chord length towards the tip. We shall discuss these effects more fully in the following chapter.

In order to avoid the deleterious effects discussed above, one should transform the sail planform, shown in Fig 2.120A, into a more efficient form, illustrated in Fig 2.120B, by filling the cut-away portions indicated by the cross-hatched zones, and making the foil semi-elliptic instead of elliptic. The complementary semi-elliptic planform of the mirror-image sail will supplement the first, thus producing a final form resembling the full elliptical planform, but of double the span. With such a configuration no additional deleterious aerodynamic effects should be incurred. Nature appears to avoid inefficient forms of wing, and there are no soaring birds with wings resembling the form given in Fig 2.120A. The same applies to subsonic aircraft wings after a long period of development.

As already mentioned, the harmful effects that may arise due to an improper kind of sail-hull junction, can be mitigated if necessary cuttings are located near the trailing edge. This is demonstrated in Photo 2.30. The catamaran *US-2* has a large cutting at the trailing edge, but the part of its wing sail adjacent to the deck has a smooth and straight leading edge. An even better junction, from a purely aerodynamic point of view, is depicted in Photo 2.31. In conclusion it can be said that the problem of the most effective junction between the sail and hull, particularly in the case of high speed craft such as C-Class catamarans, is still a potential improvement area waiting for explorers.

In so-called One Design classes, where the sail planforms are strictly controlled by the class rules, the modifications to sail shape that might lead to higher AR_{ef} can only be done through proper tuning and sail setting. Figs 2.111 and 2.113 indicate that there is much room for improvement. In so-called development classes, the sail planforms, and even the method of sail construction (soft sail, semi-rigid sail, or wing sail) are not controlled by rules, leaving only the sail area fixed. The possibilities for improvements, particularly on the designer's part, are much higher in this case. Consequently, the aerodynamic perfection already observed in C-Class catamarans (Photo 2.31) is very high and is apparently stimulated by aerodynamic theories and sophisticated aeroplane technology.

Almost all conclusions derived so far in this chapter apply equally well to hull appendages.

(5) Foil-shape effects

> 'What the eye doesn't see,
> the heart doesn't grieve over.'

From what has been said so far, it is clear that lift and drag are not entirely dependent on foil section (camber and thickness), they are also controlled to a greater or lesser extent by other geometrical features. Three of these have already been mentioned, namely–aspect ratio, taper ratio, and twist–but there are more ways in which the three-dimensional foil shape can be varied. Figure 2.121 depicts, for example, the basic planforms that are commonly used in man-made air- and water-borne machines, and again some of them are also popular in Nature. Figure 2.122 illustrates for instance Tabarly's famous *Pen Duick III*, in which all basic sail-forms are incorporated, with foresail No 2 of rectangular form, foresail No 3 of tapered planform, yankee No 1 and No 2 and genoa of triangular forms and finally a mainsail of nearly semi-elliptical planform. The sails shown are of different aspect ratio and different sweep or rake angle.

If we reserve the term planform for the contour of the foil, as shown in Fig 2.121, the essential geometrical features of the three-dimensional foil shape may be listed as follows:

   a. planform
   b. aspect ratio
   c. taper ratio
   d. twist
   e. camber distribution (in chordwise and spanwise direction)
   f. sweep angle (sweep-back or sweep-forward)

Although some people may think that by far the most important of possible foil-shape modifications is the change of aspect ratio, and the penalty for an excess of AR in IOR formula may substantiate such a feeling, it may not necessarily be true, particularly in the case of soft, highly tapered mainsails. It will be demonstrated that

Fig 2.121   Basic planforms:

     Rectangular Planform   1
     Tapered Planform   2
     Triangular Planform   3
     Elliptical Planforms   4, 5 and 6

Aspect ratio is defined as the ratio of span b to average chord c_{av}

$$\mathrm{AR} = \frac{b}{c_{av}} \quad \text{or} \quad \mathrm{AR} = \frac{b}{c_{av}} \times \frac{b}{b} = \frac{b^2}{S_A}$$

since average chord c_{av} is frequently a less convenient measure than foil area S_A.

$$\text{Taper ratio} = \frac{\text{tip chord}}{\text{root chord}} \text{ ratio} = \frac{c_t}{c_r}.$$

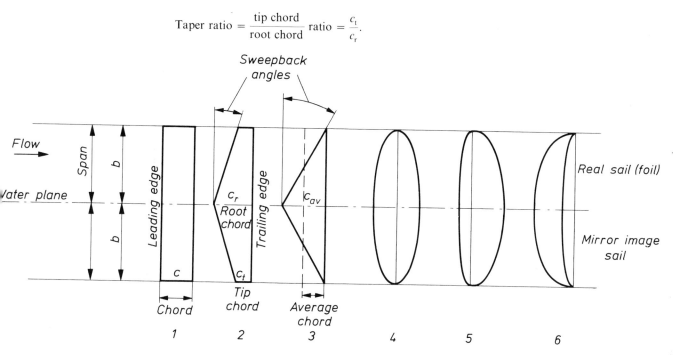

sails of the same aspect ratio and camber distribution but with different other geometrical features (mast diameter and shape, for example) may manifest different aerodynamic characteristics. In the course of numerous tests it has been established that certain combinations of taper ratio, twist and sweep angle, may produce very poor aerodynamic characteristics while the other combinations may result in high efficiency of the foil of given AR.

Figures 2.111 and 2.113 illustrate this point. The measurements clearly indicate that there is a close link between twist and additional induced drag, but the question as to what is the mechanism of this link is as yet open.

Which combination of various shape components a to f has to be incorporated into the sail in order to obtain the desired lift at minimum expense in terms of drag and heeling moment is the gist of the whole problem. In order to find the answer to

Fig 2.122

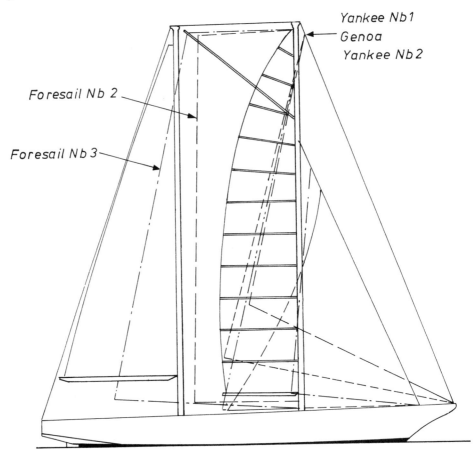

Yankee Nb1
Genoa
Yankee Nb2

Foresail Nb 2

Foresail Nb 3

Pen Duick III
original rig
Mainsail: 342 sq. ft.
No 2 Foresail: 674 sq. ft.
No 3 Foresail: 1005 sq. ft.
Genoa: 860 sq. ft.
No 1 Yankee: 714 sq. ft.
No 2 Yankee: 571 sq. ft.
Genoa jib: 408 sq. ft.
No 1 jib: 275 sq. ft.

Fig 2.123   Relationships between the load, lift and downwash distribution of the elliptic planar (untwisted) foil of uniform camber.

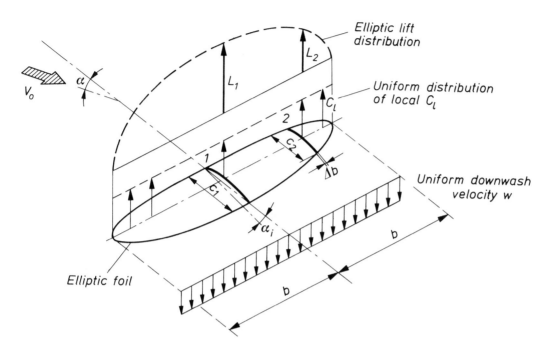

this one must determine first the relationship between the loading (spanwise distribution of lift) and the effective angle of incidence along the foil span called into play by a given planform.

(a) *Planform effects*

The easiest case from an aerodynamic point of view is to consider a foil, elliptical in planform, which has all sections geometrically similar and their angles of incidence equal along the span, as shown in Fig 2.123. With these restrictions the question of finding the load distribution curve can be solved by using tentatively Eq 2.10:

$$\frac{L}{b} \sim \rho V_0 \Gamma \qquad \text{Eq 2.10 repeated}$$

according to which lift per unit span L/b, i.e. the lift developed by a very narrow strip of the foil area $\Delta b \times c$, shown in Fig 2.123, is proportional to fluid density ρ, flow velocity V_0, and circulation Γ.

In turn, according to Eq 2.12

$$\Gamma \sim V_0 c \alpha \qquad \text{Eq 2.12 repeated}$$

by substituting Γ given by the above equation into Eq 2.10 yields

$$\frac{L}{b} \sim \rho V_0^2 c\alpha \qquad \text{Eq 2.36}$$

In this equation, which is derived basically for two-dimensional flow conditions, it is implicitly supposed that the geometric incidence angle α is the effective angle of incidence α_{ef} as, in fact, it is. In the simple lifting line theory for a three-dimensional foil it is assumed that every section of a foil of finite span acts exactly as the section of a two-dimensional foil except only that the effective angle of incidence:

$$\alpha_{ef} = \alpha - \alpha_i \quad \text{(see Fig 2.100)}$$

is used in place of the geometric angle of incidence α. Introducing α_{ef} instead of α, which is apparent from mere inspection of the three-dimensional foil attitude relative to flow direction V_0, Eq 2.36 can be rewritten:

$$L/b \sim \rho V_0^2 c\alpha_{ef} \qquad \text{Eq 2.36A}$$

If those very narrow strips into which the foil shown in Fig 2.123 is subdivided are of the same width Δb the local lift generated by each strip along the span (distinguished by vectors L_1 and L_2) will be directly proportional to the local chord lengths c_1 and c_2 respectively. The elliptical lift distribution can therefore be realized by making a foil elliptical in planform, since if the chords are elliptically distributed the lift distribution should also be elliptical. By this special choice of foil form the straightforward dependence of lift distribution on planform, i.e. chord distribution, is plainly exposed. The tacit assumption incorporated in Fig 2.123, namely that an elliptic foil produces uniform downwash w, which gives uniform induced angle α_i and therefore a uniform effective angle of incidence α_{ef} for every section, is justifiable for this particular untwisted planform only (Ref 2.79). This was mentioned earlier and the reader should perhaps be reminded yet again that by virtue of uniform downwash the elliptical planar (untwisted) foil produces minimum induced drag for given total lift. Another important feature of the elliptical foil of geometrically similar sections is that it produces uniform distribution of local lift coefficient C_l along the span. This is just a consequence of uniform effective angle of incidence α_{ef}.

Since an elliptical foil has the same lift coefficient across the span one may expect that it will stall evenly along the foil span, i.e. the separated flow should evenly and progressively spread, beginning from the trailing edge towards the leading edge, when incidence angle gradually increases and reaches the stalling angle. These expectations are verified by the experimental results displayed in Fig 2.124A (Ref 2.89). The wool strands used for observing the nature of the flow over the suction side of the foil were stuck onto one end of the foil along spanwise lines spaced roughly one inch apart. Flow diagrams were made by observing the strands through a window in the wind tunnel wall and sketching on a plan of the foil the boundaries between the different types of flow conveniently distinguished from each other as 'streamline', 'disturbed', or 'stalled'. The diagrams were made by drawing in the

Fig 2.124   Stalling pattern of an elliptical and a rectangular foil. Single
cross-lines indicate disturbed foil area. Double cross-lines
indicate stalled portion of the foil.

The elliptical, untwisted (planar) foil has constant local lift
coefficient along its span, therefore it stalls evenly. Uneven stall
of the rectangular foil indicates uneven distribution of local lift
coefficient along the foil span (Ref 2.89).

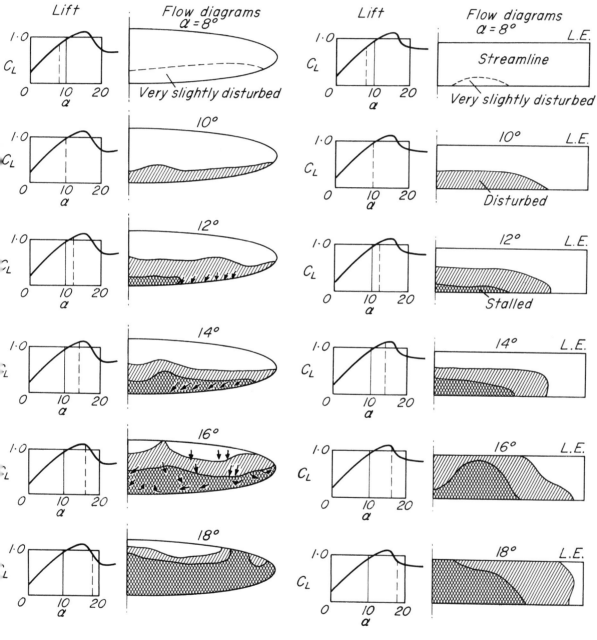

boundary on one side of which the flow was definitely streamline and undisturbed and then putting in another line on one side of which the flow was definitely violently turbulent or stalled and calling the region in between the two lines disturbed. The diagrams representing the nature of the flow given in Fig 2.124A are accompanied by lift coefficient C_L versus incidence α curves, thus showing the relationship between them.

One of the most tempting methods of reducing the heeling moment or foil root bending moment is to concentrate the lifting surface near the foil root corresponding the sail foot, and with most conventional sails this implies a high taper ratio. So, the question of aerodynamic efficiency of highly tapered foils becomes of immediate interest.

Returning for a while to Fig 2.98, which shows a perspective view of induced velocities in the immediate neighbourhood of a rectangular untwisted foil, it is seen that the distribution of downwash is different when compared with that for an elliptical foil. It is not uniform and this implies that the distribution of:

trailing vortices
downwash w
induced angle α_i
lift or load L, and finally
local lift coefficient C_l (see Note 2.90)

across the foil span must somehow be mutually dependent and controlled by the foil planform. It has been established, both by analytical methods as well as by experiments, that the concentration of trailing vortices near the foil tips, as observed for instance in the case of a rectangular untwisted foil, produces also a concentration of downwash near its tips. This, in turn, affects the remaining aerodynamic characteristics listed above, i.e. distribution of α_i, L, and C_l. Greater downwash will result in larger induced angle of incidence α_i. This is shown in Fig 2.98, where it can be seen that the induced angle midway between the tips at station 1 is smaller than that nearer to the foil tip at station 2. Therefore, the effective angle of incidence α_{ef} near the tip will be smaller than that near the centre, so although the geometric angle of incidence α is uniform, the effective local angle of incidence α_{ef}, which controls the local lift, is not. For this condition, the sections near the centre, operating at higher effective incidence, stall first. Figure 2.125B based on a series of photographs illustrates the approximate lines of flow separation at gradually increasing geometric angle of incidence α. The shaded areas correspond to a stalled condition on the portion of foil at each particular angle of incidence. Those portions which stalled early are therefore heavily shaded, while those which stalled only when a large geometric angle of incidence had been reached are lightly shaded. The unshaded areas represent portions of the foil which remained unstalled at the largest angle indicated on the contour lines (Ref 2.91).

Exactly the opposite pattern of behaviour is observed in the case of the highly tapered, untwisted foil shown by sketch A in Fig 2.125. Here, the downwash de-

Fig 2.125 Contours giving approximate boundaries of flow separation
for tapered and rectangular planform. Shaded areas correspond to a stalled condition. The numbers attached to
contour lines give the angles of incidence at which separated
flow was observed (Ref 2.90).

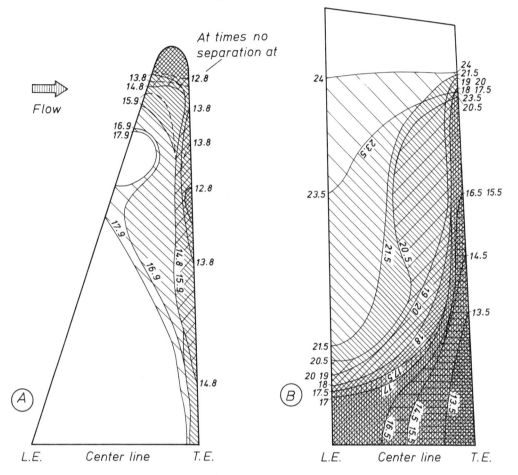

creases from the midspan or root section towards the tip so that the effective angle
of incidence is higher at the tip than near the midspan, causing the portions of the foil
near the tip to stall first. Thus, the higher the taper ratio the more liable is the foil to
develop an early tip stall. This tendency is of fundamental practical significance and
should not be underestimated.

These apparently complex mutual relationships between:

planform
lift distribution L
local lift coefficient distribution C_l and
induced angle distribution α_i

Fig 2.126   Qualitative distribution of lift load, local lift coefficient C_l and downwash w (or induced angle of incidence α_i) for three basic planforms. Foils are assumed to be planar (untwisted).

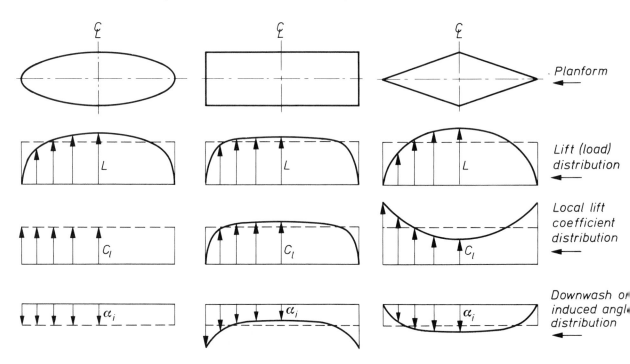

for three basic planforms are shown qualitatively in Fig 2.126. It should be noted that:

a. The elliptical foil has a uniform distribution of local C_l and α_i across the span, which is the result of elliptical lift distribution (see Fig 2.123).
b. The rectangular form has a somewhat more uniform distribution of lift L in the centre than the elliptical foil. Hence the maximum values of downwash w, reflected in the distribution of α_i, are shifted from the centre towards the tips. The middle parts of the foil are therefore operating at higher effective incidence $\alpha_{ef} = \alpha - \alpha_i$ than the portions of foil closer to the tips. For this reason the local lift coefficients C_l, attained by the middle parts of the foil are higher than those further out towards the tips.
c. The highly tapered foil of taper ratio $c_t/c_r = 0$ carries a heavier load in its median part where the foil area is concentrated. Hence, an educated intuition may suggest that the downwash, reflected in distribution of the induced angle of incidence α_i, has to be larger where the foil leans more heavily on the supporting stream and decreases where the push of the foil on the passing fluid is lightened. Photograph 2.28 substantiates such a feeling. Accordingly the local lift coefficients C_l of that part of the triangular foil which operates at lower effective angle of incidence $\alpha_{ef} = \alpha - \alpha_i$ must be smaller than the local C_l

Fig 2.127   Variation of local lift coefficient C_l for three basic planforms of the same geometric aspect ratio $AR = 6$. It is assumed that the total lift coefficient developed by each foil is $C_L = 1.0$. Distribution of local lift coefficient is shown for half of the foil only; the remaining half will produce identical complementary distribution.

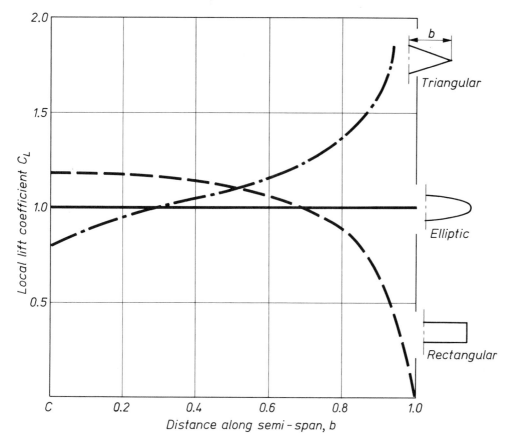

developed by the tip-part of the foil which operates at higher effective incidence angle

Figure 2.127, which is based on Ref 2.92, illustrates the variation of local lift coefficient C_l along the foil span for the three basic planforms of the same geometric $AR = 6$. The total lift coefficient developed by each foil is the same, i.e. $C_L = 1.0$.

Figure 2.128 supplements Fig 2.127 by giving variation of the effective angle of incidence α_{ef} in terms of α_{ef}/α ratio for three foils of different planforms. The peculiar upturn of one of the curves requires explanation.

As shown earlier in Fig 2.114B, the induced velocity w in the case of a triangular foil becomes an upwash at the tip, thus causing an increase of the effective angle of incidence there. Evidently the induced angle of incidence α_i must be added to the

Fig 2.128   Variation of the effective angle of incidence along the semi-
span b, for three foils of different taper ratio c_t/c_r and
AR = 6.0. Distribution of incidence is given in the form of the
ratio of effective incidence α_{ef} to the geometric incidence α. It is
seen that, for the triangular form its effective incidence angle
α_{ef}, as distinct from other forms, is greater than the geometric
incidence angle α near the foil tip.

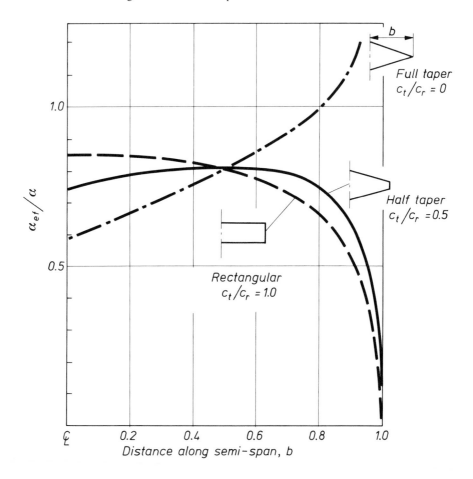

geometric incidence α for this part of a sail or foil where the upwash exists, i.e.

$$\alpha_{ef} = \alpha + \alpha_i$$

And this explains why the ratio of α_{ef}/α in Fig 2.128 becomes greater than 1 for the tip
part of the triangular planform.

Such an upwash at the foil tip should be expected if one anticipates a possibility
that the span S between the tip-vortex cores, shown in Fig 2.96, becomes smaller
than the foil span. This is exactly what happens in the case of triangular planforms
and Photo 2.28 demonstrates it convincingly. As can be seen, the cores of the trailing

vortices are shifted inboard. Such a shift is much less pronounced in the case of an elliptical or rectangular foil.

Several qualitative conclusions of the utmost practical significance can be derived from Figs 2.127 and 2.128. Consider, for instance, an untwisted, triangular sail tapering uniformly to almost a point at the head, as depicted in Fig 2.114. It is assumed that the gap $g = 0$ and the combined AR of the sail including its mirror-image is 6. The sail develops say, a total lift coefficient $C_L = 1.0$ at the geometric angle of incidence $\alpha = 16°$. Using the data incorporated in Fig 2.128 it can be estimated that the sail sections close to the boom operate at the effective angle of incidence α_{ef} of about $16 \times 0.6 = 9.6°$; while the sections near the sail top operate at the effective incidence α_{ef} of about $16 \times 1.2 = 19.2°$, or even more. Thus, due to the presence of the sail, the local airstream is twisted in a non-uniform manner over the whole mast height. In consequence, the sail sections from the boom upwards are subjected to increasingly greater effective incidence angle, and without geometric twist the sail would operate at considerably higher effective angles of incidence at the top than at the boom.

Since usually there is a gap between the boom and the deck the estimated value of effective incidence angle at which sail foot operates is modified depending on the gap g (see Fig 2.114). Such a modification does not however change radically the general trend in the effective angle variation given in Fig 2.128.

Remembering that only the effective angle of incidence α_{ef}, and not the geometric incidence α which can be estimated from mere visual inspection, really matters, one should not be surprised to find that the upper part of the sail in question may frequently operate in a fully stalled condition; Fig 2.129A demonstrates this. An untwisted single sail is seen obliquely from a bird's-eye-view and the miniature pictures of polar diagrams plotted next to the sail contour illustrate the local resultant sail coefficients C_{t1}, C_{t2} and C_{t3}, developed at three different sail-sections 1, 2 and 3. According to the so-called strip theory it is legitimate to regard, at least tentatively, each section 1, 2, or 3 as a very narrow chordwise strip of sail area. As such, these elementary strips or sections may be treated as isolated aerofoils and the resultant total coefficient C_T of the whole sail may be found by summation of the local coefficients C_{t1}, C_{t2}, C_{t3}, developed by each strip. For the sake of simplicity there are distinguished in Fig. 2.129A only three representative elementary strips.

It can be seen that the local resultant force coefficient C_t is relatively low at section 1. However, it rises sharply towards the sail top. Subsequently, somewhere near section 3, the stall and maximum C_t values are reached. Further towards the head of the sail, as indicated by dots, a fully separated flow occurs.

For the sake of definition, one may say that a sail or foil is stalled when any section stalls. By such a definition the practical upper limit to the optimum foil efficiency occurs immediately before this local stall takes place. For the stall, once initiated, may spread, and besides, the foil rapidly loses its efficiency in terms of available driving force. To make this point clear let us consider the direction of action of the local resultant force at section 2 in Fig 2.129A. This is shown by the

Fig 2.129A   Distribution of aerodynamic forces (in terms of local coefficients) at three different sail sections. In this presentation it is assumed that the effect of Reynolds Number and wind gradient are negligible. In fact they may be of some significance.

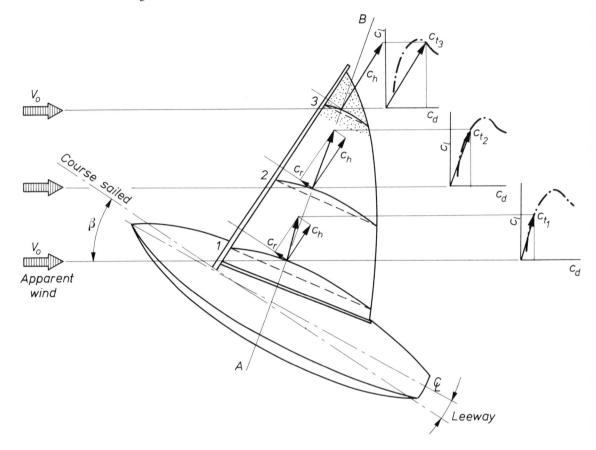

direction of the arrow marked C_{t2} which gives the value of force coefficient plotted also in the adjacent polar diagram. If one resolves the C_{t2} coefficient by means of the usual parallelogram of forces into the local driving and heeling force coefficients, C_r and C_h respectively, one may assess how much this part of the sail contributes towards the driving and heeling forces. If, in similar manner one resolves the C_t coefficient at section 3 it becomes evident that the resultant force developed by this part of the sail does not contribute at all towards the driving force. It produces, however, a large undesirable heeling force component C_h and may even produce a negative C_r component. Such a harmful effect of stall depends on two parameters:

   a. geometric angle of incidence of the sail,
   b. local camber of the sail or foil.

Fig 2.129B   If camber increases towards the sail head the stall of the upper part of the sail can be delayed or avoided. This preventive measure is effective when combined with appropriate twist.

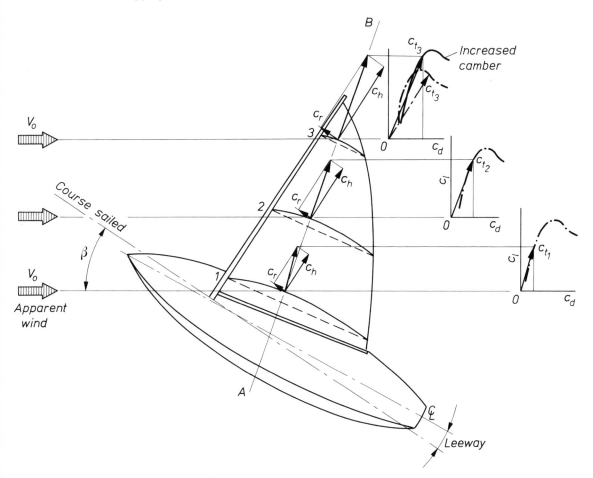

The larger the geometric angle of incidence the earlier is the onset of stall and separation and the more dramatic is the reduction in driving force. In the example discussed it was assumed that the geometric incidence of the untwisted sail was $\alpha = 16°$ and as a result of it the effective incidence α_{ef} of the tip part of the sail was about 19°. If the geometric incidence α is increased to say, 20° the α_{ef} will increase to about $20 \times 1.2 = 24°$ and hence one may rightly expect that the stalling conditions will spread down, from section 3 towards section 2 (Fig 2.129A). Vice versa, if the geometric angle of incidence is sufficiently reduced one may reasonably expect that the stalling of the upper part of the sail could be avoided. In such a case, however, the total lift coefficient C_L of the sail will be low.

Increased camber of the sail may also serve as a means of delaying an early stall. A glance at Fig 2.70 reveals that the greater the camber ratio f/c, the higher the incidence angle at which stall occurs. Thus if the camber ratio $f/c = 0.06$, the stall angle is in the order of $10°$, while for a foil of camber ratio $f/c = 0.1$ this angle increases to about $13°$.

An introduction of variable sail camber, in such a way that camber gradually increases towards the sail head, may be assumed to be equivalent to twisting of uniformly cambered sections along the sail span. Evidently, as seen in Fig. 2.70, the camber affects the zero lift angle of incidence α_{L_0}. Therefore, by increasing the camber towards the top sections, the effective twist is increased. Conversely, by reducing the camber towards the top part of the sail, the effective twist is reduced.

From the above discussion we may draw two important practical conclusions:

1. The twist (top of the sail falling off to leeward) can be used as a method of preventing tip stall. Certainly, the twist or washout, as it is called in aeronautical parlance, may reduce the effective angle of incidence of the top part of the sail below the stalling angle. Designating the angle of washout by $-\varepsilon$ the Eq 2.23 which defines the effective angle of incidence can be written as:

$$\alpha_{ef} = \alpha \pm \alpha_i - \varepsilon \qquad \text{Eq 2.37}$$

The $\pm$ signs at α_i indicate that according to Fig 2.128 the induced angle of incidence, usually negative, may in the case of triangular form such as a conventional sail, be positive near the foil tip.

Figure 2.128 implies that the necessary amount of washout or twist $-\varepsilon$ should increase slowly at first and then more rapidly towards the upper part of the sail. By gradually twisting the sail it is possible, at least theoretically, to reach the stalling conditions instantaneously, along the whole sail height. It would mean that every sail section, from the sail foot to the head, will reach the stalling angle at the same moment. If the twist is properly adjusted then the direction of action of the local resultant force at section 3 in Fig 2.129A will be turned to the left, bringing the arrow marked C_{t3}, which reflects the direction of the local resultant aerodynamic force, as close as possible in parallel to line AB. The optimum sail setting would then be achieved when all the local resultant forces represented by the arrows C_{t1}, C_{t2}, and C_{t3}, developed at different sail sections 1, 2, and 3, are roughly parallel to each other. In the case of uniform vertical distribution of sail camber this would imply the same effective angle of incidence at each sail section as distinct from different geometric incidence angles. Thus, contrary to popular belief, the sail washout or twist is not harmful but indispensable, provided of course that its amount is strictly controlled according to Eq 2.37. Photograph 3.22 may clarify this point further.

So far, it has been assumed that the effect of wind gradient is negligible. In fact, it may be of appreciable significance and requiring an additional twist in excess to that implied by the induced effects. The Reynolds Number effect may also be of some importance.

Although the stalled flow conditions are not directly visible, properly situated wool streamers may greatly help to detect stalling so that the correct sail twist is achieved.

2. This preventive effect of twist on an early stall may advantageously be augmented by the non-uniform spanwise distribution of camber in which camber increases towards the sail top. Figure 2.129B illustrates this point in a self-explanatory manner; by increasing the camber at section 3 the total aerodynamic force, represented by its coefficients C_{t3}, will change its direction of action advantageously. The two polar diagrams of coefficients plotted next to section 3, one for smaller camber and another for the increased one, demonstrates this sequence.

Such a distribution of camber, shown for example in Photo 2.33, is relatively easy to maintain on boats with non-bendy masts and fully battened sails with a large amount of roach near the top. However, in the case of flexible rigs, such as are fitted to the Finn or Star class yachts, the mast bend tends to flatten the top part of the sail. This tendency, apparent in Fig 2.111, and Fig 3.18 Part 3, depends primarily on the mast flexibility, and also on sail cut, and kicking strap or sheet tension, etc. If the sail cut is not matched to the bending characteristics of the mast, or the sail shape and the mast bend are wrongly adjusted, the sail cannot possibly be set to achieve optimum conditions. As a consequence, a large part of the sail may not contribute at all to the driving force but to the heeling force only.

It should perhaps be added that twist (washout) and increased camber as stall preventers are most effective for foils with moderate taper and are relatively ineffective for highly tapered planforms in which the c_t/c_r ratio approaches 0. This is partly due to the fact that the foil sections of progressively shorter chord towards the tip are operating at progressively smaller Reynolds Numbers. The available local $C_{l\,max}$ of the sections in question is then reduced and this may provoke an abrupt premature stall of a large part of the foil near its tip. A similar effect may, in the case of a sail, occur due to the presence of the mast which is particularly acute when the ratio of the mast size (diameter) to the sail chord increases substantially towards the sail head. Consequently, a large part of the sail may suffer separated flow at the incidence angle at which the remaining part of sail is operating well below the stalling angle.

Figure 2.130 demonstrates for example a bermudan rig of 12-Metre type tested in the wind tunnel. Clearly, substantial areas of both sails are subjected to separated flow at relatively low heading angles $(\beta-\lambda) = 20°$ and $25°$. Both sails had uniform vertical camber distribution and their twist (washout) was in the order of 3–4°, which presumably might account for a wind gradient effect in the close-hauled condition. Evidently, however, this amount of twist was not adequate to compensate for the increased effective incidence due to induced flow effects.

From what we have said so far it can now be seen that the common practice of using the masthead indicators of wind direction as a basis for judging the angle of incidence of a sail can be deceptive. It gives an incorrect indication of the effective incidence angle at different sail section along the mast height. By way of illustrating

Fig 2.130   Flow over the leeward side of a bermudan sloop. Sheeting angle of mainsail $\delta_m = 5°$, genoa $\delta_f = 10°$. Figure adapted from *Sailing Theory and Practice.*

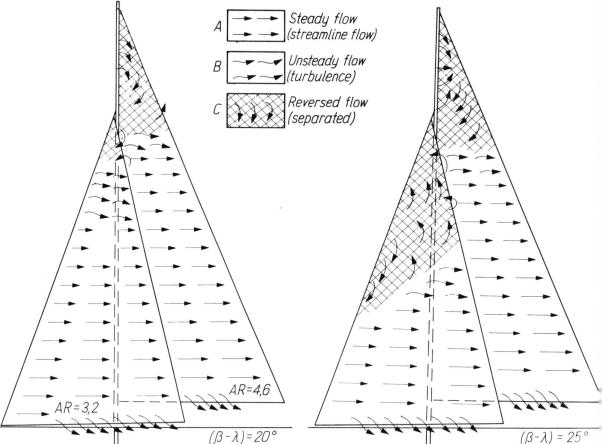

how widespread this illusion is we may quote an excerpt from an article published by a leading magazine, in which a well known author states: 'The foot of the average mainsail is trimmed at angles of attack of 20° or more. Thus, the flow over the lower chords is potentially separated. In the case of cat-rigged boats this is precisely what happens.' And 'as we move up the mast we see that the chord angles of attack gradually decrease to values which are below the stall point. Concomitantly, the thrust-to-heel ratio becomes more favourable.'

One more glance at Fig 2.130 may convince us that, contrary to these popular suppositions, the upper not the bottom part of the sail is potentially prone to an early stall. The old saying, 'what the eye doesn't see the heart doesn't grieve over', partly explains why the induced flow effects are underestimated by sailing people. Besides, one must agree that certain aspects of induced flow effects may not be easy to grasp, particularly when there are considerable preconceived and well established but wrong opinions in this respect. Once again referring to our senses and admitting

Fig 2.131   Loading distribution for rectangular and triangular planforms.

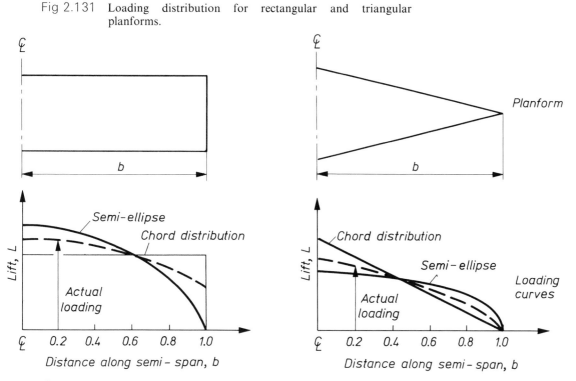

that we may sometimes be victims of deception and illusion, let us quote an apt remark made by Sir Peter Medawar (Ref 2.93) that may lead to the next chapter–'For all its aberrations, the evidence of senses is essentially to be relied upon, provided we observe nature as a child does–without prejudices and preconceptions, but with·that clear and candid vision which adults lose and scientists must strive to regain.'

(b) *Influence of taper ratio and twist on foil efficiency*

'The wind bloweth where it listeth,
and thou hearest the sound thereof,
but canst not tell whence it cometh
and whither it goeth.'

St John 3:8

Since the intensity of loading, as represented by the lift L distribution curves in Fig 2.126, can be confused with intensity of local lift coefficient C_1, also represented in the same figure, the meaning of these two terms should be clarified further. It is true that in the case of an untwisted, triangular planform the load L is concentrated in the widest part of the foil. Thus in the case of a sail the lift forces are concentrated in its lower part adjacent to the boom, but this does not mean that the local lift coefficient is also large there. Just the opposite is true. Figure 2.131 is intended to clarify this distinction between the local load distribution and local lift coefficient C_1

distribution. What follows is an attempt to look, from yet another point of view, on what has already been said when discussing Fig 2.126.

If, ignoring for the moment induced effects, the rate of change of loading (lift) across the foil span were proportional to the relevant change of chord c, as might be suggested by Eq 2.36,

$$\left(\frac{L}{b} \simeq \rho \times V^2 \times c \times \alpha \right)$$

then the intensity of local lift would be determined in a simple manner by the foil planform. So that the local lift at each foil section along the span would be proportional to the product $c \times \alpha$ for a given incidence angle α. Assuming that α, which designates the geometric incidence, is constant, the load distribution would simply be given by the chord distribution, i.e. the shape of the load curve would directly reflect the foil geometry. Due to induced flow effects, such a distribution of load is, however, unrealistic for any other planar forms but the elliptic one. The reason is rather obvious, the elliptic untwisted planform is the only form which produces a uniform downwash and this results in uniform effective incidence angle across the foil span. For other planforms, notably triangular ones, the effective incidence is strongly affected by the non-uniform downwash and is therefore necessarily non-uniform (Fig 2.128). Consequently, the actual shape of the load distribution curves, depending not only on the chord distribution but on the non-uniform incidence distribution as well, will differ from those given by the foil geometry alone.

Although no conclusive mathematical explanation is readily available, the inherent tendency of lift distribution is to approximate an elliptic form marked in Fig 2.131 as 'semi-ellipse'. In fact, according to the findings of Ref 2.94 the actual loading distribution lies between an ideal distribution given by the semi-ellipse and a distribution determined simply by the chord length distribution, i.e. foil planform. It can be seen in Fig 2.131 that actual load distribution is represented by the curve approaching an ellipse drawn half way between, as the arithmetical mean between the two curves.

It is now evident that local lift intensity varies from station to station–at some stations rising above the value corresponding to an ideal elliptic distribution of loading, and at other stations falling below that value. For the triangular foil in Fig 2.131 the spanwise decrease in chord is faster than the relevant decrease of lift. For this reason, the local load per unit foil area $\Delta b \times c$, i.e. a load which is sustained by a very narrow strip of foil area of width Δb (as shown earlier in Fig 2.123, and to which the local lift coefficient C_1 is proportional) is larger near the foil tip where chord length c is short, than that near the widest part of the foil where chord c is long. Conversely, for a rectangular foil the local lift coefficient C_1 near the foil root is larger than that near the foil tip.

The effect of taper is then to increase the local lift coefficient C_1 in the tip portion of the foil. This feature of taper is of the most important practical significance as far as trimming or tuning the sail to its best performance is concerned. It appears that there are no more important factors affecting the sail performance than the taper ratio and the associated induced phenomena. The taper ratio may be regarded as the key factor and one can hardly overestimate its significance. Let us illustrate it in the following example.

In order to decrease the induced drag of a triangular sail we should more nearly approach elliptic loading and this, according to Fig 2.131 B, would mean making the upper half of the sail more heavily loaded. With this end in view, it would be necessary to have higher angles of incidence near the sail top than near the boom. However, with current sail control arrangements this is very difficult, if not impossible, to achieve. It would require having the twist of the sail reversed, or, using aeronautical terms, to have washin instead of washout. For an ordinary, highly tapered, triangular sail such setting is impossible. But even if we could attain control over the sail shape to the extent that arbitrary washin could be achieved, there is a contradictory aerodynamic reason for not doing so. The upper part of the untwisted, triangular sail or foil, due to the induced effects discussed earlier at some length, is operating at much higher lift coefficients than the remaining part of the sail and this may lead to an early stall and separation, that in turn can only be partly cured by washout and increased camber. This constitutes a typical vicious circle of conflicting requirements inseparable from the triangular planform.

In fact, the only sailing condition in which the triangular sail may approach relatively closely the desirable semi-elliptical loading depicted in Fig 2.131 B is that encountered in close-hauled sailing in heavy winds, when sails are usually set at low incidence angles and the resulting lift coefficients are small. A glance at the sketch marked 'strong wind' in Fig 1.9 Part 1 (right upper corner) may help the reader to visualize a condition in which the incidence angle and hence the total lift coefficient C_L, largely restricted by the available stability, are low. The sailing dinghy crew of sufficient weight to give adequate righting moment may be able to afford to reduce the sail twist to the minimum possible, thus shifting the load of the sail upward and so decreasing the induced drag of the rig. This reduction, it must be stressed, can only be achieved if the effective incidence angles are sufficiently low, so that the upper part of the sail has not yet reached the stalling angle. In the same conditions the lightweight crew, commanding smaller available righting moment, will be forced to spill the excess of load from the top part of the sail, by freeing the upper leech. Hence the penalty incurred in terms of additional induced drag will be paid earlier by the lightweight crew and will be higher than that for the heavy crew.

If one attempts to come to grips with the frequently argued problem of light versus heavy crew and its effect on boat performance, one should not forget the close relationship between the stability or power to carry sails effectively, the induced drag, and the boat's performance. In heavier winds low stability is inseparable from higher induced drag which is the most important part of the total drag of any sail. It appears that by their very nature triangular sails penalize light crews greatly in conditions when the 'spilling' wind technique cannot be avoided. By applying the

Fig 2.132

25 ft *Jester* with the Chinese junk type rig introduced by 'Blondie' Hasler, originator of the Single-handed Transatlantic Race. She finished second in 1960 race. The rig consists of a single sail of 240 sq ft.

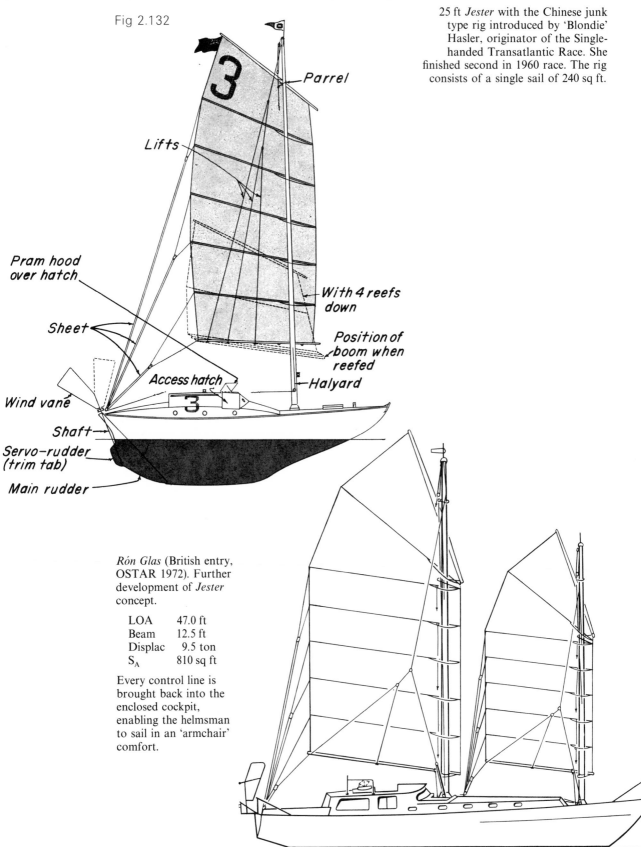

Parrel

Lifts

Pram hood over hatch

Sheet

With 4 reefs down

Position of boom when reefed

Access hatch

Halyard

Wind vane

Shaft

Servo-rudder (trim tab)

Main rudder

Rón Glas (British entry, OSTAR 1972). Further development of *Jester* concept.

LOA	47.0 ft
Beam	12.5 ft
Displac	9.5 ton
S_A	810 sq ft

Every control line is brought back into the enclosed cockpit, enabling the helmsman to sail in an 'armchair' comfort.

spilling routine the lower part of the sail becomes more loaded, thus the lift distribution departs further from the elliptic loading and the leads to rapid deterioration of boat performance, as demonstrated earlier in Figs 1.23 and 1.23A of Part 1.

The bermudan rig shown in Fig. 2.130, consisting of triangular sails having a maximum chord at the foot and tapering almost uniformly to a point at the sail head, although very simple from a practical point of view is, from the standpoint of aerodynamics, inferior when compared with planforms marked 4, 5 and 6 in Fig. 2.121. Very severe tapers may lead to aerodynamic characteristics much inferior to those of rectangular sails such as, for instance, that incorporated in the rig of *Pen Duick III* shown in Fig 2.122, or even the Chinese junk type sail, shown in Fig 2.132.

One must expect that in conditions when high values of total C_L are required, i.e. in light winds or when sailing on courses other than close-hauled in a strong wind, the inferiority of triangular sails relative to other planforms shown in Fig 2.121 will become more evident. In such conditions, the penalties paid for premature stall of the upper part of the highly tapered sail or for excessive twist, if stall is to be avoided, may be so high that triangular sails have no chance of competing with other planforms on equal terms.

The examination of Figs 2.126 and 2.131 reveals, for example, that the rectangular foil as compared with triangular one has more uniform distribution of the local lift coefficient C_l and its tip parts, operating at lower effective incidence angles, produce lower C_l than the remaining part of the foil. By allowing a certain amount of washout, to which every sail has a natural tendency, a relatively close approximation to the elliptical loading may, in the case of a rectangular sail, be easily achieved. These characteristics of a rectangular planform are exactly opposite to those of a triangular one.

Although–to quote an expert opinion from Ref 2.95–'…the study of everything connected with the Chinese junk is complicated by contradictions…and…no sooner is an apparent solution found, or a rule permitting of a classification arrived at, than along comes an exception so formidable as to wreck all previous conclusions…', in one respect the junk type rig is remarkable. This is the sheeting arrangement shown in Fig 2.132, an independent sail shape and sail incidence control system. The first function is performed by thin lines which may well be called 'sheetlets', running from each batten to and fro through a common adjusting device called the *euphroe*, which was frequently used in setting up rigging in the sailing ships of an earlier time. Such an arrangement permits accurate changes in camber and sail twist independently of changes in incidence angles. It is believed that '…if only poor quality material and workmanship are available the junk rig would certainly be the most efficient one could make for fore-and-aft sails' (Ref 2.95).

As already mentioned, while referring to Figs 2.126 and 2.131, the extreme foil planforms–rectangular and triangular–manifest in some respects opposite aerodynamic characteristics. From Figs 2.124B and 2.125 it may be inferred that one of the principal planform characteristics affecting the manner in which the foil will stall

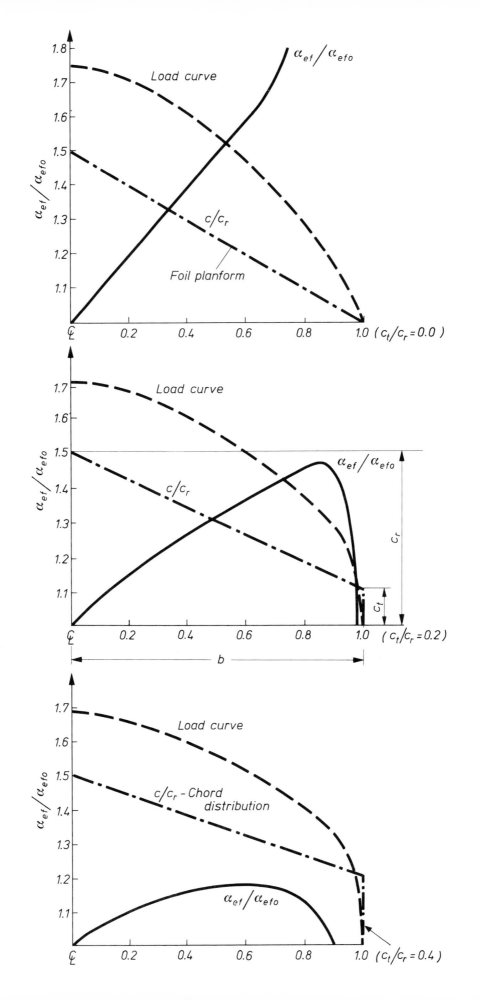

Fig 2.133   Effect of taper ratio c_t/c_r on loading distribution and variation
of the effective incidence angle along the foil span.

is the taper ratio. Figures 2.126 and 2.127 may further suggest that there must be an
intermediate planform which may approximate closer the semi-elliptical loading
than either of the extreme forms without using twist or camber variation as a means
of achieving this desirable load distribution.

Some experimental data presented in Fig 2.133 indicates that, indeed, a certain
amount of taper brings the load curve much closer to the semi-elliptical form than
does the triangular shape (Ref 2.96). In the three sketches of Fig 2.133 there are
given:

a. The foil planform (the root chord c_r, of course, being a maximum at the foil
midspan, which is equivalent to the length of the sail foot).
b. The load curve for which no scale is plotted since only the shape of the curve is
of interest.
c. Ratio of the effective angle of incidence α_{ef} for a given section along the foil
span to the effective angle of incidence α_{ef0} at the root section c_r.

The tendency for rapid increase of the effective incidence α_{ef} towards the tip when the
taper ratio c_t/c_r approaches 0 is very plain. By increasing the taper ratio c_t/c_r, i.e. by
making the foil trapezoidal, one can change the α_{ef}/α_{ef0} curve appreciably.

When the taper ratio $c_t/c_r = 0$, i.e. the planform is triangular, as given by the top
sketch in Fig 2.133, the effective incidence angle α_{ef} at the distance $0.8\,b$ from the
centre line of the foil is about twice as large as the effective incidence angle α_{ef0} at the
root section. By increasing the chord of the tip section c_t from 0 to 0.4 (bottom sketch
in Fig 2.133) the differences between the effective incidence angles along the foil span
can be reduced dramatically. Thus the effective angle of incidence α_{ef} at the same
station $0.8\,b$ is now only about one-tenth higher than α_{ef0}. This indicates that the
downwash is much more uniform in the case of the trapezoid form and therefore the
induced drag will be much lower as compared with that generated by the highly
tapered, triangular planform. The best result seems to be obtained when the chord at
the tip c_t is about 0.4 of c_r although the c_t/c_r ratio may be anything between 0.3 and
0.5 without appreciable change of effect.

Bearing in mind that the elliptic planform is aerodynamically the best planar
(untwisted) form, Fig 2.108 in which a trapezoid of one half taper has been
superposed on an ellipse shows that the experimental results, pointing at the
trapezoid form as the second best, are just about what ought to have been expected.

While the minimum induced drag of the planar foil of given span occurs with the
elliptic planform, some changes in this type of drag might be foreseen if the foil is
twisted (non-planar).

A large amount of work has been done on the determination of tapered and
twisted foil characteristics assuming that the two-dimensional section characteristics
C_l and C_d are known (Refs 2.7, 2.11, 2.12, 2.91, 2.92). The object was to answer the

Fig 2.134A   Graph for determining the induced drag factor u. For elliptical planform $u = 1.0$.

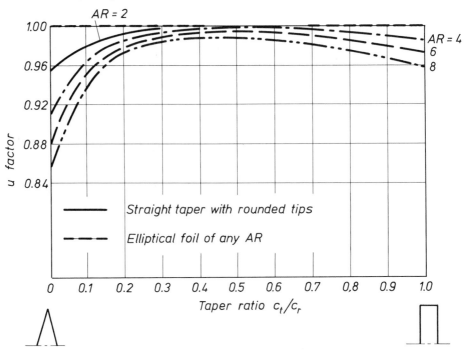

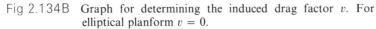

Fig 2.134B   Graph for determining the induced drag factor v. For elliptical planform $v = 0$.

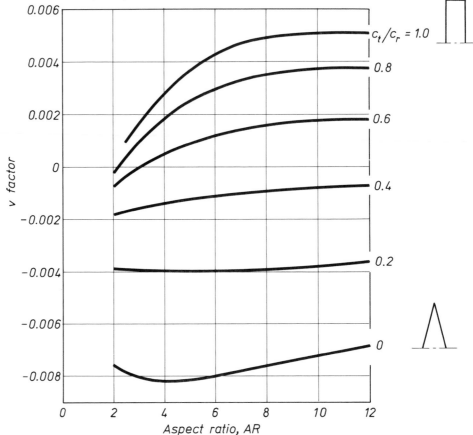

Fig 2.134C   Graph for determining the induced drag factor w.

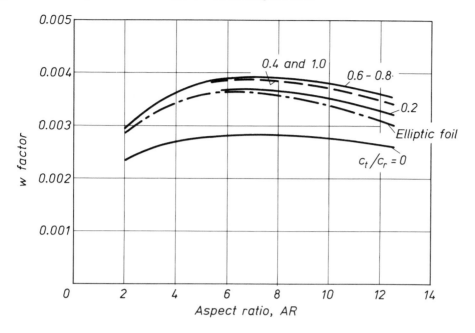

practical problem: what is the penalty, in terms of additional induced drag, if, for some reason, the actual foil shape deviates from the ideal one? Anderson (Ref 2.92) presented a solution to this problem in the following formula:

$$C_{Di} = \frac{C_L^2}{\pi AR u} + C_L \varepsilon a_0 v + (\varepsilon a_0)^2 w \qquad \text{Eq 2.38}$$

Although this may look complicated at first sight it is quite easy to apply using the graphs of Fig 2.134A–C. In order to find the induced drag C_{Di} of an arbitrary foil for a given total lift coefficient C_L one must know, or assume first:

AR–geometric aspect ratio $(AR = (b^2/S_A)$
c_t/c_r–taper ratio
    ε–aerodynamic twist in degrees from root to tip measured between the zero-lift direction of the root and tip sections.

Twist is taken positive for washin $(+\varepsilon)$ and negative for washout $(-\varepsilon)$. If camber of the foil sections along the span is uniform the aerodynamic twist can simply be measured between the chords of the root and tip sections. One may call such a measure the geometric twist. If camber is not uniform but, say, increases towards the tip section, the aerodynamic twist will be smaller than the geometric. The reason can be found in Fig 2.70 according to which the angle of zero lift α_{L0} increases with camber. Conversely, the aerodynamic twist ε will be greater than the geometric twist in the case where the camber decreases towards the tip.

When developing the C_{Di} formula 2.38, Anderson considered only the case of a linear distribution of twist between the root section and the tip section. It would be equivalent, when thinking in terms of sail geometry, to the straight imaginary leech-line when looking at the sail from behind the leech-line towards the mast or forestay. In fact, in most cases the leech-line is concave. Such a departure from the assumed linear twist distribution may not be negligible if appreciable twist is combined with a highly tapered triangular planform. In such a case, the induced drag may be higher than suggested by Anderson's formula.

The remaining terms incorporated in the formula are as follows:

u–induced drag factor. This can be estimated from Fig 2.134A which gives its dependence on aspect ratio AR and taper ratio c_t/c_r.

a_0–is the already known section lift-curve slope per degree. It can be estimated from Fig 2.70, being approximately 0.11.

v–is the second induced drag factor. Its variation, depending on aspect ratio AR and taper ratio c_t/c_r, is presented in Fig 2.134B.

w–the third induced drag factor, depending on AR and c_t/c_r, can be estimated from Fig 2.134C.

The breakdown of Anderson's formula, which is on page 435, may be of some help to the non-mathematical reader in appreciating the meaning of the factors ε, u, v and w. Such a presentation may facilitate better understanding of this new, rather long equation, and may prevent the reader from feeling confused. In order to clarify further Eq 2.38 and the associated graphs let us examine them step by step.

The first, boxed, term in the formula represents the already known Eq 2.28A, which gives the basic minimum induced drag coefficient C_{Di} for an elliptical, planar untwisted foil. Factor u in the denominator reflects the penalty or first drag correction for the departure of the actual planform from the elliptic, for which $u = 1.0$. For all other planar forms this factor is smaller than 1.0 and Fig 2.134A explicitly demonstrates that the penalty is greatest for the triangular planform. Thus, for the triangular foil of AR = 6 the factor u is just below 0.88, i.e. the induced drag coefficient C_{Di} will be about 14 per cent greater than the possible minimum. For the rectangular planform of the same AR the factor u is about 0.98 and therefore the respective increase in drag coefficient will be in the order of 2 per cent only. For any other tapered planform the drag increase will be even smaller than 2 per cent and the penalty increases with aspect ratio.

The second term in the formula may be regarded as a second drag correction for twisted, non-elliptic forms. Since, according to Fig 2.134B, v equals 0 for the elliptic planform, the second term in the formula has significance only for tapered and twisted forms. As seen in Fig 2.134B the factor v can be positive or negative depending mainly on taper ratio c_t/c_r and to some extent on AR. It is always negative for the triangular form. If such a foil is twisted in the sense that there is washout at the tip the twist angle must enter the formula with negative sign $-\varepsilon$. Consequently, the combined effect of high taper in conjunction with washout will result in the positive

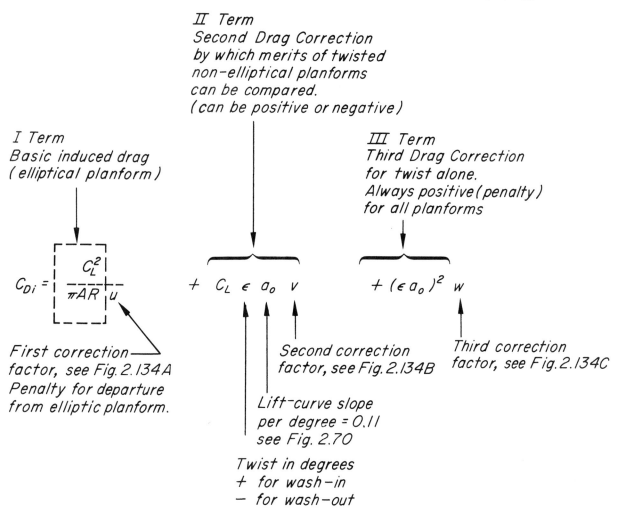

II Term
Second Drag Correction
by which merits of twisted
non-elliptical planforms
can be compared.
(can be positive or negative)

I Term
Basic induced drag
(elliptical planform)

III Term
Third Drag Correction
for twist alone.
Always positive (penalty)
for all planforms

$$C_{Di} = \frac{C_L^2}{\pi AR}\,u \quad + \quad C_L\,\epsilon\,a_0\,v \quad + \quad (\epsilon a_0)^2\,w$$

First correction—
factor, see Fig. 2.134A
Penalty for departure
from elliptic planform.

Second correction
factor, see Fig. 2.134B

Third correction
factor, see Fig. 2.134C

Lift-curve slope
per degree = 0.11
see Fig. 2.70

Twist in degrees
+ for wash-in
− for wash-out

value of the second term since clearly the product of $(-v) \times (-\varepsilon)$ is positive $(+)$. Thus, the drag penalty already incurred and given by the first term of the formula will be increased. For the rectangular planform the reverse is true. In this case the v factor is positive $(+v)$. In conjunction with washout, which carries a negative sign $(-)$, it results in a negative value for the second term, decreasing the drag penalty given by the first term of the formula.

The above discussion confirms in quantitative terms the conclusion reached earlier that washout (or in other words falling off to leeward of the top part of the sail)–the rather natural response of most sails particularly evident in strong winds, when combined with taper ratio tends to penalize the highly tapered and triangular forms. While for rectangular and trapezoidal forms the washout may be advantageous.

It can also be inferred from Fig. 2.134B and the formula 2.38 that the triangular planform, with washin instead of washout, might result in a negative value of the second term of the formula. However, as mentioned earlier, such a combination is rather unrealistic in practice and might be advantageous only in the case where the total lift coefficient C_L of the sail is very low. In such a condition only, the top part of the sail might operate at the effective incidence α_{ef} which is smaller than the stalling angle proper for sections of given camber.

Finally, the third term in the formula supplemented by the graph in Fig 2.134C gives the correction for twist alone for all planforms, including the elliptic one. This correction is always positive, i.e. indicating drag increase since both the factor w and $(\varepsilon \times a_0)^2$ are also always positive. Thus, the third term will invariably be added to the second term of the formula whenever one wishes to estimate the induced drag of twisted foil. The second and the third terms are, of course, discarded if the foil in question has no twist.

The following example is intended to illustrate the method of using the formula and the graphs in Fig 2.134A–C. The results of the induced drag C_{Di} calculations are given in Table 2.11 below, for three hypothetical sails having the following characteristics:

sail planforms: elliptic, triangular and rectangular; twist (washout) $\varepsilon = -30°$; the total lift coefficient $C_L = 1.0$ and $AR = 6$ are the same for all forms.

TABLE 2.11

Planform	u	I term	v	II term	w	III term	Total C_{Di}
Elliptic	1.0	0.0530	0	0	0.0036	0.0392	0.0922
Triangular	0.88	0.0603	−0.0082	0.0271	0.0028	0.0305	0.1179
Rectangular	0.98	0.0541	0.0043	−0.0142	0.0038	0.0414	0.0813
		Untwisted sails			Twisted sails		

The above numerical values give a good indication of what is the order of merit of twisted sails of different planforms in, at least, a qualitative sense. It is seen, for example, that the twist (washout) of the triangular planform, although desirable in most conditions as a means of avoiding stall of the upper part of the sail, is aerodynamically very expensive in terms of induced drag. It is also seen from Table 2.11 that twisting the elliptic foil causes it to lose its superiority when compared with a rectangular foil, if both foils are twisted to the same angle ε. The triangular form is always the worst, no matter whether it is twisted or planar, the other shape factors being equal.

It should perhaps be added that Anderson's method of estimating the aerodynamic characteristics of foils of various shapes has been developed further by other researchers, since initially published in 1936. This is understandable; no scientific analysis, no matter how clever and deep in scope, is self-terminating. However, Anderson's method was selected for its simplicity. It is still within the intellectual capability of the average sailor to appreciate, while the other methods, applicable to sophisticated aircraft wings, are well beyond our present requirements.

The conclusion we have just derived, that the elliptic foil loses its superiority when compared with the rectangular one if both foils are twisted to the same washout angle, may come as a surprise to some readers. To give this conclusion a physical meaning we shall briefly interpret Fig 2.135. It shows the variation of lift loading along the semi-span b for three foils: elliptic, triangular and rectangular. There is good reason to expect that a certain amount of washout on the rectangular planform will improve its lift loading curve by easing the intensity of load on the outer part of the foil. Subsequently, the load curve will be shifted towards the semi-elliptic one,

Fig 2.135   Variation of lift loading along the semi-span b for elliptic, triangular and rectangular planforms, $AR = 6$, total lift coefficient $C_L = 1.0$.

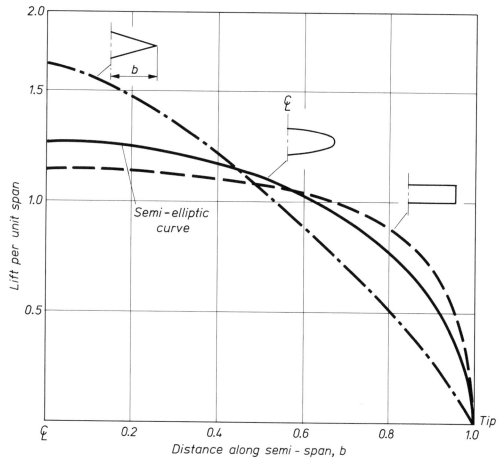

Fig 2.136   Three customary methods of panel design in fully-battened
sails–see also Photos 2.33 and 1.5A.

A. Parallel panels (batten pockets fanned).
B. Fanned panels and batten pockets.
C. Fanned panels with wide seams forming batten pockets.

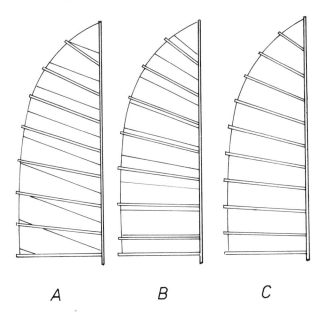

A   B   C

with which is associated the minimum drag. The washout in the rectangular foil may
be beneficial in the same sense as tapering of the chord towards the foil end. On the
other hand, here is no reason to expect any reduction of induced drag from the twist
introduced into a foil that has already attained, through its taper, a substantial
approximation to the ideal elliptic distribution of lift. Thus, according to Eq 2.38 (III
term) any washout introduced into the foil of elliptic planform is bound to increase
the induced drag in proportion to the square of the twist angle ε. Figure 2.135 may
help in anticipating the effect of the washout in the elliptic planform. Any twist will
flatten the tip part of the lift load curve making it similar to the triangular loading
curve, and this, as we already know, is an expensive type of load distribution in terms
of induced drag.

From what has been discussed so far it would seem that a substantial reduction in
sail drag can be obtained by re-shaping of the sail planform with the object of
making the upper portion more and the lower portion less effective. Development in
the catamaran sails may serve as an indication in this respect. As reported by Farrar
(Ref 2.97), their shape evolution illustrated by Fig 2.136 and Photo 2.33 owes a lot to
the International 10 sq m Canoe Class shown in Photo 1.5A, in which class J
Aumonier's *Wake* sported a fully battened sail in 1938. To quote reference 2.97
directly:

Photo 2.33   Increased camber towards the headboard together with sail
              twist may serve as a dodge in delaying an early stall of the
              upper part of the sail which is particularly susceptible to stall.
              An opposite trend in camber distribution is shown in Photo
              2.30 (Tornado cat); its effect must be compensated by larger
              twist which in turn incurs higher penalty in terms of induced
              drag.

'Introduced initially as a light weather sail, it had a higher centre of effort than the conventional sail because the battens held out a large rounded "shoulder" of leech and the foot was short to keep the area right. In light weather Aumonier's Chinese sail, as it was called, stood ready shaped to catch the first of a new wind. When caught out in a blow it paid an *unexpected bonus* [the italics are not in the original] as it could be eased away without flogging so causing much less drag and making it possible to feather a canoe to windward pointing higher than with soft sails!'

The explanatory facts, which clearly escaped the attention of those who made these experiments were, no doubt in the first place, an essential reduction in induced drag due to redistribution of the sail chord and possibly camber. To quote further from the same source (Ref 2.97): '...no question arose of gaining extra area because every square inch of a canoe's sail is measured, but the sail was more efficient area for area.' Subsequently, when the early catamarans emerged they soon turned to the same line of thought and Fig 2.136 illustrates the development in panel cutting which lead to better shape control and smooth airflow.

Photograph 2.34 depicts another way of approximating an elliptic planform, by means of curved spars. The same effect can also be achieved by employing more efficient planforms than triangular, such as rectangular and trapezoidal shown in

Photo 2.34 Curved spars–another way of approximating an elliptic planform. The Chesapeake Bay log canoe. The curved spars rig is a new version but the typical boat ahead and to leeward with highly tapered sails dates back to the 1880s.

Photo 2.35   A. Rig designed by L F Herreshoff for R class boat about 50
                  years ago.
              B. Recent application of similar concept on land-yacht.
                  (*Showing off*, Brittany France).

Fig 2.122 and Photo 2.35, or the gaff-headed sails depicted in Fig 2.137. It is rather a pity that gaff-headed sails have become almost completely ousted from the sailing scene. Certainly, the rating rules have in this respect a more profound effect on the shape of sails than the aerodynamic requirements or wind in all its moods. The penalty incurred for example by the sail measurement system on the width of headboard of the mainsail or length of its top batten is so high that it virtually precludes any attempt to improve the aerodynamic effectiveness of the modern tall rig. Those curious prohibitions, which after years of enforcement became part of sailing tradition, effectively discouraged ocean racing people from making experiments with unorthodox rigs which could have led to the development of less tall but more efficient rigs. So triangular sails prevail.

Fig 2.137   Gaff-headed sails.

A. Traditional Dutch fishing vessel which became a commodious cruising yacht. An extra wide headboard may well improve sail efficiency.

B. Variation on the same theme. Remarkable small yacht designed by L F Herreshoff for shoal water cruising. The total sail area of 456 sq ft is kept low by providing short gaffs.

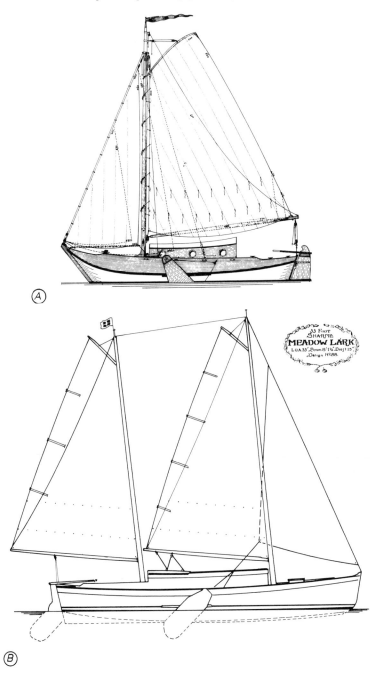

(c) *Effect of aspect ratio on maximum lift*

With some reservations expressed earlier (aspect ratio effect should be considered in conjunction with taper and twist), one must agree that high aspect ratio sailplan is efficient for the round-the-buoys racing, where close-hauled performance is of primary importance. It would be even better, however, if the triangular planform could be modified as discussed previously. However, many people question the advantage of a high AR rig in the case of boats intended solely for cruising where close-hauled performance may not be considered to be of primary importance and the sail design alternatives are not restricted by the rigid rules.

If the potential possibilities in developing more power with relatively low heeling moment are appreciated, the long passage ocean racing of 'Around the World' type might well lead to an entirely different concept of offshore racing rig–something along the *Pen Duick III* line (Fig 2.122) with lower aspect ratio sail of various planforms. In Tabarly's account of his experiences with *Pen Duick III* its rig is described as, '...undoubtedly the best of all rigs for ocean racing and for singlehanded races', and '...it is a great pity that a schooner rig suffered so badly in the recent revision of handicap rules' (Ref 2.98).

Where the close-hauled course is not the most important performance feature and the maximum available lift coefficient becomes of pre-eminent value, low aspect ratio gaff-headed or even square sails may prove superior.

Figure 2.138, which is reprinted from the author's *Sailing Theory and Practice*, substantiates such a judgement. It shows in a qualitative sense the effect of aspect ratio on the maximum lift and driving force coefficients for various rigs. It is evident that there is no ideal aspect ratio or ideal type of rig superior for all points of sailing. The prevailing or expected sailing conditions–winds and courses sailed–should be regarded as significant factors when evaluating the merits and demerits of different rigs.

When discussing the differences between the two-dimensional and finite span foils it was stated that the C_L coefficient of a foil of finite span is, at a given incidence angle, smaller than that of a two-dimensional foil (see Fig 2.84). This can be explained by taking into account the pressure losses due to the end effect at the foil tips. One may rightly expect that progressive reduction in aspect ratio should result in marked drop in $C_{L \max}$ compared with the two-dimensional value, i.e. sectional maximum c_l coefficient corresponding to an aspect ratio of infinity. With reduction in AR the loading at the foil median line becomes relatively higher, with the local c_l in excess of the foil total or mean lift coefficient C_L, so that the stall occurs at a lower mean C_L. This explanation is fairly adequate, except for foils of low AR below, say, about 2 and also foils with very thin sections (Refs 2.99, 2.100). As shown in Fig 2.139, a decrease of AR below 2 introduces the curious feature of a rapid increase of maximum lift, a phenomenon first discovered with the flat plate in the earliest days of heavier-than-air flight.

The effect of AR on maximum lift is displayed by plotting the measured increment $\Delta C_{L \max}$ above $C_{L \max}$ of the flat plate of infinite AR which is around 0.7. The C_L

Fig 2.138   Polar diagrams of four foils of the same camber 1/13.5 (7.4 per
cent) but different planform (aspect ratio). Note conspicuous
differences in the total force coefficients C_T when foils operate
at the same incidence angle α.

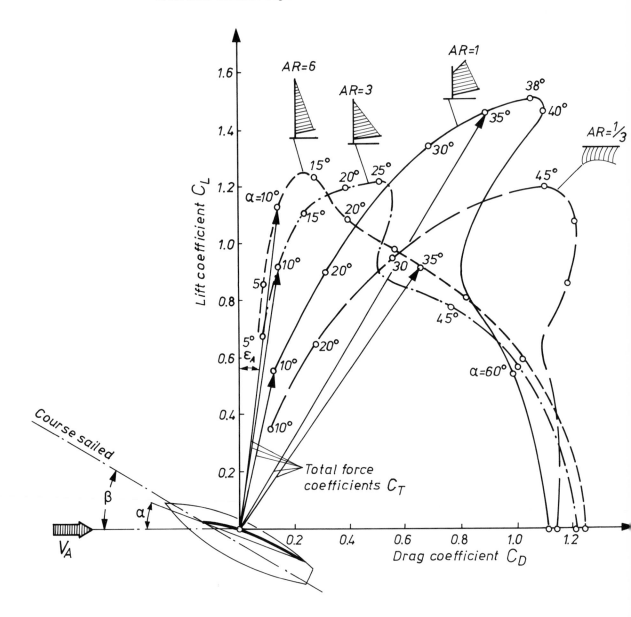

Fig 2.139   The effect of aspect ratio on the maximum lift of a flat plate. Increase in $C_{L max}$ may not occur in the case of streamline foils of low AR such as rudders or keels. The Reynolds Number or surface roughness effects can be more significant than AR effect (see Figs 2.64 and 2.82). Flat plates or very thin foils which develop a leading edge separation are relatively insensitive to Reynolds Number effect.

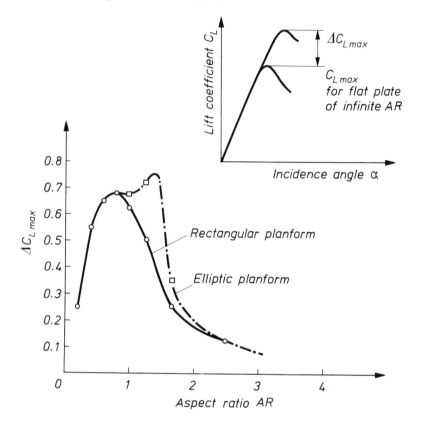

coefficient reaches its maximum for most planforms where the span is roughly equal to the root chord. For foils of elliptic planform, for instance, the highest $C_{L max}$ is obtained on a circular foil. And, as aptly remarked by one commentator, it makes one wonder whether the flying saucer may not have its advantages.

In the course of experiments on low aspect ratio foils (Ref 2.99) the following conclusions were reached:

1. There is a range of AR extending approximately from 0.5 to 2.0 wherein the tip vortices cause a marked delay in the breakdown of the flow as the angle of incidence of the foil is increased. In other words, for the very low AR foils the boundary layer flow is considerably affected by the strong tip vortices and separation is delayed to a high angle of incidence (see Fig 2.140), thus the maximum lift coefficient is bound to increase.

Fig 2.140   Lift characteristics of thick foils of low aspect ratio (Ref 2.99).

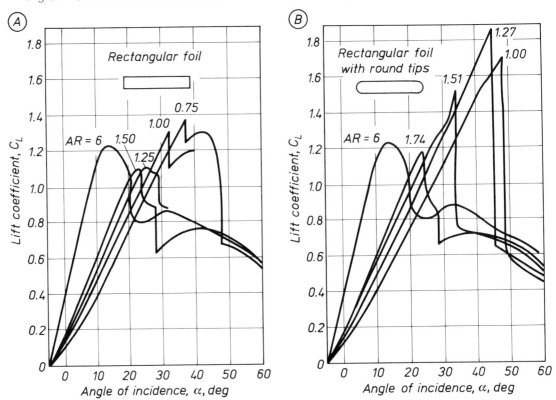

2. It is possible, within the particular range of AR indicated in Fig 2.139, to obtain a maximum lift coefficient C_{Lmax} considerably higher than can be obtained for a foil of the same section having AR = 6. Thus, the highest maximum lift coefficient obtained in a series of tests represented in Fig 2.140B was 1.85 at $\alpha = 45°$ for a foil with semi-circular tips as compared with $C_L = 1.24$ at $\alpha = 14°$ for foils of AR = 6 with semi-circular or square-cut tips.

3. The tip shape is, in the case of low aspect ratio foils, of paramount importance among the factors affecting the foil efficiency at large angles of incidence. The foils with semi-circular tips were found to be much superior to those having square-cut or faired tips.

   Since it is believed that the increases in C_{Lmax} are the results of the delay of separation caused by the tip vortex flow near the foil ends, it is apparent that tip form plays an important part in this phenomenon.

4. The low AR, thick and thin foils behave somehow differently in this respect. On thick foils the rise in ΔC_{Lmax} is delayed until the aspect ratio is below 1.5 and the extent of the rise shown in Fig 2.140B is consequently less. There appears, as for the flat plates, to be a significant difference in the results for foils with round-tips and square ones.

Thick foils with square-cut ends are appreciably worse than flat plates of the same AR. This seems well established, at least qualitatively, by the data presented in Figs 2.139 and 2.140A; the reason for it is, however, obscure.

It has already been mentioned that the Prandtl–Lanchester theory, called frequently the lifting line theory, by which the lift and drag generated by a foil can be predicted contains, as does every theory, certain limitations:

a. It is applicable with a reasonable satisfaction to straight unswept foils of high aspect ratio provided the flow is attached to the foil surface, i.e. separation has not yet occurred. This non-separation condition constitutes the first limitation.

b. The lifting line theory is not satisfactory in the case of the low aspect ratio foils just discussed. When the span becomes shorter and shorter as compared with the mean chord length, the tip vortices increasingly influence the flow round the foil section. The flow pattern along the foil span cannot therefore be approximated by the uniform two-dimensional flow shown in Figs 2.84, 2.94 and 2.96. This is the second limitation.

c. The lifting line theory does not give good approximation for swept foils (usually swept-back foils), a feature which has been adopted for different reasons both for high-speed aeroplanes and sailing yacht keels as well (see Fig. 2.141).

d. Finally, as mentioned earlier, the lifting line theory fails to some degree in cases where a foil is placed in non-uniform flow velocity. Sails operating in a strong vertical wind gradient are good examples of such a non-uniform flow condition.

(d) *Sweep angle effects and low* AR *foils*

'The study of fluid dynamics, and indeed of all sciences, is like a tree trunk whose root is physical experience, drawing for its strength on our powers of observation. It stems into many branches, some leading to regions of thought far removed from the root of physical phenomena; if it were otherwise, much of its fascination would fade away.'

B. THWAITES, *Incompressible Aerodynamics*

So far, we have been mainly concerned with straight unswept foils, in which the leading edge or the quarter chord line is, at least roughly, perpendicular to the direction of undisturbed flow ahead. Foils such as sails, keels and rudders are, however, frequently raked bodily backward or, much more rarely, forward through some angle called sweep-back or sweep-forward respectively. The basic features of the raked geometry are defined for convenience in Fig 2.141.

Foil sweep may lead to considerable changes in aerodynamic or hydrodynamic

Fig 2.141   For different reasons sweep-back foils have been adopted both for high-speed aeroplanes and sailing yacht keels as well. Sweep angle Λ is an important factor of foil efficiency.

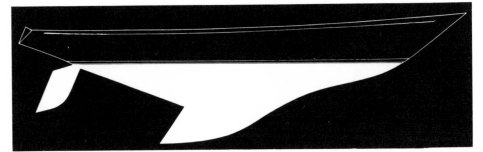

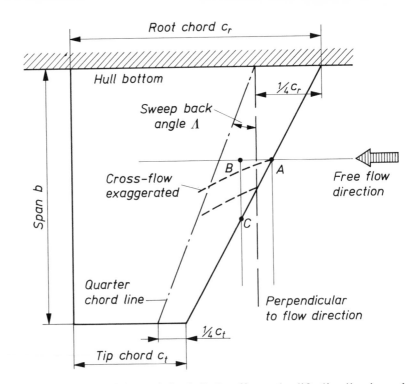

characteristics as compared with straight foil. It affects the lift distribution, drag, stalling pattern, etc. Despite this important fact, keel or rudder shape design, for example, is often influenced by eye appeal or a desire to cut its wetted area, rather than some definite knowledge on the part of the designer regarding its efficiency.

Although the results of tests on keels or rudders are available, as described later in Part 3, they are unfortunately not fully understood by many yacht designers; and certainly some of the tests appear contradictory, so that they cannot be used for designing purposes with full confidence. It seems therefore to be instructive to examine more closely what happens if, for some reason, the foil is given a sweep angle.

Experimental evidence indicates that a foil with sweep-back behaves similarly to a foil with more taper, while a swept-forward foil acts like a foil with less taper. These characteristics are illustrated by the results depicted in Fig 2.142, which is adapted from Ref 2.89, already quoted earlier. The character of flow on the suction side of the two foils is presented in the same manner as that in Fig 2.124. Foil A, with straight leading edge and of taper ratio $c_t/c_r = 0.25$, behaves like an elliptical planform, i.e. it stalls almost evenly across the span while the stall pattern of a similar foil, but

Fig 2.142A, B Stalling pattern developed by two tapered foils with different sweep angles. AR = 7.2 and $c_t/c_r = 0.25$ in both cases. Foil B is more heavily loaded on the tip part than foil A.

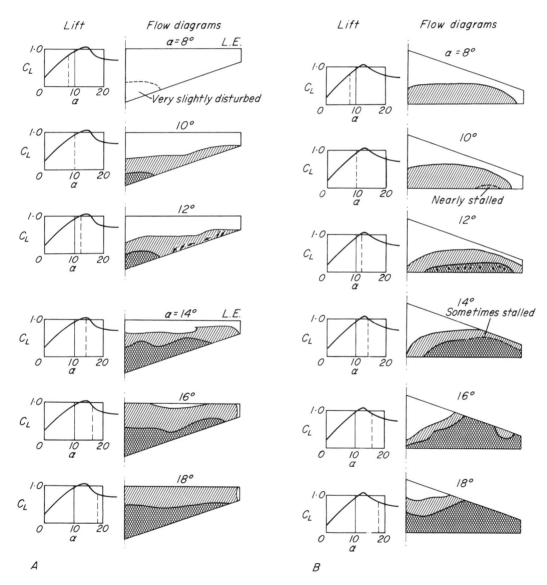

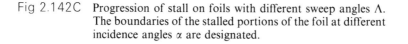

Fig 2.142C Progression of stall on foils with different sweep angles Λ.
The boundaries of the stalled portions of the foil at different
incidence angles α are designated.

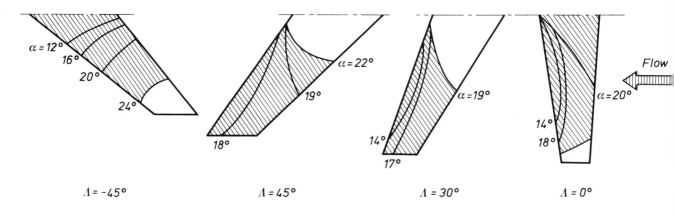

with sweep-back angle about 15°, resembles that of highly tapered form shown in Fig 2.125. It stalls first at the tip.

In the case of a foil with sweep-back the pressure gradient, existing along the foil span, causes a cross-flow towards the tip in the boundary layer. Why? In Fig 2.141 the pressure on the suction side at point A is sensibly equal to that at point C but there is a pressure difference along the foil surface lines perpendicular to the flow direction and the pressure at point B is higher than that at point C. Such a spanwise variation in pressure, i.e. the existence of a spanwise pressure gradient, even a comparatively small one, will tend to produce a sideways cross-flow of boundary layer along the foil surface. The fluid, be it air or water, adjacent to the foil surface has lost most of its momentum (mass × velocity) and therefore more readily flows towards the region where pressure is lower than straight in the chordwise direction where pressure increases. Consequently, the boundary layer gradually thickens while approaching the tip of a swept-back foil, thus encouraging premature separation and stall, depicted in Fig 2.142B. Photograph 2.21B illustrates the fact that steep rake of a keel encourages water to flow downwards; and the steeper the rake or larger the incidence of leeway, the more conspicuous the cross-flow.

These cross-flows can occasionally be observed on headsails with the wool-tufts placed near the luff of the sail. In some conditions, determined mainly by the shape of the headsail, i.e. its distribution of camber along the sail height, twist, sag in forestay, etc., the tell-tales may not stream horizontally in some places but instead they are inclined diagonally upwards, as shown in Fig 3.54. Their behaviour indicates the presence of a vertical pressure gradient, i.e. a decreasing of the pressure towards the sail head, and this should be regarded as a warning that a premature tip stall is imminent. If this is to be avoided, which may be desirable, particularly in light weather conditions when sails operate at high lift coefficients and therefore the tip stall is highly probable, the sail twist should be increased. Reduction of forestay sag, which may be excessive, can also, for the reason given in Part 3, be of some help in this respect.

As far as the design of the moderately swept keel, centreboard or rudder of higher aspect ratio is concerned the proper distribution of maximum thickness of their symmetrical sections along the span may to some extent serve as a means of delaying an early tip stall. Thus the tip sections should have the position of maximum thickness nearer the leading edge than those sections closer to the root, i.e. the hull bottom. Such a distribution of thickness brings the suction peaks all along the foil span closer to the same perpendicular line to the flow direction. This effect of thickness distribution can slightly be augmented by an inverse taper of the foil thickness along its span, that is, the tip sections are thicker than the root sections. As was explained earlier in the chapters dealing with the boundary layer, the larger nose radii associated with the thick foils generally produce a more gentle adverse pressure gradient with an attending increase in local $C_{L\,max}$. This requirement for fatter tip sections happily coincides with the hydrostatic stability requirement which implies that the mass of the keel, regarded as a ballast container, should be concentrated close to the keel bottom.

Tests have shown that it is almost hopeless to prevent tip stalling by correct selection of foil section along the span if the angle of sweep is large. Twisting of the foil becomes necessary in order to counteract the characteristic increase in the foil loading occurring near the tip. Such a preventive measure, although practical in the case of sails, cannot be easily used on keels or rudders.

There are other means which have been employed in aeronautics to prevent the sideways motion of the boundary layer, thus delaying premature separation and so allay the increase in drag. Amongst them are the boundary layer fences similar to those shown in Photo 1.25 Part 1, fitted to the surface of the foil and running fore and aft parallel to the flow direction. They are quite effective in checking any spanwise flow along the foil surface which is likely to cause a breakaway of the flow and so lead to tip stalling.

The three-dimensional cross-flows of boundary layer, shown in Fig 2.141 and earlier in Photo 2.21B, become (according to Ref 2.31) particularly marked under conditions approaching separation. The flow of this low-energy air or water from one section to another tends to delay separation in some places and to promote it in others. The result is that not only the lift but also the drag characteristics of swept foils, particularly of those with large sweep angle, depart seriously from that of unswept foils where the two-dimensional flow pattern prevails.

It is worth while bearing the induced-drag factor in mind when variation in planform and sweep angle of hull appendages of higher AR are being considered. The provision of large sweep-back or sweep-forward can give rise to a severe increase in induced drag at given lift. Experimental results from various sources have been evaluated in Ref 2.101 and the resulting curve in Fig 2.143 illustrates this point. The experimental drag characteristics of the swept foil are presented in relation to similar but straight foils and the ratio of the induced drag of a swept foil to that of an

Fig 2.143 Effect of sweep angle on drag of foils of larger aspect ratio and elliptic or nearly elliptic planform.

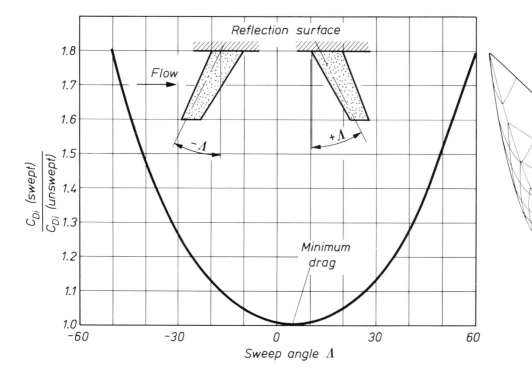

unswept foil $C_{Di\,swept}/C_{Di\,unswept}$ serves as a measure of drag increase due to sweep. It is seen that the minimum induced drag is reached when the angle of sweep Λ is in the order of $+5°$, i.e. the foil is slightly swept-back. It appears that sweep-back angles greater than $15°$ can hardly be justified from the hydrodynamic point of view in the case of hull appendages of higher AR, such as the deep wing-like centreboard sketched in Fig 2.143, which is really distinct from the body of the hull.

The magnitude of drag penalty due to sweep given in Fig 2.143 and valid for unspecified but presumably elliptic or nearly elliptic foils of higher aspect ratio has been questioned by some (Ref 2.103). Evidently, the amount of additional drag depends critically on aspect ratio. An increase in aspect ratio magnifies the effects of sweep on induced-drag increment. Conversely, the drag increments quoted in Fig 2.143 tend to become smaller when aspect ratio decreases. Finally, when aspect ratio is in the order of 1–1.5 and the mechanism of circulation reaches certain limiting conditions; where the actual flow pattern round the foil changes radically as compared to the near two-dimensional condition, which prevails on higher aspect ratio foil, the induced drag becomes much less sensitive to sweep angle variation. From Fig 2.143 and Table 2.12 it can be inferred, for example, that variations of sweep angle from $-8°$ to $22.5°$ for a constant taper ratio of 0.45 do not significantly affect the hydrodynamic characteristics of foils. It appears that relatively small

Photo 2.36   Hull appendages tested in wind tunnel.

A. Rudder mounted against simulated hull in presence of the water surface simulated by the wind tunnel floor. Photograph from *Principles of Naval Architecture* copyrighted by The Society of Naval Architects and Marine Engineers and included herein by permission of the aforementioned Society.

B. Mirror-image model of the underwater part of 5.5-Metre yacht hull.

A

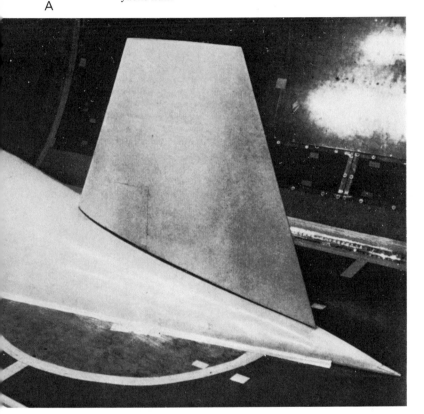

B

sensitivity of induced drag to even extreme planform change is characteristic of all low aspect ratio lifting surfaces. According to Ref 2.106A it can be concluded, nevertheless, that the highly swept keel is a less efficient lifting surface, having a lower lift-curve slope and higher induced drag.

This conclusion must be taken with reservation. It would be valid on the assumption that the water surface can be regarded as a rigid reflection surface in which the underwater portions of the hull, including appendages, are reflected to give what is called in professional parlance a *double model* or *mirror-image* model, as shown in Photo 2.36B. This concept, discussed in some detail in section 4, was employed for calculation of the effective aspect ratio and associated induced drag of sails. In relation to the airflow round sails the water surface can be regarded as a rigid reflection surface. However, as far as the underwater part of the hull is concerned, the mirror-image analogies might be considered with due respect to peculiar conditions in which the optimum sweep angle may largely be affected by the hull–fin keel interaction and associated wave pattern, generated both by the hull and its appendages. At higher speed, when wave drag becomes dominating, the eventual losses in terms of induced drag of the fin keel, due to sweep, may more than outweight the gains arising from lower wave drag.

Swept-forward hull appendages are rarely applied or considered in the yacht design. As demonstrated in Fig 2.143, their induced drag increases with the angle of sweep for the same reason as outlined while discussing the sweep-back foil characteristics. Namely, the drag increase is due to non-uniform distribution of loading along the span. In swept-forward foils the lift concentration is reversed as compared with swept-back foils and the maximum load is now nearer to the root sections. Thus the tip sections are less loaded than in the case of swept-back foil. This difference in load distribution, which is indirectly visualized in Fig 2.142C, has a pronounced effect both on the maximum lift and the shape of the lift curve beyond the angle of stall. This is shown in Fig 2.144 (Ref 2.9). Sweep-forward produces

Fig 2.144   Effect of sweep on lift characteristics of a straight rectangular foil of effective AR = 6 (semi-span foil of geometric AR = 3 was tested in the wind tunnel and mounted over ground board which simulated the mirror-image plane) (see Photo 2.36A).

A. 20° sweep-forward
B. 0 sweep angle
C. 20° sweep-back

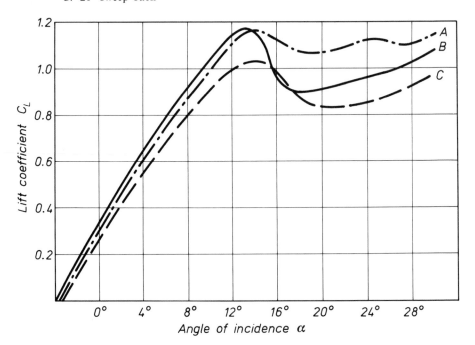

TABLE 2.12

Effect of aspect ratio, sweep angle and tip shape on the free-stream characteristics of all-movable control surfaces (all control surfaces tested against groundboard with gap $= 0.005\,\bar{c}$)

(Section shape in all configurations (NACA 0015))

Item No	Profile	Direction	Mean chord $\bar{c}$ (ft)	AR	Λ (deg)	$\dfrac{c_t}{c_r}$	Tip shape	$Re \times 10^{-6}$ of test	$\left(\dfrac{\Delta C_L}{\Delta\alpha}\right)_{\alpha=0}$ (per deg)	C_L @ 10°	C_L @ 20°	$C_{L\max}$	Stall angle (deg)	α for $L/D_{\max}$ (deg)	$L/D_{\max}$	L/D stall angle
1		↑	2.0	1	−8	0.45	Square	2.25	0.023	0.25	0.59	1.24	39.4	8	8.0	2.0
		↓					Square	3.00	0.023	0.30	0.67	0.96	31.2	8	5.4	1.7
		↑					Faired	2.28	0.021	0.23	0.53	1.03	36.3	8	8.2	1.8
2		↑	2.0	2	−8	0.45	Square	2.72	0.04	0.42	0.91	1.33	28.7	6	12.2	4.0
		↓					Square	3.00	0.021	0.40	0.64	0.64	21.0	7	6.8	2.3
		↑					Faired	2.72	0.039	0.41	0.86	1.24	28.8	5	13.0	4.0
3		↑	2.0	3	−8	0.45	Square	2.26	0.056	0.56	1.09	1.13	21.0	4.6	15.8	6.8
		↓					Square	3.00	0.047	0.46		0.62	17.4	5	8.1	2.7
		↑					Faired	2.26	0.049	0.51	1.02	1.21	24.0	4.4	15.4	5.4
4		↑	2.0	1	0	0.45	Square	2.70	0.023	0.27	0.60	1.26	38.5	7.0	8.0	2.2
		↓					Square	3.00	0.023	0.30	0.68	0.93	29.2	7.8	5.5	1.9
		↑					Faired	2.29	0.020	0.24	0.54	1.11	36.4	8	8.2	2.0
5		↑	2.0	2	0	0.45	Square	2.72	0.04	0.44	0.93	1.33	28.7	5.5	12.4	4.0
		↓					Square	3.00	0.039	0.40	0.62	0.63	19.2	6.8	6.8	2.7
		↑					Faired	2.72	0.04	0.42	0.87	1.17	26.8	5	13.0	4.2
6		↑	2.0	3	0	0.45	Square	2.70	0.054	0.55	1.10	1.25	23.0	4.5	16.0	6.0
		↓					Square	3.00	0.042	0.46		0.59	15.5	5.2	7.5	3.3
		↑					Faired	2.26	0.052	0.53	1.05	1.14	22.0	4.6	16.2	6.4
7		↑	2.0	1	11	0.45	Square	2.28	0.024	0.26	0.60	1.40	43.4	7.0	8.2	1.7
		↓					Square	3.00	0.023	0.29	0.67	0.84	28.2	8.0	5.7	1.8
		↑					Faired	2.28	0.021	0.25	0.55	1.21	39.4	7.0	8.3	2.0
8		↑	2.0	2	11	0.45	Square	2.72	0.042	0.45	0.94	1.33	28.8	4.0	13.2	4.0
		↓					Square	3.00	0.026	0.39	0.61	0.65	18.2	8.0	6.8	3.1
		↑					Faired	2.72	0.043	0.44	0.90	1.18	26.8	4.6	14.0	4.4
9		↑	2.0	3	11	0.45	Square	2.26	0.050	0.52	1.05	1.25	24.2	5.0	15.5	5.6
		↓					Square	3.00	0.046	0.48		0.57	13.4	5.6	7.4	3.6
		↑					Faired	2.26	0.054	0.54	1.05	1.08	20.9	4.0	16.8	5.6
10		↑	3.0	2	22.5	0.45	Square	3.00	0.045	0.46	0.96	1.46	31.8	5.2	10.6	3.3

Table partly adopted from *Principles of Naval Architecture*. Copyrighted by the Society of Naval Architects and Marine Engineers and included herein by permission of the abovementioned Society.

higher maximum lift and, what is more important, there is a marked flattening of the peak of the C_L curve. Tapered foils seem to show it more strongly than straight ones. The two other curves, B and C, indicate that a substantial amount of lift is rapidly lost immediately after the foil reached the stalling angle, a rather undesirable characteristic when the foil is used as a rudder, which unlike a keel frequently operates at high C_L. The results presented in Fig 2.144 cannot be generalized and extrapolated to larger sweep angles.

As remarked by the author of Ref 2.9–'It would need a very impressive demonstration of an extraordinary aerodynamic virtue in the swept-forward wing to make designers reconcile themselves to the use of so odd-looking a shape.' This 'odd-looking a shape' was, however, successfully used no less than about 200 years ago, strangely enough on the rudders of the so-called 'coble' type boats of the NE Coast of England, depicted in Fig 2.145. To quote Ref 2.102, 'Cobles have always had a good reputation for speed and seaworthiness.' One can imagine that on account of their deep fore-foot a strong tendency to broach was rather inevitable in conditions

Fig 2.145   The Tyne coble, about 27 ft LOA. This type of boat, about 200 years old in conception, descended from much older Scandinavian ancestors, is still built today. This is probably because of the qualities that made them seaworthy and suitable for beaching in surf.

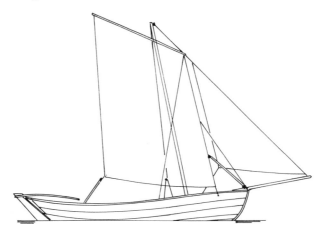

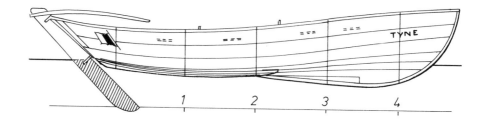

when the relatively full stern was lifted by a wave, depressing the bow. Only the powerful rudder operating effectively beyond the angle of stall, as illustrated by curve A in Fig 2.144, could possibly prevent the impending broach.

Swept foils might, at least theoretically, have the same induced drag as straight unswept ones of the same aspect ratio, provided that the load distribution along the span is the same. However, in swept foils the lift is concentrated either near the foil tip (swept-back foil), or vice versa near the root sections (swept-forward foil). Such a harmful concentration of lift can be alleviated by proper tapering of the foil planform. According to Ref 2.103, for each angle of sweep there is an optimum taper ratio for which the theoretical loading is practically elliptic and the induced drag is at a minimum. Thus the planforms conforming to these optimum configurations should not give induced drag appreciably higher than those for an unswept foil. These optimum relations between the angle of sweep and taper ratio c_t/c_r are given in Fig 2.146. For instance, a taper ratio c_t/c_r of about 0.2 is needed to make the lift

Fig 2.146   Theoretical relationship between the sweep angle and taper ratio to achieve minimum drag. For large angles of sweep-back the BL flow aggravates the tip stalling tendencies and for large angles of sweep-forward the BL flow promotes root stalling. Note: Table 2.13 is attached to Fig 2.151.

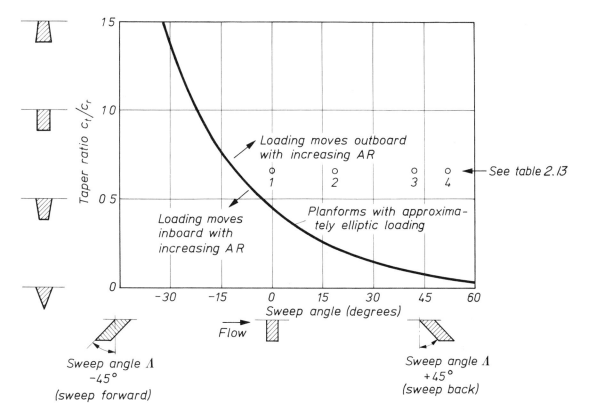

Fig 2.147 Variation of lift-curve slope $\Delta C_L/\Delta\alpha$ with sweep angle and aspect ratio. Taper ratio $c_t/c_r = 1.0$. For moderately tapered foils the peaks of curves are bodily shifted more towards the positive values of sweep angle.

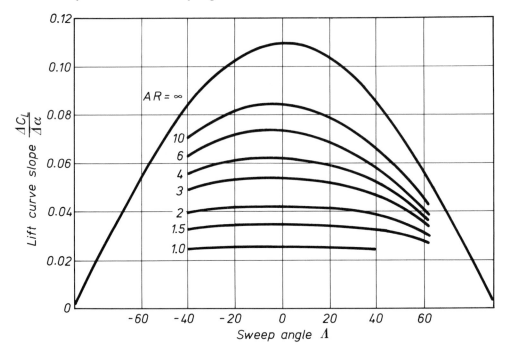

distribution of a 30° swept-back foil near-elliptical. For the 30° swept-forward foil the recommended taper ratio c_t/c_r is in the order of 1.4, which means the taper is reversed. In other words, as the foil taper ratio c_t/c_r is increased from 0, the angle of sweep, at which the minimum value of induced drag is expected, changes progressively from positive to negative angles of sweep. The small sketches along the vertical axis in Fig 2.146 illustrate this tend in c_t/c_r variation, which in turn indicates that optimum taper ratios, from a hydrodynamic point of view, are less than would be desired from considerations of ballast effectiveness.

Sweep angle affects also the slope of the lift curve $\Delta C_L/\Delta\alpha$, which as we may recollect is a measure of the rate of lift increase with the incidence angle (see caption, Fig 2.103). The theoretical variation of $\Delta C_L/\Delta\alpha$ with sweep angle and AR for taper ratio $c_t/c_r = 1.0$ is shown in Fig 2.147 (Ref 2.104). It is seen that, for foils of higher AR, the angle of sweep has a marked effect on the lift-curve slope, the greatest effect occurring for foils of infinite aspect ratio (AR = ∞). As the AR approaches low values, the lift-curve slope for unswept foils is greatly reduced and the effect of sweep becomes increasingly smaller, except for very large angles of sweep.

Figure 2.148 illustrates the separate effects of aspect ratio and taper ratio. It can be seen clearly how much an increase in the angle of sweep affects the variation of lift-curve slope with aspect ratio. It shows further that while the taper ratio, as compared

Fig 2.148   Effects of aspect ratio, sweep-back angle and taper ratio on lift curve slope $\Delta C_L / \Delta \alpha$.

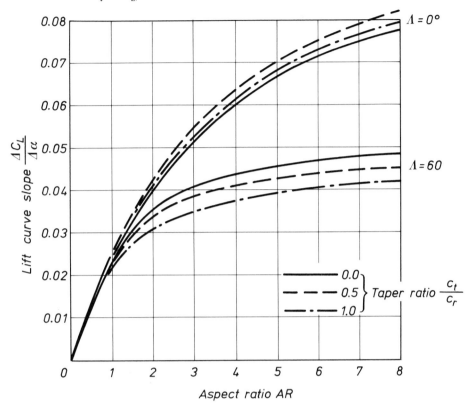

to aspect ratio, has only a small effect on the lift-curve slope of an unswept foil, the taper ratio has a predominant effect on the lift-curve slopes of highly swept foils of moderate to high aspect ratios. For very small aspect ratios AR below 1.5, however, the lift-curve slopes of all the foils converge and become almost a linear function of aspect ratio, being essentially independent of the effects of sweep and taper.

At the end of the previous section it was mentioned that the lifting line theory has certain limitations in its applicability. As a matter of fact this theory proved inadequate when used to predict the characteristics of foils having appreciable angles of sweep and/or low aspect ratio. 'Low aspect ratio' implies a foil with values of AR normally less than 2.5. Referring back to Fig. 2.87 we should perhaps remind ourselves of the fact that in the development of the lifting line theory it was assumed that each section of the large AR foil all along its span acts independently of its neighbouring sections except for the induced downwash. Strict compliance with this assumption would require the chord or streamwise two-dimensional flow, as presented in Fig. 2.84, which implies no variation of section shape and hence pressure along the lifting line, which is in principle perpendicular to the undisturbed flow direction. When these requirements are satisfied, as in the case of an untwisted elliptical foil of large AR, the local lift coefficient C_l along the foil span is uniform and the two conditions of constant local C_l coefficient and minimum induced drag go together. This is no longer true on the swept, low aspect ratio foils for the following reasons:

Firstly–the powerful tip vortices increasingly influence the flow round the whole foil when the aspect ratio becomes progressively smaller. With foils of higher aspect ratio the cross-flow affects only a relatively small tip portion of the span, thus the decrease in suction over the foil caused by this flow is almost negligible. With very low aspect ratio foils, however, the relative effect of the cross-flow becomes dominating.

Secondly–the cross-flows of the boundary layer developing on low aspect ratio foils and amplified on swept foils, in which case the lifting line is not straight but makes an angle at the root section, destroy the basically two-dimensional flow existing on high aspect ratio straight foils.

Thirdly–it is not correct to consider a foil or, more specifically, a keel, entirely exclusive of hull, because the existence of the hull at its root section contributes to the lift and also affects the flow round the appendage.

(e) *Slender body theory: implications and shortcomings*
To cope with these apparent deficiencies of the lifting line theory a number of complementary theories have been developed, such as 'extended lifting line theory', 'lifting surface theory' or 'slender body theory' to cover foils of small aspect ratio and arbitrary planform. A well referenced summary of these theories is given by B Thwaites (Ref 2.15). The common feature of all these concepts is that, inevitably, they are more complicated than the simple lifting-line theory and one must admit that their inherent mathematical difficulties render them useless from the viewpoint of the yacht designer. For our purpose it is sufficient to say that the so-called slender body theory, notably that recently developed by Newman and Wu (Ref 2.105) considers quite an interesting configuration of a slender foil, such as that shown in Fig 2.149B, intersecting a relatively large yacht-like body. Happily, some effort has been made by Kerwin, Herreshoff and others (Ref 2.106) to utilize the selected bits of complicated and still deficient 'science' incorporated in slender body theory directly for design ends.

Let us now look more closely at some remarkable properties of slender, delta-like foils of the type shown in Fig 2.149B, a practical application of which is demonstrated in Fig 2.149C. These contrast sharply with those of straight foils of large aspect ratio. If the slender body theory is applied to a highly tapered foil with swept leading edge attached to the hull it will be found that the lift arises only on the forward triangular front part of the foil-hull combination, i.e. that part which is to the right of the dotted line marked X in Fig 2.149C. It is presumed that lift contribution of the remaining wetted area to the left of the line X is negligible. According to theory, the lift force is given by the expression:

$$L = C\rho\pi V^2 b_0\alpha \qquad \text{Eq 2.39}$$

Fig 2.149 Comparison of local lift coefficients C_l at various spanwise
stations throughout angle of incidence range. Re = 2.4 × 10⁶
based on mean chord.

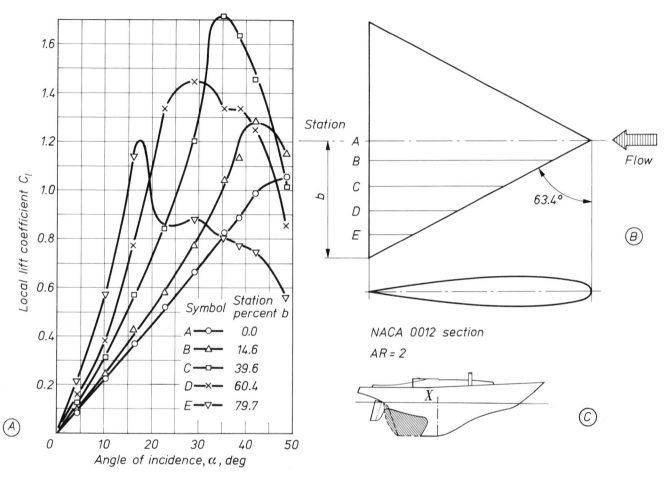

where C = lift coefficient depends on the ratio r_0/b_0, r_0 being the depth of the hull
and b_0 the draft, as shown in Fig 2.150A.
ρ = water density
π = 3.14
V = boat speed
α = angle of incidence (leeway)

Evidently, the magnitude of lift is proportional to the square of the keel draft and
not to the amount of its lateral plane. Numerous tests, amongst them those
published in Ref 2.106, indicate that indeed the slender, low aspect ratio appendages
operate somewhat that way. However, the tests also revealed that by no means all the
hidden mechanism which determines the character of the flow, and therefore the

Fig 2.150   Lift coefficient C of a yacht hull-like body with single fin (no rudder).

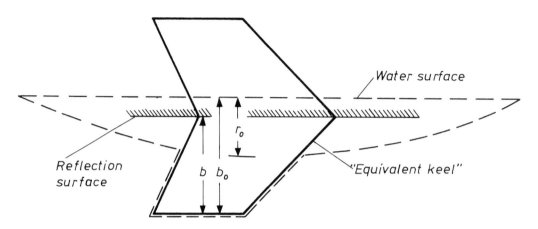

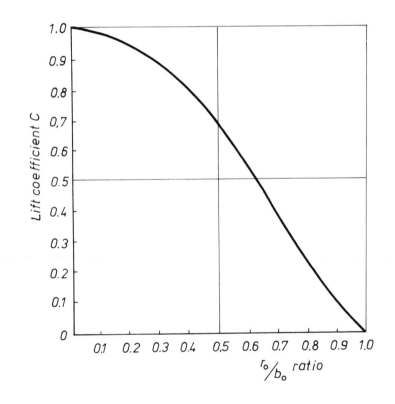

Ⓐ   *Underwater view of yacht showing "equivalent keel" for $r_o/b_o = 0.5$*

Ⓑ

properties of a slender foil interacting with a hull has been uncovered.

Figure 2.150B discloses the theoretically established dependence of lift coefficient C on r_0/b_0 ratio, by which the mutual contribution of the fin and hull to lift is reflected. In the limiting case when r_0/b_0 becomes 0, which corresponds to a flat-bottom hull of negligible depth r_0, the value of coefficient C would be 1. This decreases to a value of 0.69 for $r_0/b_0 = 0.5$, a rather common configuration illustrated in Fig 2.150A. In the same figure there is also shown the concept of the so-called 'equivalent keel', which was developed in order to facilitate both the theoretical and experimental study of the separated keel but accounting also for the hull contribution to lift. The argument, given in Ref 2.106A, is as follows. Suppose that we use the slender-body theory to find the draft b of an 'equivalent keel' which produces the same lift as the combination of hull and keel for a particular value of r_0/b_0, as shown in Fig 2.150A. For $r_0/b_0 = 0.5$ we can find from Fig 2.150B and Eq 2.39 that:

$$b/b_0 = \sqrt{0.69} = 0.83$$
so $$b = 0.83\,b_0$$

As seen in Fig 2.150A, the equivalent keel of span b protrudes into the hull to a point approximately two-thirds of the distance from the hull-keel juncture to the water surface.

The induced-drag coefficient of a slender foil is given by the expression:

$$C_{\text{Di}} = \frac{C_{\text{L}}^2}{K\pi\text{AR}} \qquad \text{(Eq 2.28A repeated)}$$

The value from the above equation is often used as a standard of comparison for the induced drag of a foil which is not so shaped as to produce minimum induced drag. The efficiency factor K given by the theory and never greater than 1.0, depends on foil geometry. Efficiency $K = 1.0$ would indicate a keel with minimum induced drag corresponding to elliptical spanwise loading. Apart from the K factor the theory makes it possible to calculate the lift curve slope $\Delta C_{\text{L}}/\Delta\alpha$ which in turns allows one to plot a graph of C_{L} versus α and estimate the lift increase with incidence angle.

As an example, Fig 2.151 demonstrates the results of computations based on the lifting surface theory and equivalent keel concept (Ref 2.106). In this case the effective aspect ratio of the keel AR = 1.62 obtained by reflecting its planform about the root section (rigid reflection surface), as shown in Fig 2.150A, was held constant. The taper ratio $c_t/c_r = 0.66$ was also constant and the only variable investigated was the sweep-angle Λ–ranging from 0° to 51°. Table 2.13, next to the four sketches in Fig 2.151 illustrating the planforms of keels, gives the computed values of the lift-curve slope $\Delta C_{\text{L}}/\Delta\alpha$ and efficiency factor K. In addition there are tabulated the friction drag D_f, the induced drag D_i and total drag D_t in kilograms for a specific choice of keel area of 3.84 sq m (41.3 sq ft) and lift force corresponding to an Admiral's Cup yacht sailing to windward at a speed V_s of about 6.5 knots. The leading dimensions of this yacht, together with its hull lines, are also presented in Fig 2.151.

Fig 2.151 Hydrodynamic properties of four keels with different sweep angles from 0° to 51°, at constant taper ratio $c_t/c_r = 0.66$ and constant effective aspect ratio AR $= 1.62$ (twice the geometric aspect ratio).

The drag components are calculated on the assumed keel area $A = 3.84$ sq m (41.3 sq ft) and boat's speed $V_s = 6.74$ knots in close-hauled conditions.

Table 2.13

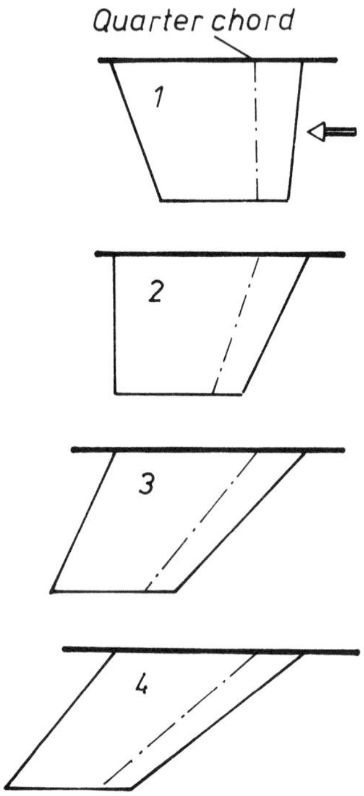

Quarter chord

Λ degrees	c_t/c_r	$\dfrac{\Delta c_L}{\Delta \alpha}$	K	D_f kg	D_i kg	D_t kg
0	0.66	0.0368	0.998	23.55	29.73	53.28
20.5	0.66	0.0368	0.996	23.55	29.79	53.34
41.0	0.66	0.0359	0.993	23.55	29.88	53.43
51.0	0.66	0.0342	0.987	23.55	30.06	53.61

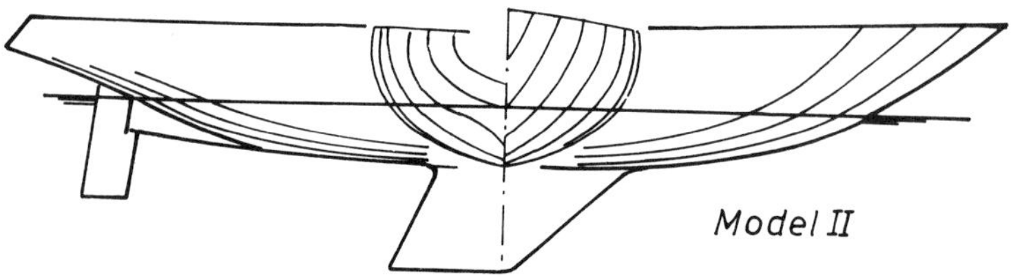

Model II

The frictional drag D_f does not come from the aforementioned theory but was calculated on the assumed drag coefficient

$$C_f = 0.0085 + 0.0166 \, C_L^2$$

based on the NACA data of Ref 2.31.

By comparing the results of Table 2.13 one might easily jump to the conclusion that although the hydrodynamic differences between the four keel configurations appear to be trivial in magnitude, the planform 1 with no sweep is the best. The calculations reveal that the lift-curve slope $\Delta C_L / \Delta \alpha$ is reduced by about 8 per cent and induced drag is increased by about 1 per cent as the sweep angle increases from $0°$ to $51°$. This trend agrees with Fig 2.146 from which it can be seen that the keel configuration 1 approaches the optimum combination of c_t/c_r and Λ more closely than other keel planforms 2, 3 and 4.

But does such a conclusion hold true for the complete keel-hull combination moving on a free air-water interface? A confrontation of the above results with some towing tank and full-scale tests seems to point to the general conclusion that, in reality, the keel does not always respond to the above theoretical predictions. It must be borne in mind that the results presented in Fig 2.151 rest on the assumption that the fin action is confined to the fin itself; no matter whether the actual fin geometry is corrected by means of the equivalent keel to take the hull into account, or not. The whole concept of equivalent keel is based on the assumption that the keel-hull junction is a reflection surface, therefore the flow at the root section of the keel is two-dimensional. In fact, the hull-keel junction may not be flat, moreover the water surface above is a deformable surface and the waves generated by both the hull and keel interact. The presence of the curved bottom of the hull modifies the flow at the root section of the keel through a change in local speed and direction. The resultant pressure effects of the hull and its appendages manifest themselves as surface waves which are different at each hull attitude and speed. All these facts make the problem of a slender body-foil combination, developing lift and drag in a free-surface flow, exceedingly complex.

The towing tank tests have revealed that the resulting resistance of the fin-body combination is much more sensitive to sweep angle variation than suggested by the theory and furthermore in the opposite sense to that implied in Table 2.13. One should not feel too despondent regarding this, however; after all, it is a well established fact in aerodynamics, as well as in other branches of science, that the whole is more than the sum of its parts. The yacht hull is a system of interdependent parts and every change in any one part influences the other parts and also the whole.

Despite these difficulties, which were recognized quite early in the history of the scientific investigation of yacht performance, they are still unsolved and much open to speculation. The idea that the water surface can be regarded as a mirror in which the underwater portions of the hull and appendages are reflected, to give a double model of an effective aspect ratio which is twice the geometric aspect ratio, as illustrated in Photo 2.36B, was apparently first suggested by Davidson (Ref 2.109). If accepted, the most apparent characteristics of the yacht motion, the bow and stern waves shown in Photo 1.7A and B in Part 1 and the following wave train which spreads out in a characteristic pattern behind the stern, would have to be ignored. In the discussion following the presentation of Davidson's paper objections were raised by Von Karman, who pointed out that the discrepancies in measured and calculated drag of model yachts can only be explained by assuming that a part of the wave drag is the induced drag effect due to keel action.

The original wording of Von Karman on this subject, published more than 40 years ago, is worth quoting more extensively since it is still relevant.

'Considering the hull as an aerofoil with small aspect ratio, Professor Davidson found that the induced drag is equal to about twice the amount predicted by the aerodynamic theory. The discussor believes this discrepancy is due primarily to the fact that the author considers the immersed portion of the hull as the half of an aerofoil; as a matter of fact, the aerodynamicists would agree to this assumption only in the case of a fluid jet between rigid walls. Because of the presence of the free water surface, the case analogous to the yawing boat hull is that of an aerofoil submerged into a jet of infinite cross-section limited by a free surface with constant pressure. Calculating the induced drag for this case, the theoretical value is slightly larger than twice the value given by Professor Davidson, and is in accordance with the experiment.'

The two assumptions are illustrated by Fig 2.152 and, to continue Von Karman's comments,

'That it is incorrect to substitute a "half aerofoil" for the boat hull is easily seen in the limiting case of a boat of "infinite draft". Professor Davidson's formula [which used the standard aerodynamic expression for the induced drag in the form of Eq 2.28A, i.e. $C_{Di} = C_L^2/\pi AR$ (evidently, when the AR in the denominator gradually increases approaching infinity AR $= \infty$, the C_{Di} value approaches zero)], would give zero induced drag in this case because of the infinite span. However, it is evident that the elevation and depression at the water surface at the two sides of the yawing boat must be connected with some kind of resistance corresponding to the kinetic energy produced per unit time. It is rather a matter of terminology whether this resistance be called induced drag or wave resistance. The physical fact is that dynamic lift cannot be maintained without transferring momentum permanently into the fluid; hence in every second new kinetic energy must be produced and stored up in the wake. This energy is equal to the work done by the drag. Considering the motion in the wake as a kind of wave motion, we call the drag corresponding to the work done wave resistance; considering the motion as circulation motion we call the same resistance induced drag.'

It was reported by T Tanner (Ref 2.110) that investigations carried out in an attempt to clarify this matter appear to indicate that the assumption of the 'double

Fig 2.152   *View of trailing vortices from behind:* both foils A and B are set at an incidence angle at which they develop a lift force.

Case A represents the double model concept, in which the water surface is regarded as a rigid wall. This idea is similar to the mirror-image presentation in Fig 2.114.

Case B represents the trailing vortices configuration in a free-water surface flow. The surface of the water is now being induced to travel up (elevation) and down (depression) while the fin passes through. The 'depression' in the sketch represents the water level on the suction side of the foil and the 'elevation' illustrates the water level on the pressure side. This disturbance in the water level on the suction side is clearly seen in Photo 1.7B. The surface waves or up-and-down pendulum-like motion can be set up whenever the water particles are displaced above or below their at-rest or datum position by any disturbance.

The general nature of the fin action piercing through the water surface can be well demonstrated by holding a sheet of glass partly immersed in a stream. If held at a small incidence angle the difference in the level of the water surface can easily be distinguished.

The disturbances of the free-water surface due to the passage of a body, be it hull or fin, can be resolved into two kinds–(1) local disturbance, such as that shown in Photo 1.7B, whose dimensions (contour) are determined by the size and shape of the body. (2) wave train which spreads behind the body in a characteristic manner, and is also visible along the hull, as depicted in Photo 1.7A. The existence of a keel underneath the hull implies some modification to the wave pattern generated by the hull itself. It should be expected that this modification to the wave shape and its amplitude, as seen along the hull side, will be strongly dependent on the keel loading, i.e. lift generated and heel angle of the boat. Photograph 2.37 illustrates this point.

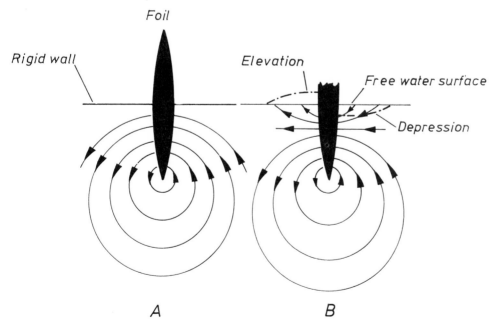

Photo 2.37   Model of a yacht being tested at Davidson's Laboratory, USA, at 3 different angles of heel: 10°, 20° and 30°. The resultant wave pattern of the hull is mainly due to sources of disturbance, one situated near the bow, second near the stern, and third at the midspan where the keel is attached. These are the regions where the water flow velocity and pressures change rapidly. Any modifications to the shape of these parts of the hull, for example, by adding a 'bustle' to the afterbody, or by fore-and-aft shift of the keel, or by changing keel rake, may quite dramatically affect the resulting wave pattern and wave drag. A study of wave contour along the hull at different speeds and heel angle can be very helpful in assessing the advantages of eventually introduced modifications to the hull shape.

model' can lead to grossly underestimated values of both lift-curve slope $\Delta C_L/\Delta \alpha$ and induced drag.

The calculated theoretical values of the efficiency factor K in Table 2.13 are very close to unity and suggest that foils of small aspect ratio possess elliptic, spanwise loading and thus minimum induced drag, no matter whether the sweep angle is large or not. However, according to Ref 2.15, it is seldom found that measured drag values conform to the theoretical prediction. The reason is that the theory does not take viscous effect, and therefore premature separation, into account. In real flow conditions the separation normally occurs not only along the trailing edge, as assumed in the theory, but also locally along the leading edge, and this implies a higher local lift in some places, so that the drag is higher too.

The measurements of local lift coefficient C_l at several stations along the span of highly swept slender foil of AR $= 2$ demonstrated in Fig 2.149A lend more than adequate support to this point. The five curves show the variation of local lift coefficient with incidence angle for each of the sections indicated in sketch B. In contrast to the linearity of lift versus α curves for larger aspect ratio foils, shown for example in Figs 2.103 and 2.104, the measured local lift curves for low aspect ratio foil are non-linear; and even at small angles of incidence they bend upwards. The non-linearity is greatest at the outboard section where the rate of increase of the lift coefficient with incidence, reflected by the lift-curve slope $\Delta C_L/\Delta \alpha$, is also greatest. For instance, at zero lift the slope of the lift-curve varies from about 0.05 at the outboard section to about 0.02 at the root section. At, say, incidence $\alpha = 15°$ the section E nearest the tip is much more loaded than the other sections towards the root, and is the first to stall. It can be found in Fig 2.149A that while the tip section E stalls at just above 15° incidence, the stall angle of the root section A is about 45°. The previously mentioned cross-flow in the boundary layer is apparently responsible for this favourable, one may say, anti-stalling effect, which determines so strongly the local lift produced by the inboard sections. The boundary layer entrainment is presumably drained off these sections, thus delaying separation and allowing them to support a higher load than that they would be able to support if the flow were two-dimensional, i.e. chordwise only.

The effects of the three-dimensional flow are so great that neither the chordwise pressure distributions nor the lift characteristics of the foil with large sweep-back are predictable from the two-dimensional data, such as that given in Ref 2.31.

To make this point of vital practical importance clear; one might expect, for instance, that a keel of a given NACA 0015 section and aspect ratio, say 2, will have the lift and stalling characteristics of that given by the two-dimensional data after correction for aspect ratio. The collected data in Table 2.12 item numbers 2, 5, 8, may even be felt to justify this expectation. It can be seen from it that the stall angle is reached when the lift coefficient C_{Lmax} is in the order of 1.2–1.3. This information may, however, be misleading if the sweep angle effect is forgotten and only the aspect ratio is taken into account. As a matter of fact, the foil of the same aspect ratio but of triangular planform with high angle of sweep, as depicted in Fig 2.149B, stalls when the total lift coefficient C_L is in the order of 0.8 only. The tendency towards reduced C_{Lmax} for the foil with sweep-back already demonstrated in Fig 2.144 is apparently exacerbated by the larger sweep angle. This is due to the already mentioned differences in spanwise distribution of lift having its maximum value farther and farther out towards the tip, as the sweep angle increases.

In conclusion, one may say that as far as delta-like foils are concerned the three-dimensional flow effects appear to be so powerful that any attempt to exploit the relatively small differences in section characteristics, such as those demonstrated for instance in Fig 2.58, becomes meaningless. Since the shark-fin type of keel of small wetted area, as shown in Fig 2.149C, is similar to the delta planform, one should expect a similar kind of response, an early stall and high drag once the stall took place. When downwind rolling develops, such keel characteristics become of enormous practical importance.

The theory of the slender foil, initially developed in aeronautics by Jones (Ref 2.108) and so distinctively different from the lifting line theory, according to which lift depends on the foil area, certainly stimulated radical modifications in the underwater part of sailing yachts. This theory gives a clear hint that there is no point in paying an 'unnecessary' friction drag penalty for the large wetted area of a traditional long-keel planform since lift is independent of the area. By carrying this argument to its apparent logical conclusions the shark-fin configuration, shown in Fig 2.149C, was bound to be invented. In fact, in the past 25 years the relative size of keels and rudders has shrunk quite dramatically. In some cases of the fashionable separated keel-rudder configuration, which superseded the traditional long continuous keel-rudder planform, the wetted area is reduced to one-third of the traditional planform.

Of course, there is a limit in reducing the wetted area of a low aspect ratio, slender appendages for a given draft or span. In Ref 2.106 we find clarification of this point–'If we were to test a hull having a keel with an extremely long chord and gradually shorten the chord, we would find experimentally that the lift would initially be practically unchanged. However, as the keel aspect ratio increased sufficiently as a result of shortening the chord, the lift would eventually decrease. This contradiction is due to the fact that the assumptions of the slender body theory are violated if the aspect ratio of the fin becomes too high...', and then the high aspect ratio theory begins to be valid so the foil area becomes again an important factor determining the available lift.

No doubt, in some circumstances drastic reduction in the wetted area of a keel may be advantageous. It may lead to better performance in rather light weather conditions, when steady motion prevails and the friction drag matters most. However, one must expect some unpleasant side-effects when tacking in waves, or when potential rolling conditions are expected. An over-small keel of shark-fin type can then easily lose its advantages. The reason is apparent, while in irons the boat decelerates rapidly so her actual speed V_s may become a fraction of the normal speed before tacking. In order to regain speed on the new tack quickly the keel must develop an ample lift at the minimum possible induced drag. This is not an easy task in a condition when the forward speed V_s has been reduced substantially and lift according to Eq 2.39 depends on the velocity squared V^2. Sufficient lift can, according to the same Eq 2.39, only be developed at high incidence angle (leeway) α, at which, in turn, stall may occur incurring a heavy induced-drag penalty and limiting available lift. Both factors delay acceleration of the boat on the new tack causing a typical vicious circle due to the conflicting requirements.

Without being excessively nostalgic or apologetic towards the past generation of sailors and yacht designers, one must recognize the fact that the separated keel-rudder concept was also quite popular at the end of the 19th century. Figure 2.153

Fig 2.153 Separated keel-rudder configurations were quite popular at the end of the 19th century.

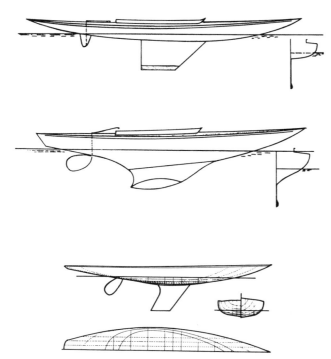

shows some of the almost traditional planforms of the hull of that time. These boats had the reputation of being '...very quick on the helm, a quality which however useful in racing vessels is undesirable in cruising boats' (Ref 2.111). Apparently, those boats with good steering qualities had rather poor course-keeping and in some extremes, poor anti-rolling characteristics and this may be very tiring for the crew on a long run. Subsequently, following the cruiser-racer idea, which gradually gained impetus, and in which speed, manoeuvrability and seaworthiness were regarded as the trinity of a vessel's virtues, the separated rudder-fin configuration was abandoned, until its recent re-introduction.

References and notes to Part 2

2.1   *Yachting Monthly* F E M Ducker, *About Fins*, 1943.

2.2   Pressure measurements on the earth's surface, which is at the bottom of our atmosphere, can only be made relative to the atmospheric pressure. This pressure, called sometimes 'barometric pressure', changes depending on the dominating weather pattern. The total variation in atmospheric pressure is in the order of ± 5 per cent of the average pressure. For the sake of having a reference base for pressure measurements a 'standard' or 'normal' atmospheric pressure was established at sea level,

     1 atm = 14.7 lb/in^2 (psi), which is equivalent to 2116.2 lb/ft^2.

Pressure gauges and other pressure measuring devices indicate pressures called 'gauge pressures' relative to this atmospheric pressure regarded as 0 psi gauge, often abbreviated psig.

The gauge pressure (psig) is positive if it is greater in magnitude than atmospheric, and is negative if less than the atmospheric reference. Negative gauge pressure may be called vacuum. The condition of no pressure at all, which is equivalent to negative pressure of -14.7 psig, is called absolute zero. Thus we have a second reference from which pressure can be measured, absolute zero or 0 psia, so the atmospheric pressure is actually 14.7 psia.

2.3   *Conference on Yacht Design and Research* T Tanner, ACYR–March 1962, University of Southampton. Section 2. *Basic Principles of Aerodynamics and Hydrodynamics.*

2.4   *On the Triple Origin of Air Forces* Max Munk, J of Aeron Sciences, 1938. (M Munk is regarded as pre-eminent scientist in history of aerodynamics.)

2.5   *Mastery of the Air* Sir G Sutton, Hodder and Stoughton, 1965.

2.6   *Essentials of Fluid Dynamics* L Prandtl, 1952, Blackie & Son.

2.7   *Modern Developments in Fluid Dynamics* Vol II, Ed S Goldstein.

2.8   *The Flettner Rotor Ship, Engineering*, January 1925.

2.9   *Aeroplane Design–Performance* E P Wagner, McGraw-Hill, 1936.

2.10   Frederick W Lanchester, born 1868, educated at the Hartley Institution at Southampton, now the University of Southampton, made several extraordinary discoveries in the field of aerodynamics. In 1897 he presented to the Physical Society of London a paper in which the concept of the origin and nature of lift generated by the aerofoil (a term coined by him) was outlined. Unfortunately,

Fig 2.154   Definition of positive and negative pressure.

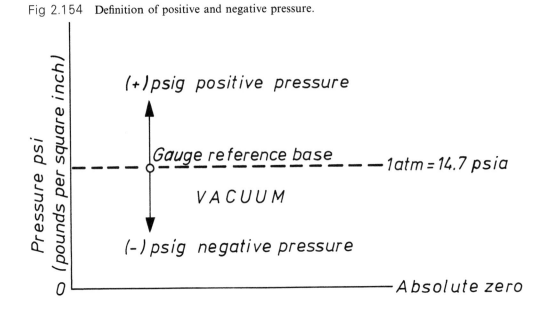

the paper was rejected by the referees. In 1907 Lanchester managed to publish his book *Aerodynamics* in which he developed the vortex theory of the finite aerofoil.

Strangely, his originality and pioneering approach were recognized in Germany well before his theories were accepted in England. L. Prandtl of Göttingen, Germany, paid tribute to Lanchester's work when delivering the Wilbur Wright memorial lecture to the Royal Aeronautical Society some 50 years ago. The following words are taken from Prandtl's lecture: 'Lanchester's treatment is difficult to follow, since it makes a very great demand on the reader's intuitive perceptions, and only because we had been working on similar lines, were we able to grasp Lanchester's meaning at once.'

2.11   *Basic Wing and Aerofoil Theory* A Pope, McGraw-Hill, 1951.

2.12   *The Elements of Aerofoil and Airscrew Theory* H Glauert, Cambridge University Press, 1959.

2.13   *Growth of Circulation about a Wing and Apparatus for measuring Fluid Motion* P B Walker, R and M No 1402.

2.14   ARC Reports and Memoranda, Aeronautical Research Council, England–No 1353, 1931.

2.15   *Incompressible Aerodynamics* Editor B Thwaites, Oxford, 1960.

2.16   We shall occasionally deal with changing quantities, and it seems desirable to introduce a shorter way of writing 'change of...with respect to...', or rate of change. Choosing the symbol Δ, a capital Greek letter D, pronounced *delta*, which conventionally stands for the 'difference' (increment or decrement), we may express the pressure gradient, according to its definition, rate of change of pressure Δp with distance Δx by:

$$\frac{\text{change of pressure}}{\text{change in distance}} = \frac{\Delta p}{\Delta x}$$

read 'delta' p over 'delta' x.

Inspecting the sketch given in Fig 2.155 below we may notice that between the leading edge of the foil and the point C at which the peak of suction (negative pressure) occurs, the pressure is falling, as indicated by the pressure distribution over the back surface of the foil. Such a pressure drop due to higher flow velocity has been already demonstrated in Fig 2.16C. Considering the two points A and B along the back surface of the foil and the relevant negative pressures p_1 and p_2

respectively operating there, we may express the pressure difference between those points using 'delta' notation,

$$-p_2 - (-p_1) = -p_2 + p_1 = -\Delta p$$

Minus in front of Δp reflects the fact that the numerical value of p_2 is higher than p_1, i.e. $\Delta p < 0$.

In this condition, the pressure gradient, i.e. the rate of change of pressure Δp between two points A and B, separated by a small distance Δx, is also negative $(-\Delta p/\Delta x)$ and as such will facilitate the fluid flow from higher pressure region A towards the lower pressure region B. It is customary to *call such a negative pressure gradient, a favourable or an accelerating gradient*. The Ancients used to say that nature abhors a vacuum, but they really had in mind the idea that nature loves uniformity. If there are any inequalities of pressure the fluid will rush to smooth them out. We meet examples of pressure gradient in operation every day. For instance, when the meteorologist announces that a depression, or area of low pressure, is moving across the Atlantic and the winds will soon increase to gale force over the country, he means that a large pressure gradient exists over this part of the globe and that as a consequence, air is rushing from regions of high pressure to those of low pressure.

Referring back to Fig 2.155, it can be seen that the slope of the straight line passing through the ends of pressure vectors p_1 and p_2, and tangent to the pressure distribution curve rising in the direction the fluid flows, can be regarded as a graphical representation of pressure gradient. The steeper the slope of this tangent-line the higher is the pressure gradient, i.e. higher acceleration of the flow may be expected. The slope of this tangent-line is not always the same but depends on the point chosen along the pressure distribution curve. The pressure gradient is highest near the leading edge of the foil section and the tangent-line is there vertical. The gradient then gradually decreases downstream reaching zero when the suction peak occurs at point C and the tangent-line there is parallel to the axis (chord) of the foil.

In a precisely analogous fashion as before we may define the positive pressure gradient $\Delta p/\Delta x > 0$, which operates downstream from the point C. As fluid particles continue their journey, passing the peak of pressure drop towards the tail, they are facing, apart from retarding viscous force, another kind of decelerating force, now produced by the rising pressure.

Inspecting the conditions at points C and D, we find that $\Delta p = -p_4 - (-p_3) = -p_4 + p_3$ is positive since the numerical value of p_3 is higher than p_4. The pressure difference Δp is now positive. It indicates a pressure rise along the path C–D-trailing edge. The tangent-line to the pressure distribution curve, falling in the direction of fluid flow, is a graphical representation of a positive pressure gradient. It is customary to call such a positive pressure gradient an unfavourable, adverse or retarding gradient.

As before, the steepness of the tangent-line represents the magnitude of the positive gradient. The higher the pressure rise (higher pressure gradient) and steeper the tangent-line, the more pronounced will be the retarding action of the gradient, resulting in rapid deceleration of the flow velocity inside the boundary layer.

2.17 In fact, the hull of a sailing yacht, for example, experiences other kinds of resistance to motion. The wave drag is perhaps the most conspicuous, but one may also distinguish the so-called induced drag, produced by appendages such as fin-keel and rudder. The induced drag occurs whenever lift or side-force is generated on the appendage, therefore it is induced by the lift force. Evidently, for objects which exhibit no lift, the induced drag will be zero.

There is also interference drag. An interference at the hull-keel discontinuity, hull-rudder, and other such junctions leads to modification of the boundary layer flow over the isolated hull and appendages.

2.18 *The modern System of Naval Architecture* London, 1865.

2.19 The theory of boundary layer assumes a non-slip condition, i.e. the velocity of a fluid immediately adjacent to the solid surface is presumed to be zero, so that with a surface having even a glossy finish the fluid motion is entirely one of slippage of fluid over fluid. A simple physical demonstration of the essential truth of what may, at first sight, be a rather startling assumption, is afforded by the collection of layers of dust on the body of cars driven very fast. This fact provides a clear indication that the boundary layer concept is not merely an invention of the theoreticians, but a physical reality. The velocity in a fluid flow increases very rapidly outward along a

Fig 2.155   Definition of pressure gradient.

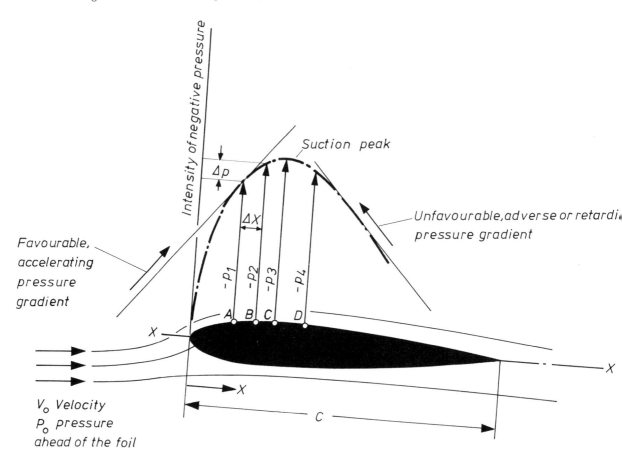

perpendicular (normal) to a solid surface, until it attains the full free-stream value V_0, at a certain distance from the wetted surface. This distance or boundary layer thickness increases gradually downstream.

One common definition of the thickness of boundary layer, δ, is the distance from the surface for which the velocity V within the boundary layer is 99 per cent of the velocity V_0 outside of BL.

For further information concerning boundary layer and viscous friction the reader is referred to:

Boundary Layer Theory–Dr H Schlichting, McGraw-Hill.

It is one of the most comprehensive publications on the subject.

2.20 'Viscosity' as a quantity, which used to concern only physicists in the past, became quite popular as a term in everyday language. It determines one of the most important characteristics of a motor oil. To become more familiar with viscosity not merely as a number specifying an oil, but as a very real physical property, let us anticipate a simple experiment. Move a thin metal plate rapidly edgewise, first through water and then through a lubricating oil, and note the difference in the forces required for the purpose in the two cases. The force, be it large or small, is called skin friction, and experiment leaves no doubt that it is a function of viscosity that can also be regarded as a measure of the ease with which a fluid will flow.

Viscosity, or internal friction, is large for heavy-bodied lubricating oil, and small for watery fluids, like water itself or air. A term combining both the viscous and density properties of a fluid is defined as kinematic viscosity:

$$v = \frac{\mu}{\rho} \text{ square ft/sec}$$

where μ = coefficient of viscosity (lb sec/ft^2)
ρ = density of a fluid (Tables 2.1 and 2.2).

2.21 It was reliably reported that a barracuda of about 4.0 ft in length can swim at 27.0 m.p.h.
A dolphin of about 6.5 ft length, swimming close to the side of a ship, was timed at better than 22 m.p.h. The relevant speed-length ($V/\sqrt{L}$) ratios, for barracuda and dolphin are of order 13.5 and 9.0 respectively. The barracuda seems to be the fastest known swimmer. A man, however, can only swim at 4 m.p.h. Further details can be found in:
Fish propulsion in relation to design R W Gawn, RINA, 1949.

2.22 *Wind Tunnel Tests of Rigging Wires for 12-Meter Yachts* J W Hollenberg, Davidson Lab Technical Mem 140.

2.23 *Aerodynamic Theory* Vol. IV, Ed W F Durand, Dover Publications.

2.24 *Thoughts on Windage Rigging and Spars* B Chance Jr. *The Land's End Yachtsman's Equipment Guide,* 1967.

2.25 *Shape and Flow* A H Shapiro, Heinemann, London.
A splendid introductory account of the basic concepts and terms of aerodynamics and hydrodynamics.

2.26 *Aerodynamics* T von Karman, McGraw-Hill, 1963.

2.27 *Boundary Layer and Flow Control* Vol 2, Ed G V Lachman, Pergamon.

2.28 *Vortex Generators: their Design and their Effects on turbulent BL* ARC Rep 16487, 1954.

2.29 *Symposium on Sailing Yacht Research* MIT Rep No 6 68–10, November 1966.

2.30 *Airfoil Section Characteristics as affected by Variations of the Re Number* NACA Rep 586, E N Jacobs and A Sherman.

2.31 Extensive information about aerofoil sections can be found in: *Theory of Wing Sections* I H Abbott and A E Von Doenhoff, Dover Publications, Inc, New York.

2.32 *An Experimental Investigation of the Circumstances which determine whether the Motion of Water shall be direct or sinuous, and the Laws of Resistance in parallel Channels* O Reynolds, Phil Trans Roy Soc London, Vol 174, pp 935–982 (1883).

2.33 *Laminar BL Oscillation and Stability of Laminar Flow* G Schubauer and H. Skramstad, J A Sc Vol 14 pp 69–78 (1947).

2.34 *Laminar BL Oscillations and Transition on a Flat Plate* G Schubauer and H Skramstad, NACA Rep 909 (1943).

2.35 *Visualization of the Effect of some Turbulence Stimulators* I van den Bosch and I Pinkster, Rep 293, Technische Hogeschool Delft, 1971.

2.36 *The Prediction of Yacht Performance from Tank Tests* W A Crago, RINA, 1962.

2.37 *Low drag aerofoils* L G Whitehead, Journ Roy Aer Soc, 1946.

2.38 *An Experimental Study of a Series of Flapped Rudders* J Kerwin, P Mandel, S Lewis, *J of Ship Research,* December 1972.

2.39 *Viscous–Viscosity.* Water is a viscous fluid, and when it is set in motion by any system of forces and then left to itself, it comes to rest by virtue of an internal friction which tends to resist the sliding of one part of water over another. This internal friction is called viscosity which is a measure of the resistance to flow.
If a state of steady motion of any water-borne craft is maintained, the viscosity of the water (viscous friction) tends to oppose the motion and leads to dissipation of energy as heat, just as does the friction between solid bodies.

2.40 *Polymer–Polymerization.* Polymerization in chemistry is considered to be any process in which two or more molecules of the same substance unite to give a molecule (polymer) with the same percentage composition as the original substance (monomer) but with a molecular weight which is an integral multiple of the original weight of the monomer. The length or diameter of most molecules is so small that they are not visible. Some polymers, such as the Polyox WSR 301

Poly(ethylene-oxide), are large enough to be seen with a microscope. The size of Polyox long-chain-molecules is a sphere roughly 1500 Å in diameter; if pulled out, the length of a Polyox molecule would be 750,000 Å, i.e. 0.0075 cm (1 Å = 1 Ångstrom = 1/100,000,000 = 10^{-8} cm).

Molecular weight of polymers is very large compared with a molecule of water which has a molecular weight of 18 times the weight of a hydrogen atom. Some polymers may have a molecular weight greater than one million times the weight of a hydrogen atom.

2.41 *A Towing Tank Storm* K Barnaby, RINA Vol 107, 1965.

2.42 *Model Experiments using Dilute Polymer Solution instead of Water* A Emerson, North East Coast Inst of Eng and Shipbuild, Vol 81/4, February 1965.

The Skin Friction Bogey J Paffett, *The Naval Architect*, April 1974.

2.43 *Polymers and Yachts* Private correspondence–A. Millward, Southampton University, 1968.

2.44 *Turbulence suppression and Viscous Drag Reduction by Non-Newtonian Additives* T Kowalski, RINA, 1967.

2.45 *Swimming speed of a Pacific Bottle-nose Porpoise* T Lang and K Norris, *Science* 151, 1966.

2.46 *Hydrodynamic of the Dolphin*
M Kramer, Advances in Hydrosc 2, pp 111–130 (1965).

2.47 *Swimming of Dolphins* G Steven, *Science Progress* 38 (1950).

2.48 *Boundary layer Stabilization by Distributed Damping* M Kramer, J of Aerospace Science 24 (1957).

2.49 *Turbulent Damping by Flabby Skins* D Fisher and E Blick, J Aircraft Vol 3, March/April 1966.

2.50 *Turbulent BL Characteristics of Compliant Surfaces* E Blick and R Walters, J Aircraft Vol 5, January/February 1968.

2.51 *Evaluation of High Angle of Attack Aerod Derivative Data and Stall Flutter Predict Techn* R Halfman, H Johnson, S Haley, NACA TN 2533.

2.52 *Examples of three representative types of Airfoil-section Stall at low speed* G McCullough and D Gault, NACA TN 2502.

2.53 *Low-speed Flows involving Bubble Separation* I Tani, *Progress in Aeronaut Science* Vol 5

2.54 *On the Theory of Wing Sections with particular reference to the Lift Distribution* Th Theodorsen, NACA Rep 383, 1930.

2.55 *Wind Tunnel Tests on a 3-dimensional Curved-plate (sail-type) Aerofoil* Terrence Miller, Part of the program for BSc(Eng), communicated by T Tanner.

2.56 *Zur Aerodynamik de kleinen Reynoldszahlen* Jahr 1953 der WGL.

2.57 *Aerofoil Sections* Dr Fr W Riegels, Butterworths, 1961, London.

2.58 *Class C Racing Catamarans* Maj Gen H J Parham, A Farrar and J R Macalpine-Downie, RINA, 1968.

2.59 *Wind Tunnel Test of a 1/3rd scale model of an X-One Design Yacht's Sails* C A Marchaj, Southampton University Rep 11, 1962.

2.60 *Wind Tunnel Tests on a Series of Circular-arc Plate Airfoils* R A Wallis, Australian Aer Res Lab, Note 74, 1946.

2.61 *Padded Sails* J Nicolson, *Yachts and Yachting*, January 1972.

2.62 *A review of Two-dimensional Sails* A Q Chapleo, Southampton University, Rep No 23, 1968.

2.63 *A Survey of Yacht Research at Southampton University* T Tanner, Journal of the Roy Aer Soc, October 1962.

2.64 *Generalisation of the Condition for Waviness in the Pressure Distribution on a Cambered Plate* E A Boyd, Journal Roy Aer Soc, August 1963.

2.65 *Further Tests on Two-dimensional Sails* W G Robbins, BSc(Eng) Thesis, University of Southampton.

2.66 *Theory of Flight* R Mises, Dover Publications, Inc., 1959.

2.67 *The Aerodynamics of Yacht Sails* E Warner, S Ober, Transaction SNAME Vol 33, 1925.

2.68 *The Aerodynamic Theory of Sails* I, Two dimensional sails. B Thwaites, Proced Royal Society A, Vol 261, 1961.

2.69 *Etude Aérodynamique d'un élément de pale d'hélicoptère* P Poisson-Quinton, AGARD Conference Proceeding No 22, September 1967.

2.70 *The Aerodynamic characteristics of Airfoils as affected by Surface Roughness* R W Hooker, NACA TN No 457.

2.71 *Sails* J Marshall, *One Design and Offshore Yachtsman*, December 1969.

2.72 *The Tacit Dimension* M Polanyi, Garden City NY, 1966.

2.73 *Hazard for Aircraft* by Wetmore, NASA TN D-1777.

2.74 *Theoretical Hydrodynamics* L M Milne-Thomson.

2.75 *Hydrodynamics in Ship Design* Vol I H Saunders, The SNAME, NY, 1957.

2.76 *The Rolling Up of the Trailing Vortex Sheet and its Effect on the Downwash Behind Wings* J Spreiter, A Sacks, JAS Vol 18, p 21.

2.77 One may be surprised to learn that according to experiments performed by N Piercy and reported in his *Aerodynamics* (The English University Press, 1964), the centre of the trailing vortex core may revolve at over 18,000 rpm. Such a high speed of the vortex core implies a very low pressure inside the vortex. This in turn may produce beautiful effects in air (under special conditions of air humidity) appearing somewhat like threads of silvery steam seen behind the wing tips of aircraft flying high.

2.78 See series of articles written by A Gentry, published in 1973 *Sail* Magazine.

2.79 *Applied Hydro-aerodynamics* L Prandtl, O G Tietjens, Dover Publications, Inc., 1957.

2.80 The meaning of Eq 2.24, $L = m \times 2w$, discussed in section 3a and derived by implication of the momentum concept is that the dynamic lift cannot be maintained without transferring momentum permanently into the fluid. Hence in every second new kinetic energy must be taken from the fluid stream (it is the wind energy in the case of a sail) and stored up in the wake. The quantity of kinetic energy ΔE deposited in the wake per unit time can be expressed in the conventional way as:

$$\Delta E = \frac{m \times v^2}{2} \qquad \text{Eq R.1}$$

where m is mass of fluid
v is velocity

Referring again to Fig 2.101 and applying the same notation it may be found that the mass of fluid affected by the foil each second is

$$m = \rho A V_0 \qquad \text{Eq R.2}$$

This mass of fluid is pushed downward at right angles to the original direction of flow V_0 reaching velocity

$$v = 2w \qquad \text{Eq R.3}$$

Substituting expressions 2 and 3 into Eq R.1 gives

$$\Delta E = \frac{\rho A V_0 (2w)^2}{2} = \frac{\rho A V_0 4w^2}{2} = 2\rho A V_0 w^2 \qquad \text{Eq R.4}$$

This energy is equal to the work done by the induced drag Di per unit time, i.e.

$$\text{Work}_{\text{(per unit time)}} = Di V_0 = \Delta E$$

In other words, the work done on the foil in three-dimensional flow conditions or energy given by the flow stream is equivalent to the kinetic energy lost in the wake in the form of downward momentum. Induced drag found from the above expression is

$$Di = \frac{\Delta E}{V_0} = \frac{2\rho A V_0 w^2}{V_0} = 2\rho A w^2 \qquad \text{Eq R.5}$$

Substituting expression R.2, i.e.

$$m = \rho A V_0$$

into Eq 2.24

$$L = m 2w$$

yields

$$L = \rho A V_0 2w \qquad\qquad \text{Eq R.6}$$

hence the ratio of

$$\frac{Di}{L} = \frac{2\rho A w^2}{2\rho A V_0 w} = \frac{w}{V_0} = \alpha_i \text{ (in radians)} \qquad\qquad \text{Eq R.7}$$

and this is confirmed by the geometrical relationships already drawn in Fig 2.100.

It is known from observation of the generation of momentum $m \times v$ in a fluid that the total kinetic energy of a group of fluid particles with a given average velocity will always be smallest when all the particles individually have velocities equal to that average, i.e. when the velocities of all the particles are the same.

This observation, let us quote Rauscher (Ref 2.81), '...is of interest in the study of propelling devices whose action involves the generation of momentum in a fluid. Clearly, a propeller handling a given mass of fluid in unit time, and deriving from this fluid mass a given thrust by imparting to it a given average velocity, will require the least input of work when that average is the common velocity of all the elements in the slipstream.'

Foils such as sails work on a similar principle and the relation between the work done by induced drag and the energy extracted from the fluid is of particular interest in connection with sail action which involves the creation of momentum. In figure 2.156, the sketch A illustrates the two different distributions of downwash, *one* for elliptic loading in which case the downwash is uniform and *another one* for non-elliptic loading (in fact for a highly tapered foil) in which case the downwash is non-uniform but its average value w is the same as for the elliptic loading. Since the average downwash is the same in both cases, the lift will be identical, however the induced drags will be different.

As shown in Eqs R.4 and R.5 both the induced drag as well as the kinetic energy extracted from the stream are proportional to the downwash velocity squared. Sketch B illustrates this point graphically. The area of rectangle ABCD represents the flow energy lost in the wake when loading of the foil is elliptic. The crossed area of the figure ABE represents the energy lost when loading is non-elliptic. This area is greater than that of ABCD by the amount given by the dotted area FGE which represents the additional energy lost due to non-uniform distribution of downwash. Since drag is a measure of energy taken out of the flow, the induced drag will increase in the same ratio as the energy lost. The fact that uniform downwash results in a minimum induced drag can be understood through the observation that the downward momentum imparted to the passing stream by the lift reaction carries with it a minimum of kinetic energy. Hence, it requires a minimum of work to be done by the foil as a deflecting agent, when the mass of fluid, be it water or air, experiencing the deflection receives uniform downwash velocity. The distributions of downwash and energy in the two sketches A and B are given for half of the foil span. Provided that there is no gap between the two halves the other half will have identical distributions.

2.81 *Introduction to Aeronautical Dynamics* M Rauscher, J Willey, 1953.
2.82 *The Aerodynamic Characteristics of a 2/5 scale Finn Sail and its Efficiency when Sailing to Windward* SUYR Rep No 13 by C A Marchaj.
2.83 *The Aerodynamics of Sails* J H Milgram, 7th Symposium on Naval Hydrodynamics, 1968.
2.84 *The Application of Lifting Line Theory to an upright Bermudan Mainsail* T Tanner, SUYR Rep No 16.
2.85 *Effect of a Chordwise Gap in an Aerofoil of Finite Span in a Free Stream* B Lakshminarayana, Journal Roy Aer Soc, April 1964.
2.86 *Wind Tunnel Tests on a 1/4 Scale Dragon Rig* C A Marchaj and T Tanner, SUYR Rep 14, 1963.
2.87 *Aerodynamic Theory* Vol IV, Ed W Durand, Dover Publications, Inc.
2.88 *Aerodynamics of High-Performance Wing Sails* J O Scherer, *Marine Technology*, July 1974.
2.89 *Some Aerodynamic Characteristics of Tapered Wings with Flaps of Various Spans* H Irwing, A Batson, I Warsap, H Gummer, R and M, N 1796, 1937.
2.90 The term 'local lift coefficient', designated by C_l as distinct from the total lift coefficient C_L, requires some explanation. The total lift coefficient C_L of the foil of a given shape (defined by its planform, aspect ratio, taper ratio, twist, camber distribution, etc.) is the mean or average

Fig 2.156 Distribution of downwash *w* and kinetic energy lost in the
wake of the foil. The area FGE crossed with dots represents the
additional energy lost in the wake: this is the case when the
load is not elliptic. Note that the areas ABCD and ABGFD are
equal.

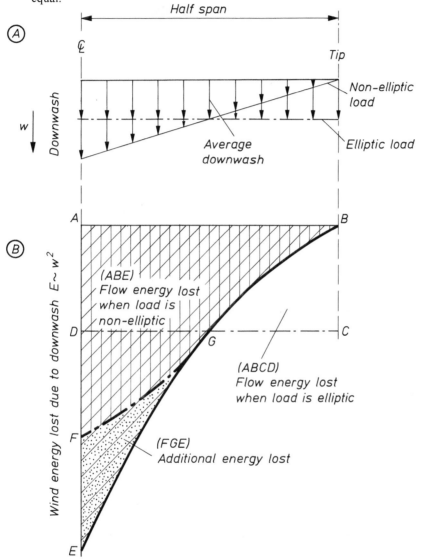

coefficient for the complete three-dimensional foil. Its value will depend upon the local lift
coefficients C_l of each section of the foil. In the case of an elliptical planar foil depicted in Fig 2.123
the local lift coefficient C_l is uniform across the span, therefore the toal lift coefficient C_L is equal to
C_l.

For other foil shapes this may not be the case simply because the local lift coefficient depends
upon the effective angle of incidence α_{ef}, the camber of each section of the foil and also upon the
other induced effects, as illustrated in Fig 2.119. It can be seen for example that at some sections

along the foil span the local lift coefficient C_l reaches the value in the order of 1.2. However, the total lift coefficient C_L of the whole foil, which can be roughly estimated from Fig 2.119, will not exceed 1.0.

What has been said about the differences between the local C_l and total C_L lift coefficients applies equally well to other foil coefficients, namely:

drag coefficient,
driving force coefficient,
heeling force coefficient,
and resultant force coefficient.

In order to differentiate between the coefficients that refer to a local section of the foil and those that refer to the complete foil, the lower-case subscripts are used for the former, i.e. C_l, C_d, C_r, C_h and C_t (see Fig 2.129) while the upper-case subscripts are used for the latter, i.e. C_L, C_D, C_R, C_H and C_T respectively.

2.91 *On the Stalling of Highly Tapered Wings* C Millikan, Journal of Aer. Sc. Vol. 3 No 5, 1936.
2.92 *Determination of the Characteristics of Tapered Wings* R F Anderson, NACA Rep 572.
2.93 *Induction and Intuition in Scientific Thought* P B Medawar, Methuen and Co Ltd, 1968.
2.94 *A Simple Approximation method for obtaining the Spanwise Lift Distribution* O Schrenk, NACA TM 948.
2.95 *The Chinese Junk and the Chinese Rig* Brian Platt, *Yachts and Yachting*, July 1961.
2.96 *Design Charts relating to the Stalling of Tapered Wings* H Soule and R F Anderson, NACA Rep 703.
2.97 *Catamaran Sails* A Farrar, *Yachting World*, January 1966.
2.98 *Ocean Racing* E Tabarly, W W Norton, New York.
2.99 *Characteristics of Clark Y Airfoils of small AR* C H Zimmerman, NACA Rep 431.
2.100 *Maximum Lift Data for Symmetrical Wings* T Nonweiler, *Aircraft Engineering* January 1955.
2.101 *Fluid Dynamic Drag* S F Hoerner, 1958.
2.102 *The Cobles* I W Holness, *Yachting World*, March 1969.
2.103 *A Review of Aerodynamic Cleanness* E J Richards, Journal Roy Aer Soc, 1950, pp 137–144.
2.104 *Aircraft Stability and Control* A W Babister, Pergamon Press, 1961. See also NACA Rep 921–*Theoretical Symmetric Span Loading at Subsonic Speeds for wings having arbitrary plan form* J De Yong and Ch Harper.
2.105 *A generalized Slender-Body Theory for Fish-like Forms* J N Newman and T Y Wu, Journal of Fl Mech Vol 57, p 4 (1973).
2.106 A–*Sailing Yacht Keels*
    J E Kerwin, H C Herreshoff,
    HISWA, 1973
    B–*Yacht Hull Research*
    J E Kerwin,
    MIT, Rep No 68–10
    C–*The influence of Fin Keel Sweep-back on the Performance of Sailing Yachts*
    W Beukelman and J Keuning,
    HISWA, 1975
2.107 *Chordwise and Spanwise Loadings measured at Low Speed on a Triangular Wing having an Aspect Ratio of two and an NACA 0012 Airfoil Section* B H Wick, NACA TN 1650.
2.108 *Properties of Low-aspect Ratio pointed Wings at Speeds below and above the Speed of Sound* R T Jones, NACA Rep 835, 1946.
2.109 *Some Experimental Studies of the Sailing Yacht* K S Davidson, SNAME, 1936.
2.110 *Yacht Sailing Close-Hauled* T Tanner, Correspondence–Journal Roy Aer Soc, July 1965.
2.111 *Yachting* The Badminton Library, Vol. II, London 1894.

PART 3

Research on sails: practical implications

'Experience is the name everyone gives to their mistakes'

OSCAR WILDE

Lady Windermere's Fan

'Our whole problem is to make the mistakes as fast as possible...'

J A WHEELER

American Scientist

A Speed performance prediction: scope and limiting factors

The late Lord Brabazon of Tara, at one time British Minister of Aircraft Production and also member of the Advisory Committee for Yacht Research, expressed an opinion that '...the designing of aircraft is child's play compared with the difficulties of the sailing craft.' This view, held some years ago, may be used today to accentuate the complexity of the task and the difficulties the modern sailmaker or sail designer is facing.

Unlike the aeroplane, whose development resulted from the close co-operation of scientists and technologists, the modern racing yacht has been evolved almost entirely by the concerted efforts of enthusiastic skippers, designers and sailmakers with little direct reference to basic scientific principles. In fact, yacht designing and particularly sailmaking, have been working to a rule of thumb–a very good rule of thumb based on hundreds of years of accumulated experience. The only disadvantage to this method being that reasons *why* certain factors contribute to successful design remain somewhat obscure or misplaced.

In order to understand the fundamental factors governing yacht performance one may reasonably look at the sailing yacht as:

1. A combination of two distinct systems, one of which is the *aerodynamic* (rig), and the other the *hydrodynamic* (hull with appendages), whose performances can be considered and measured separately, or

2. A complex dynamic system consisting of two interdependent parts, aerodynamic and hydrodynamic, in which case each part is the cause and effect of the other part, and of the whole system; and the system itself is the cause and effect

Photo 3.1 Which is the best hull?
Tank test models of 12-Metre type yacht from the period just before Sceptre-Columbia Challenge, 1958. To build the representative models and to test them in a towing tank is, no doubt, less costly than an analogous procedure with full-size hulls. Picture presented with kind permission of the British Hovercraft Corporation Ltd, Cowes, Isle of Wight, England.

of its parts. In other words, the whole system's characteristics are more than just the sum of the characteristics of its parts. Hence, the resulting performance cannot be estimated directly from model results by a simple, straightforward process.

The first approach is more tempting than the second, since it suggests the possibility of relatively uncomplicated, direct comparison between different rigs, or different hulls that can, for example, be developed and measured separately in the wind tunnel and towing tank respectively. As a matter of fact, the standard method of assessing the likelihood of success by limited tank testing and used in the development of more mundane cruising yachts is nothing but a relative comparison.

An immediate practical purpose of such limited testing (see, for example, Photos 3.1 and 3.2) is to determine whether one proposed design is better than another, either proposed, or already constructed. In this case, advantage is taken of the knowledge that for any but mediocre sailing performance, the hull in the close-hauled attitude must have a resistance R no greater than somewhere between a third and a quarter of the hydrodynamic side force F_s, generated at a given angle of leeway

Photo 3.2   A. Above–*Intrepid* (1971) modified by B Chance.
            B. Below–*Intrepid* (1967) originally designed by O Stephens.
              Tested in the towing tank at the same angle of heel.
            Differences in the stern wake are apparent; they are due to the
            modifications shown in the two sketches. Shaded hull indicates
            Intrepid–1971, the dotted line shows contour of the original
            Intrepid–1967.

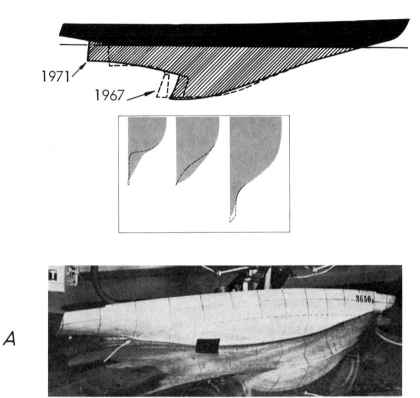

(Ref 3.1). The attainable F_s/R ratio in the range of hull attitudes, corresponding to close-hauled sailing, can be regarded as one of the criteria of hydrodynamic efficiency of the hull.

Figure 1.12, Part I, depicts the relevant hydrodynamic forces, as usually measured in the towing tank. The model of a yacht is run in the tank at constant speed over a range of leeway and heel angles and the resulting variations in side force F_s and resistance R are recorded. In some cases the ratio of side force to resistance may not even reach the minimum desirable value of 3, so that means must be sought to increase the side force, say by deepening the keel, by reducing the resistance, or both. In other cases, an acceptable ratio may be achieved, but at a high angle of leeway. In this instance, the yacht designer may choose to allow some development to take place in an effort to improve the design. While no absolute measure of the competitive performance is obtainable by this means it is at least possible to ensure that a *non-sailer* is not produced.

If sailplan and its characteristics are not considered with the above simplified technique, the tank tests cannot, except by good fortune, predict the true full-size speed performance of the yacht. Even the comparative merits of two hulls can only be correctly determined when they are driven by identical sails. However, one can sensibly argue that, even if two identical hulls are driven by the same inventory of sails (and this is quite large in number on modern ocean racers), those identical yachts may manifest different performances in the same wind and sea conditions. This is because the available sails can be set in numerous combinations each with shapes (defined by camber and twist) which can be altered substantially by sheeting, halyard tension, Cunningham holes, mast bend, zippers etc; thus (as we will see in later chapters) producing different aerodynamic driving and heeling force components. Inevitably, the same hulls are bound to respond differently to unequal aerodynamic input yielding unequal speed performances.

The recent development of methods of measuring the aerodynamic characteristics of sails in the wind tunnel, as illustrated in Photo 3.3, prompts the question: what are the comparatively simple means of making a relative assessment of the results of such tests, even if the answers so obtained do not represent the best possible attempt at estimating the merits of different rigs in an accurate quantitative sense? Various methods have been suggested, notably those by Spens (Ref 3.2) and Tanner (Ref 3.3), which understandably involve special simplifying assumptions. They explicitly indicate that, as in the interpretation of towing tank tests, progressing too far with reasoning based on wind tunnel testing alone, divorced from detailed characteristics of the complementary part (the hull), may bring disappointing results. To take an extreme case, it might be demonstrated, for instance, that good all round performance cannot be obtained with plain, rigid or semi-rigid sails set on a traditional, displacement-type hull, although such rigs, shown in Photos 2.26, 2.30 and 2.31, Part 2, may successfully be applied for specific types of hull, designed for specific purposes.

As mentioned earlier, the essence of the second alternative approach to the

Photo 3.3   Model of 12-Metre rig in the wind tunnel at Southampton
University.
    P Spens of Davidson Laboratory USA, poses the question–
'How can one decide from their aerodynamic characteristics
whether one rig or another will give the better performance
to windward, on a hull of known hydrodynamic characteristics?'

estimate of sailing yacht performance is that the craft is regarded as one complex
aero-hydrodynamic machine, so that the boat speed cannot be determined from the
characteristics of its aerodynamic and hydrodynamic components alone, i.e.
without going through performance calculations in which the nature of the
interconnection or interaction of both parts of the yacht is reflected. The ultimate
purpose of it would be to predict the absolute, as opposed to the relative,
performance of a given design. As might be expected, to achieve such a goal a great

deal of sail and hull data are required. Furthermore, the speed estimate is complicated by the fact that hull behaviour is affected by a multiplicity of other than design factors, of ever-changing comparative dominance and associated with unsteady character of the real wind, the state of the sea, the course sailed relative to waves etc. When every relevant factor is taken into account, the prediction of speed performance would become, in the mathematical sense, a laborious, if not impossible, task. Some simplification must be made and the many methods of performance prediction, which have been proposed since Davidson's method, differ only in their simplifying assumptions and the particular techniques employed in solving the problem (Ref 3.4). All of these methods have points for and against them, and it must be emphasized that in no case has an entirely satisfactory check been made against the actual behaviour of a boat.

After all, one should realize that any performance prediction, based on towing tank and wind tunnel results, is subject to inevitable error and uncertainty. They arise both from difficulties in scaling-up the model results, and from the fact that the test conditions do not simulate the actual full-scale conditions in all respects, and therefore the influence of dissimilarity cannot be foreseen with certainty. By its very nature, a good workable model can be constructed only upon a limited number of specific postulates and empirical assumptions. Some factors or conditions must be neglected on a relative basis as leading to small or negligible effects, otherwise the model and associated methods of testing and speed prediction programmes might easily become misleadingly precise in the guise of *exact* knowledge.

Since predictions and conclusions derived from model testing are valid only with the particular set of assumptions stipulated, a change in any of the assumptions, or unintentional omission of the relevant factor (simply because of lack of sufficient knowledge), will affect the final performance estimate. Considering now the impressive room for disagreement in assessing the full-scale performance of sailing craft from model testing, one must admit that the generally good agreement found is rather satisfactory.

On the other hand, one must also admit that there is evidence which came to light in the course of much-publicized design difficulties in the sophisticated 12-Metre class, that towing tank tests may be all too easily misinterpreted. It has been claimed, for instance by Kirkman and Pedrick (Ref 3.5), that one can observe severe scale effect on the hydrodynamic side force which the filled-out afterbody can create on small models; tank tests may thus lead some designers to select a keel that is too small. It has been reported that the 1970 crop of 12-Metres all had extremely full bustles, and some designers, it is said, '...did not believe that model scale performance was reproduced on the race course'. This experience helped Olin Stephens to change his thought when designing *Courageous* and move backwards in the evolutionary process towards his 1967 *Intrepid* designs, shown in Photo 3.2.

However, one cannot help but wonder whether, in tank tests on boats with a fat stern underbody, any effort was made to determine the possibility of flow separation in the afterbody, which can be done with tufts attached to the hull surface as shown

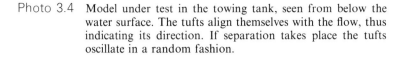

Photo 3.4   Model under test in the towing tank, seen from below the
water surface. The tufts align themselves with the flow, thus
indicating its direction. If separation takes place the tufts
oscillate in a random fashion.

in Photo 3.4. Such separation is, as mentioned earlier in Part 2, section C, dependent
on Reynolds Number and therefore affected by scale. This means that if a model is
so formed that the separation is so well aft as to be insignificant, then premature
separation should not occur on the full-size hull.

When assessing reliability of the model-test results, one must bear in mind both
the accuracy of the required answer and the inherent test errors, which largely
depend on the scale of the model used. For example, the *America's* Cup trials are
frequently '…decided by one-half per cent margin of the elapsed time with two to
three per cent being considered as a rout!' Many tests on the commercial model sizes
indicate that 5 per cent test error is not infrequent, and this is significant considering
the tiny margin by which the races are won or lost. One feels intuitively that the
larger the model the more reliable the results; Fig 3.1 which is taken from Ref 3.5
supports this view. It can be seen that as the model size is decreased the uncertainty
band becomes wider and for, say, a model 8 ft long, the correlation error may be in
the order of ±5 per cent. So the best model may not in fact be the best prototype.
However Fig 3.1 takes no account of the wide ranging possibility for differences in
sophistication of measuring apparatus, methods of turbulence stimulation (to which
reference was made in Part 2), correction for stimulator drag penalty, blockage
correction for oversized models and other factors which affect the final estimate.

Fig 3.1   Variation of correlation error with size (defined by its LWL) for
          upright resistance. Similar trend can be observed in the error
          variation for the hydrodynamic side force.

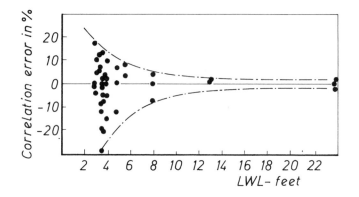

Besides, as pointed out by Pierre De Saix in discussion of Ref 3.5, full-scale data
should be treated with the same healthy scepticism as model tests. And since the full-
scale results obtained outside the laboratory generally show scatter of ± 20 per cent,
any attempt to demonstrate a correlation error using full-scale data as a base that
can scatter so much is not valid. It should be added that the observed large scatter in
measuring the full-size characteristics is no reflection on any particular investigator
but is a comment on the difficulty of making such tests outside the laboratory.

It is perhaps worth mentioning that, because of the cost involved, the present
practice is to confine the use of performance prediction methods to such craft as
12-Metres or highly competitive boats such as Admiral's Cup contenders. Usually,
their performance is expressed in terms of speed made good to windward, as shown
in Fig 1.29, or more rarely in the form of a complete speed polar diagram, such as
that presented in Fig 1.28, Part 1.

Moderate optimism with respect to the possibility of speed prediction based on
specific model tests is justified by past records. However, the claim made by the
scientists who have recently developed the USYRU/MIT computer program
designed, to quote from Ref 3.6, '...to predict the speed of an ocean racing yacht of
any size and almost any type, beating, reaching or running in winds ranging from
light air to a moderate gale', sounds incredibly over-optimistic. They believe that
this goal can be achieved by use of the Massachusetts Institute of Technology's
electronic device for recording hull lines. These, with some information concerning
the sailplan, stability and displacement of a given yacht, are sufficient to be
introduced as an input into the computer. Subsequently, the Velocity Prediction
Program (VPP) of the computer, which 'sails' any yacht, of any design, under
various wind conditions, carries out the task of predicting speed on any course
relative to the wind of any strength.

As explained in Ref 3.6, the data base of VPP consists of results obtained both by theoretical means and by model and full-scale experiments on hulls and sails. As far as the hydrodynamic part is concerned, the characteristics of the hull used in the VPP originated from experiments made with a systematic series of scale models of about 7 ft long in a towing tank. The models were all derived mathematically from a single parent hull, which was considered to be a good design with average proportions. The derivative models were then designed with a variety of modifications to the basic shape, that is, they were of either heavier or lighter displacement, wider or narrower beam, shoaler or deeper etc. From these tank data the mathematical equations were derived for the resistance of each hull, which can be readily scaled up to any size yacht of the same configuration. Figure 3.2 shows, for example, a typical plot of hull resistance, including that due to side force, drawn by computer for a yacht of 35 ft waterline length and of a given set of hull lines. The other information needed by the computer to carry out a speed performance prediction are the height of the centre of gravity, the sail area and the height of the centre of effort.

Since the aerodynamic theory alone could not provide all the necessary information about the complementary sail forces, these were derived by assessing the full-scale sailing performance on the assumption that they must be equal and opposite to the hull forces. The authors of this VPP project believe that the speed of ocean racing yachts can be predicted with reasonable precision; what is meant by this is not defined. However, one may expect that the results obtained from the VPP computer will be less precise than the speed predictions based on specific model tests of a particular yacht, particularly one which possibly incorporates breakthrough design features.

An interesting by-product of the VPP is that '...the computer predicted speed differences between a well-sailed and poorly-sailed version of the same boat were comparable with the speed differences among yachts of widely varying design.' This leads us to the frequently-asked question as to which is the crucial race-winning factor: an outstanding boat, or a first rate crew? Probably, as in One Design classes, level rating racing is bound to put a high premium on a good crew. In handicap racing a good crew also matters, but to a lesser degree. Not infrequently, mediocre prototype boats sailed by an outstanding crew are capable of scoring conspicuous successes in regattas just by virtue of crew expertise. Success obviously serves well to advertise the new design which, once it becomes a production boat, often shows disappointing qualities. Conversely, the true worth of a boat may be masked by bad handling.

It is intended to use the VPP '...to analyse the existing rating rules and time allowance formulae, thus providing one means of guiding future modification. In addition, it can be used as a rational starting point in developing entirely new rule formulation.' This new approach to the old problem of estimating speed potential of racing boats stems from much of the discontent with the International Offshore Rating Rule (IOR). Particular disgust is directed against expensive stripped-out

Fig 3.2   Typical plot of hull resistance for two-degree increments of heel
angle θ. The LWL of the yacht is 35 ft. The curve corresponding to
zero heel angle represents the case of running downwind. Looking
at other curves which include the induced drag due to side force
one may notice that the resistance increases not only at higher
boat speed V_s but in the lower speed range as well. The
explanation is as follows: when tacking, the boat may lose her
forward speed V_s substantially. To generate the same side force on
a new tack at much lower speed may require such a large leeway
that the keel will stall and the hull will merely drift to leeward and
eventually be slowed by it.

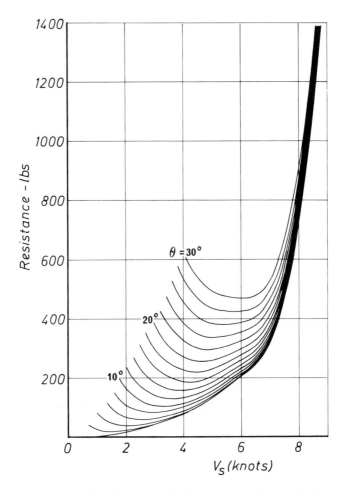

boats crammed with pricey hardware including electronics and a large wardrobe of
sails which has, it is said, been encouraged by the IOR.

As reported by Peter Johnson in *In the Offing* (*Yachting World* 1976), the intention
of the new USYRU rule backed up by the VPP is to:

1. Reduce obsolescence;
2. Avoid another revolution by handicapping existing yachts fairly;

3. Encourage dual purpose cruising and racing boats;
4. Adjust rating factors within each yacht to varying weather, and thus discourage extreme types of racing boats;
5. Match each factor as nearly as knowledge permits to its effect on speed;
6. Provide for quick rule changes without new measurements;
7. Control excessive costs.

No doubt, referring to the conclusions expressed in the introductory chapter in Part 1 of this book, the above praiseworthy package appears to contain some factors which can hardly be reconciled. As to whether computers, which helped put men on the moon, will make yacht racing people happier remains to be seen. However, when searching for future trends, before costly decisions are taken and offshore owners submit themselves to the computer centres, one should not ignore past experience and, particularly, the changing psychological attitude of the sailing fraternity towards yachts and racing.

Offshore racing is and will remain a game played to certain rules, and if full satisfaction and enjoyment is to be experienced by the competitors, the rating rule should estimate accurately enough the relationship between rating and a boat's potential speed performance, thus providing good, even racing for one crew against another in direct competition. However, an estimation of a boat's potential performance cannot be done accurately with any simple formula. The known speed-producing and speed-reducing factors are too numerous to be taken into account and, unfortunately for rule makers, their number is progressively increasing as our knowledge of sailing improves.

It may come as a surprise to some readers to learn that, at this stage in the development of mathematics and hydrodynamics, the resistance of even an ordinary ship form cannot, as yet, be calculated accurately enough by applying mathematical formulae. In fact, most of today's knowledge about ships and yachts too is based upon experiments, and therefore the solution of practical problems is still largely empirical.

From past experience one may conclude that, however much rule makers or handicappers have striven after equitable offshore racing, its accomplishment was and still is not feasible by simple and cheap methods. This is why a large proportion of yachtsmen have become tired of racing against another man's ability to wriggle through measurement rules and handicaps. Those who have tried handicap racing have not been convinced by all the flights of their handicappers' mathematical fancy; with some exceptions perhaps–those who are actually winning. It was recently reported that '...currently the yachting cult in America is the cruiser. The word racing is fun but race boat becomes a dirty word.'

Dissatisfaction with handicapping gave an enormous impetus towards level or fixed rated class racing, in which there is no need for a handicap. A notable stimulus for this boat-for-boat racing was given by the One-Ton Cup concept, shortly followed by the Half- and Quarter-Ton Cup competitions. It appears that level racing is more rewarding: a length ahead of a rival is a length ahead, without any magic number or computer in between the crews involved in the duels.

The level rating concept is however not a new remedy; history repeats itself somewhat in this respect. About fifty years ago the unhappy offshore men, dissatisfied with the unjust rating and handicap formulae, invented what they called Metre classes, which are equivalent to the fixed rating classes so much advocated nowadays. When those Metre classes were accepted with the blessing of the IYRU, the racing people abandoned them almost immediately and returned to the unjust handicap system. Apparently, there must be a certain gulf between the dreams of rationally minded people and the people who only pretend that they think in a rational way. Often, unfortunately, they are the rule makers.

Going through the writings of the famous American yacht designer L F Herreshoff (*The Common Sense of Yacht Design*) one may find the following explanation of this almost chronic offshore malaise:

> 'One-design classes may be all right for young democrats who are labouring under some inferiority complex or who are scared that someone with better mental equipment will take advantage of them, but as a matter of fact their only hope of winning is to race in an open class where, if they pick the right designer, they will have an advantage that will make up for their other deficiencies...in an open class the young democrat can blame his defeat on the designer, and this often gives him some satisfaction, but in a one-design class he has no redress whatsoever.'

Apparently, the problem of equitable racing was quite acute in Herreshoff's time too. Recently, when discussing Herreshoff's explanation with American friends, the author of this book was assured that nowadays '...rule beating in the United States has not been the exclusive province of either Democrats or Republicans.'

Undoubtedly, the principal source of the trouble seems to be the disregard of reality or, more precisely, of human nature. If the amateur sport of yachting is plagued with cheating or rule-beating, it is mainly due to the psychological fact that perhaps every racing sailor, as a human being, is torn apart by a dilemma: everybody wishes to adhere to an ideal concept of fair play, yet surreptitiously very many are prepared to take any action to put their hands on the tiller of a breakthrough or underrated boat. Cheating, perhaps one of the most frequently used words in yachting magazines, as pointed out on some occasion by Bernard Hayman, editor of *Yachting World*, is a somewhat unkind word because of the association with, say, cheating at cards. By the *Yachtman's Guide* definition 'Rule cheating is not so much cheating as trying to obtain best advantage from the legal wording or rule. The result is a reduction in rating or increase in speed, out of all proportion to the related measurements' (Ref 3.7). So, by this definition, no one can suggest, for example, that J Milgram, when designing *Cascade*, was keeping the spare jokers up his sleeve. This outstanding specimen *Cascade*, shown in Photo 3.5, deserves to be remembered as the most spectacular rule-cheater of our time. In a way she is a symptom of revolt against the tacitly agreed set of norms and against the

Photo 3.5 The extraordinary *Cascade* designed by J Milgram threw the rule-makers into a state of considerable confusion. She rated 22 feet on 30 feet waterline length. Fast for her rating, she was slow for her length.

establishment. This is the reason why her success infuriated many of the old guard yachtsmen, and stirred up a kind of 'mass hysteria' in some quarters, as reflected in Refs 3.8 and 3.9.

Cascade rated 22 ft on her waterline length of 30 ft. She was fast for her rating; not surprisingly since she was about 25 per cent longer than any boat of the same rating with a conventional rig carrying headsails, and therefore rated for the foretriangle. *Cascade* had no foretriangle, but she was carrying enormous unrated sail area. This enabled Milgram to keep her rating low enough to offset the disadvantage due to the intrinsic miserable windward performance of this cat-ketch rig. In fact, *Cascade* was a pretty poor performer for her size measured by her actual waterline length and, as such, cannot be regarded as heralding an improved breed of yacht. *Cascades* do not make the lives of IOR men easy, although they may delight those revolutionary, loop-hole minded and competitive enthusiasts who think that the sole object of sailing is to beat their rivals by all means, or to ridicule the establishment to gain notoriety.

In his book *The Act of Creation* Koestler says: 'Man cannot inherit the past; he has to recreate it.' In other words, it appears that every generation has to rediscover the expectations and disappointments, answers and methods of the past. Judging by the recent development in modern offshore racers, our generation of sailors seems to be as confused as that living at the end of the 19th century. The only consolation is that perhaps we are confused at a higher level of understanding of the factors involved. Some excerpts from the Badminton Library volume *Yachting*, published in 1894, may serve to illustrate one more point.

By the end of 1892, a group of famous yacht designers, amongst them Fife, Nicholson and Watson, gathered in London to discuss the undesirable, extravagant, and costly trends in yacht evolution, induced by the rating rule in operation. The discussion resulted in a letter sent to the Council of the Yacht Racing Association, in which the following view was expressed.

'We take it that the general yachting public require in a yacht: that she shall be safe in all conditions of wind and weather; that she shall combine the maximum of room on deck and below with the minimum of prime cost; and that she shall be driven as fast as may be with the least expenditure of labour– i.e. that she shall have a moderate and workable sail area. Therefore, as but few men can afford to build for racing, and for racing only, and as the racer of to-day is the cruiser of a few years hence, any rating rule should by its limitations encourage such a wholesome type of vessel.'

The Council of the YRA, however, took the view that '...what the yacht-owning public want in a racing yacht is speed, and speed at any price'. Subsequently,

'...1893...saw new boats in the classes, fast it is true, in fresh breezes, but undesirable from anything but a racing point of view...In America [continues the author of the book] where money is spent like water, when the national honour is at stake, 85 ft machines were built on the off-chance of their being successes.'

Is the tone of the controversial articles about recent trends in ocean-racer development and the rating rules, which largely control this development, very much different from that recorded in the 80-year-old volume of the Badminton Library?

B   Sail design in general

Despite the fact that mathematics, computers and wind tunnel testing are playing an increasing part in the designing of sails, sailmaking as well as sail tuning are still strongholds of art based on a hit-or-miss technique rather than on science. Numerous and frequently controversial articles published in yachting magazines, telling the racing crew how to bend spars and pull various controls to attain optimum sail trim and tune, are good examples of confusion in this field. This is quite understandable. After all, unlike the aeroplane wing, which can be regarded as a rigid structure whose shape is unaffected by variation in incidence and speed, sail shape is a function of both; in which case the shape of the sail affects the pressure distribution and vice versa, in a rather unpredictable manner. Moreover, as mentioned earlier, while sailing the whole sail geometry changes continually due to fabric stretch, which may or may not be recovered after a spell of leaving the sails in their bag. Besides, the sail shape, as potentially predetermined by the sailmaker, depends on how the sail is set by the crew, i.e. by the tensions applied along the sail edges by means of the halliard, out-haul, sheet, kicking strap, Cunningham hole, etc.

To illustrate this point in more practical language, let us quote Austin Farrar, a prominent sailmaker himself, when discussing the sheeting of stretch-luff genoas (Ref 3.10)

'After 20 minutes or so of sailing close-hauled, the sheet will need tightening to keep the sail correctly trimmed even though the wind strength has not increased; and the clew will have come back and down several inches...As the wind gets up a bit more, the luff will need stretching more and the sheet tightening again...Another increase in wind and more stretch still is required on the luff; but it already fills the forestay. However, the sail has a Cunningham hole part way up the luff like a reef cringle. If a tack purchase is provided, hook it into the Cunningham hole and heave down till the sail section looks right again.'

This procedure of tuning and trimming goes on with further increase in wind; and as a result, a 50 ft luff may stretch as much as 5 ft, while the clew goes back 3 ft. And farther on, from the same source,

'Now 10 per cent stretch on the luff sounds a lot, but it does not mean that a 30 ft luff will stretch 3 ft, since the stretch is a function of the area of the sail and not just the length of the luff; and the wider a sail of a given luff length the more stretch will be needed. This stretching does not mean that area has been created in the sail, the actual length of the yarns increases only minutely; but area has been transferred from one place to another and the leech become more hollow.'

The above description of sail behaviour in actual sailing conditions should be sufficient to give some appreciation of the difficulties associated with the unstable characteristics of the fabric that sails are made of.

It seems reasonable to raise the question as to whether any sailmaker, employing the most cleverly programmed computer and using the available fabric, can design an ideal, or optimum soft sail for all weather. Not infrequently, the readers of sailing magazines find a positive answer to this question. As a matter of record, some sailmakers proudly advertise that their computers can do it. Whether their claim is justifiable is another story.

The art of sailmaking is based on a not necessarily firm belief that the sailmakers know what aerofoil shape they wish to achieve. Their most difficult problem, which the majority of them will honestly admit, is to interpret into a sail the required shape and make it retain that shape. The frequently lamented difficulty arises from the sailcloth, since for various reasons, mostly commercial, the desired standard is not achieved consistently. There is much variation in this respect, and even in one consignment there may be quite appreciable variations. In fact, one can find variation between two halves of the same bolt of cloth. This unpredictability of fabric behaviour under stress makes logical or analytical sail design extremely difficult, if not impossible. As rightly pointed out by E Venning in the discussion on one of the SNAME papers (Ref 3.11), 'The elusive feature of sail design is the fact that it is problematic whether or not any sail will actually assume the shape one wishes it to take, and whether it ever will be used in exactly the wind conditions for which one has decided to design.'

Computers are of little help in this respect. They can perform tedious computation in an incredibly short time, but their output is completely dependent on the input data and these necessarily include the clear-cut assumptions as to specific course sailed, wind conditions, available righting moment at a given angle of heel, aero-elastic properties of the sailcloth etc., to be met. And even if the sail design difficulties, related to the recoverable and irrecoverable stretches in the sailcloth, were minimized by using a fabric having negligible stretch, the accurately computed ideal sail could only manifest its optimum performance over a very narrow range of wind speed and course relative to the apparent wind. In the manufacture process, such a sail would be forced into the required shape and, when set on spars, it would readily assume its design shape regardless of the wind speed. Needless to say, apart from the obvious disadvantages of having an optimized sail which can only operate efficiently in strictly specified conditions, such a necessarily rigid sail would be difficult to handle and could not be stuffed in a bag. It is thus not a very practical proposition.

To illuminate the sail designing dilemma from yet another point of view, let us quote some remarks written by J Milgram, and based on his own experiences in this field (Ref 3.12)

'...if a maximum heeling angle is specified, a good design criterion is obtained by designing the sails such that the forward force is maximized, under the restriction that the heeling moment should not exceed the value resulting in the maximum allowed heeling moment. Once the sail plan is designed or specified, it remains to determine the camber distributions of the sail. Now, I cannot sit down and directly design the optimum camber distribution for a suit of sails. *If anyone tells you he can do this, I advise you not to believe him* [italics are introduced by the author]. This is no basis for doing such a thing. However, what I can do is design the pressure distribution, which if attained on the sails in use will result in the highest boat speed. Then I can calculate the camber distribution needed to attain these pressure distributions. There is an accurate basis for doing this...The determination of the sail shapes needed to attain specified pressure distributions must be done on a digital computer because many millions of arithmetic operations are involved.'

So far so good, but to continue Milgram's remarks,

'...As opposed to the situation with the design of sail shapes, there is no precise scientific way to design a sail cut such that the result will attain the accurately computed shape. There are many reasons for this deficiency.'

And further on '...Given that there is no precise way to design sail cuts that yield the designed shapes, *the sail designer must do the best he can with the available information.*' Finally

'...In saying that we do not know how to achieve our computed shapes precisely, it is wise to state what we mean by being precise. In this case, a reasonable limit of tolerable error in local angle of incidence or camber ratio would be ten per cent. That might sound like quite a lot, but actually it is very restrictive. For example, typical camber ratios are about 12 per cent and aerodynamic angles of local incidence are about 5°. Thus a ten per cent error limit would restrict the acceptable error in camber ratio to 1.2 per cent and in angle of incidence to one half of a degree. *Unless these limits of error are obtainable, there is little sense in going through a precise shape design process.* I have checked many sails which were supposed to be built to computed shapes, many of which I built myself, and only very rarely did the actual shape resemble the computed shape within the forementioned error limit.'

Summing up, the above considerations justify a rather conservative or sceptical position in regard to the availability of sufficient knowledge for truly scientific sail design. So, beware of confounding the precision of computers and advanced aerodynamic theories, which facilitated the development of aeroplanes of startling performances, with actual sail design achievements if those precise methods or analytical tools are applied to material not suited to them. The soft sail in common use being an elastic membrane susceptible to deformation and therefore having, for lack of a better term, 'floating' aerodynamic characteristics is very different from a rigid aeroplane wing of 'fixed' aerodynamic characteristics. Besides, sailing craft are not constant cruising speed vehicles, while for practical purposes, aeroplanes in fact are. 'Mathematics are only of value', says the empirically-minded Uffa Fox,

'to the person who has the sense to use the right formula and start with the true value. Too many mathematicians today multiply an unknown quantity by an illogical factor, and arrive at proportions that a man with discerning eyes can see are wrong, even though the mathematicians believe the answer to be correct if the mathematics are correctly worked' (Ref 3.13).

It must be said in defence of some sail designers that they are by no means the only people who '...multiply an unknown quantity by an illogical factor'.

What we really need is not an ideal sail of predetermined ideal shape for predetermined wind speed and course of sailing, but an *infinitely adjustable sail*, which can be trimmed and tuned effectively to cope with a great variety of wind speeds and course conditions. This requires a different shape for near calm and for gale conditions, different for close-hauled work and different still for reaching. To achieve this goal a certain amount of elasticity in the fabric is essential, provided the stretch is fully recoverable. Then it is very likely, let us stress it again, that the many adjustments in the sail shape, which the crew can intelligently introduce in the process of tuning and trimming by means of various control gadgets, are more important than having the best computerized sails.

Since distortion of sails, resulting from stretch in the fabric under stress, is a major factor affecting efficiency, the material needs to be chosen with great care. What are the important parameters which distinguish a good sailcloth from a bad one? Why are some fabrics more suited to one type of sail than another? How big are the stretches in sail fabric and how are they distributed? These are just a few of the questions which yachtsmen and sailmakers have been discussing for a long time.

Undoubtedly research and tests are needed for at least two practical purposes, which would assist sailmakers.

1. To discover and establish a set of criteria by which the sailcloth quality could be assessed;
2. To agree upon the method of measuring fabric properties to answer the question as to whether a particular sample of cloth is up to specification.

(1) Loading

The development of a method which enables the stretch of a sail to be measured under wind loading is an important step forward. Scientists working for ICI Fibres Limited, and independently at Southampton University (Ref 3.14), have devised methods of measuring the distortion which occurs in a sail at the time that the sails are in this condition. Photographs 3.6 and 3.7, together with Fig 3.3A and B depict the technique used by ICI, and Photo 3.8 shows a genoa with attached strain gauges in the wind tunnel at Southampton University.

Textiles deform in different modes according to the direction of the principal stresses. It is convenient to resolve these loads into tensile and shear stresses along warp and weft axes, and to consider three modes of deformation:

warp extension,
weft extension and
shear

Figure 3.3 displays the variation of strain with direction relative to weft and warp at a number of points on the sail shown in Photo 3.6. The two sails shown are made of different fabric. At each point, the magnitude of the cloth extension in a given direction is represented by a line joining the point to the edge of a 'clover leaf', the length of which is proportional to the stretch. The shape of the clover leaf portrays, in a way, the distortion at a given point and at a given wind velocity. The direction of maximum stretch is then along the axis of the leaf. It can be seen that in some directions the fabric contraction occurred at right angles to the stretch, although in some regions the large apparent contractions were in fact due to creasing of the sailcloth.

Another solution to the problem of measuring sailcloth properties is offered by the Automate/Yendell fabric testing machine which was developed as part of the sail research programme of the University of Southampton; the broad principles of the machine are illustrated in Photo 3.9. It provides a method of carrying out simple tests to determine the relationship between load and extension of fabrics under more realistic conditions than has hitherto been possible. Basically it simultaneously loads the warp and weft by means of a simple system of levers and hand operated screw jacks. The magnitude of the load is measured by spring weighing machines through which the screw jacks act. The corresponding fabric extensions are indicated by micrometer dial gauges.

With the fabric test panel loaded as described above, a bias or shear load can next

Photo 3.6 An experimental rig erected on the Yorkshire Moors. Sails previously marked with accurately spaced crosses are set on a 35 ft Dragon mast. These series of crosses form reference points distinguished in Fig 3.3 as row numbers and line numbers when photographing the sail to obtain a stereoscopic picture of the movement of the cloth under load.

Photo 3.7 Photographs of the sail, taken on a pair of high-precision cameras, are analysed on a stereoautographic plotter, using a technique similar to that for preparing maps from aerial photographs. In this picture, a pantograph is drawing out a true-to-scale elevation view. From these drawings the distortion of the sail under various conditions can be measured accurately.

Photo 3.8 Photograph showing a genoa with attached strain gauges used to assess the forces within the sail from measurements of stretch, made in the wind tunnel at Southampton University.

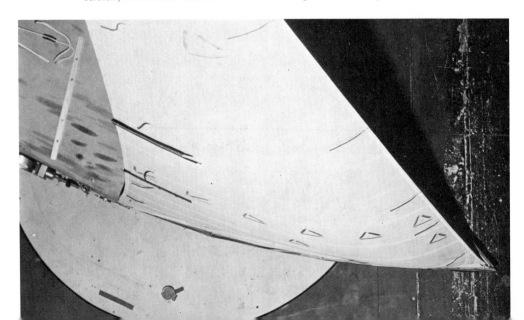

Photo 3.9   The Automate/Yendell fabric testing machine.

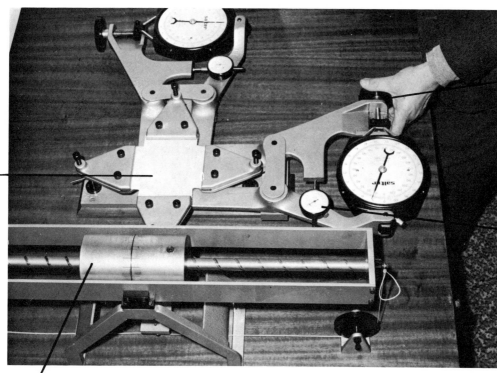

Hand operated screw jacks to apply warp and weft loads

t square fabric

Dial gauges indicate extension on warp and weft

Weighbeam to apply bias load to fabric

be applied by means of a weigh-beam arrangement in which a weight is moved along a pivoted lever. The resulting distortion is measured by a simple pointer and scale.

The bias loading facility can conveniently be used to fatigue the fabric so as to loosen any resins added during processing so that 'before and after' tests can be carried out.

The use of a low power microscope enables the detailed behaviour of the thread geometry to be observed as the load pattern is changed.

Referring to Fig 3.3, a point of interest is that the stretch in the sail observed with a relatively low wind velocity of 6.5 mph was greater than that observed with wind velocities of 17 and 27 mph. This apparent anomaly is explained by the fact that the observations at the higher wind velocities were taken shortly after the sail had been hoisted, whereas those at the lower velocity were recorded after several hours of exposure to a wide range of conditions. This indicates that the extension of the sail is not immediately reduced as soon as the wind velocity falls. When the sails were finally lowered, however, allowed to recover and the dimensions between the crosses remeasured, it was found that the sail had recovered to a residual extension of rather less than 1 per cent.

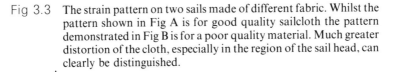

Fig 3.3 The strain pattern on two sails made of different fabric. Whilst the pattern shown in Fig A is for good quality sailcloth the pattern demonstrated in Fig B is for a poor quality material. Much greater distortion of the cloth, especially in the region of the sail head, can clearly be distinguished.

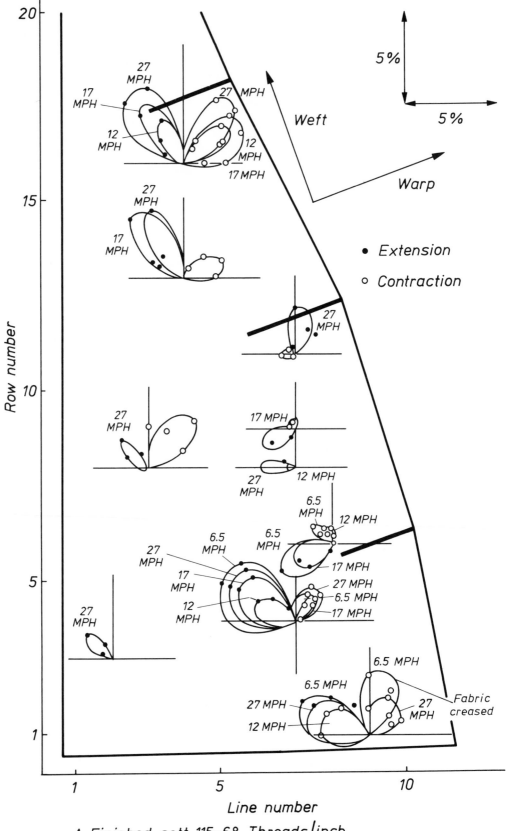

A. Finished sett 115×68 Threads/inch.

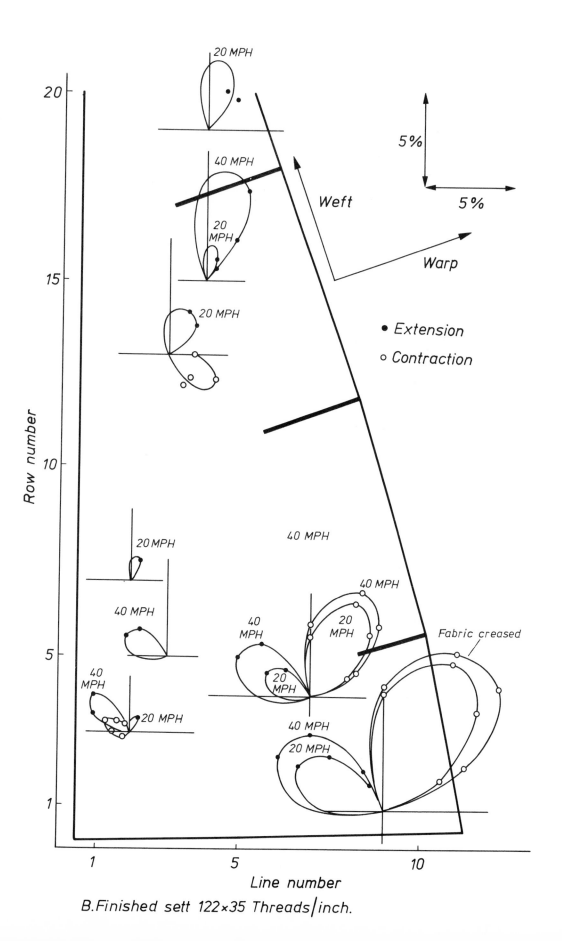

B. Finished sett 122×35 Threads/inch.

Photo 3.10   Picture A above shows a loosely woven fabric in the unsheared state; Photo B below shows the same fabric sheared through 25°, × 100 magnification.

From these studies the following conclusions can be drawn. If a sail is stretched or distorted intentionally, by use of Cunningham holes for example, it *should* recover its original shape when the load is removed. A material which creeps or progressively manifests non-recoverable stretch may be good for a few races but sooner or later, even after recutting, will have to be scrapped. In other words, a good cloth must be sufficiently stable both in elasticity and bias-distortion.

Shear distortion generally accounts for the major changes of shape in a sail. When a loosely-woven fabric is sheared, most of the deformation is caused by the fibres slipping over one another at the intersections. If the shearing force is removed, the major part of this distortion will remain and not recover. Photo 3.10A and B, taken for demonstrating the problem, shows a loosely woven fabric (not a sailcloth) in both unsheared and sheared states (Ref 3.15).

If the fabric is more tightly woven, the shear due to this slippage becomes smaller. At low shear angles, the shear stiffness of a fabric is due to the yarns bending, while being held in place by frictional forces at the intersections. As the shear force increases more and more of the fibres slip at the intersections until the warp yarns are jammed against one another, as in Photo 3.10B, at which point the shear stiffness increases markedly.

Resin fillers added to the sailcloth during the finishing process have a large effect on shear stiffness, stabilizing the material. However, many of the resins used at present are not sufficiently elastic to survive the large extensions within the sail, to ensure consistent shear performance through a long life. Rough calculations indicate that when a fabric deforms in shear, the resin between neighbouring warp threads must expand from a thickness of perhaps $\frac{1}{1000}$ in to something of the order of $\frac{1}{200}$ in, which is 500 per cent, while the fabric shears 10 per cent. Being rather brittle and stiff in relation to threads of the cloth, the resin fractures under working conditions. It gradually comes out and the material becomes a loose cloth which pulls all over the place. Due to the failure of the resin filler the material is liable to non-elastic extension and cannot go through the same load-extension cycle repeatedly. Photo 3.11 shows the resin filler peeling from the surface of the fibres due to breakdown in adhesion between resin and woven fibres.

Unless better fillers are invented, the sensible way to obtain long-lasting shear stiffness seem to be a tighter weave, which will become jammed at the smallest possible shear angle. The present trend towards fabrics with no filler seems to be correct.

It is common knowledge that sail fabrics are being improved by a calendering process, where the cloth passes through heated rollers under high pressure. By compressing the fibres onto one another at the intersections, flattening them to some extent, the shear response can be improved; at the same time the air porosity is reduced. The problem is somewhat controversial as to whether the effects of calendering wear off quickly. Photographs 3.12 and 3.13, taken through an electron microscope, may throw some light on this question. They show, at two different magnifications, the damage which is done to fibres by excess calender temperature and pressure.

Photo 3.11   Photograph shows resin peeling from the surface of the fibres in Terylene sailcloth, × 1100 magnification.

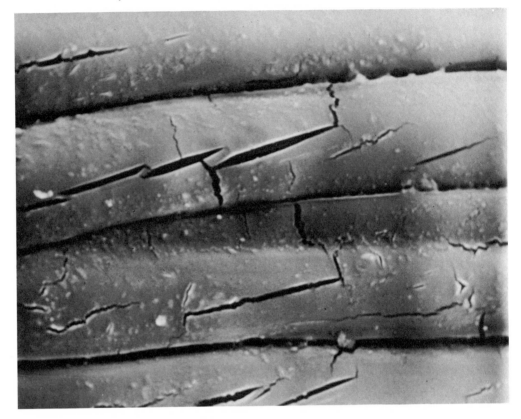

Photo 3.12   Photograph shows the damage to the fibres caused by excessive calendering temperature and pressure, × 110 magnification.

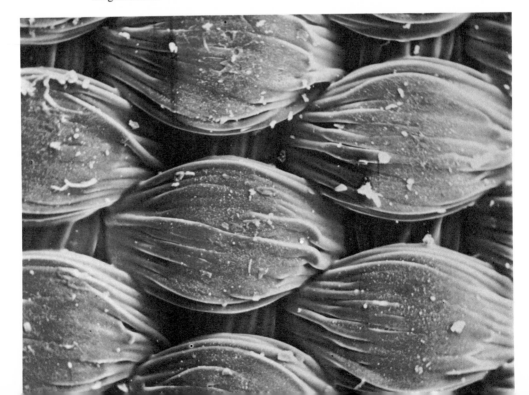

Photo 3.13   Photograph shows individual fibres of the fabric in Photo
3.10, × 1100 magnification.

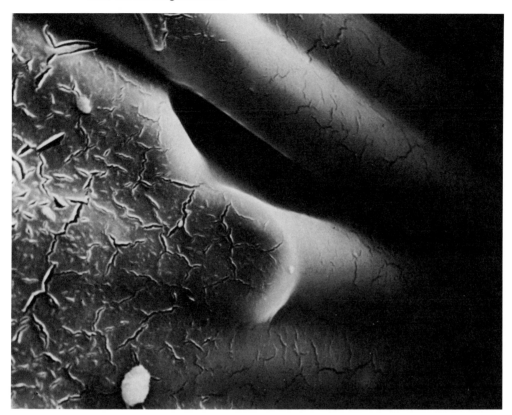

Air porosity is another important parameter in the design of a sail fabric. Wind
tunnel tests carried out by M Yendell (Ref 3.15) showed an increase in the driving
force produced by a model genoa of up to 15 per cent, due to sealing the pores in the
fabric. The air porosity of a fabric may be tested by a variety of standard methods,
the most popular of which involves timing the passage of a known quantity of air
through the fabric under a given pressure. The routine method applied in
Southampton University involves measuring the air through-flow velocity V_p with a
given pressure difference Δ_p across the fabric.

The tester used in these experiments consisted of a hollow thin-walled cylinder,
closed at one end, falling in a cylindrical tube filled with liquid. The trapped air is
forced through a known area of sailcloth by pressure created by the weight of the
falling cylinder. Since the cylinder has thin walls, the air pressure changes only
marginally due to their displacement as the cylinder falls. Knowing the test area and
the area of the cylinder, the through-flow velocity V_p may be calculated from the
time taken for the cylinder to fall a certain distance.

This tester used a pressure difference of 5 lb/ft^2, which corresponds to a pressure experienced by genoa in a 30 ft/sec wind (Beaufort force 5), this being a feasible value for the pressure peak in an efficient genoa.

A variety of genoas, as set out in Table 3.1, were used for the different portions of the experiment; in each case they were rigged on a yacht model which had a mast height of approximately 8 ft (Photo 3.32). They were of similar outline shape but had different porosity and area. Sails 2a and 2b were cut from Sail 2 and they were used to assess the effects of sail size.

TABLE 3.1
The model sails

Sail No	Area (ft^2)	Porosity (V_p, ft/sec)
1	10.6	0.16
2	21.2	0.13
2a	4.6	0.13
2b	4.7	0.13
3	21.2	0.05
4	11.7	0.02

Sail forces were measured and Table 3.2 sets out the driving force coefficient C_x and side force coefficient C_y (see Fig 3.14) both before and after the sail pores were sealed by spraying the sail with Ethylene Glycol, which has a high boiling point and is therefore slow to evaporate.

TABLE 3.2
Tests on sails on equal porosity
$(\beta - \lambda) = 25°$, $V_A = 31$ ft/sec

Sail No	Porous		Pores sealed		% Change	
	C_x	C_y	C_x	C_y	C_x	C_y
2	0.321	1.31	0.350	1.37	9	5
2a	0.195	1.28	0.222	1.33	14	4
2b	0.121	1.08	0.141	1.13	16	5

This test shows that porosity becomes more important for windward sailing when a yacht is propelled by a smaller sail than usual.

A series of tests was carried out, using Sail No 1 (Table 3.1), which was more porous than any of the others, to investigate the effects of porosity at various wind speeds and course angles. Sail forces were measured at four wind speeds, 33.2, 30.7,

21.7 and 15.3 ft/sec, at angles of 25°, 45°, 90°, 135° and 180° between the yacht's heading and the apparent wind. In the case of the directly downwind run, the sail was boomed out to create a realistic running situation. Table 3.3 gives the results of these tests, from which it is apparent that the effect of porosity increases considerably with wind speed, and that it has little effect off the wind. Since this fact became obvious in the experiment, no runs were made at the lower speed for apparent wind angles of 90°, 135°, or 180°.

TABLE 3.3
Tests on Sail 1 at various wind speeds and course angles

β-λ	V_A (ft/sec)	Porous		Pores sealed		% Change	
		C_x	C_y	C_x	C_y	C_x	C_y
25°	33.2	0.329	1.14	0.361	1.20	10	5
25°	30.7	0.332	1.13	0.353	1.19	6	5
25°	21.7	0.336	1.14	0.350	1.15	4	1
25°	15.3	0.332	1.15	0.334	1.15	1	1
45°	33.2	0.664	1.44	0.742	1.49	12	3
45°	30.7	0.686	1.47	0.740	1.52	8	3
45°	21.7	0.738	1.52	0.772	1.54	5	1
45°	15.3	0.745	1.51	0.753	1.50	1	0
90°	33.2	0.872	1.04	0.894	1.04	3	0
90°	30.7	0.873	1.06	0.890	1.05	2	−1
90°	21.7	0.872	1.05	0.877	1.05	1	0
135°	33.2	0.937	0.52	0.948	0.52	1	0
135°	30.7	0.951	0.53	0.949	0.53	0	0
135°	21.7	0.966	0.54	0.966	0.54	0	0
180°	33.2	1.253	0.25	1.253	0.25	0	0
180°	30.7	1.267	0.25	1.267	0.25	0	0
180°	21.7	1.287	0.26	1.292	0.26	0	0

Having established that porosity has most influence on the driving force when close-hauled and at high wind speeds, the data from all available tests have been collected in Table 3.4 to compare the effect of porosity on sails with different porosity levels. Although some of these sails were of different size, with the exception of Sails 2a and 2b, the ratio of hull size to sail size was approximately constant. Results are compared at an angle (β-λ) of 25° to the apparent wind of 30.7 ft/sec.

While the figures in the per cent change column do not bear a direct relationship to the porosities of the sails, this is hardly surprising due to the errors possible in C_x and C_y. However, at no time did Sail 4 show a measurable change in forces when the pores were sealed, this sail having one-eighth the porosity of Sail 1 and less than half the porosity of Sail 3.

TABLE 3.4
Effects of varying porosity level at $(\beta-\lambda) = 25°$

Sail No	V_A (ft/sec)	Porous		Pores sealed		% Change	
		C_x	C_y	C_x	C_y	C_x	C_y
1	30.7	0.332	1.13	0.353	1.19	6	5
2	30.7	0.322	1.31	0.350	1.37	9	5
2a	30.7	0.195	1.28	0.222	1.33	14	4
2b	30.7	0.121	1.08	0.141	1.13	16	5
3	30.7	0.342	1.34	0.362	1.37	6	2
4	30.7	0.427	1.57	0.424	1.56	−1	−1
1	33.2	0.329	1.14	0.361	1.20	10	5
4	33.2	0.420	1.55	0.416	1.55	−1	0

An acceptable level of porosity for sails used in windward sailing conditions, in winds of up to 33 ft/sec (force 5), is therefore that of Sail 4, $V_p = 0.02$ ft/sec at $\Delta_p = 5$ lb/ft^2.

In stronger winds, porosity is unlikely to be an important factor determining a yacht's performance, since changes in sail force of the order of 10 per cent are unlikely to make a big difference in performance when the hull is working on the steeper portion of the resistance curve.

It is important to realize that sail fabric porosity can change considerably after a period of use, due to the fibres moving within the weave and to the failure of the resin filler. Of course, a more porous cloth allows greater transfer of air through the fabric into the boundary layer developing along the leeward side of the sail, thus accelerating separation and reducing the pressure differential between the leeward and windward sides of the sail.

C How and why sail forces are determined

'When you can measure what you are speaking about, and express it in numbers, you know something about it.'

LORD KELVIN

Reverting to Fig 1.2, Part 1, illustrating the simple case of a boat sailed in close-hauled conditions, in which the disposable crew weight is sufficient to keep the boat nearly upright, we found that the boat's speed performance is controlled by nine basic variables. They were given in Table 1.3, and are repeated here for convenience.

TABLE 1.3
(From Part 1)

Geometry of sailing velocity triangle Fig 3.4	Sail aerodynamics	Hull hydrodynamics
V_s	V_A	V_s
V_A	β, δ_m	λ
β	F_R, F_H	R, F_S

Developing a step further our discussion which led to Table 1.3, and simplifying to bare essentials the problem of boat performance, we can write a *Basic Performance Equation* as follows:

Yacht speed performance = Sail aerodynamics + Hull hydrodynamics
+ Geometry of sailing
+ Mutual desirable and undesirable interference effects
between the sails and hull.

Eq 3.1

This incorporates the 3 groups of variables given in Table 1.3, plus one more which will be explained shortly. Some of the variables can largely be controlled by the crew and some others are almost independent of crew will or action. Clearly, the boat once built has a hull configuration of fairly rigid shape and therefore of fixed hydrodynamic characteristics, which can hardly be affected by the crew, or a little by changes in hull trim and quality of its wetted surface smoothness. In contrast to the fixed hull characteristics the boat may have a large wardrobe of sails, which can be set in various combinations, each with infinitely varying shapes and hence infinitely varying aerodynamic characteristics that can deliberately be played by the crew in the course of tuning and trimming.

The importance of the variables belonging to the group called 'geometry of sailing' becomes evident if we examine Fig 3.4. Any change in the course sailed β leads to changes in apparent wind V_A for a given V_T, and this will affect both the sail aerodynamics and subsequently the hull response. Finally, considering the fourth group of variables in the Basic Performance Equation, an example may explain what is meant by the mutual interference effects between sail and hull. For a given sail area, the greater the aspect ratio, the more efficient the sail in terms of windward ability–a desirable effect. But a taller sail produces greater heeling and this in turn usually increases the hull resistance, causing at the same time a deterioration of sail efficiency, which is clearly an undesirable effect. Another example, concerning desirable interference effect is that due to closing or reducing the gap between the sail foot and the deck, which can lead to an increase in effective aspect ratio of the rig and therefore to reduction of the induced drag (Photo 3.14).

If the boat cannot be sailed upright, which means that the 'overturning' or heeling moment becomes significant, one more variable must be added to Table 1.3–this is the heeling angle Θ. Up to about 20° its effect on performance is relatively small but beyond 25–30° it becomes increasingly dominant, modifying in a detrimental sense the sail and hull data indicated in Table 1.3. The relationship between the heeling moment, angle of heel, driving force and course sailed β is an extremely complicated game, but in general, the greater the heel angle, the broader is the β angle. This effect of heel angle will be demonstrated on an experimental basis when discussing the so-called Gimcrack coefficients in the following chapter.

Concentrating on the aerodynamic aspect of boat performance, it is now pertinent to inquire what determines the magnitude of the sail forces. Figure 3.5 gives an indication of the complexity of factors influencing the forces developed on a sail. It shows only the main relationships and much has been omitted for the sake of simplicity. One such omission is that of *feed-back*, the way in which one factor

Fig 3.4   The geometry of the velocity triangle and aerodynamic forces on a
boat sailing close-hauled on the port tack.

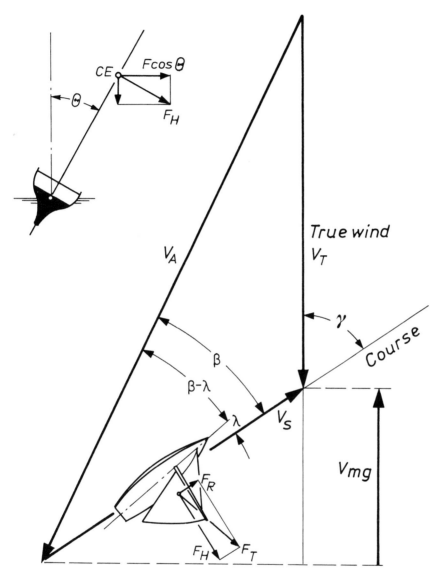

affecting another is in turn affected by it. Some of these factors, such as 5 and 6, are
determined by the class rules, some others, 7 and 8, depend on the sailmaker's skill;
many among them, 1, 2, 9, 10, depend on the crew, and unfortunately not all are
sufficiently well documented scientifically.

As an example of how the block diagram may be interpreted, let us consider the
effect of the mast. Its presence profoundly modifies the flow over the whole mainsail.
In addition, when the mast bends under the combined action of sheet, kicking strap

Photo 3.14   This shows a badly set genoa on a Flying Dutchman. A gap
              between the deck and the foot of the sail makes it possible for
              air to flow from the windward high pressure side to the
              leeward low pressure side. The arrow shows direction of the
              undesirable flow. The wrinkles apparent near the leading
              edge of the genoa are a symptom of too little tension along
              the luff of the sail.

Fig 3.5   Main factors affecting aerodynamic forces.

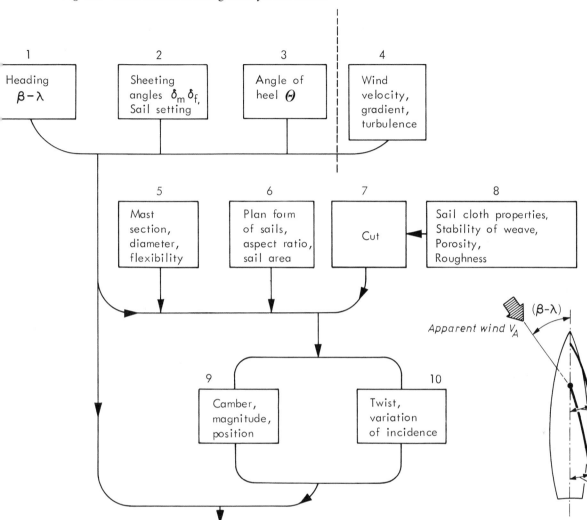

and wind, it can further influence the shape of the sail, i.e. its twist, the magnitude of camber and its chordwise and spanwise distribution. All those geometric parameters change the airflow round the rig and hence they lead to changes in the pressure distribution from which the aerodynamic forces come, and so to variations in the forces themselves.

Although the diagram in Fig 3.5 may be regarded as instructive, it gives no quantitative information to really appreciate the effects of all these factors on boat performance. To solve the speed/performance equation quantitatively, the numerical values for the terms given in Table 1.3 must be known and these can be obtained wholly or partly from measurements of sail characteristics. There are a number of different ways of deducing sail forces and of investigating the influence of essential factors on which those forces depend. The whole problem of testing and how the tests are conducted is closely allied to what one hopes to gain from them. Some testing techniques are more fruitful than others but all of them have certain limitations in reliability and accuracy. A short account of possible methods of sail force testing should be helpful in appreciating the difficulties and limitations that every researcher may face.

(1) Determination of sail forces by strain gauging the rig

If one wishes to determine the forces on an actual sail, under normal sailing conditions, then the logical thing to do is to go on a yacht, whilst sailing, and measure these forces. At first sight this proposal appears attractive, but the problems involved in such an undertaking are very considerable. The only serious attempt to do something of this nature was an effort on the part of Sir Geoffrey Taylor. His experiments were not entirely successful and despite technical progress during the intervening years many of his problems remain unsolved (Ref 3.16).

One of the principal difficulties is inherent in the geometry of the rigging. Large loads would have to be measured in shrouds and stays inclined at relatively small angles to the mast. The simplest case of a dinghy mast with mainsail alone and one shroud per side, as shown in Fig 3.6, may serve to illustrate the problem. Let T_t be a transverse component of shroud tension due to an aerodynamic loading in close-hauled condition. One can find that:

$$T_t = T_S \times \sin A \cos B$$

where T_S is a tension in the shroud. Hence

$$T_S = \frac{T_t}{\sin A \cos B}$$

Assuming for example that $A = 11°$
$$B = 20°$$
$$T_t = 100 \text{ lb}$$

one finds that the shroud tension:

$$T_S = \frac{100}{0.191 \times 0.94} = \frac{100}{0.18} = 560 \text{ lb}$$

If T_S is measured by a strain gauge, then the transverse component T_t due to sail action can easily be calculated. So far so good, but flexing of the rig, say bending of the mast by means of swinging spreaders shown in Fig 3.6C, will result in changes of both angles A and B, and hence in the magnitude of the component forces which

Fig 3.6   By using free-swinging spreaders the mast can be forced to bend
forward and sideways. Amount of fore and aft mast movement
will depend on the length of spreaders, on the β angle at which
they are angled and on the initial tension in the shrouds.

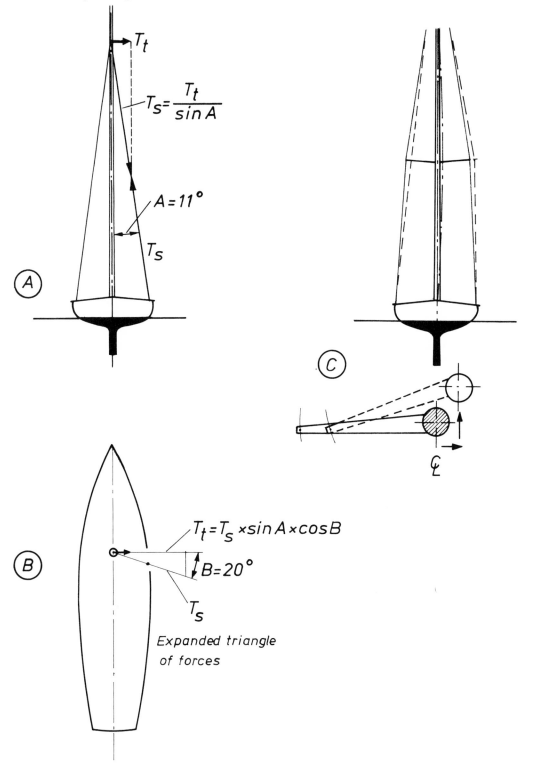

T_t

$T_s = \dfrac{T_t}{\sin A}$

$A = 11°$

T_s

$\textcircled{A}$

$\textcircled{C}$

$\unicode{0x2104}$

$\textcircled{B}$

$T_t = T_s \times \sin A \times \cos B$

$B = 20°$

T_s

*Expanded triangle
of forces*

would be resolved from the measurements. One may argue that the angular changes could be measured and suitable allowances made, but the significant values would be so small as to make practical estimates inaccurate. Another difficulty is found to be associated with the determination of relative zero for the sailing loads, since they may be measured in the presence of the relatively large static tensions, depending on initial tightening of the shrouds. The effect of ambient temperature variations, elongation of the shrouds, or strain gauge drift, can easily upset the whole calibration procedure.

Having obtained the data referring to one yacht with a strain-gauged rig, it would not be too difficult to extrapolate the data to another yacht with a similar sail plan but, because of the difficulties mentioned, the method has not been used successfully. It was claimed by the late A Robb, the designer of a prospective *America*'s Cup Challenger, in discussion of Ref 3.16, that extensive strain-gauge tests were carried out by him in 1961 on *Norsaga*, the British 12-Metre test-boat shown in Photo 1.2, and from which useful data were obtained. However, nothing more has been disclosed since, and it seems probable that these measurements were ultimately only of value for stressing purposes.

(2) Determination of sail forces by measuring mooring loads

An alternative to strain gauging the rigging would be to erect the complete sail system on a platform, let the wind blow through it, and measure the forces required to balance the aerodynamic loads. Two forms of this proposal have been considered and the first uses the hull as the platform. The yacht would be moored by means of two ropes and dynamometers in between connections. Relevant tensions in ropes could be recorded and evaluated in terms of aerodynamic sail forces. With this concept in mind, McLaverty (Ref 3.17) attempted to measure the sail shapes and forces developed on a full-scale Dragon, moored in a static water pond, as shown in Photo 3.15. A further intention was to reproduce a full-scale shape of the sail on a $\frac{1}{4}$ scale wind tunnel model, which had been specially designed to make the shape adjustment possible so as to obtain correlation between model tests and full scale measurement.

In the event, the full-scale tests were largely abortive, due to the extremely unsteady nature in both magnitude and direction of the natural wind. Whereas yachts which are sailing close-hauled, even in a rather disturbed sea-state, are remarkably steady in roll, maintaining an almost constant angle of heel, the yacht moored in the pond was remarkably unsteady in roll. On a day when the wind appeared to be reasonably steady, at one moment the Dragon would be almost upright and a few seconds later the deck edge would be immersed.

The reason for this violent rolling motion was subsequently found (Ref 3.18), and is associated with the roll damping characteristics of the rig, which are closely related to the boat's speed through the water and the course sailed relative to the wind.

Professor Bradfield of New York State University modified McLaverty's testing

Photo 3.15   Full-scale Dragon moored in a static water pond. The two
              mooring steel-wire ropes attached to dynamometers are
              visible.

technique by carrying out the full-scale sail investigations towing the rig, suitably mounted on a balance system, along a straight track or a runway in calm weather conditions. In this way he hoped to eliminate the unsteadiness which caused McLaverty so much trouble.

The sail force components or pressures could also be measured by mounting the complete full-size rig on a specially constructed plan-form. The apparatus might be situated in some place exposed to natural winds, the velocity of which together with the representative wind gradient could be recorded. Photograph 3.6 depicts such an experimental rig erected by ICI on the Yorkshire Moors. As mentioned earlier, the basic object of these experiments was the measurement of distortions occurring in the sail at the time when the sail is actually under the wind load.

Fig 3.7 *Gimcrack* sail coefficients.

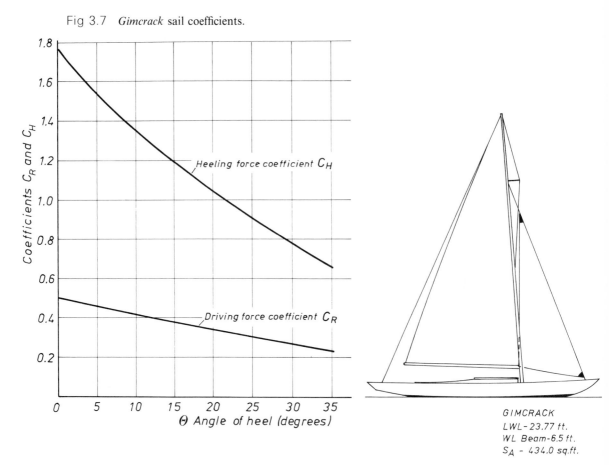

GIMCRACK
LWL - 23.77 ft.
WL Beam - 6.5 ft.
S_A - 434.0 sq.ft.

(3) Determination of sail forces by correlating the results of full-scale trials and model tank test (Gimcrack sail coefficients)

This method had been tried on several occasions and in particular it was employed on *Gimcrack* USA (Ref 3.4), *Yeoman* and *Norsaga* GB (Refs 3.4, 3.19). The pioneer work leading to this technique was undertaken by Professor Davidson at the Stevens Institute of Technology, Hoboken, USA, about 1932. In his historic paper *Some Experimental Studies of the Sailing Yacht*, presented in 1936 (Ref 3.4), he described this method and some other tests, by which he proved that very useful and reliable information can be obtained by testing in the tank relatively small models of yacht hulls about 4–6 ft long, provided the correct boundary layer flow is simulated using sand strips, studs, or trip wires near the bow. From a series of full-scale trials of the yacht *Gimcrack*, shown in Fig 3.7, supplemented by towing tank model tests, K S M Davidson produced in 1933 the well-known set of *Gimcrack* sail coefficients and thereby provided the first systematic method of using tank test results to predict yacht performance. He argued that, since in steady conditions the wind and water forces must be numerically equal and opposed in direction, the wind forces could be determined by measuring the water forces by a model hull, when towed at corresponding speeds, and at the correct geometrical attitudes.

The full-scale water speed V_s, the heel angle Θ, relative wind speed V_A, and apparent wind angle β, as shown in Fig 3.4, were all obtained by direct measurement on the *Gimcrack*, a 6-Metre type yacht. She was sailed by an experienced helmsman, and measurements were recorded when it was judged that the boat was sailing at optimum speed made good to windward. Table 3.5 gives in lines 1, 2, and 3 these recorded values of V_A, β and V_s, at different heel angles Θ ranging from 0° to 35°. Subsequently, that information was used to set up the model tests that were carried out in the Stevens Tank. The hull resistance R and hydrodynamic side force F_S, being equivalents of driving force F_R and heeling force F_H, were ascertained for the upright and inclined conditions. Figure 1.12 in Part 1 illustrates these relevant air and water forces at one particular wind speed and course sailed.

By repeated trials and interpolation it was possible to establish the actual heeling force F_H and resistance R that should occur in the full-size yacht, while sailed at the same angle of heel and corresponding speed. Figure 3.8 depicts for example the heeled resistances of the *Gimcrack* hull, as predicted from the model tests, for short ranges of speed, which include the actual sailing speed at each heel angle. The curve of upright resistance is also drawn for comparison. These curves show the large increases of resistance for which the heel is responsible. The predicted values of F_S = F_H and $R = F_R$ are given in lines 4 and 5 in Table 3.5. With these data, which supplement the full-scale measurements given in lines 1, 2 and 3 in Table 3.5, and the known sail area $S_A = 434$ sq ft, it became possible to calculate the set of sail coefficients in terms of C_R and C_H as a function of the heel angle alone. The results of such calculations are given in Table 3.5 lines 7 to 13, and also graphically in Fig 3.7. The most striking feature of the graph is rather pronounced reduction of both coefficients C_R and C_H with an increase of heel angle.

With these coefficients at hand it became possible to reverse the procedure used for *Gimcrack*, that is to work forward from known hull forces to wind and boat speeds, rather than backward to the coefficients themselves. If we wish now to know the actual sail forces that would be generated on a geometrically similar sail, we multiply the relevant coefficients C_R or C_H from Fig 3.7 by given sail area S_A and selected dynamic pressure of the apparent wind $0.00119 \, V_A^2$. Thus, say, the driving force F_R can be calculated from the customary expression:

$$F_R = 0.00119 \times C_R \times S_A \times V_A^2 \qquad \text{Eq 3.2}$$

and likewise

$$F_H = 0.00119 \times C_H \times S_A \times V_A^2 \qquad \text{Eq 3.3}$$

The *Gimcrack* sail coefficients provided thereby a valuable means of establishing the comparative merits of different hulls from the results of towing tank tests. In a

TABLE 3.5
Gimcrack sail coefficients

	Angle of heel Θ	0°	5°	10°	15°	20°	25°	30°	35°
Line		Values from full-scale trials							
1	Apparent wind V_A (knots)		6.22	9.33	11.87	14.33	16.97	19.70	22.50
2	Apparent course β	25.8°	26.1°	26.5°	27.0°	27.6°	28.6°	29.7°	31.0°
3	Boat speed V_s (knots)		3.32	4.50	5.18	5.60	5.87	5.97	5.97
		Values from model tank tests							
4	Driving force F_R (lb)		26.0	53.4	78.5	103.0	130.0	154.0	175.0
5	Heeling force F_H (lb)		87.6	172.0	248.0	317.0	383.0	444.0	496.0
6	Ratio $\dfrac{\text{Heeled resistance}}{\text{Upright resistance}}$		1.2	1.2	1.2	1.2	1.22	1.35	1.53
		Aerodynamic coefficients to reproduce F_R and F_H							
7	Driving coeffic C_R		0.457	0.417	0.378	0.341	0.307	0.269	0.234
8	Heeling coeffic C_H		1.54	1.345	1.195	1.045	0.902	0.778	0.666
9	Total air force coeffic C_T	1.835	1.60	1.405	1.253	1.090	1.037	0.825	
10	C_R/C_H ratio	0.280	0.296	0.310	0.317	0.331	0.340	0.345	
11	Lift coeffic C_L	1.814		1.397		1.079		0.809	
12	Drag coeffic C_D	0.286		0.203		0.157		0.137	
13	C_L/C_D ratio	6.35		6.88		6.82		5.92	

Note: Driving force F_R is in the direction of V_s, Heeling force F_H is at right angles to the F_R and to the mast centreline plane. Horizontal component F_{lat} of heeling force equals $F_H \times \cos \Theta$. See Fig 3.4.

Table adapted from *Some Experimental Studies of the Sailing Yacht* 1936 (Ref 3.4) and Technical Memo No 17, 1936. Stevens Institute of Technology, USA.

slightly modified form, this famous set of sail coefficients is still in use today. Obviously, as we shall see, the accuracy of the quantitative estimate of V_{mg} values for given wind speed V_T must necessarily depend on the degree of resemblance of the rig actually used with the hull tested to that original *Gimcrack* rig which can be described as a three-quarter rigged sloop with non-overlapping headsail. Bearing in mind all these limitations the *Gimcrack* sail coefficients have been applied with considerable success in the past 40 years, although the temptation to read too much into their numerical results has, on certain occasions, resulted in the drawing of misleading conclusions by the uninformed.

In the light of more recent wind tunnel experiments the validity of the *Gimcrack*

Fig 3.8   Heeled Resistance of *Gimcrack* as predicted from model tests with leeway.

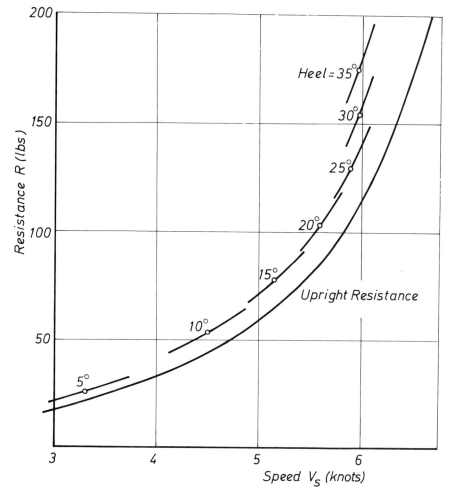

coefficients seems open to question. One should not forget that they were the result of an incredibly clever mixture of tank testing, intuitive approach and the practical genius of Davidson who derived these coefficients from some rather doubtful data. 'There was little reason'–Davidson stated–'to suppose that details of both the tests and calculation procedures might not have to be modified, or that the *Gimcrack* sail coefficients could be considered more than a first approximation.' An extremely modest statement, as subsequent history has shown, but one which we should not forget.

Why? Let us examine the two basic full-scale measurements, namely: the apparent wind V_A and the course sailed β, the accuracy of which certainly affected the reliability of the *Gimcrack* coefficients. As admitted by Davidson himself, the sailing tests of *Gimcrack* had included comparatively rough measurements of both

V_A and β. The suspected accuracy of measured values of V_A and β was partly due to the poor instrumentation of that time and partly due to inherent difficulties in measuring these two quantities and associated with the distortion of the wind-field by the presence of the sails together with an undefined wind gradient. These measuring difficulties could not easily be overcome either in Davidson's experiments or today. In spite of the fact that the masts of modern racers are '...sprouting an increasing load of instruments aimed at replacing intuition with science'—the accuracy of V_A and β measurements have not substantially improved since Davidson's attempt. Electronics has not helped very much in this respect.

(a) *Instruments*

Since nowadays the peak of a cruiser-racer's performance is frequently judged by reference to instruments, practical sailors might like to know what their precise electronic instruments really measure and what possible errors may occur. In order to establish the influence of sails on β angle reading an investigation was undertaken by Kamman (Ref 3.20) with the object of measuring the local deflection of the apparent wind at various points near the sails of a model in the close-hauled attitude in a wind tunnel. The tests were carried out on a $\frac{1}{6}$ scale model of a Class III ocean-racer, a masthead sloop with large genoa, heeled to 20° (Photo 3.32). The angular deflections of the wind were measured with the Brookes and Gatehouse wind direction indicator type *Hengist*, shown in Fig 3.9. A wind speed of 25 ft/sec was used throughout with the model set at a heading angle $\beta\text{-}\lambda = 25°$ to the axis of the wind tunnel, i.e. to the apparent wind direction. The wind-vane, supported by a tripod similar to a camera stand, was put in a fixed position relative to the model, and the wind direction reading without any sail set was taken. Then sails were hoisted and a new reading was taken without altering the position of the vane. The difference between these two readings is called the deflection of the wind. The measuring positions and the wind deflection in direction and magnitude are shown in Figs 3.9 and 3.10. The arrows show the direction in which the wind is deflected due to the presence of the sails. These measured wind deflections are caused by the powerful trailing vortex operating at the sail tip. Its action was discussed at length in section D, Part 2 and a glance at Photo 3.27A and D may remind the reader that the trailing vortex is not an illusory invention.

The maximum error in the reading of the measurements is estimated to be less than $\pm 2.0°$ taking into account error due to inaccuracies of the measuring instrument stated by the makers to be $\pm 1.0°$, error due to fluctuating wind direction in the wind tunnel making the wind angle indications unsteady $\pm 0.2°$, error due to different deflection of the supporting structure caused by the change in wind direction between two measurements, error due to the size of the vane and to the restraining effect of tunnel walls.

The results seem to indicate that the usual position of the wind-vane at the masthead is not entirely satisfactory as the error in the measured apparent wind angle can be in the order of 7°, which in performance prediction is too large if

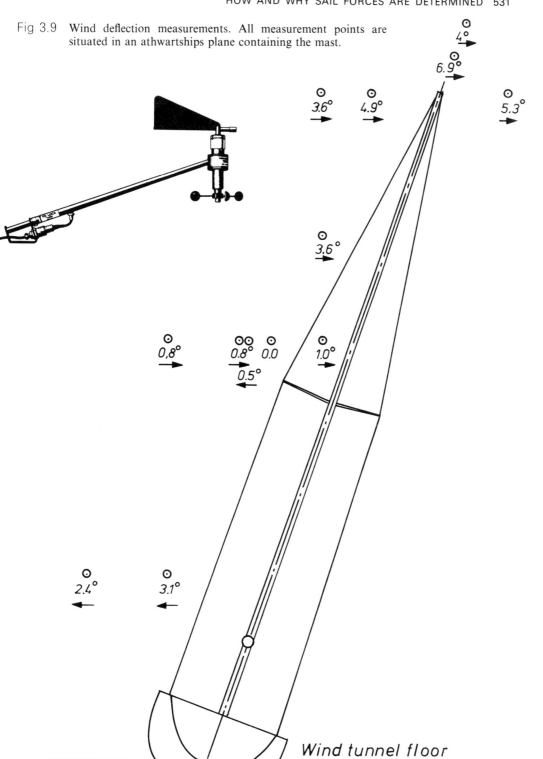

Fig 3.9   Wind deflection measurements. All measurement points are situated in an athwartships plane containing the mast.

Wind tunnel floor

5'
Full scale

Fig 3.10 Wind deflection measurements. All measuring points abeam or forward of the mast are placed in a horizontal plane 1 in above the mast top of the model. The five points above the transom are in a plane through the centre line of the model containing the mast, the lowest at boom height, the highest at mast top height and the rest evenly spaced in between.

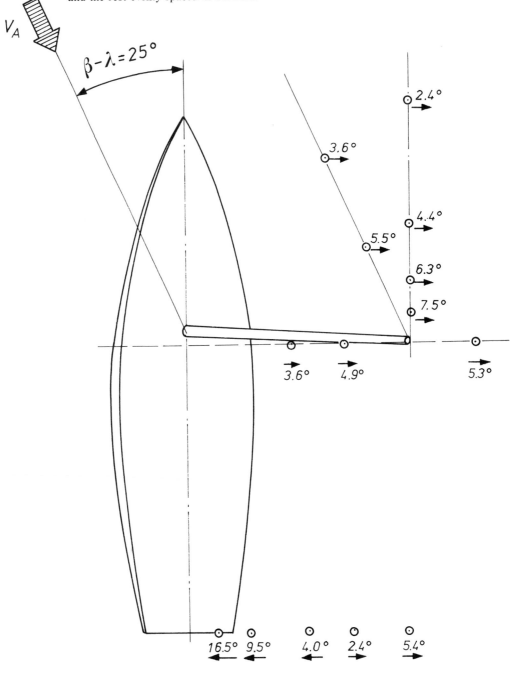

worthwhile calculations of V_{mg} are to be made. The data of Fig 1.9D in Part 1, which gives the variation of V_{mg} at different β angles and different apparent wind speed V_A, together with some approximate calculation, indicates that even a 5° error in selecting the optimum heading angle may easily result in a difference of 80–100 ft made good to windward for every nautical mile sailed. This is not a negligible distance, bearing in mind that racing deals with very small differences indeed.

As already mentioned in Part 2, to extract the reliable values from wind speed and direction indicating instruments it is necessary to keep them far enough from the sails. Photographs 3.6 and 3.15 demonstrate, for example, how investigators have attempted to reduce this interference effect of sail on the wind-measuring instruments in two different experiments. An alternative practical position for the wind-vane on a racing yacht is difficult to suggest, but for close-hauled work it might be sensible to put vanes outboard of the spreaders, one on each side where, as shown in Fig 3.9, the wind deflection is relatively small. Overlapping foresails and spinnakers, however, are likely to cause trouble, and this position might not be suitable for sailing off the wind.

There is also another factor which complicates the measurements of both the direction and velocity of the apparent wind. This is the wind gradient, the existence and effects of which have been debated for years by sailing enthusiasts. The wind gradient, definable as the rate at which the true wind V_T increases with the height above the water level, has been discussed by the author in Ref 3.21; it is sufficient to repeat here that its magnitude depends on a number of factors. Essentially the wind gradient reflects itself in a variation of β angle and apparent wind velocity V_A along the mast height, and this is shown in Fig 3.11A. It can be seen that the β angle is smaller near the sail foot than at the sail head. This difference for, say, the Dragon or Soling class may be in the order of 3–4° in the close-hauled condition. Such a twist in β angle is accompanied by an increase in wind velocity V_A towards the mast top. The differences in V_A, as shown in Fig 3.11A, may be in the order of 20–30 per cent, or much less, depending to a large extent on the modifying influence of the hull and the actual gradient of the true wind. The hull induces a contraction of the airflow above the deck, thus accelerating the wind speed there and diminishing the effect of the true wind gradient on the apparent wind gradient. This rather strong modifying influence of the hull on the apparent wind gradient is responsible for the conflicting opinions in this respect. In extreme statements it has been argued that wind gradient does not exist. A sketch B in Fig 3.11 illustrates Francis Herreshoff's idea of how the air-flow is distorted by the hull superstructure. To be on the safe side, we should assume that it is rather likely that at least some apparent wind gradient will often occur. Therefore we should expect that, depending on the position of the wind-measuring instruments relative to the deck level, their readings may indicate different values of both V_A and $(\beta-\lambda)$ although we may not know how big those differences are.

Needless to say, the measurement errors in $(\beta-\lambda)$ and V_A can upset any calculation, no matter whether the object is to establish a set of sail coefficients or to use the

Fig 3.11 True wind gradient effect on apparent wind angle.

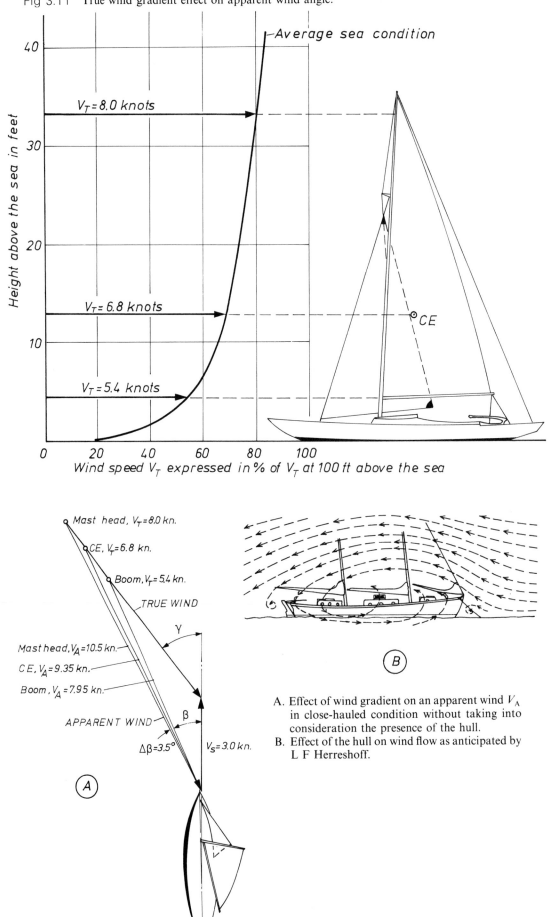

A. Effect of wind gradient on an apparent wind V_A in close-hauled condition without taking into consideration the presence of the hull.
B. Effect of the hull on wind flow as anticipated by L F Herreshoff.

measured (β-λ) and V_A values as an input to a 'black box' V_{mg} meter circuit. As a matter of record the values of V_A, shown in line 1 in Table 3.5, established in the course of Davidson's tests on *Gimcrack* are corrected values of V_A for wind gradient (Ref 3.4). They were corrected for the relative heights of the points at which the measurements were made (9 ft above the deck) and of the centre of effort CE (15.75 ft) according to the relationship:

$$(V_A)h = 0.464 \sqrt[6]{h} (V_A)100 \qquad \text{Eq 3.4}$$

where h = height above water surface, in feet.

  $(V_A)h$ = wind speed at height h, in knots.

  $(V_A)100$ = wind speed at height of 100 ft, in knots.

The question as to whether the corrected values of V_A, which were subsequently used to calculate the *Gimcrack* coefficients, were in fact the real, average values of V_A at the CE height, cannot be answered now. However, it seems highly improbable that the experimenters were just the lucky ones and the expression for the wind gradient given above reflected accurately the wind structure on every day when experiments were performed.

Considering now sailing by instruments, the helmsman should be presented with the data that are essential for the best control of the helm. It can be argued that this is the heading angle (β-λ) alone, since the determination of the best value of (β-λ) to match the actual sailing conditions is properly the helmsman's responsibility. Bearing in mind what has been said about the difficulties in measuring the heading angle (β-λ), the helmsman or navigator should be aware of the fact that the data transmitted by the instruments is only information which is somehow related to the true value of (β-λ) but is not the true value itself. The same criticism can be applied to other instruments, the sensors of which are operating in distorted air- and water-flow fields. So the instruments, although they may play an important part in tuning or trimming a racing yacht to her best potential, cannot replace the personal skill or intuition of the helmsman. Moreover, the effective use of sailing instruments becomes itself an additional skill, which can hardly be perfected by those who do not dare to brave the realms of sailing theory.

(4) Analytically derived sail coefficients

> 'I could have done it in a much more complicated
> way,' said the red Queen, immensely proud.
>
> LEWIS CARROLL

Within the framework of existing aerodynamic theory J Milgram derived sets of sail coefficients by an almost completely analytical method. The result was presented in the SNAME Report (Ref 3.22). It consists primarily of tables, an example of which is demonstrated in Table 3.6A and B and reproduced here from the aforementioned

SNAME publication. The sail force coefficients are given for systematic variation in a sloop-rig geometry. It includes as variables:

jib overlap,
jib span,
boom height,
course angle.

There are also given the coefficients for a cat-rig of different aspect ratio, as illustrated in Tables 3.6A and B, to which reference will be made in the next chapter dealing with wind tunnel tests.

These analytical sail coefficients were determined by a computer program written in accordance with a mathematical model developed by application of the so-called lifting surface theory, which is a modification of the lifting line theory. The details of this application were reported in the 1968 Transactions of the SNAME (Ref 3.23), to which the reader is referred for more precise descriptions. As might be inferred from Table 3.6, those sail coefficients cannot be presented either in the familiar form of polar diagram C_L versus C_D, or in a form similar to the *Gimcrack* coefficients in Fig 3.7. Instead, there are given the numerical values for pairs of coefficients, namely, forward and side force coefficients and the related lift and drag data distinguished in Tables 3.6A and B by a rectangle and applicable to two distinct close-hauled conditions.

The first set of coefficients demonstrated in Table 3.6A is regarded as appropriate for *light to moderate winds*, in which the heeling moment generated by the sails is easily balanced by the stability of the boat. This is the so-called *high lift condition*. The relevant true wind angle 42.9° and the apparent wind angle 27° are assumed, rightly or wrongly, to be typical.

Table 3.6B gives the relevant coefficients for the same rigs as in Table 3.6A, but now the condition considered is that of *moderate to strong winds*, in which the stability requirements yield lower lift than in the previous case. Accordingly, the condition is called the *reduced lift condition*. It can be seen in Table 3.6B that in these circumstances the relevant true wind angle is 45° and the apparent wind angle equals 32.5°, both being larger than those in the previous condition.

Apart from the force coefficients the Tables give information concerning the position of the Centre of Effort. These together with known hull characteristics enable calculation of forward force, side force, heeling moment, and heel angle. Thus, at least in principle, the water sailing speed can be predicted if the hydrodynamic characteristics of the hull are known from the tank tests.

As compared with the *Gimcrack* coefficients, which are in fact applicable for only one particular rig geometry, this new set of analytical coefficients in which a variety of rig planforms is taken into account may look more promising, even in spite of the fact that their application is limited to strictly predetermined sailing conditions, given by the selected true and apparent wind angles, and therefore the full performance calculation in terms of V_{mg} at various wind speeds V_T cannot be made.

A question that inevitably arises is, how well does this analytical data agree with experimental measurements, such as are obtained from wind tunnel tests, bearing in mind that a number of practical as well as theoretical problems make it difficult to obtain the absolute, in a quantitative sense, wind tunnel measurements of the aerodynamic characteristics of sails. As stated by Milgram in Ref 3.22, results of the wind tunnel tests recently carried out on analytically designed sails indicate that '...the only quantity in which agreement is not excellent is the pitching moment.' It sounds encouraging. However, an analysis of available wind tunnel data obtained in the University of Southampton and reported in Ref 3.21, raised doubts on this score. Let us consider the simplest case of a cat-rig of the three aspect ratios shown in Fig 3.12. This gives polar diagrams of C_L versus C_D for the three models, which had

Fig 3.12 Combined effect of aspect ratio and mast diameter on rig efficiency.

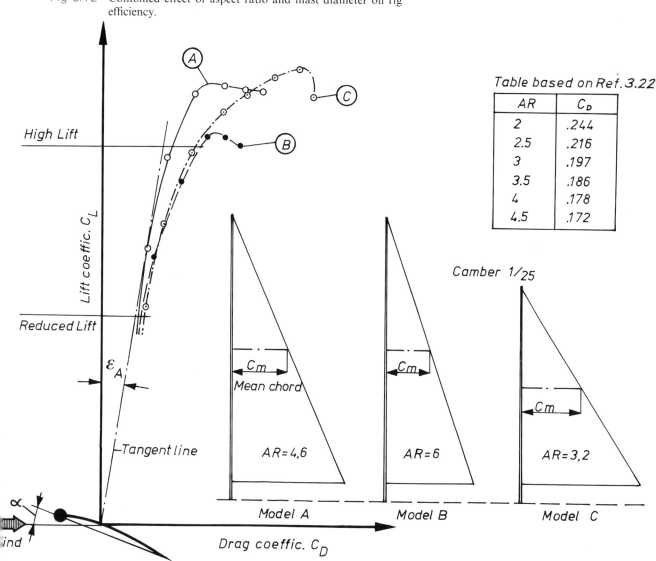

Table based on Ref. 3.22

AR	C_D
2	.244
2.5	.216
3	.197
3.5	.186
4	.178
4.5	.172

TABLE 3.6A

High lift catboat series. Aspect ratio = 2.0

CDF = 0.04: Mast CD = 0.40 based on mast diameter of 0.01 × (Span + Foot)
Main aspect ratio = 2.00 Foretriangle aspect ratio = 0.0
Jib span/main span = 0.0 Jib foot/main foot = 0.0
Foretriangle/boom length = 0.00 Jib foot/foretriangle = 0.0
Boom height/main span = 0.10 LP/foretriangle = 0.0
True wind angle = 42.9° Apparent wind angle = 27.0°

Apparent wind/true wind = 1.50   Boat speed/true wind speed = 0.60

Force coefficients (actual area)

	Jib	Main	Total		Jib	Main	Total
						(1969 new int rule area)	
Forward	0.000	0.426	0.426		1.000	0.608	0.608
Side	0.000	1.366	1.366		1.000	1.952	1.952
Lift	0.0	1.410	1.410		0.0	2.015	2.015
Drag	0.0	0.244	0.244		0.0	0.348	0.348

Sail area ratios

	Jib	Main	Total
RU/AC	1.000	0.700	0.700
Vert side force centre/mast ht	0.0	0.481	0.481
Vert forward force cent/mast ht	0.0	0.522	0.522
Hor side force cent/(Boom + Fortri)	0.0	0.354	0.354

High lift catboat series. Aspect ratio = 2.5

CDF = 0.04: Mast CD = 0.40 based on mast diameter of 0.01 × (Span + Foot)
Main aspect ratio = 2.50 Foretriangle aspect ratio = 0.0
Jib span/main span = 0.0 Jib foot/main foot = 0.0
Foretriangle/boom length = 0.00 Jib foot/foretriangle = 0.0
Boom height/main span = 0.10 LP/foretriangle = 0.0
True wind angle = 42.9° Apparent wind angle = 27.0°

Apparent wind/true wind = 1.50   Boat speed/true wind speed = 0.60

Force coefficients (actual area)

	Jib	Main	Total		Jib	Main	Total
						(1969 new int rule area)	
Forward	0.000	0.453	0.453		1.000	0.581	0.581
Side	0.000	1.358	1.358		1.000	1.741	1.741
Lift	0.0	1.416	1.416		0.0	1.815	1.815
Drag	0.0	0.216	0.216		0.0	0.276	0.276

Sail area ratios

	Jib	Main	Total
RU/AC	1.000	0.780	0.780
Vert side force centre/mast ht	0.0	0.481	0.481
Vert forward force cent/mast ht	0.0	0.517	0.517
Hor side force cent/(Boom + Fortri)	0.0	0.341	0.341

TABLE 3.6B

Reduced lift catboat series. Aspect ratio = 2.0

CDF = 0.04: Mast CD = 0.40 based on mast diameter of 0.01 × (Span + Foot)

Main aspect ratio = 2.00	Foretriangle aspect ratio = 0.0
Jib span/main span = 0.0	Jib foot/main foot = 0.0
Foretriangle/boom length = 0.00	Jib foot/foretriangle = 0.0
Boom height/main span = 0.10	LP/foretriangle = 0.0
True wind angle = 45.0°	Apparent wind angle = 32.5°

Apparent wind/true wind = 1.32   Boat speed/true wind speed = 0.40

Force coefficients (actual area)

	Jib	Main	Total	(1969 new int rule area) Jib	Main	Total
Forward	0.000	0.428	0.428	1.000	0.611	0.611
Side	0.000	1.073	1.073	1.000	1.533	1.533
Lift	0.0	1.134	1.134	0.0	1.621	1.621
Drag	0.0	0.218	0.218	0.0	0.311	0.311

Sail area ratios

RU/AC	1.000	0.700 0.700
Vert side force centre/mast ht	0.0	0.433 0.433
Vert forward force cent/mast ht	0.0	0.473 0.473
Hor side force cent/(Boom + Fortri)	0.0	0.383 0.383

Reduced lift catboat series. Aspect ratio = 2.5

CDF = 0.04: Mast CD = 0.40 based on mast diameter of 0.01 × (Span + Foot)

Main aspect ratio = 2.50	Foretriangle aspect ratio = 0.0
Jib span/main span = 0.0	Jib foot/main foot = 0.0
Foretriangle/boom length = 0.00	Jib foot/foretriangle = 0.0
Boom height/main span = 0.10	LP/foretriangle = 0.0
True wind angle = 45.0°	Apparent wind angle = 32.5°

Apparent wind/true wind = 1.32   Boat speed/true wind speed = 0.40

Force coefficients (actual area)   (1969 new int rule area)

	Jib	Main	Total	Jib	Main	Total
Forward	0.000	0.451	0.451	1.000	0.578	0.578
Side	0.000	1.065	1.065	1.000	1.365	1.365
Lift	0.0	1.140	1.140	0.0	1.461	1.461
Drag	0.0	0.194	0.194	0.0	0.248	0.248

Sail area ratios

RU/AC	1.000	0.780 0.780
Vert side force centre/mast ht	0.0	0.434 0.434
Vert forward force cent/mast ht	0.0	0.467 0.467
Hor side force cent/(Boom + Fortri)	0.0	0.368 0.368

The Tables, reprinted from Ref 3.22, give the values of sail force coefficients for a cat-type rig as drawn. The aspect ratio (span length/foot length) is 2.0 and 2.5. The pairs of coefficients, (forward, side) and (lift, drag), are enveloped by a rectangle for the sake of easier identification.

identical camber, mast diameter and twist, the latter being in the order of 4° between foot and head.

According to Milgram's analytical sail coefficients, one should expect that, by virtue of decreasing induced drag when aspect ratio increases, the windward efficiency of the taller rig will be better than that of the lower one. The table of drag coefficients, for cat-rigs of various aspect ratios but the same lift, shown in the right upper corner in Fig 3.12 and taken from Milgram's Tables, substantiates such an expectation; the wind tunnel test does not, however, confirm this. It can be seen from the curves in Fig 3.12 that by increasing the aspect ratio of the sail one should not necessarily assume that the windward efficiency of the rig will automatically be improved. It would be so if the aspect ratio were the only factor controlling the aerodynamic properties of a sail, but it is not. It appears that the ratio of mast diameter to mean chord of the sail c_m is of considerable significance. More specifically, model A with aspect ratio of 4.6, which is equivalent to AR = 2.3 in Milgram's Tables, is aerodynamically better for windward work than model B of higher aspect ratio. Even model C of the lowest aspect ratio is, in the range of high C_L coefficients, superior to model B, although the merits of rigs B and C are reversed in the reduced lift range.

This powerful effect of mast diameter, its section and rigging on boat performance is common knowledge amongst racing people. As a matter of fact, there are two distinct schools of thought which favour one or the other of the two ultimately contrasting solutions. In one, a large diameter mast section held up by as few shrouds as possible is preferred, whilst in the opposed concept, a tiny mast section, possibly of the so-called delta shape, is argued to be the best in the close-hauled condition, even if it is supported by a complex but well engineered network of rigging. Anyway, extensive evidence in the form of racing results and the tests on various mast sections support the view that the effect of the mast on air-flow may be of greater importance than that of aspect ratio. Taking these two effects into consideration, the geometrically identical planforms can be aerodynamically different enough to render satisfactory quantitative comparison practically impossible. In this respect the analytical sail coefficients, as presented in Ref 3.22, can only be regarded as misleading.

Referring once again to the analytical sail data in Tables 3.6A and B, the aerodynamic lift/drag ratios, which can be calculated from the information for, say, an AR = 2.0, is found to be about 5.8 for the high lift condition and about 5.2 for the reduced lift condition. This implies that, as the wind speed increases, the L/D ratio decreases. Such a trend is hardly in accord with experience obtained in the course of full-scale trials such as the *Gimcrack* tests and others. It is also contrary to the results of performance calculations, to which we shall refer later. As far as the variation of L/D ratio with lift is concerned, all aerofoils manifest the same trend and the sails are no exception in this respect. It can be seen, for example in Fig 3.12, that when the lift coefficient is high the L/D ratio, as reflected by the magnitude of the drag angle ε_A, is relatively low. By reducing lift, or in other words by reducing the incidence angle, the L/D ratio gradually increases. It reaches a certain maximum value (minimum ε_A) and then decreases again shortly before the sail is flogging. This pattern is well presented in Table 3.5, which represents the *Gimcrack* coefficients. 'Thus,' to quote Davidson, 'while there may be instinctive objections to the idea that the lift/drag ratio increases with the heel angle [which is associated with higher wind velocity–author's remark] it is difficult to find tangible evidence against it' (Ref 3.4).

This seeming improvement in sail performance with heel, provided the heel angle does not exceed 20–25°, should not of course be regarded as a mysterious effect of heel angle increase. The explanation is rather straightforward. According to Eq 2.30, Part 2, i.e. $C_D = c_d + (C_L^2/\pi\,AR)$, the total drag C_D of a sail is, through its induced drag component, a function of lift squared. One must therefore expect that in a certain range of incidence angles, at which lift is high, the induced drag increases at a higher rate than lift. Consequently, the L/D ratio is bound to decrease. Conversely, one might expect that by decreasing the incidence angle, and so the lift, a certain optimum in variation of lift and drag is reached and L/D becomes a maximum. As we already know, this optimum is given by the tangent line to the C_L versus C_D polar curve, as plotted in Fig 3.12.

A glance at the two sketches in Fig 1.9, Part 1, illustrating the light and strong wind conditions, may help the reader in refreshing his memory and grasp the physical meaning of the L/D ratio in relation to windward performance. The observed trend in L/D variation is clearly opposite to that which is implied in Tables 3.6A and 3.6B.

Concluding, it appears that the analytical sail coefficients, as demonstrated, cannot possibly be accurate enough to be applied with confidence, and they can be misleading. Their accuracy, strictly conditioned by the theoretical assumptions made, necessarily depends on the exactness of those assumptions in reflecting the actual sails as they operate in real conditions. While describing the method of deriving the analytical sail coefficients J Milgram says:

> 'Almost all the aerodynamic quantities of interest on sails are determined by the pressure distribution on the sails. The effects of the pressure distribution are such that a natural method for the design of sails is to first design the desired pressure distribution, and to then use lifting surface theory to complete the sail shapes necessary to attain these pressure distributions' (Ref 3.22).

Unfortunately, however, there is no assurance that the sail shapes so designed will sustain the desired pressures and in addition that the sail shapes developed theoretically can be obtained even by the most skilled sailmaker. In fact, we already know from section D (5), Part 2 and section B, Part 3 that for various reasons it cannot practically be done. That is why '...the ugly empirical facts', to use T Huxley's expression, as manifested in real sail behaviour are bound to be stubborn and so hardly be ordered to be compliant to the postulates of the theoretical sail coefficients.

Let us now complete this chapter by quoting Einstein's somewhat consoling message, taken from his *Special and General Theory*: 'There could be no fairer destiny for any...theory than that it should point the way to a more comprehensive theory in which it lives on, as a limiting case.'

(5) Determination of sail forces by wind tunnel tests

> 'Be as clear as you can about the various theories you hold, and be aware that we all hold theories unconsciously, or take them for granted, although most of them are almost certain to be false.'
>
> K R POPPER
> *Objective knowledge*

For as long as yacht racing has been a popular sport, the waters of the Solent in Britain (according to the *Badminton Library* 1894, '...clear as crystal, with the air healthy as Switzerland and the scenery nearly as beautiful') or the waters of Long Island Sound in the USA, have been regarded as natural laboratories, where the products of the best designers could be compared in direct competition. At first sight these waters appear to provide a splendid substitute for the towing tank and wind tunnel. The wind, its gradients and squalls, the waves and tides, and in fact everything, is simulated *naturally*. Unfortunately, their effect on yacht behaviour can hardly be measured accurately enough for the purpose of confident systematic development by analysing racing results. On the other hand, in the wind tunnel, for example, everything seems to be *unnatural*, but the forces generated by the sails can be measured relatively easily, allowing the effect of systematic variations of the geometrical and physical factors held under control to be found. This is much more difficult to do while experimenting with a full-scale yacht in natural conditions, even in what would appear to be a carefully controlled trial.

Needless to say, to make meaningful measurements the most rigorous control of the factors involved is essential, whether the experiments are conducted on a full-scale yacht or on a model. In this respect the wind tunnel clearly offers great advantage. Tests can be carried out relatively cheaply on scaled model-sails, made of flexible material, as shown in Photo 3.3. It perhaps does not need emphasis that the largest possible models, or even full-size rigs, provide the best objects for test. This is possible for a set of sails designed for small boats, but the size of wind tunnel required for a 12-Metre rig probably does not exist as yet. Whatever may be said of the limitations of wind tunnel tests, they provide us with the only data source on sail behaviour which we can rely upon for systematic quantitative knowledge of how the various adjustment and geometric factors, available to the designer and sailor, affect the rig efficiency.

It is not proposed to enter here into a detailed discussion of how the wind tunnel tests are made, neither is it pretended that the sail coefficients presented in the following chapters are accurate in an absolute sense. The purpose of the described tests is merely to show some trends in the variation of aerodynamic characteristics of a given rig, when the sail shape is modified in a controlled manner, say, by altering the kicking strap or halliard tension and so on. From the viewpoint of a practical racing sailor such trends are more important than precise numerical values and are of particular significance when considering so-called One Design Classes, such as the Finn, Dragon or Star, in which the hulls may be regarded as identical and thus have fairly constant hydrodynamic characteristics. In these cases, differences in the attainable speed made good to windward V_{mg} will depend almost entirely on the rig efficiency, which in turn depends on tuning, trim, sail-setting, etc.

As far as the art of sail tuning or trimming to give the best performance is concerned, the knowledge of what variables in sail shape are most sensitive or what features of the sail shape are most accessible to intervention, is of primary importance. In establishing priorities in these respects wind tunnel experiments can be of great help.

An inspection of Fig 3.13A and B should help the reader to interpret wind tunnel results, which are frequently presented in standard aerodynamic terms, such as lift L and drag D. Figure 3.13A shows the principle of measuring the aerodynamic forces L and D, from which the total force F_T can be found by means of a vector diagram. Figure 3.13B illustrates how subsequently the total aerodynamic force F_T can be resolved into: the driving force F_R which makes the boat sail, and the heeling force F_H which is responsible for heeling and drift (leeway).

Bearing in mind that the windward performance of a sailing boat depends partly on its leeway λ, which in turn depends on the hydrodynamic characteristics of the hull, the wind tunnel tests alone cannot give more than an approximate indication of sail efficiency, since leeway is not represented in them. In other words, it is not possible to resolve the total aerodynamic force F_T in terms of components F_R and F_H parallel and perpendicular to the direction of motion of the hull through the water, i.e. in the direction of boat speed V_s (course sailed) as illustrated in Fig 3.14A. These in fact are the directions commonly used in determining the forces measured in towing tank tests. Since the leeway angle is not the same for every boat and its value depends on the course sailed β and speed V_s, it became common to present the wind tunnel results in a slightly different way to that demonstrated in Fig 3.14A. This is illustrated in Fig 3.14B, where the components F_x and F_y of the total force F_T are given parallel and perpendicular to the hull centreline, i.e. boat heading $(\beta-\lambda)$. The relevant trigonometrical relations between those new components and lift L and drag D are given by the following equations that can be easily derived from Fig 3.14B.

$$F_x = L \sin(\beta-\lambda) - D \cos(\beta-\lambda) \qquad \text{Eq 3.5}$$

$$F_y = L \cos(\beta-\lambda) + D \sin(\beta-\lambda) \qquad \text{Eq 3.6}$$

consequently the relevant coefficients of forces F_x and F_y are given by:

$$C_x = C_L \sin(\beta-\lambda) - C_D \cos(\beta-\lambda) \qquad \text{Eq 3.5A}$$

$$C_y = C_L \cos(\beta-\lambda) + C_D \sin(\beta-\lambda) \qquad \text{Eq 3.6A}$$

Fig 3.13 Principles of wind tunnel measurement.

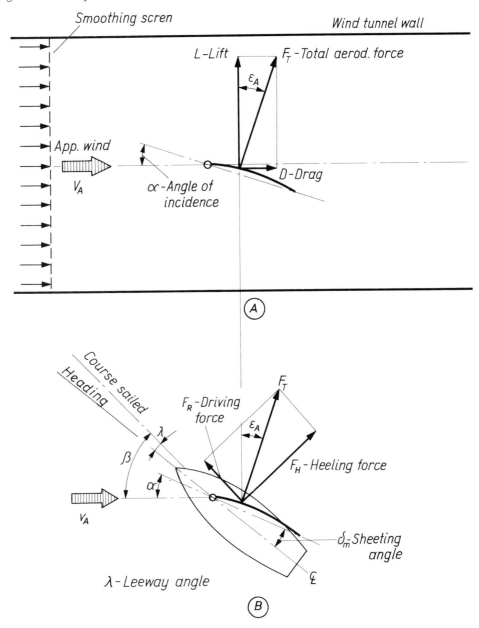

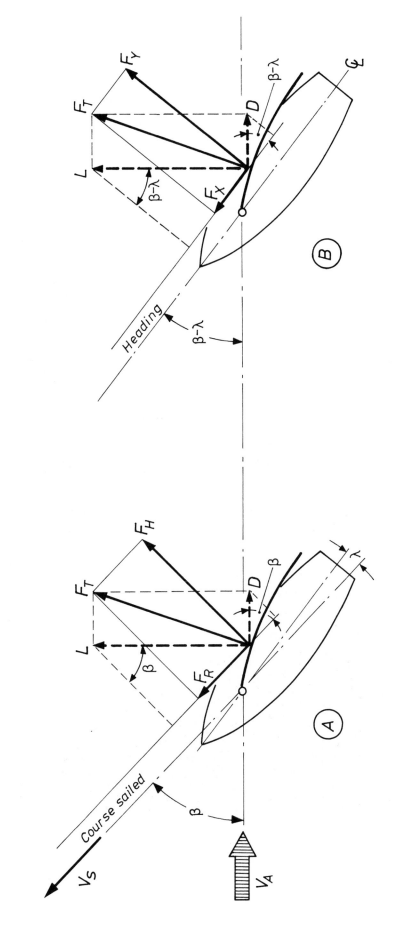

Fig 3.14 Definition of aerodynamic force components (F_R, F_H) and (F_X, F_Y).

Photo 3.16   Above–$\frac{2}{5}$ scale Finn sail in the wind tunnel at Southampton University. Bendy mast is entirely unstayed. To enable the camber and twist of the sail to be defined under various test conditions black lines were painted on the sail one foot apart and parallel to the boom. The sail configuration was recorded using a camera mounted on the wind tunnel roof directly above the masthead and photographing downwards.

Below–a typical photograph yielding sail camber and twist at $\alpha = 25°$. The crossing line from the boom to the sail head gives mid-chord points.

A

Wind

$\alpha = 25°$

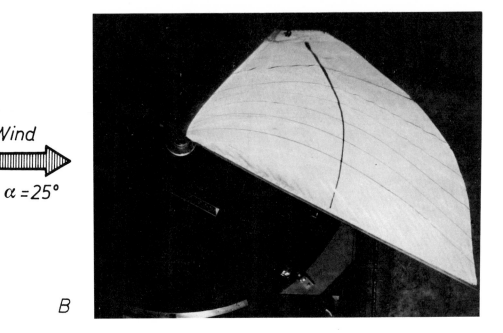

B

The usual procedure with the balance system installed in the large wind tunnel at the University of Southampton, described in Ref 3.24 and partly shown in Photographs 3.3 and 3.16, is to rig the model so that its centreline is at known angle to the axis of the tunnel. This angle is represented by $(\beta-\lambda)$ in Fig 3.14B and the wind speed in the tunnel is represented by V_A. The horizontal force components L and D are measured for known sheeting angles δ_m (mainsail) and δ_f (foresail) and the heading angle $(\beta-\lambda)$ is then altered so that ultimately graphs may be plotted to show how lift or drag vary with heading angle. The whole procedure is then repeated to determine the effect of changing δ_m and δ_f separately. Obviously, once the lift and drag are known, it is possible to present the wind tunnel results in any convenient form, such as the familiar polar diagrams C_L versus C_D, or in terms of F_x and F_y against $(\beta-\lambda)$, as demonstrated earlier in Fig 2.118, Part 2. Subsequently, the sail coefficients derived in the wind tunnel can be used to estimate the boat's speed performance.

D   Wind tunnel results: factors affecting the sail forces and their effects on boat performance

(1) Finn sail tests

Let us now present some measurements of the aerodynamic characteristics of the Finn-type rig, supposedly the simplest rig ever invented. Photographs 1.5B, Part 1, and 3.16, Part 3, illustrate the full-scale rig and its $\frac{2}{5}$ scale model, as tested in the wind tunnel at the University of Southampton (Ref 3.25). The single sail is set on this particular boat on a grooved mast and boom, the mast being entirely unstayed. The position of the boom relative to the mast is fairly fixed and maintained by a form of gooseneck, while the whole rig can be rotated for sail trim.

The mast of the Finn is flexible so that under combined aerodynamic, sheet halliard and kicking strap loading, the whole rig is subject to elastic deformation which, as might be expected, affects the sail shape, the sail forces and hence the boat performance. As seen in Photo 3.16, no attempt was made to represent the hull. However, the height of the model boom above the floor of the tunnel was the scaled height of the boat's boom above the sea. In order to enable the twist and camber of the sail to be recorded under various test conditions the black lines were painted on the sail one foot apart and parallel to the boom. The model was then photographed using a camera in the roof of the tunnel.

The object of the investigation was to determine:

a. The effect of gradually increased kicking strap tension and associated mast flexure on the aerodynamic characteristics of a given mast-sail combination;
b. The effects of varying the tensions in the luff and foot of the sail;

c. The aerodynamic effects due to varying the height of the rig boom above the 'surface of the sea';

d. The effects of varying the wind speed on the sail shape, i.e. twist and camber distribution along the mast height.

In each test run, the angle of incidence α between the sail boom and the wind direction was varied from 2.5° to 40.0° in increments of 2.5°. At each incidence angle the values of the horizontal components of wind force–lift L and drag D–were recorded. All tests were carried out with the model unheeled and, with the exception of item d, at a constant wind speed of 29.3 ft/sec (about force 5 on the Beaufort scale).

(a) *Influence of kicking strap tension*

Consider now the first series of experiments, the object of which was to establish the influence of kicking strap tension and associated variation in camber and twist distribution. The four tests marked Run V, VI, VII, and VIII, in which the kicking strap tension was gradually increased, were carried out with the positions of the head and clew of the sail on the spars left unaltered. The tack of the sail was maintained at a distance of 10 in above the wind tunnel floor, a distance equivalent to the height of the sail tack above the sea surface on the full-scale Finn. Figure 3.15 gives the variation of measured C_L and C_D coefficients with incidence angle α and the sketch attached depicts the sail with the boom pulled vertically downward to four different positions determined by a distance X. Figure 3.16 shows the lift and drag coefficients plotted in another familiar form of polar diagrams with angles of incidence inscribed along the curves. Figure 3.17 demonstrates the variation of L/D ratios over the range of incidence angles used in this series of experiments. The various sail shapes, resulting from the four boom positions, and the consequent flexure of the mast, are shown in Photo 3.17 and Fig. 3.18. The camber and twist of the sail, at each vertical station in each test condition together with mean values of camber, are given in tabular form in Table 3.7. It can be seen from the Table and Fig 3.18 that the basic sail geometry factors, namely camber and twist, change together under the action of the kicking strap in a manner which, as we shall see later, may or may not be the most desirable from the standpoint of sail efficiency but, as demonstrated in Figs 3.15, 3.16, and 3.17, their effect on sail characteristics is profound.

Taking for example the L/D ratio as one of the criteria of windward performance potential of each rig configuration, it can be seen in Fig 3.17 that by hardening the kicking strap it is possible to improve the L/D ratio from 3.6 (Run V) to 5.6 or so (Run VIII), i.e. by 55 per cent!

Inspecting Photo 3.17 it should be noticed that the sail shape investigated in Run VIII, which produces much better L/D ratio than that by the sails marked Run VI or VII, is not the nicest one from a purely aesthetic point of view. We shall see that the best-looking smooth and crease-free sails are not necessarily the most efficient

Fig 3.15 Lift and drag coefficients of Finn-type rig at different kicking strap tension. Incidence angle was measured relative to the boom (see Photo 3.16).

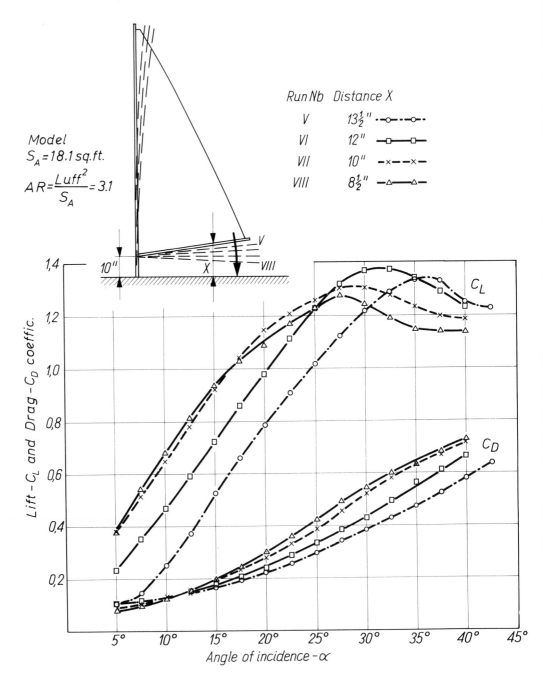

Fig 3.16   Polar diagram of sail coefficients of Finn-type rig at different
kicking strap tension (see Fig 3.15).

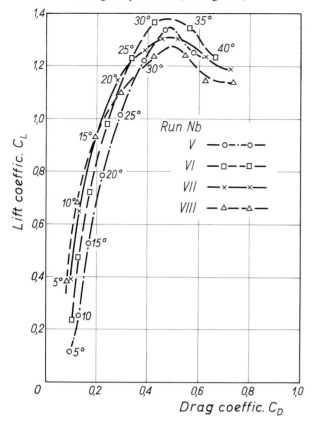

Fig 3.17   *L/D* ratio variation for Finn-type rig (see Fig 3.16).

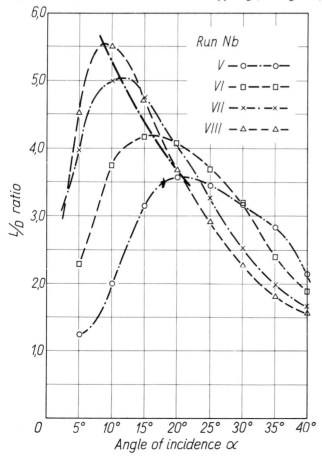

Fig 3.18   Variations in sail twist and camber recorded at different kicking strap tensions. Wind speed $V_A = 29.3$ ft/sec and the incidence angle α of the boom relative to V_A equal to 25° was kept constant (see Fig 3.13A).

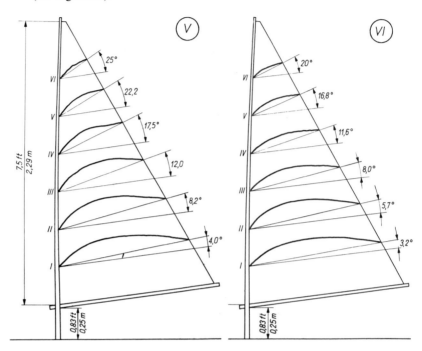

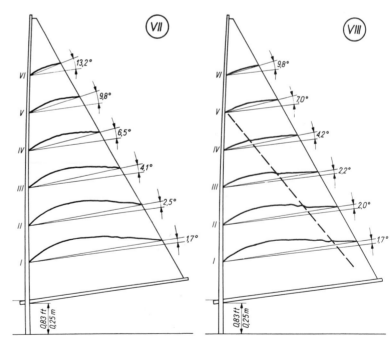

Photo 3.17   View of the Finn rig under different kicking strap tensions;
(no Cunningham hole used). Sail appearance judged from an
aesthetic point of view can be misleading as a criterion of sail
efficiency. The best looking, smooth and relatively crease-
free sail, marked Run VI, is not necessarily the most efficient;
much depends on wind strength and course.

TABLE 3.7

Section	Run V Camber (per cent)	Run V Twist (degrees)	Run VI Camber (per cent)	Run VI Twist (degrees)	Run VII Camber (per cent)	Run VII Twist (degrees)	Run VIII Camber (per cent)	Run VIII Twist (degrees)
I	12.5	4.0	12.3	3.2	10.7	1.7	9.7	1.7
II	15.2	8.2	14.1	5.7	11.6	2.5	9.9	2.0
III	14.0	12.0	13.5	8.0	11.0	4.1	8.2	2.2
IV	12.8	17.3	12.2	11.6	9.2	6.5	6.5	4.2
V	11.7	22.2	11.2	16.8	7.8	9.8	4.7	7.0
VI	10.1	25.0	9.6	20.0	6.7	13.2	3.8	9.8
Mean camber	12.7		12.0		9.5		7.1	

In all cases recorded in Table 3.7 the geometric angle of incidence between the boom and
the wind V_A was constant and equal to 25° (see Fig 3.13A).

Fig 3.19   Effect of changes in luff and foot tensions on L/D ratio.

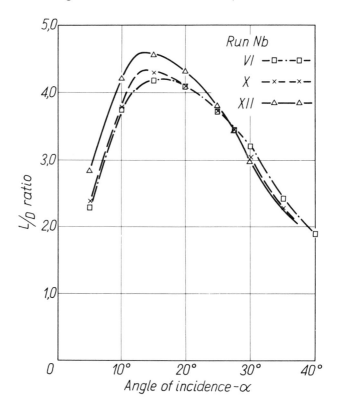

aerodynamically. It will also be shown in quantitative terms that the right sail shape for one wind strength may be quite the wrong shape for another.

(b) *Influence of changes in luff and foot tension*
In this investigation the basic sail-configuration used was that of Run VI of the kicking strap series. Increasing the tension in the clew, outhaul and luff was found to have a similar effect on the L/D ratio as flattening the sail by hardening the kicking strap with its associated bending of the mast. The influence of changes in the sail foot and luff tensions on the L/D ratio is shown in Fig 3.19. The curve marked Run X illustrates the L/D variation when the foot of the sail was stretched a further $\frac{1}{2}$ in beyond the control position (Run VI). Equivalent data is given in the same figure when both luff and foot were stretched $\frac{1}{2}$ in beyond their control positions (Run XII). Study of Fig 3.19 reveals that increasing the foot tension improves the L/D ratio from 4.2 to 4.35, i.e. by about 3.6 per cent. With both luff and foot tension increased, the L/D ratio increases further to 4.6, i.e. 9.5 per cent more than the original L/D ratio given by the curve marked Run VI.

Since the extension of the sail along the luff and foot was comparatively small,

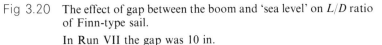

Fig 3.20   The effect of gap between the boom and 'sea level' on L/D ratio of Finn-type sail.

In Run VII the gap was 10 in.
In Run IX the gap was 6 in.

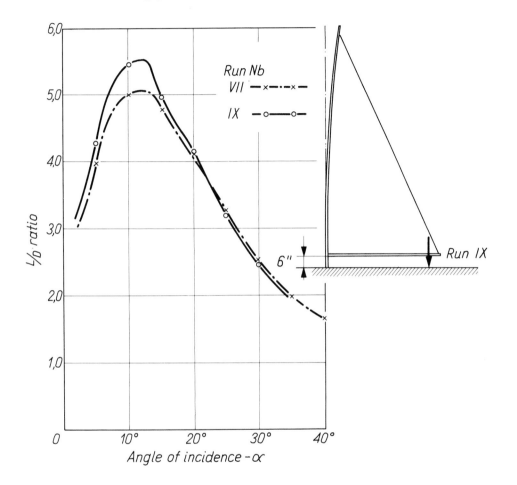

about 0.5 per cent and 1.0 per cent respectively, the foregoing results imply that the setting of the sail on spars is an important factor in sail efficiency attainable from a given sail. This suggests that the ability to vary these tensions easily when under way would be advantageous.

(c) *Influence of rig height above sea level*

In this investigation, the sail form designated VII of the kicking strap series was used as the basic sail configuration and the only change made on the rig was to reduce its height above the wind tunnel floor, which simulated the flat sea surface, by 40 per cent, i.e. from 10 in to 6 in. The variations of L/D ratio with incidence angle for the two cases is presented in Fig 3.20. It can be seen that by reducing the gap between the sail foot and the tunnel floor or eventually the deck of the hull, the maximum L/D ratio increases some 10 per cent from 5.0 to 5.55. This quite measurable effect may be attributed to the diminution of the end-losses due to the trailing vortex shed underneath the boom. In general, except in very light winds where strong wind gradient cannot be ignored, the lower the rig is mounted in the boat the better, and the above experiment confirms our previously reached conclusion, when discussing Fig 2.115, Part 2.

(d) *Influence of wind speed on sail shape*

Figure 3.21A, B, C illustrates the changes in the camber and twist combinations adopted by the sail for tunnel wind speeds of 19.5, 25.0 and 33.5 ft/sec respectively. Table 3.8 gives numerical values of camber and twist recorded at stations I–VI.

TABLE 3.8

Section	Run A $v = 19.5\,\text{ft/sec}$		Run B $v = 25\,\text{ft/sec}$		Run C $v = 33.5\,\text{ft/sec}$	
	Camber (per cent)	Twist (degrees)	Camber (per cent)	Twist (degrees)	Camber (per cent)	Twist (degrees)
I	10.0	1.0	10.3	1.6	11.4	2.7
II	11.1	1.0	11.9	2.0	13.1	4.3
III	11.1	0	11.6	2.0	12.0	6.3
IV	10.4	0	10.9	2.6	11.2	8.0
V	10.0	3.0	11.4	6.0	11.0	12.5
VI	9.5	5.0	9.6	8.0	9.5	15.0
Mean camber	10.3		11.0		11.4	

Study of these data indicates that variation of the wind speed, on which the pressure experienced by the sail depends, modifies the shape of the sail to one which is considerably different from that determined by the sailmaker's cutting. Table 3.8 shows that camber is less subject to variation due to wind speed than is the twist, but both show a tendency to increase with wind speed. It should be pointed out here that this variation of shape is not only due to stretch in the sail cloth under increasing pressure forces, but also includes the effects of the increasing bend in the mast due to these same forces. The results are not therefore applicable to stayed masts which remain effectively straight.

We have seen from our previous discussion on Figs 2.111–113 and on Anderson's formula in Part 2, that excessive twist is detrimental to windward performance since it incurs a heavy drag penalty in terms of additional induced drag. In the following chapter we shall discuss in some detail how heavy those penalties might be in terms of speed made good V_{mg}.

Fig 3.21   Effect of wind speed on camber and twist.
$A–V_A = 19.5$ ft/sec   (5.9 m/sec)
$B–V_A = 25.0$ ft/sec   (7.6 m/sec)
$C–V_A = 33.5$ ft/sec   (10.2 m/sec)

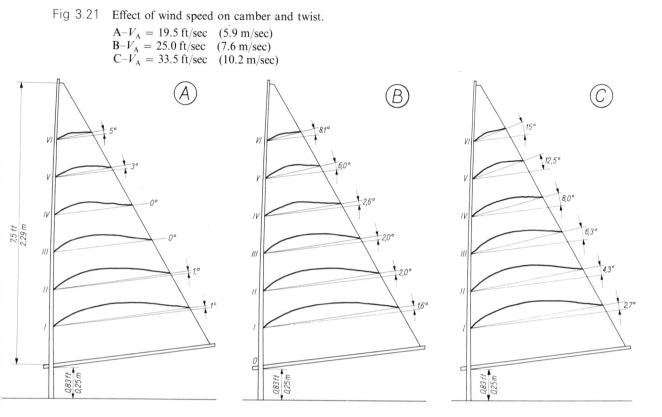

(2) Effects of sail shape on boat performance

'We are working together to one end, some with
knowledge and design, and others without knowing
what they do.'

MARCUS AURELIUS

It is realized that for a rigorous quantitative assessment of the performance of a sail
there can only be one approach, namely, to do a full performance calculation of a
yacht, using known hull data. In other words, it would be necessary to solve
numerically the whole Basic Performance Equation, mentioned at the beginning of
section C. This, however, is not always convenient, since one may wish to consider
sails, albeit in very general terms without reference to a particular hull, and be
satisfied even if the answers so obtained do not represent the best possible attempt at
estimating the absolute performance.

The question arises as to whether this approach is feasible, and if so, what criteria
one should adopt when making a relative assessment of the wind tunnel results. In
previous chapters we have used the L/D ratio as one of the criteria of sail efficiency.

Now we should consider the meaning attached to the term 'sail efficiency' in more precise terms. By analogy, applying the customary engineering routine, the sail efficiency might tentatively be expressed as a ratio of the actual effective sail power which could be used to drive a boat, to that power which the sail captures from the wind, at a given incidence angle. According to Fig 3.14A this actual *effective sail power* may be represented by the driving force F_R, while the total aerodynamic force F_T might be regarded as an equivalent of the captured wind energy. Since in close-hauled conditions the total aerodynamic force F_T is not much greater than either lift L or heeling force F_H, we shall use those two forces as more convenient equivalent quantities of F_T.

Now one can argue that if at a particular course sailed β an alteration in sail sheeting, twist or camber is accompanied by an increase in the driving force component F_R without a corresponding increase in the heeling force F_H, then a better performance to windward will result, i.e. the sail efficiency is higher. In other words, the aerodynamic comparison between rigs of a given plan form, but different in shape, in terms of twist and camber distribution, should be made at the same heeling force for each, so that the better rig will be the one with the higher driving force. The comparison should extend over a range of lift forces or lift coefficients which the rigs are expected to develop at various wind speeds.

Let us apply this measure of sail efficiency to the results we already have at hand, namely the Finn rig data. Figure 3.22 illustrates the set of four measurements–the kicking strap series, demonstrated earlier in Fig 3.16. The assumed course β to windward is 30° and two wind conditions are considered. In the first, marked 'strong winds', the heeling force that can be balanced by a helmsman of a given weight is represented in Fig 3.22 by the relevant coefficient C_{H1}. One can see that depending on the kicking strap tension the available driving coefficient C_R changes considerably from C_{RB} to C_{RA}, some 60 per cent, the heeling coefficient C_{H1} being constant. Clearly, sail configuration V is the worst one and configuration VIII is the best.

Let us now assume that another, heavier or more athletic, helmsman is capable of balancing a greater heeling force than that defined by a heeling coefficient value of C_{H1}, i.e. he sets exactly the same sail of configuration VIII but at a higher incidence angle. This is equivalent to shifting the relevant coefficients along the polar curve from point A to A'. The gain is about 20 per cent more driving power than in the previous case, denoted by point A on the polar curve.

In *light winds*, when the tolerable heeling force coefficient is now given by C_{H2}, i.e. a much larger value than C_{H1}, the merits of the sail configurations are reversed, when compared with those in the strong wind condition. The configuration VIII is now the worst one and configuration VI becomes superior. Considering point C on the polar curve VI it becomes evident that this particular configuration produces more driving force at a given heeling force than any other configuration.

We may express the above tentative findings concerning the sail efficiency in terms of L/D ratio. It can legitimately be argued that, if the C_L versus C_D curve for a given

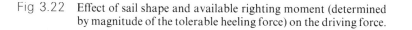

Fig 3.22   Effect of sail shape and available righting moment (determined by magnitude of the tolerable heeling force) on the driving force.

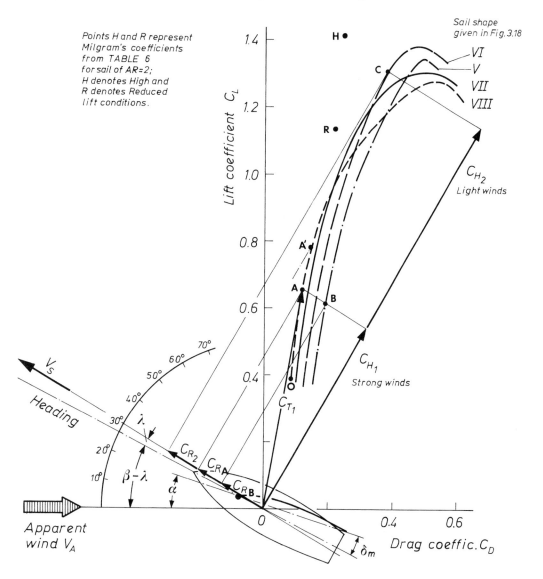

rig is bodily shifted to the left, towards the lower drag relative to the other curve regarded as a *reference curve*, the sail efficiency is improved. This assumption is already represented graphically in a self-explanatory manner in Fig 3.22 and can be expressed as follows: when the two different sail test points, for example A and B, have the same C_H value, then the better is simply the one having the higher L/D ratio. The above criterion of sail efficiency can be expressed in yet another way: if comparison is made between the two different rigs, the estimate should be done at the same lift for each, so that the better rig will be the one with the lower drag.

From the following equations, which can be derived in a similar manner to Eqs 3.6 and 3.6A from Fig 3.14A:

$$F_R = L \sin \beta - D \cos \beta \qquad \text{Eq 3.7}$$

$$F_H = L \cos \beta + D \sin \beta \qquad \text{Eq 3.7A}$$

we can conclude immediately that the drag D not only lowers the driving force F_R but also increases the harmful heeling force F_H. Hence one may infer that particularly in *high wind velocities* a reduction in drag should be an ultimate aim, when tuning or adjusting the sails to their most effective functioning. The largest possible L/D ratio, in the whole range of C_L coefficients applicable in windward work, appears to be a sufficient criterion of sail efficiency. In other words, the comparison of potential merits of various sail configurations should extend over the whole practical range of C_L coefficients. In the case of our Finn rig, this range begins somewhere at C_L about 0.4 extending upwards to C_L about 1.3. These figures may be different, particularly the upper value of C_L, depending on the type of rig in question.

The physical meaning of the increasing demand for the higher L/D ratio when the wind is blowing harder, should become clear if we refresh our memories by examining once again Figs 1.9 and 1.21, Part 1, together with Eq 1.6, which reads:

$$V_{mg}/V_T = \frac{\cot \gamma}{\cot(\gamma-\beta) - \cot \gamma} = \frac{\cot \gamma}{\cot(\gamma - \varepsilon_A - \varepsilon_H) - \cot \gamma}$$

From Fig 1.9 it can be seen that when the boat speed V_s increases the ε_H angle increases, at first slowly and then more rapidly. At higher boat speeds, associated with a high wind, the course β can only remain unchanged if the sail is tuned and trimmed in such a way that its ε_A angle decreases by the same amount as the ε_H angle increases. This significance of ε_A variation on the true course γ sailed to windward is well reflected in Fig 1.21. Another way of considering the problem is to insert just two numerical values in place of β angle, say 25° and 30°, into the above equation and find out that it is advantageous to maintain the β angle low to obtain a high V_{mg}/V_T ratio in strong winds.

Cautious scientists will ask the question: if the driving force F_R increases, does this gain offset the losses in terms of additional hull drag due to higher induced drag associated with an increase in the heeling force? Besides, they may argue, an increase in V_s, which accompanies an increase in the driving force, may result in increase of the course angle γ and this may, according to the equation $V_{mg} = V_s \cos \gamma$, impair the V_{mg}. Such a question cannot be answered for certain without complete performance calculations. But those calculations which were carried out by the author, or are known to him, indicate beyond doubt that in the conditions assumed when discussing Fig 3.22 and sail efficiency criteria, an increase in driving force more than offsets the eventual losses in hull drag. A reservation should perhaps be added: provided the hull itself is not a bizarre freak.

The results of the Finn performance calculations, presented in Figs 1.23 and 1.23A, Part 1, for a light and a heavy helmsman, illustrate this point in a quantitative sense. Photographs 1.5A, 1.6, 1.14, 1.15 in Part 1 illustrate this problem in a different way. They show the attempts of crews in various classes to increase the tolerable heeling force F_H as much as is humanly possible in order to gain more driving force and so improve the speed performance. If the losses, in terms of higher hull drag due to higher heeling force, were not offset advantageously by greater driving power from the sail, it would be difficult to persuade the crew to hang, with no purpose, outside the gunwale. Another example along the same line is the IYRU rule referring to soaking wet garments, which it is stated should not weigh more than 15 kg (initially the limit was 20 kg). Soon after, people began to conjure up strange water-jackets with self-drainers and pumps, which would be filled upwind, emptied off-wind, and so encourage sailors to wear as little as possible to make full use of the permitted 15 kg as water ballast. More recently, rumours have spread that some eager devotees to the cult of high performance '...have been slipping mercury into their water-jackets'.

Now, it remains to check whether the relative merits of the sail configurations of Fig 3.22, established tentatively by the above criteria of sail efficiency, agree with the performance predictions calculated quantitatively by including the hull data. Figure 3.23 shows the results in terms of V_{mg} plotted against true wind V_T for the four Finn rig characteristics V, VI, VII and VIII, defined earlier in Figs 3.18, 3.22 and Table 3.7. The calculations were based on the assumptions that the light helmsman sitting outside the gunwale balances about 57 lb of the heeling force F_H developed on the sail. If the heeling force exceeds this limit the boat cannot be sailed upright and her performance deteriorates. The same graphical method of performance prediction was used as in the example demonstrated in Figs 1.9A–F, Part 1. This method, employing a minimum of hull and sail data, and in which the crudity of simplifications introduced is somewhat offset by the ease and rapidity of the calculations, is described in detail in Ref 3.21. Thus the following restrictions and simplifications were introduced: the boat is sailed upright, the sail camber and twist are assumed to be dependent on the kicking strap tension only, but not on the wind force. The effects of wind unsteadiness and rough water are not taken into account, neither are accelerations in the yacht's motion, whether caused by wind, water or crew dynamic action, trying, say, by means of pumping or rocking to get the best out of the craft. These limitations may seem drastic, but there is little to learn from our investigation of variable sail shape until steady sailing conditions have been studied to the extent of reaching a fair measure of understanding.

One can see from Fig 3.23 that the aerodynamic properties of a given sail, the shape of which is modified by kicking strap tension, have a powerful effect on the attainable V_{mg}. Sail configuration V definitely produces the worst windward performance in the range of assumed wind speeds. Sail VIII is superior in the range of true wind speed V_T above 6 knots and, say, in a moderate breeze–force 4 on the Beaufort scale–the difference in V_{mg} resulting from the worst sail (V) and the best

Fig 3.23   Performance curves V_{mg} versus V_T of a Finn dinghy driven by
sails of four different shapes as shown in Fig 3.18 (see
complementary Fig 3.22).

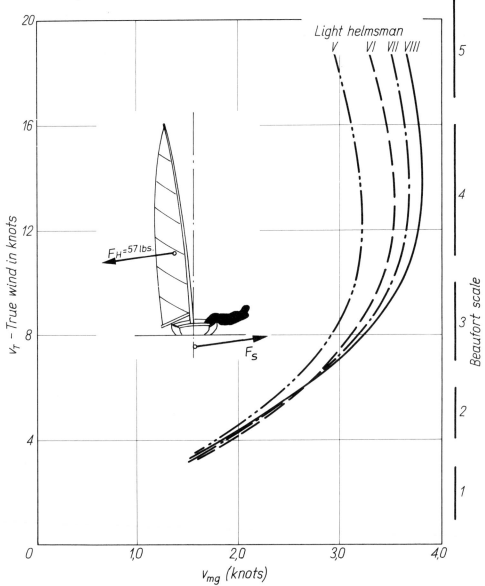

configuration (VIII) are about 20 per cent. In light wind speeds, V_T below 6 knots,
the sail configuration marked VI becomes more efficient than that marked VIII,
producing, according to Fig 3.22, the highest C_L coefficient and at the same time
higher L/D ratio than any other sail shape. However, no one sail shape is superior
over the whole range of wind velocities.

Comparison between Fig 3.23 and Fig 3.22 shows good correlation between the attainable L/D ratio and V_{mg}. In other words, our previous estimate of sail efficiency, based purely on a tentative analysis of Fig 3.22, is valid and there is no apparent reason why this analysis procedure should not also be valid for any other boat which is sailed upright, or nearly upright, within the limit set by the weight and agility of crew.

It should be pointed out that the performance curves plotted in Fig 3.23 represent the optimum performance, which means that both the sheeting angle δ_m and the course sailed β were correctly adjusted to the wind strength V_T for each sail configuration. It may be instructive to examine now the basic trends in variations of these optimum values of:

course sailed β
and sheeting angle δ_m
with wind speed V_T.

This is shown in Fig 3.24 and the two curves refer to the sail VII only. However, their character is similar to that which could be drawn for sails V, VI, and VIII. The curve illustrating the optimum sheeting angle δ_m is particularly interesting. As we may remember (see Fig 1.10, Part 1), the sail incidence α, which determines the aerodynamic forces, is related to the sheeting angle δ_m and course sailed β and so can be changed by altering either δ_m or β. It can be seen that over the lower wind speed range, up to force 2–3 on the Beaufort scale, the optimum δ_m remains fairly constant and at a relatively low value, but as the wind speed V_T increases so does the sheeting angle, quite sharply. When the wind force increases, the incidence angle α changes from that yielding the maximum driving force F_R, which is coupled with a high value of lift force L, to that which is coupled with highest L/D ratio. This sequence is illustrated in Fig 3.22 and the shift in sail trim, from point 0 on curve VIII (relevant to very heavy winds) through points A, A' to point C on curve VI (relevant to light winds), may be interpreted as the helmsman's attempt to obtain the maximum speed V_{mg} to windward from the available righting moment, i.e. tolerable heeling force coefficient C_H.

The optimum β angle varies, according to Fig 3.24, quite considerably too. In light winds, the best β is just above 30°, then it gradually decreases to its minimum which is about 25° at wind force 2–3 on the Beaufort scale, and increases again when the wind blows harder.

Figure 3.25 demonstrates the variation of L/D ratio versus incidence angle α for optimum sail trim at different wind speeds. The arrows above the curve point to the actual value of L/D at a given wind speed V_T. This figure confirms our conclusion reached earlier, when analysing the data of Fig 3.22 and also when discussing Milgram's analytical sail coefficients, that in light winds, when the sail generates a high lift coefficient, the L/D ratio is relatively low. It gradually increases with wind speed, reaching its maximum in moderate winds, after which it decreases again due to the sail being trimmed to a lower and lower incidence as the wind strength

Fig 3.24   Variation of optimum sheeting angle δ_m and course sailed β with wind speed V_T. The relationship between δ_m, β and incidence angle α is shown in Fig 1.10.

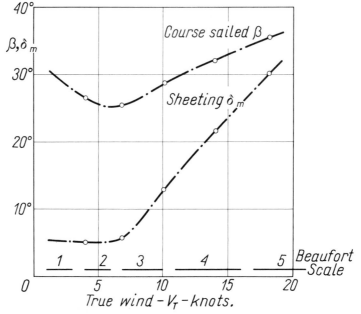

increases. Finally, at the end, sail flogging commences and this sail trim is denoted in Fig 3.25 as the 'luff lifting condition'.

In Fig 3.22, there are marked two points H and R, which are estimates of aerodynamic characteristics of a similar aspect ratio sail derived by Milgram and discussed in section C (4). One point H refers to the light weather condition, in which the heeling moment is of no significance and the other point R refers to heavy winds, in which the heeling moment is of primary importance. Four significant points arise from a comparison of Milgram's data with the results of the wind tunnel tests:

1. The estimated sail characteristics, based on analytical methods, are much more optimistic than are suggested by the results of wind tunnel tests.
2. Point R, which denotes the reduced lift conditions applicable to strong winds is, in fact, close to the wind tunnel dates which are appropriate to light wind conditions, in which the measured lift coefficients are high.
3. Point H indicates an unrealistically high lift coefficient and L/D ratio in that it appears to bear no relevance to sail data obtained by wind tunnel testing. The recorded discrepancy is too large to be possibly bridged by even the most cleverly designed set of corrections.
4. Lastly, from sail characteristics, such as are given by single points H or R it is difficult to extract useful information on how a sail really operates, or how a given rig or sail shape should be corrected if there are some faults or mistakes in their design. These are the problems which we intend to investigate in the next chapter.

Fig 3.25   Variation of L/D ratio versus incidence α for optimum sail trim
at different wind speeds V_T.

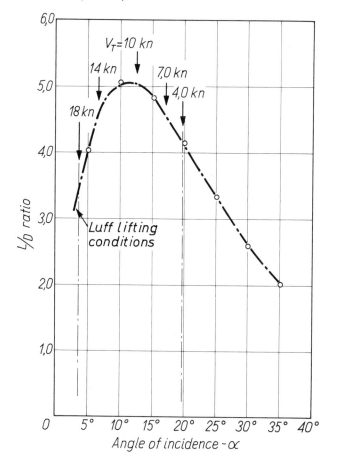

(3) Sail trim and tuning

'It cost much labour and many days before all these
things were brought to perfection.'

DANIEL DEFOE
Robinson Crusoe

The fact that the very same sail can produce entirely different performance levels on
the same boat, as evidenced in Fig 3.23, has become widely known since the startling
racing results demonstrated by W Huetschler, when sailing his Star in one of the
Olympics just before the Second World War. By introducing a flexible rig, in which
the mast bend and hence the sail shape could be controlled by quick adjustments to
the stays, shrouds and runners, the Star class gave an enormous impetus towards

Photo 3.18   The picture above illustrates the relatively simple mainsheet system. The lower picture, which bears some resemblance to a rather unkempt telephone exchange, gives an idea of the development made from simple to complex control systems. It certainly puts a heavy demand on the crew.

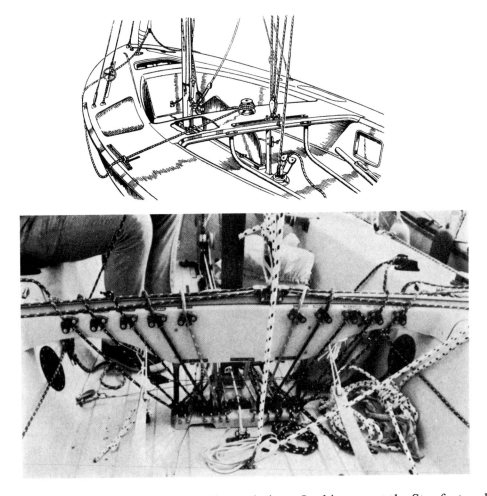

what might be called the modern sailing technique. In this respect the Star fostered more progress in developing new methods of tuning and trimming than any other racing class, and still offers opportunity to exercise the finest racing techniques. The Star can in fact be regarded as a classic Olympic boat.

The simple mainsheet and jib sheets, originally the only means of controlling the shape and incidence angles of the two sails, developed rapidly, in the life span of one generation, into the endless-line system used by modern sailors as a more sophisticated means of precise adjusting of the sails to instantly changing wind and course conditions. Since, in most racing classes, the development of fittings, gadgets and tuning devices is not inhibited by the class rules, the top helmsmen have been stretching their imaginations to invent uncompromisingly expensive go-fast equipment. Some of them are shown in Photographs 3.18–3.21.

Although ocean racing people in the past were generally the slowest to appreciate the difference that good sail control could make on boat performance, they are nowadays becoming more aware of the potential advantage which, say, a good kicking strap may offer. Many top racing keel-boats have already adopted the semi-circular Star track concept, shown in Photo 3.21, or developed powerful hydraulic vangs. Moreover, it would seem that in some cases there is even someone amongst the crew willing to play the vang constantly!

The control of sail incidence, camber and twist is nowadays extensively used by racing helmsmen in the process of sail trimming and tuning in order to improve sail efficiency, but only very few can boast of understanding what they are really doing. Understanding means here the ability to use the available tuning and trimming gadgetry with the confident knowledge of the end in view. Tuning techniques have now reached such a level of complexity that an increasing number of racing people appear to face the Chinese dilemma, so aptly defined by Lao-Tse– '...doing nothing is better than to be busy doing nothing.' In other words, non-operating is better than mal-operating. Let us examine this contemporary sailors' dilemma in some quantitative terms.

It has been mentioned when referring to Fig 3.24 that for a given sail shape, determined by its camber and twist distribution, there is an optimum angle of incidence α or sheeting angle δ_m at which the boat can attain her best V_{mg}. Now, a question of practical interest arises as to how great are the losses in a boat's performance when the helmsman does not adjust the sheeting angle δ_m for the most effective sail functioning, but maintains the correct β angle. This is answered in Fig 3.26, which shows the basic optimum performance curve for sail VII and also the two other curves touching tangentially the first one at two different points. The two other curves represent the attainable V_{mg} for the same sail, assuming that in one case the sheeting angle $\delta_m = 10°$ and in another this angle $\delta_m = 18°$. It can clearly be seen that in both cases the potential efficiency of the sail is exploited fully in only a very narrow range of the true wind V_T. Below or above the particular wind speed V_T, at which there is an optimum coincidence between V_T and the proper sheeting angle δ_m, the boat will not sail as fast as she could if the helmsman continuously adjusted the sheeting angle according to the available stability determined by the wind strength. These losses in the boat's performance can be estimated from Fig 3.26 and, for example, if the true wind speed $V_T = 12$ knots and the helmsman controls the boom in such a way that the sheeting angle $\delta = 10°$, then the best attainable V_{mg} will be about 3.4 knots instead of about 3.6 knots, which might be attained with a sheeting angle of 17°. From these results the significance of a wide mainsheet track and the ability of the helmsman to use it instantly in ever-changing wind conditions can easily be appreciated.

In Figs 1.9D, E and F, Part 1, the reader will find more detailed information on the

Photo 3.19   Flying Dutchman class. Top picture–control consoles became common not only in relatively small Olympic classes in which they originated, but also in Ocean racers. To the left–one-of-a-kind worm gear to tension the genoa luff. Although it looks large, the fittings are fairly light. Alongside one can see numbers for setting tension. To the right–the British FD used this Brook and Gatehouse apparent wind indicator which costs a large fraction of the boat's total price.

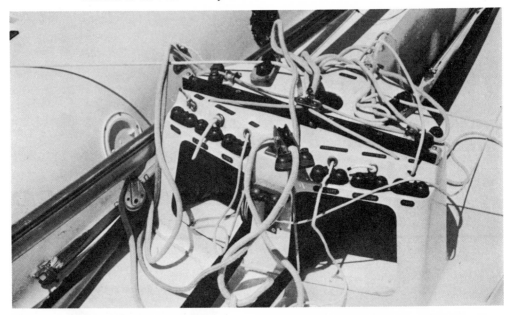

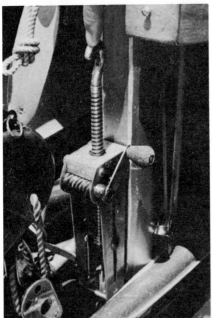

Photo 3.20   The interior equipment of a FD hull may break the most dedicated sailor. It is believed that up to 1000 hours of work is needed to get all of the control gear working reasonably well and in a fool-proof manner.

Photo 3.21 Examples of the powerful kicking strap or vang systems developed in the Star (A) and Finn (B) racing classes. In both cases the vang tension remains constant with the boom in any position. Photograph C shows hydraulic kicking strap used aboard modern offshore racer.

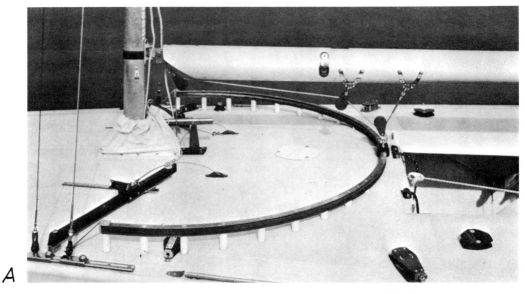

A

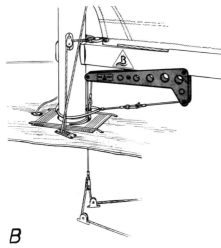

B

C

Fig 3.26 Effect of sheeting angle δ_m on V_{mg}.

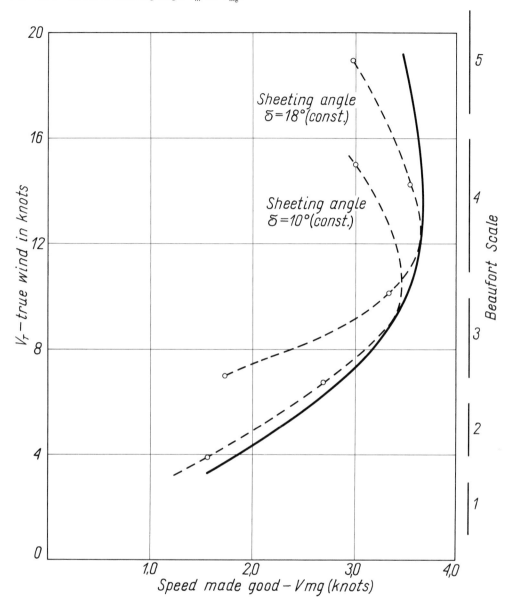

Finn performance sensitivity to maladjustments in:

apparent course β

sheeting angle δ_m

and true course γ

The above type of presentation of boat performance, as well as that in Figs 3.23, 3.24 or 3.26, may answer a number of questions related to the art of sailing, such as–what course must one steer to make the best speed upwind, how much do we lose by pinching or sailing free or if we bear away without easing the sheets, etc. The answers given apply to the Finn-type boat; nevertheless the general pattern is true for every sailing boat. From these demonstrations of trends in yacht behaviour one may not be able to prove the Lao-Tse point–that non-operating is better than mal-operating. What can be proved, however, is that it is possible to attain large improvements in terms of V_{mg} by relatively small intervention in sail trim or tuning. These data do not, however, answer the question as to why one sail configuration is better than another. And, moreover, they do not give a clear-cut clue as to what is the real cause of differences in sail efficiency and what should be done in order to optimize the rig in given conditions, determined by the wind and course sailed. To talk meaningfully about the optimum sail shape one has to consider the following groups of parameters or factors listed in Table 3.9.

TABLE 3.9

1. Accessible parameters of the sail shape and its attitude relative to V_A which can be altered in the course of sail trimming and tuning.
2. Criteria by which one can judge the sail shape and hence the adequacy of the adjustments made.
3. Constraints which may not allow the helmsman to maximize fully the potential driving capabilities of the rig in question.

The accessible parameters in group 1 are specified in Table 3.10 as follows:

TABLE 3.10

1. Trim angles or sheeting angles δ_m, δ_f, for mainsail and foresail respectively, as defined in Figs 3.5 and 1.10.
2. Twist angle, i.e. the variation of the trim angle towards the sail head, as shown for example in Fig 3.18 and Table 3.7.
3. Sail curvature at the leading edge, which determines what was described earlier when referring to Fig 2.62, as the *entrance efficiency*.
4. Camber, its magnitude, chordwise and vertical distribution, as shown in Fig 3.18 in which the position of maximum camber can also be distinguished.

The above parameters can be altered by deliberately designed control devices or by indirect control means which, in the case of the mainsail alone, are registered in Table 3.11.

TABLE 3.11

1. Mainsheet
2. Mainsheet traveller
3. Kicking strap
4. Cunningham holes
5. Outhaul
6. Mast flexibility controlled in turn by the rigging wires, spreaders, jumper stays, running back stays, etc.
7. Batten stiffness
8. Zipper foot
9. Boom flexibility
10. Leech line, etc.

In an effort to satisfy a popular demand for simple answers, the sailing magazines publish numerous articles in which racing people are given recipies on how to use the means of Table 3.11 in order to achieve better racing results. Such short cuts to the answers, although desirable at first sight, are not always feasible. In fact, many answers given in magazine articles are contradictory, and in most of them there is relatively little which is not controversial. There are several reasons why those recipes should not be followed blindly but taken with caution and a comprehension of the underlying assumptions.

It is worth while to consider this point closely by analysing critically just a short excerpt from a recipe written by a prominent racing helmsman. This may help to draw the reader's attention from irrelevant or misleading issues towards the real causal relations on which the success or failure of the sail-tuning process finally depend.

Here is the passage:

'...The rig must be designed so that by adjustments to spars and sail edge tensions the same sail can be made to assume the right shape for varying conditions upwind and downwind in light airs and strong breezes. The sailmaker has to build his sail in the right material so that it can stand these varying forces and tensions without becoming distorted. The crew have to learn how the forces should be applied either automatically by the effect of rigging tensions on the spars or physically by varying tensions on kicking strap, luff and foot.

In light to moderate winds the problems are easiest to understand. The crew can counteract the heeling forces of the wind without having to ease the sails. Waves are not so large as to affect the pointing angle of the boat. Main and jib sheets can be set so that the boat sails as close to the wind as the helmsman feels she should in the prevailing conditions, and she is competitive with the best boats around her. The forces on the sail are not great enough to cause any distortion of the cloth. Luff and foot are fully tensioned to the black bands.

Under these ideal conditions it appears that a mainsail of moderate fullness performs best with a chord depth of 10 or 11 to 1 in the lower third, gradually reducing to about 15 to 1 in the upper third and with its position of greatest depth about one-third of the length of the chord back from the luff. Twist in the sail should be as little as possible (all chord sections should be nearly in the same plane) and the leech of the sail should be firm and straight without any hooking to windward.'

The only definite instruction or advice one can extract from the above excerpt concerning sail trim in light and moderate winds, is that the magnitude of sail camber should decrease from $\frac{1}{10}$ in the lower part to $\frac{1}{15}$ in the upper part of the sail, and that twist in the sail '...should be as little as possible'. This is in accordance with some pundits who think that the kicking strap should be tight down all the time.

From section (5) (a), (b) in Part 2, the reader may recall that any triangular sail is, by its very nature, prone to develop an early stall in the upper part. This was demonstrated for example in Fig 2.129. On this occasion it had been mentioned that a method of preventing such a stalling, which may profitably be employed in conjunction with a certain degree of twist (washout), is to increase the sail camber from the boom towards the head or at least over the top part of the sail. This conclusion is exactly opposite to the advice given above, so we are left in something of a quandary as to which is right.

(a) *Kicking strap control*

The correct answer to the twist problem can be obtained from the analysis of controlled experiments we already have at hand, namely the kicking strap tests on the Finn rig. In Fig. 3.27 there are plotted four sketches showing the effect of decreasing twist. These are based on the data of Fig 3.18 and Table 3.7. All four sail configurations V–VIII are plotted and the continuous lines give the variation of geometric, incidence α, as measured at various sail-sections along the mast, while the broken lines give the relevant downwash, i.e. the induced angle α_i, as calculated by Tanner (Ref 3.26).

According to Eq 2.23, Part 2,

$$\alpha_{ef} = \alpha - \alpha_i$$

i.e. the effective angle of incidence α_{ef}, which really matters most, is the difference between the value of α given by the continuous curve and the value of α_i given by the broken line. These differences, or in other words the values of effective incidence angle α_{ef} at various mast heights, can be read within the crossed areas, as illustrated in Fig 3.27. Bearing in mind that the results presented are based on measurements taken at a constant geometric angle of incidence $\alpha = 25°$, measured between the boom and the apparent wind direction V_A, one may derive the following conclusions:

Fig 3.27 Distribution of effective angle of incidence α_{ef} for sail shapes given in Fig 3.18. Valid only for one particular geometric incidence $\alpha = 25°$, measured between the boom and apparent wind direction.

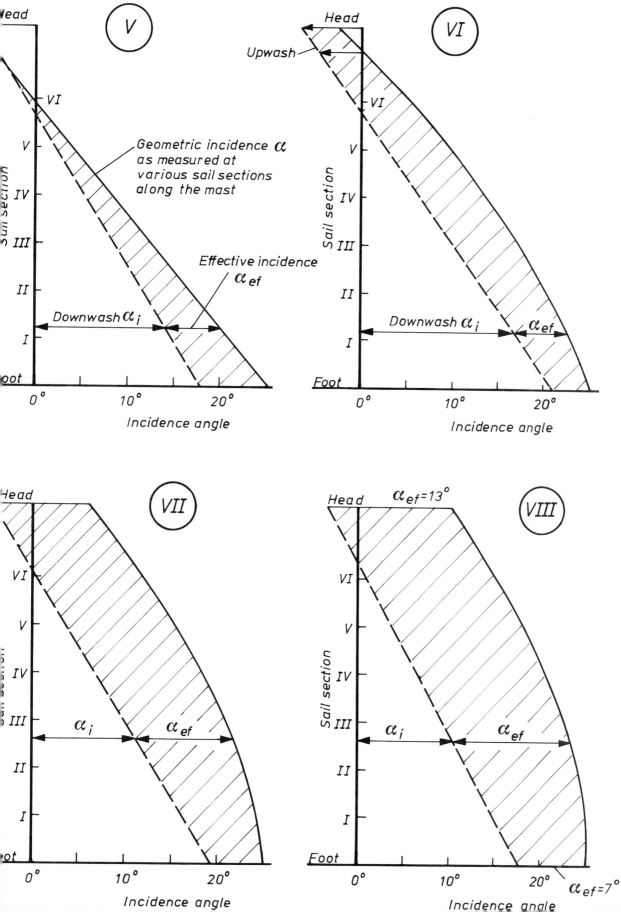

1. The influence of increased kicking strap tension, with its associated reduction of sail twist, on the effective incidence angle α_{ef} is powerful indeed. From the comparison of the four sketches it becomes plain to what extent the geometric incidence angle α, or the trim angle δ_m, measured in relation to the boom, are deceptive as indicators of the effective incidence angles at which the sail sections actually operate. At the same geometric incidence $\alpha = 25°$, the effective incidence α_{ef} for, say, sail configuration VIII, changes from about 7° at the foot to about 13° at the sail head. This increase in α_{ef} takes place in spite of the fact that the recorded twist between sections I and VI for this configuration is about 10°, which is apparently too small to compensate for the induced flow effects. Thus most of the sail area operates well beyond the stall angle, while only the lower part, close to the sail foot, operates at sufficiently low incidence to exclude stall. A glance at Fig 2.70 will remind the reader what is the range of the effective incidence angles at which the aerofoil, depending on its camber, can work efficiently without stall. It will be seen that the effective incidence angle $\alpha_{ef} = 13°$ at the sail head of configuration VIII, is far beyond the stall angle of the sail section of camber 4 per cent (0.04) as recorded in Table 3.7. Referring again to Fig 3.27, it can be seen that in the case of configuration V the effective incidence angle α_{ef} changes from about 7° at the foot to almost zero at the sail head, hence the upper part of the sail contributes practically nothing to the lift. When compared with configuration VIII, these differences in effective angle α_{ef} are quite dramatic and as such must lead to equally dramatic differences in sail performances. These were in fact convincingly demonstrated in Fig. 3.23.

2. The presence of upwash observed at the sail head, instead of the down-wash which dominates down from section VI, can be explained by referring to Fig 2.114, which shows that the downward shift of the tip-vortex core is responsible for it. This interesting effect is confirmed also in Photos 2.27C and 2.28. The variation in downwash along the sail height, associated with departure of the actual sail loading from the ideal elliptic one, leads to a non-uniform twist superimposed upon the incident air stream, so that the effective local incidence α_{ef} is bound to show variation with height up the sail. Photo 3.22 reduces in a way the level of abstraction incorporated in the last few sentences to the visually perceptible level, thus answering a question the reader may ask–how can we be sure that our theorizing is reflected in the real world? Photo 3.22 shows for instance how the air flow approaching the Finn-type sail is affected by its presence. Photograph A illustrates how light wool-streamers numbered 1–6 were attached to the wind tunnel screen ahead of the sail. The wind velocity has not reached the full test value, so the streamers are not yet fully aligned with the air flow. Photo B shows the wool-streamers, as seen by the camera attached to the wind tunnel roof above the sail. It can be observed that streamer 1 is dragged underneath the boom by the action of the powerful trailing vortex. The behaviour of the remaining streamers 2–6 gives quite a

Photo 3.22 Depicts how much the air-stream is twisted between the sail-section close to the boom and other sections towards the sail-head; $\alpha = 20°$.

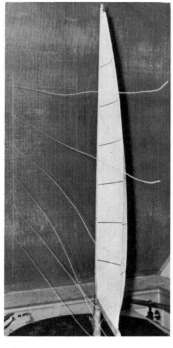

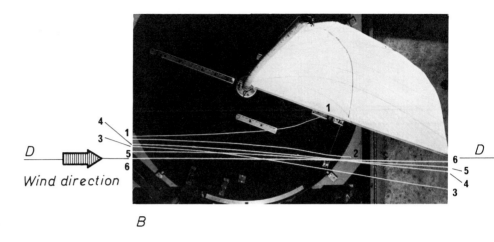

Wind direction

B

A

good idea how much the approaching air streams are twisted across the span between sail sections 2 and 6. The effect of non-uniform induced downwash is apparent; since it twists the air stream itself in the same direction as the sail is twisted geometrically, it causes the resulting local effective incidence α_{ef} to be more nearly uniform, but not quite, along the whole height of the sail. Thus, for instance, greater downwash at the height of streamer 3 is reflected by a larger induced angle α_i there. Subsequently, the effective incidence α_{ef} close to the boom is much smaller than the geometric incidence $\alpha = 20°$ measured between the boom and the undisturbed wind direction given in Photo B by the line marked D–D.

3. When discussing Fig 2.129B in Part 2, section D (5), it was stated that the optimum sail setting in the close-hauled condition is achieved if all the sail sections from the boom to the sail head are operating at the same effective angle of incidence α_{ef}. And we may add now the complementary requirement–the sail must be set at such an incidence that it produces the right amount of lift which can be tolerated, bearing in mind the available stability. Analysing from this standpoint the four sketches in Fig 3.27, one may notice that at the particular geometric incidence $\alpha = 25°$, considered as an example,

configuration VI is the only configuration which satisfies closely enough the condition of uniform effective incidence angle along the whole sail height. It would therefore seem that in this case a considerable reduction in sail drag should be expected as compared with, say, the much less twisted configurations VIII or VII, in which no such uniformity exists and a large part of the sail operates above the stall angle. Such a comparison can be made by examination of Fig 3.28. It is drawn in the same manner as Fig 2.111, but to facilitate a direct comparison, the curves of drag coefficient C_D versus lift coefficient squared C_L^2 are plotted for all four configurations in question.

Let us apply now our yardstick of sail efficiency, namely if, at given lift, the various sail configurations produce different drags, the better configuration is the one with lower drag. We can find in Fig 3.28 that at the high geometric incidence $\alpha = 25°$, at which the lift coefficient C_L is in the order of 1.3 as marked by a thin vertical line originating at $C_L^2 = 1.69$, sail configuration VI indeed produces lower drag than any other configuration. Its twist is therefore close to the optimum one for given lift, a conclusion already confirmed by the data of Figs 3.22 and 3.23. In contrast, configuration VIII, in which the twist was the smallest, produced the highest drag for the same $C_L = 1.3$.

One must remember that the above comparison of twist effects was made at a large lift coefficient and is applicable in light winds only. One may rightly expect that the optimum twist requirement will be different in stronger wind.

The approximate relationship between the desirable lift coefficient C_L, which can be generated by the sail, and the wind speed expressed in terms of the Beaufort scale, is given by thick bars plotted below the lift scale in Fig 3.28. Going to the left along these bars, towards stronger and stronger winds associated with decreasing lift coefficient and watching at the same time the behaviour of drag-lift curves, we may notice that the requirement for optimum twist changes. Now, we are in the position to establish a general rule as to how the kicking strap should be applied to control the sail twist.

The lighter the wind, i.e. the higher the geometric angle of incidence at which sail is set to produce large lift coefficients, the greater must be the amount of twist in order to prevent or alleviate an early stall of the upper part of the sail. In other words, the optimum twist in close-hauled conditions must gradually be reduced, as the wind becomes stronger, i.e. the sail operates at smaller and smaller lift coefficient. When bearing away the amount of twist allowed should gradually be increased. In short— the higher the lift, the larger the required twist for optimum sail efficiency.

Bearing in mind what has been said, it may be of some interest to look once again at Photo 3.17 in order to appreciate to what extent a helmsman's judgement, purely by eye, can mislead. The sail of configuration VIII is full of creases and although it looks ugly, is more efficient than the smooth, more pleasant looking configuration VI in conditions when the wind is strong. It is also worthy of note that configuration VIII might be improved further by the application of Cunningham holes. The sail itself could initially be better cut and also matched better to the mast's bending

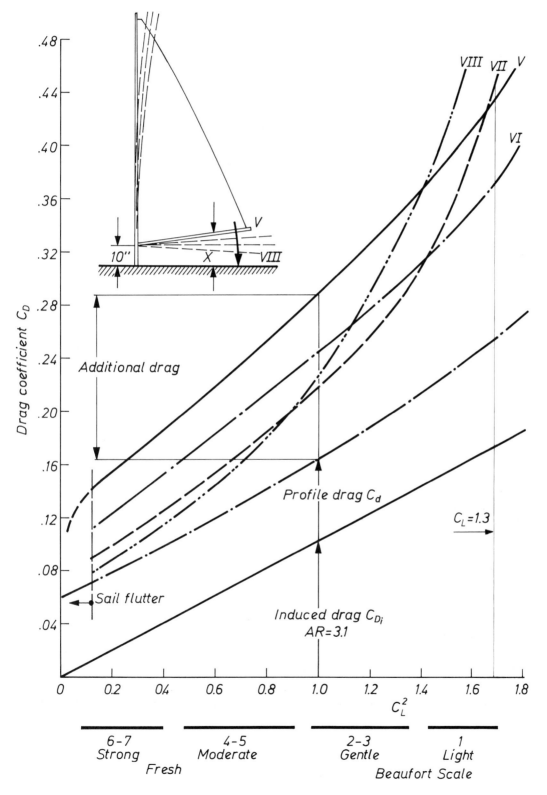

Fig 3.28 Drag variation with lift squared for four sail configurations V–VIII of decreasing twist set at the same geometric incidence α = 25°, measured between the boom and apparent wind direction.

Photo 3.23 A Lightning sail set on mast with moderate bend. Note increasing camber towards the sail head.
Photographs reproduced here by kind permission of Gary Comer of Land's End Publishing Co, Inc.

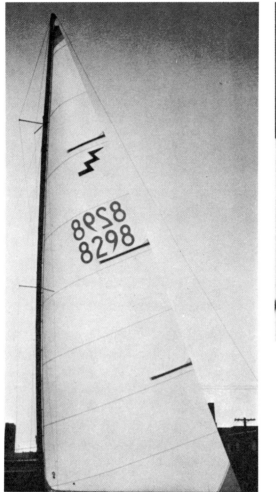

characteristics. Similar criticism can be applied to sail VI. One can easily discern excessive flatness of the upper part of the sail, as compared with camber built in the lower part of the sail, and this is opposite to the required camber distribution for the sail operating at high lift coefficient. Photo 3.23 illustrates the point. It shows a beautifully shaped Lightning class mainsail for light winds in which the camber increases towards the sail head.

Concluding, the kicking strap, if properly used, i.e. played constantly, can substantially improve the boat performance. To be effective it must be powerful enough, instantly adjustable and its tension must be separated from interference with the mainsheet, which should control the trim angle alone. Photo 3.21 illustrates correct trends in solving this problem. The kicking strap's effectiveness can be improved if its action is limited to the control of one variable in the sail shape only– namely the twist. Unfortunately, this is not always feasible because it also affects the sail camber, particularly in classes with flexible masts, and this additional effect can be detrimental and may, at least partly, offset the advantages offered by the kicking strap as a twist-control device.

(b) *Twist correction for wind gradient effect*

One more digression is necessary to complete our consideration of the effect of twist on sail efficiency. In Part 2 we derived Equation 2.37 repeated below:

$$\alpha_{ef} = \alpha \pm \alpha_i - \varepsilon \qquad\qquad \text{Eq 2.37 (repeated)}$$

where α geometric angle of incidence
   α_i induced angle
   ε sail twist

It gives the value of effective incidence angle α_{ef} at different sail sections along the mast height.

Taking now the apparent wind gradient into account, the above equation should be completed to accommodate one more correction to compensate for the effect of twist of the apparent wind which will tend to increase the effective angle of incidence towards the upper part of the sail. Otherwise the wind velocity gradient may cause partial separation near the sail head.

Equation 2.37 can finally be presented in the form:

$$\alpha_{ef} = \alpha + \alpha_i - \varepsilon + \Delta\beta \qquad\qquad \text{Eq 3.8}$$

where $\Delta\beta$ is the apparent wind twist due to the presence of wind gradient, as depicted in Fig 3.11.

As already mentioned when referring to Fig 3.11, the value of $\Delta\beta$ in close-hauled conditions is relatively small, perhaps 3–4° for Dragon or Soling size sails. It increases gradually when bearing away and, as shown in Ref 3.21, in reaching conditions the apparent wind twist may be in the order of 11° for the size of rig mentioned above. Its actual value will depend on the wind velocity profile, which may be of two distinct types resembling the curves illustrating the boundary layer flow drawn in Fig 2.23. In fact, the atmospheric wind flow near the sea or land surface differs, as compared with that of the boundary layer discussed in Part 2, in one respect–its size. In other words, the atmospheric boundary layer can be regarded as a considerable expansion of the more familiar boundary layer close to the bodies in relative motion, which we discussed at length in Part 2.

The observed differences in wind velocity gradient for the alternative light or strong and gusty wind, can be explained by taking into consideration the character of the flow close to the sea or land surface. In the case of a light wind and overcast sky, with no significant action of vertical thermal currents, there probably prevails a laminar flow inside the whole mass of air in motion. Hence the wind velocity profile, or in other words the wind variation with height, will be similar to that demonstrated in the left-hand sketch in Fig 2.23. In such circumstances, the wind gradient effect may be pronounced, and it is not uncommon, in light winds or in so-called drifting conditions, to see large yachts on the move while their little sisters lie hopelessly becalmed. It happens because the tall rigs can reach to the upper strata of the air and catch the wind aloft which may be very feeble nearer the sea surface.

When the wind is strong and/or gusty there exists a certain exchange of kinetic energy between the higher and faster moving layer and the lower and slower one. This is shown in the right-hand sketch of Fig 2.23. Due to this exchange of energy (momentum) between the neighbouring layers, the differences in flow velocities are reduced and the wind gradient is less pronounced.

To compensate for these possible differences in wind gradient and so to increase the sail efficiency over a wide range of courses sailed, the wing sail of *Patient Lady*, depicted in Photo 2.31, incorporates flaps divided into three sections. According to Ref 2.88–this sail is '...capable of operating with both small and large amounts of twist while still maintaining good cross-sectional shape along the entire span.' By setting the bottom flap at higher incidence angle than the upper flaps, arbitrary twist can be created.

(c) *Sail camber control*

'Take your choice of those that can best aid your action.'

Coriolanus

'I would I knew in what particular action to try him.'

All's Well That Ends Well

Although the effective incidence angle α_{ef} at which the sail operates can be regarded as a major factor determining the magnitude and direction of the aerodynamic forces, the significance of the sail camber, together with its spanwise and chordwise distribution, should not be underestimated. Its effect on sail forces is equally as powerful as that of incidence angle. After all, the pressure distribution and finally the lift developed on the foil depend on both the incidence and camber. Figure 3.29, in which one curve of C_L versus α is replotted from Fig 2.70 for the cambered foil of $f/c = 0.04$, illustrates this point. The lift generated, say at 4° incidence, consists of two components:

1. Lift due to camber alone, which is measured along the vertical C_L axis, equals about 0.42 at incidence angle $\alpha = 0$,
2. Lift due to incidence alone, which is about 0.39, so the total lift coefficient C_L, being the sum of these two components, equals 0.81.

Fig 3.29 Illustrates how camber and angle of incidence contribute towards the total lift or pressure developed by the foil.

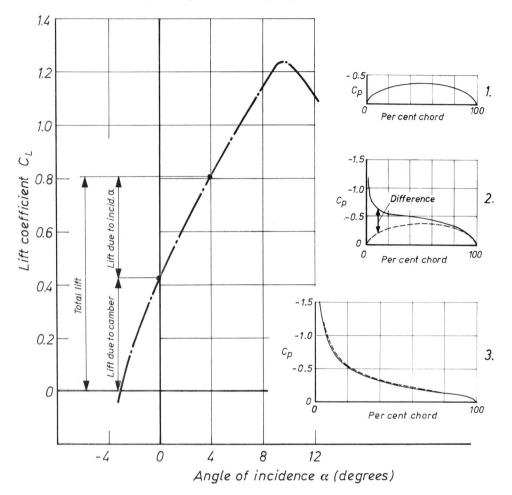

It has been found in the course of early aerodynamic studies that those two parts of lift are additive on the basis that the relevant pressures are additive. The three sketches in Fig 3.29 illustrate this concept: sketch 1 gives the distribution of pressure coefficient C_p over the suction side of the thin cambered circular arc foil at zero incidence, sketch 2 gives the distribution over the same foil set at incidence angle $\alpha = 3°$. Now, when the distribution of sketch 1 is subtracted from that of sketch 2 the remainder very closely approximates the pressure distribution over a flat plate set at $\alpha = 3°$, as shown in the slightly enlarged sketch 3. The solid line shows the pressure over a flat plate at $\alpha = 3°$. This can be directly compared with the broken line which gives the difference between the total pressure over a circular-arc foil measured at $\alpha = 0°$ and $\alpha = 3°$. Those two curves are almost identical.

Figure 3.30 may assist in establishing the two fundamental trends associated with camber alteration. The polar curves presented give the variation of lift versus drag coefficients for four untwisted rigid sail models of increasing camber ratio f/c from $\frac{1}{15}$ to $\frac{1}{4}$, tested as a mirror-image combination. Their aspect ratio was 4 and the position of the maximum camber was about $\frac{1}{3}$ of the chord from the mast. The essential intention of the test was not to determine any set of coefficients for this particular bermudan-type sail but to demonstrate, as clearly as possible, the trends in variation of C_{Lmax} and L/D ratio on the assumption that the camber distribution is both vertically and horizontally uniform and independent of the incidence angle. Such a clear demonstration of trends is impossible in the case of soft sails, where the camber distribution, twist, and camber ratio change whenever angle of incidence is altered, so that the overall picture of trends is somewhat blurred. Direct comparison of Fig 3.30 with Fig 3.22, particularly in the region of intersection of the polar curves at higher lift coefficients just before and after stall, should give an idea of what is meant by the blurred picture.

Referring to Fig 3.30 it is seen that as the camber ratio decreases the L/D ratio increases and this trend, distinguished by a thick horizontal arrow, is noticeable in the range of C_L coefficients up to about 1.2. Assuming now that the apparent course β relative to V_A is 30°, which is quite representative, one can find that there is little to be gained in terms of the driving force from the sail by increasing its camber beyond $\frac{1}{15}$. The camber ratio in the order of 1/10 appears to be the maximum acceptable camber in close-hauled conditions when there is no restriction on the upper limit of lift coefficient, i.e. the wind is light. The thin line drawn from point A on the course sailed line and tangential to the polar curves, illustrates this conclusion graphically.

When the wind speed increases and the tolerable lift coefficient C_L becomes lower and lower, there is no better way of optimizing the sail shape than to make it gradually flatter up to a drum-like membrane, before the sail area is eventually reduced by reefing or sail change. The thin line drawn from point B on the course sailed line and intersecting the three polar curves at points 1, 2 and 3, illustrates the fact that for a given tolerable lift coefficient C_L, which is close to the tolerable heeling force coefficient C_H, there is an optimum camber that produces the minimum drag and therefore the best sail efficiency.

From Fig 3.30 it can also be seen that the maximum lift increases as the camber increases, but only to a certain limiting value, which appears to be restricted by the state of the flow right at the leading edge. This trend is distinguished by a thick arrow pointing upwards. Although these high lift coefficients cannot be exploited in close-hauled conditions, they might be used on other courses relative to the wind, provided, of course, the helmsman can sufficiently increase the sail camber to obtain the largest possible lift. This, in fact, is not feasible in the case of a conventional mainsail but it is easy to achieve by hoisting another type of sail such as spinnaker or drifter-genoa, etc. The advantages of having well cambered sails, in terms of driving power, can be estimated by resolving the force coefficients in the manner shown in Fig 3.31, which depicts the beam reaching condition.

Fig 3.30   Effect of camber on L/D ratio and $C_{L\,max}$ (see Fig 2.70 Part 2).

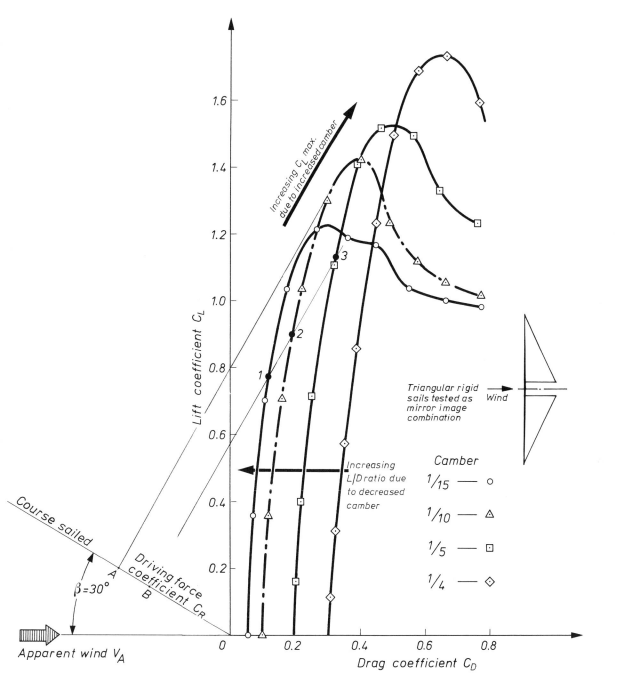

Fig 3.31 Forces developed by a single bermudan sail in beam reaching
conditions.

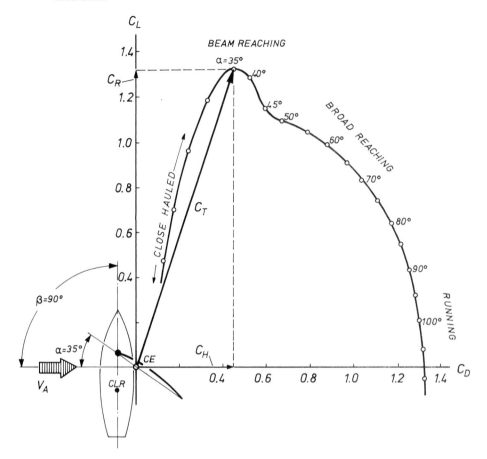

As seen from Figs 3.30 and 3.31, the general trends of the polar curves of sail forces over the range of apparent courses β, from close-hauled in strong and light winds through close reaching to beam reaching, are fairly well established. This is extended in Fig 3.32 to cover a range up to angles of incidence $\alpha = 170°$ corresponding, as indicated by the yacht silhouette, to running dead downwind and by-the-lee. The plotted polar curve represents the characteristics of a bermudan-type sail of AR $= 4$, a camber ratio in the order of $1/10$ and a twist of about $10°$ only. By comparing the magnitudes of the sail total force coefficients C_{TI} and C_{TII} at two different boom positions corresponding to the two angles of sail incidence $\alpha = 90°$ and $\alpha = 105°$, one can notice that the boat may run faster–if a little uncertainly–when sailed by-the-lee. A reservation–if a little uncertainly–arises in the situation where no kicking strap is applied and therefore an unintentional Chinese gybe may occur. The golden rule expressed by some experts that one should never let the boom out to the shrouds in anything but light airs, because of a danger of the mast

Fig 3.32 Boat may run faster, if a little uncertainly, if sailed by the lee.

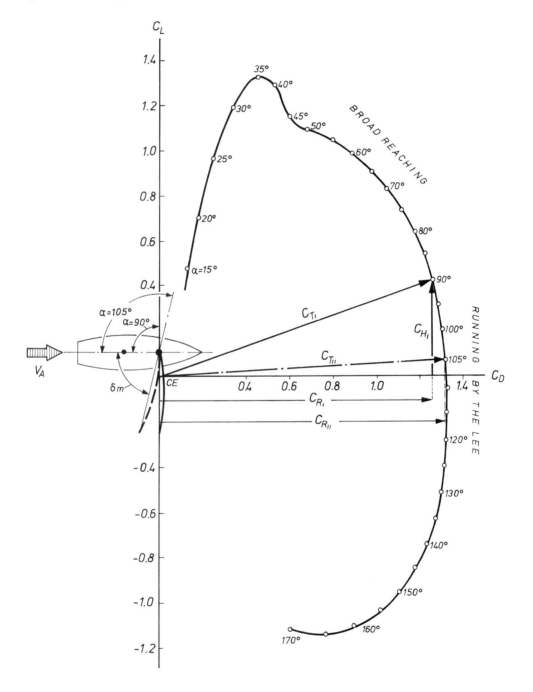

Fig 3.33   Methods of building up sail camber from flat cloth panels.

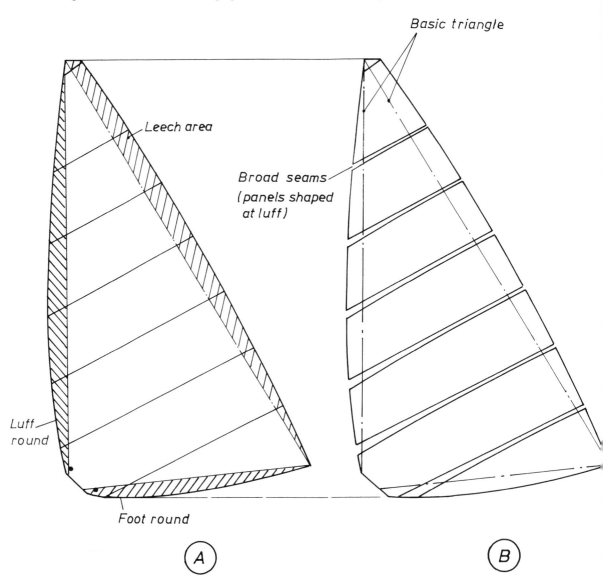

breaking, can thus be justified on the basis of the data in Fig 3.32. There is yet another reason for not allowing large sheeting angles for the mainsail even if the sail twist could be reduced to the minimum. This will be demonstrated in the section dealing with rolling downwind.

The camber in a sail built up from flat cloth panels is produced by:

a. Giving the luff and foot a prescribed surplus of material, i.e. luff and foot rounds as indicated in Fig 3.33A.

b. Tapering the panels at the luff by introducing so-called broad-seams, as shown in Fig 3.33B.

c. Leaving a certain amount of roach in the leech area beyond the basic sail triangle, in association with suitably tapered battens.

The sail camber, so predetermined, is further affected by stretch of the sail cloth caused by wind-loading and, finally, to a greater or lesser extent, by a number of control devices and means listed in Table 3.11. So the actual amount and distribution of camber which contributes substantially to the forces a given sail can produce, depends upon the separate skills of the sailmaker and the sailtrimmer. Since there is no *best* sail camber for all sailing conditions, the skill of the helmsman is shown by the way in which he modifies the sail flow to suit the particular sailing demands. This task is unfortunately complicated by the fact that practically none of the control devices given in Table 3.11 satisfies the basic requirement of an ideal control system which would enable the helmsman to influence only one parameter of the sail shape at a time by a given gadget and, moreover, to reproduce the desirable shape again whenever similar wind and course conditions occurred. It is not the case however and an adjustment with one gadget may affect, at the same time, several of the sail shape parameters given in Table 3.10, so that the result cannot be ascribed to the effect of a single one. This is the essence of the difficulties encountered during tuning trials and racing. For example, the sheets do more than just control the incidence angle; when they are eased, the boom not only increases its sheeting angle but is also lifted up, thus changing the whole vertical and horizontal distribution of sail camber, not to mention the twist. This is the reason why the complementary control systems were introduced: the wide mainsheet-track and/or kicking strap to compensate for the deficiency of the simple sheeting arrangement. Subsequently, another system (Cunningham holes) was invented to bring back the maximum camber position, whenever it travels too much towards the leech when wind load increases or the mast bends extensively. Thus, we already have 3 or 4 different control means, which to varying degrees modify the sail camber, by virtue of their rather unwanted side effects. When there are ten or more possible adjustments that can be made on each leg of the course in ever-changing wind, it would be hopeless for the helmsman to seek the best combination amongst these variable adjustments without a clear concept of what particular sail shape he is trying to achieve and why.

Owing to the great complexity and variability of modern sail tuning devices, there is little to be gained just by copying others. Tuning knowledge, like any science, advances not only by an accumulation of new devices but by continuous development of new fruitful concepts. Mastery cannot be achieved otherwise. In other words, there is little to be gained by having a highly sophisticated and infinitely adjustable rig if the crew have no idea how to use it effectively in ever-changing sailing conditions. 'The whole science'–to quote Albert Einstein–'is nothing more than a refinement of everyday thinking.' This ties in with Bridgman's (Ref 3.27) operational approach to knowledge, that is, the concepts must be constructible out of materials of human experience and workable within that experience. When concepts move beyond the reach of experience they become unverifiable hypotheses. Knowledge advances when we find how things are related and in what order.

(d) *Evolution of the Finn mast*

To illustrate this point let us consider briefly the evolution of the bendy mast as one of the means of controlling the sail camber in the Finn class since its introduction into the Olympic Games in 1952. Originally, the Finn was conceived with a mast of telegraph-pole rigidity. Consequently, the common problem of every helmsman before the race, as aptly described in an early report, was–'Shall I use the flat main or the full one? Blowing a bit fresh now but I reckon it will ease before the second round–or will it not? Eeenie meenie, minie mo–catch a dilemma by the toe!' Once the too full sail was selected, the helmsman could only survive the heavy wind either by easing the sheet so that the leech at the mast-head fell off with a consequent convulsive drag-causing flogging, or by luffing, which caused similar spasmodic sail flutter, but of the whole sail instead of its upper part only. Needless to say, both techniques lead to rapid deterioration of boat performance.

Since it was almost impossible to keep the original stiff-masted Finns upright in heavier winds, they could successfully be sailed only by tough and rather heavy guys. According to the recollection of some lightweight people who sailed the prototype craft–'They were wretched boats. The masts were stiff and the cotton sails were like bags; the effect was that you could not hold the boats up in much above Force 2, and in any wind you could not stop them broaching.' Incidentally, as stated in Ref 3.28, '...A number of Australian helmsmen who tried them out for the 1956 Olympics, when the class first appeared in Australia, have never recovered from back injuries suffered during training and the selection trials.' Paul Elvström, who won the first three consecutive gold Olympic medals in this class, weighed at that time between 180–200 lb (82–91 kg), depending on what state of training he was in; his super-athletic and absolutely professional approach to sailing undoubtedly helped him to achieve such an outstanding Olympic record. Since then racing sailing ceased to be just an art and became more and more a sport for athletes. After some six years, the rig developed enormously. A wide mainsheet track was introduced, polyester replaced cotton as the sailmaking material and above all, people learned how to make a mast that would bend and not break, thus allowing the sail to be flat or full at the helmsman's choice. In other words, a bendy spar increased the wind range in which one sail could be used effectively. Table 3.12, together with Fig 3.34 based on measurements done by Richard Hart and kindly presented to the author, illustrate the evolution in mast flexibility during a period of 10 years. The fore and aft as well as side bends were measured by hanging a 20 kg (23 kg later) weight halfway between the black control bands with the mast supported at the bands, as shown in Fig 3.35, and taking the difference in the amount of deflection in centimetres with and without the weight.

Fig 3.34 Evolution of Finn mast flexibility.

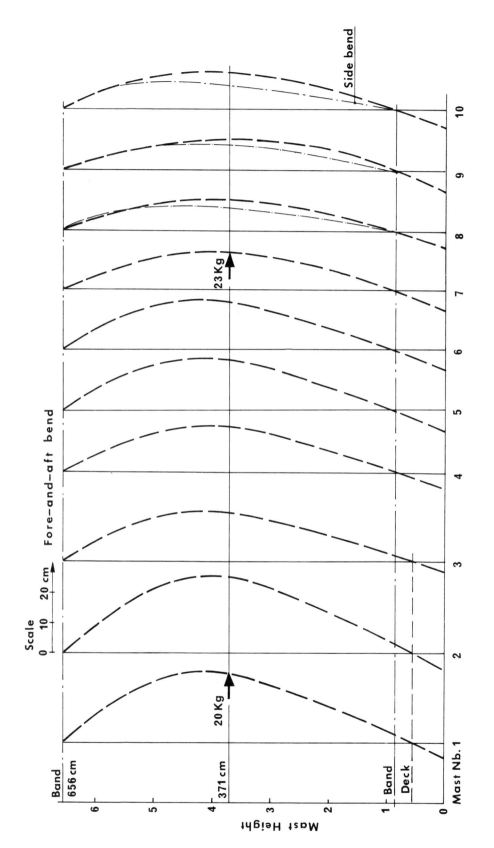

Fig 3.35   Method of measuring mast-bend. The 20 kg weight was replaced
            by 23 kg weight when masts became stiffer.

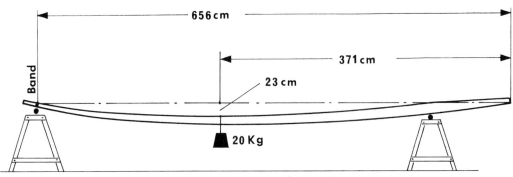

Fore-and-aft bend

TABLE 3.12

No	Name	Year	Fore and aft bend	Side bend
1	Standard Elvström	1962	23.0	11.0
2	Collar	1963	25.0	9.0
3	Standard Benrowitz	1965	16.5	9.5
4	Standard Raudaschl	1966	15.0	10.0
5	Standard North	1967	16.0	12.0
6	Collar (Oval)	1967	16.0	11.5
7	Bruder	1968	13.0	
8	R Hart	1970	11.0	8.0
9	New Bruder	1970	10.0	9.0
10	RYA Mast 1352	1971	12.0	8.0

The soft, flexible masts changed the Finn behaviour so dramatically that it appeared that from now on quite light helmsmen, just above 10 stone (63.5 kg), could compete with the goliaths on truly equal terms. This feeling is well exposed in one of the enthusiastic opinions written by R Creagh-Osborne in that time (*Yachting World*, April 1964)–'David Thomas said he found it too tough for him. Come, come! I really think this often-heard excuse ought not to be permitted any more. *Anyone above ten stone can sail a Finn, believe me!* But it has got to be tuned correctly and the harder you work, both physically and mentally, the faster it goes.'

It was believed that the differences in boat speeds attributable to crew weight and so to the available righting moment, could be ironed out simply by allowing mast flexibility in inverse proportion to the weight of the crew. A contest between heavy and light helmsmen was, in a way, reduced to the mysterious ability to tune the wooden mast by means of a plane-it and glue-it routine. A flexible spar not only allowed the flattening of the sail but also an automatic twisting of the sail by virtue of

Fig 3.36   The feathering technique became popular in most racing boats.

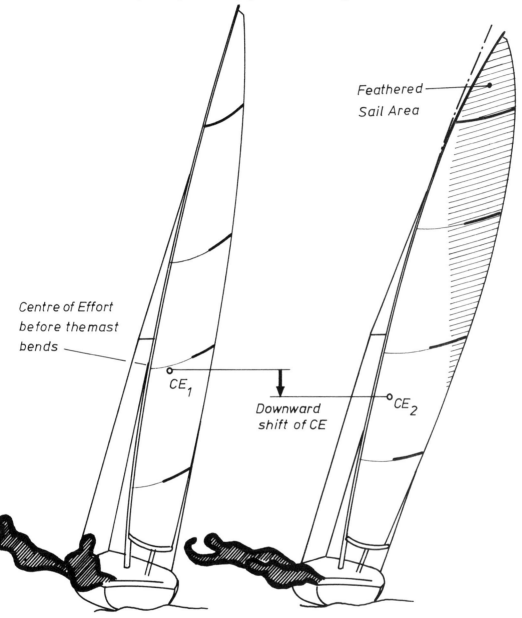

Feathered Sail Area

Centre of Effort before the mast bends

CE_1

Downward shift of CE

CE_2

easing the leech tension as the wind strength increased. This way the so-called unwanted portion of the sail is unloaded, so that it does not contribute to the heeling force and therefore the height of the centre of effort is reduced.

The feathering technique, illustrated by Fig 3.36, has become popular in all racing classes in which flexible spars make it possible and particularly in those classes in which the crew weight contributes substantially to the righting moment. If mast flexibility has been tuned to the sail shape and to the helmsman's weight and his sitting-out ability, the automatic leech release should come into action as soon as the boat becomes overpowered in a blow. The twist is self-adjusting and operates instantly as a kind of safety valve.

Although feathering the sail top is much more efficient than the old-fashioned method of flogging the entire sail area in a heavy blow, it is nevertheless quite an expensive way of lowering the actual centre of pressure to that limited by the ballasting ability of the crew. The sail twist, as discussed earlier on a theoretical basis, must be dearly paid for in terms of harmful induced drag, which, referring to Fig 3.22, is tantamount to shifting the sail aerodynamic characteristics from, say, point A on curve VIII (sail with small twist) towards point B on curve V (sail with large twist). The associated losses, in terms of the available sail-driving power, are not negligible; so, the softer the mast, i.e. the earlier the feathering commences, the greater are the losses in boat performance. This explains why finally stiff masts returned to popularity. It took however several years to fully recognize, irrespectively of the true reason, the disadvantages of an early and large twist, which is inevitable if the feathering is to be effective. The apparently retrograde step towards stiffer masts, shown in Fig 3.34, became in fact possible because of the rediscovery of the advantages associated with shifting ballast and which materialized on the contemporary racing scene in the form of heavy, soaking wet garments for light helmsmen.

It is worth recording the protest of an experienced Finn sailor, a doctor by profession, against the IYRU intention to reduce wet clothing (mainly for safety and medical reasons, especially fear for the spine). The excerpt is taken from the *Finn-Fare*, November 1976, which is the official publication of the International Finn Association:

'*Reducing wet clothing is putting the clock ten years back*. The era of the big heavy sailor will come back...Another danger is that if the IYRU reduces the weight, some sailors who are around 70 or 75 kg body weight will take anabolic steroids to increase their weight and so achieve the 95 kg weight optimum. Insiders know that in the past anabolic steroids have been taken and are still used in sailing nowadays. Let us not be hypocrites...This is not restricted to the Finn class. This happens in other classes as well and this happens also with amphetamines and other drugs. There is enough evidence for that, even if the official version is that sailing is clean! *I am sure that reducing wet clothing will provoke an increase in anabolic intake in dinghy sailing.*'

No doubt, such an evolution of additional and movable ballast in the pursuit of performance is well beyond the very much stretched limit of acceptance on the part of the establishment, although the revolutionary, progress-minded young people can hardly resist it. As usual, in every good there is ensconced some evil.

In the course of a rather slow, truth-revealing trial-and-error process it became evident that feathering is more expensive a remedy for stability deficiency than an additional weight carried on the helmsman's back; even if this additional ballast were, in some extremes, in the order of an almost unbelievable 100 lb of wet sweat clothing, as reported in *One Design and Offshore Yachtsman* by P Barrett discussing the 1969 Finn Gold Cup. Evidently, the reduction in twist and therefore a reduction of the induced drag more than offsets an increase in hull resistance due to increased displacement by the amount of the additional ballast, no matter how it is hidden.

(e) *Other methods of sail camber control*

Considering the camber variation, one should realize that by bending the spar, the sail draft induced by the surplus of material along the luff can be taken out, but the camber induced by means of shaping the panels will be little affected. Subsequently, if the luff round shown in Fig 3.33 is not adjusted to the mast-bending characteristics, which can differ substantially, as demonstrated in Fig 3.34, inevitably, whenever a certain amount of kicking strap or sheet tension is applied, a distortion in sail shape will occur. This, for instance, is shown in Photo 3.17–Run VIII. One can clearly recognize the stress in the sail cloth which manifests itself in the form of creases coming out radially from the luff to the clew. These, although they might to some extent be alleviated by the use of a Cunningham reef, indicate a poor sail-mast partnership, that is, there is an incorrect amount of the round in the cut of the sail luff. Such a distortion in the sail shape and/or wrongly distributed camber must necessarily affect the flow round the sail, and therefore the aerodynamic forces, in an undesirable manner.

Sensitivity of camber variation along the sail height when the mast bends can be estimated with the help of Fig 3.37, which is based on Ref 3.14. It can be seen from the diagram that the camber/chord ratio f/c depends primarily on the excess of curved length l over chord length c and to a lesser extent on the section shape, mainly the position of maximum camber from the leading edge. Values of camber/chord ratios for varying excess length ratios $(l - c)/c$ in per cent had been calculated for the simple section consisting of a circular arc at the leading edge, with a tangent to this circle forming a flat leech (which sail makers attempt to produce) at three different positions of maximum camber, $p = 0.3$, 0.45 and 0.5. One can anticipate that the final camber of a given sail section, shown in the sketch of Fig 3.37, will depend on the combined effect of the bendy mast movement, the leech sagging forward, an increase in curved length l, due to fabric extension in this direction, and finally on the amount of the material surplus given by the sailmaker at the section in question.

Figure 3.37 can also be of some help when considering the well known phenomenon of the camber of headsails increasing as the wind speed increases. It is due partly to the decrease in chord c, caused by the leech sagging forward and magnified by the forestay sag and partly by increase in the curved length l, due to horizontal fabric extension. It is apparent that an increase in excess length $l - c$ must either increase the camber chord ratio f/c or be absorbed in a movement forward of the position p of maximum camber.

Fig 3.37   Sensitivity of sail camber to the fabric extension or surplus along
the sail chord.

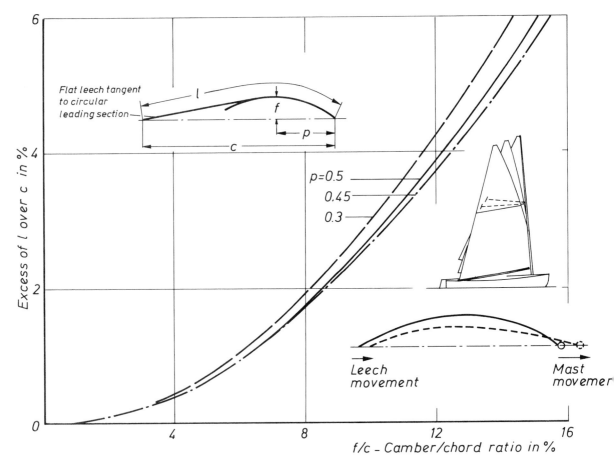

The basic effect of stretching the luff is to pull the flow forward. Stretching produces a fabric contraction in the chordwise direction but it also has the effect of pulling the leech forward. To establish the size of this effect on the sail shape and the driving force component at a given heading, two genoas of identical planform (Photo 3.8) were tested in the wind tunnel at a constant wind speed of 31 ft/sec (Ref 3.14). They were cut from the same template and intended to be fairly full. Sail No 1 was made from a sailcloth of a lightly resinated finish, while sail No 2 had a dead soft finish.

The trends in variation of the camber/chord ratio and position of maximum camber for each sail are plotted in Fig 3.38 against the five different values of luff stretch on Sail 1 and four values on Sail 2. It is seen that in the extreme case, when stretching increases, the mean position of maximum camber moves forward from 45

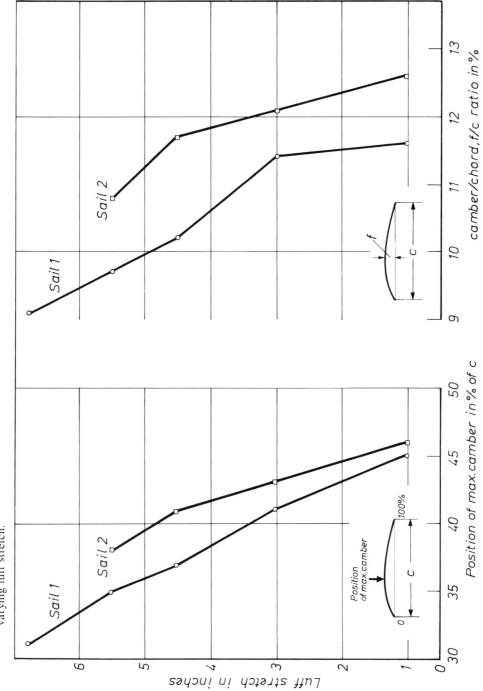

Fig 3.38   Position of maximum camber and camber/chord ratio for varying luff stretch.

to 31 per cent of the chord c. The effect on the camber/chord ratio was not so straightforward but in the more usual case, given in Fig 3.38, the camber decreases slowly with luff stretch. The results of the tests suggest that, whereas the forward half of the sail uses the fabric's shear flexibility to allow adjustment to the position of maximum camber, say, by means of a Cunningham hole, to counteract the wind-induced deformation, the after section is relatively unaffected by luff stretching. It is simply acting as a sort of pre-tensioned beam to prevent the leech sagging forward and the shear stiffness of the fabric plays a major role in this action.

The stretch luff genoa, as well as the mainsail, would thus seem to have two requirements:

1. In order to move the position of maximum camber forward, the fabric should have relatively low shear-stiffness near the luff, to allow adjustment without inflicting enormous halyard loads.
2. At the same time, to allow the sheet loading to affect reduction in camber at positions other than the foot of the sail, the fabric must have high shear-stiffness in the leech region, coupled if possible with high weft stiffness, so that a kind of stiff beam is formed to transmit horizontal forces to high points in the sail (Ref 3.14).

As reported in *Seahorse* magazine in 1974, John Oakeley developed along this line of thought the so-called compensation mainsail–a concept which is depicted in Fig 3.39. The sail is made up from varying weights of cloth, the lighter weight being along the luff. It was claimed that, since this lightweight cloth stretches at a greater rate than the heavier cloth in the leech area, the fullness does not move aft as much as it would in a conventional sail when the wind increases. Besides, the light cloth on the leading edge is more sensitive to adjustments so it gives greater control when using the Cunningham holes. The upper 12 per cent of the sail shown in Fig 3.39 should be made of stiffer material; this has the advantage of reducing the stress lines radiating from the headboard and also improves the Cunningham hole action, which usually diminishes rapidly towards the sail top. Admittedly, the finding of a correct method of joining the light and heavy cloths was quite a problem. Apparently the radial seams, shown in Fig. 3.39, solved the difficulties associated with stress differential, which otherwise manifested itself disastrously as a scallop effect, when the seams were straight. It required some experimenting to find the best matching of cloths, as there are so many different weaves, finishes and weights to choose from.

It is expected that the advantages of the compensator mainsail will be particularly appreciated in offshore classes, where the IOR rule prohibits using more than one mainsail and therefore the conventional sail cannot possibly produce maximum driving power in the full range of wind conditions. A genoa manufactured by this method may offer similar advantages.

Referring again to the result of the stretch test on genoa No 2 shown in Fig 3.38, an attempt was made to establish the relationship between the amount of luff stretch, sail shape and the magnitude of driving force coefficient for each value of

Fig 3.39 The compensator sail.

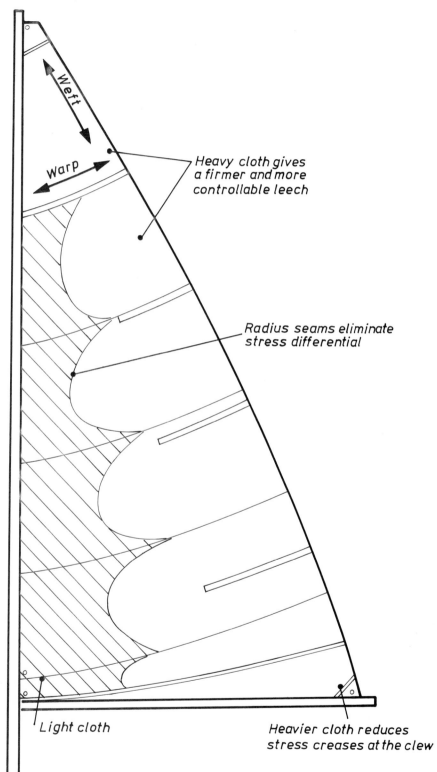

Weft

Warp

Heavy cloth gives
a firmer and more
controllable leech

Radius seams eliminate
stress differential

Light cloth

Heavier cloth reduces
stress creases at the clew

Fig 3.40   Sail No 2: the effect of luff stretch on camber.

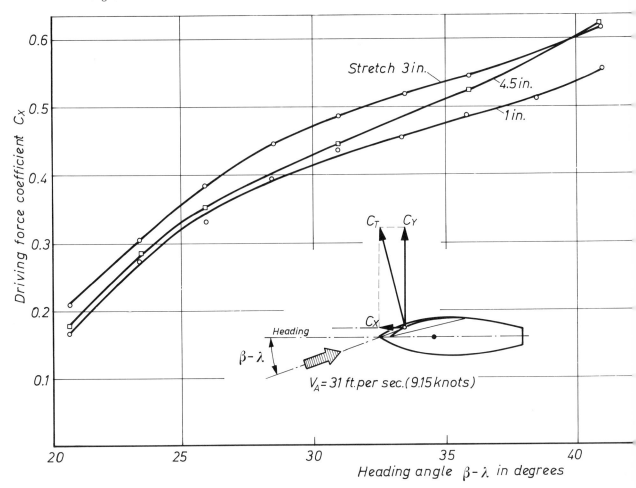

stretch at constant wind speed $V_A = 31$ ft/sec. A study of the driving force graphs in Fig 3.40, in which the magnitude of C_x coefficients is plotted against the apparent course $\beta-\lambda$, indicates clearly that the sail produced up to about 10 per cent more driving force as the luff was stretched up to a certain point. Beyond this optimum amount of stretch, which was 3 in, the driving force dropped off as the luff was stretched further. The measurements presented in Fig 3.40 should only be regarded as indicative of trends. In the experimental work it was not possible, or even realistic, to treat any of the factors listed in Table 3.13 as an independent variable when recording the effects of sail shape on performance.

As a matter of fact the variables listed in Table 3.13 were controlled in the following manner: the luff stretch was chosen as the independent variable while holding forestay sag (d) constant at a small value, with high tension in the wire. The

TABLE 3.13

a. Athwartships sheet position
b. Fore and aft sheet position
c. Sheet tension
d. Forestay sag
e. Luff stretch

sail shape was then adjusted at each value of apparent course β-λ by moving the sheet fairlead position (*b*) fore or aft, and changing the sheet tension (*c*) until the sail had no leading edge separation, as indicated by wool-tufts attached to each side of the sail about 2 in from the leading edge and depicted in Photo 3.8. This is in fact the method of sail adjustment used by an increasing number of skilled crews when racing. An interesting point is that it has been found that this method is so critical and accurate that it was possible to repeat sail settings to produce sail forces within 1 per cent difference. Thus the tufts may serve as sail-trim or still indicators of extreme sensitivity. However, one must realize what their limitations are in guiding the helmsman when tuning the sails and boat to best performance.

(4) More about headsail fairings, leading edge function and tell-tales

'I do not think I am mistaken in saying that, in sailing, head-work comes first, physical prowess second.'

MANFRED CURRY
Yacht Racing

The recent advent and rapid development of grooved luff systems for supporting headsails, called sometimes headsail fairings or briefly head-foils, is a result of an increasing realization by the offshore racing fraternity that foretriangle efficiency is the prime race-winning factor.

Inventors of those new revolutionary or ultimate groove systems claim that they offer three advantages over the headsail hanked to wire stay, by providing:

1. Faster (hankless) sail changes.
2. More rigid support for the leading edge; subsequently the sag of the stay is so small that it can easily be allowed for by sailmakers–therefore more effective shapes should result.
3. A clean aerodynamically effective leading edge.

Let us concentrate on the last claim, trying to answer the questions: do those various headsail fairings, that snap over a wire or rod forestay and swivel freely, provide better entrance to the leading edge of the sail? And if so, what are the advantages in some quantitative sense?

(a) *Tests on head-foils*

Wind tunnel experiments performed on a number of head-foil models or actual, commercially available fairings, some of them depicted in Photo 2.14, throw some light on the aerodynamic effects of fairings on sail efficiency, since, to date, there appears to be no more data than is contained in the claims of sales literature. In the course of the tests, every full-scale foil was attached to the same rectangular rigid sail made of sheet steel. Dimensions of the model sail were:

Span	6.92 ft
Chord	3.92 ft
Area S_A	27.2 sq ft
Camber	circular arc of 1/10 camber/chord ratio

Tests were conducted in approximately two-dimensional conditions, i.e. the model had its slightly smaller than the wind tunnel width by about 1 in on each side. The characteristics of a hanked-on headsail (sail + forestay + hanks) were also recorded to be subsequently used as a yardstick. Each fairing section attached to the sail was tested at the same wind velocity = 38 ft/sec, which gives the Reynolds Number (see Eq. 2.19B):

$$\text{Re} = 6370 \times V \times \text{chord} = 6370 \times 38 \times 3.92 = 0.95 \times 10^6$$

i.e. almost one million, which corresponds to average full-scale sailing conditions. The foils operating at identical Reynolds Number are therefore directly comparable.

In each test-run the angle of incidence α of the sail model was varied from 0° to about 10° in increments of about 2.5° and at each incidence angle the value of the lift and drag were recorded. Some results of tests are given in Figs 3.41A, B, and 3.42. An analysis of Fig 3.41A and B reveals:

1. That an addition of fairing to the leading edge of the sail causes a small shift of the L/D peak towards the higher angle of incidence, or in other words, the maximum L/D ratio occurs at higher lift.
2. The peaks of the L/D ratio curves for sails equipped with head-foils are flatter as compared with the sharper ones of the L/D curves of ordinary hanked-on headsails and the resulting L/D ratio is better over the whole range of measured incidence angles, at which the headsails usually operate in close-hauled conditions. This facilitates correct sail trim and steering; in other words, a good head-foil system makes the sail more tolerant to errors in sheeting or incidence angles.
3. Some head-foils are better than others in improving the L/D ratio of a sail-head-foil combination. Fairings of large thickness ratio t/c, such as Nos 1 or 4 in Fig 3.41A and Photo 2.14, cause deterioration in L/D ratio as compared with that of an ordinary hanked-on headsail. More efficient head-foil section can be developed and section 6 indicates certain trends to follow. Figure 3.42 depicts the beneficial action of head-foil 6, which results in higher lift and lower drag

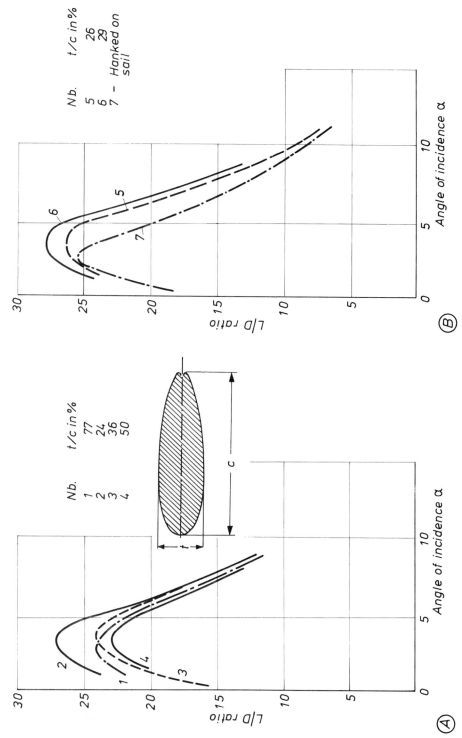

Fig 3.41 Results of wind tunnel tests on head-foils.

Fig 3.42   Results of wind tunnel tests on head-foils.

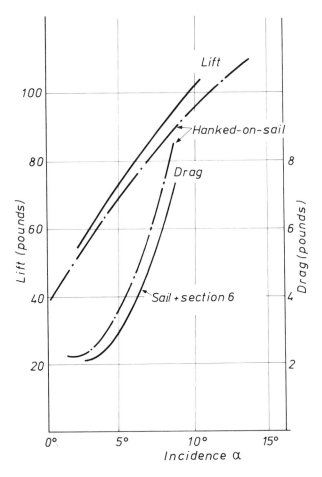

for the same angle of incidence as compared with a conventional headsail held by hanks.

It appears that the radius of curvature of the head-foil attached to the sail's leading edge, the size of the fairing in relation to the sail chord length and the foil thickness ratio t/c, are the most important parameters and the tests conducted substantiate such a conclusion. Unfortunately, the forestay penalty, under the IOR, inhibits the use of large, more efficient fairings.

The wind tunnel results disproved some manufacturers' claims that the drag of head-foils is primarily a function of their frontal area, and to a lesser extent t/c ratio. Clearly, once the sail and fairing are attached, there is no logic in considering the t/c for the fairing alone, since it merely forms the leading edge for a much larger chorded device formed by the sail and fairing combination. However, even if, for the sake of fallacious argument, the sail is ignored, there is ample evidence that in the case of streamline sections the thickness chord ratio t/c, rather than the frontal area, is of

Fig 3.43 Double luff rig patented by F Herreshoff.

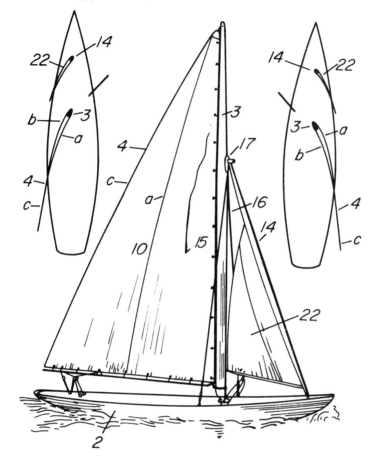

Photo 3.24 R boat *Live Yankee*, Com C A Welch 1927, showing revolving, streamlined spar forestay.

dominant significance. The tests conducted in Göttingen (Germany) and NPL (England), the results of which are shown in Figs 2.28, 2.29, and 2.30 in Part 2, substantiate this point. In particular, Fig 2.30A illustrates dramatically the advantage of streamlining.

Some readers may be interested to know that headfoils are neither new nor a revolutionary invention. They are simply a new answer to an old problem. According to Herreshoff's memoirs quoted from Ref 3.29:

'A few years later I designed three or four rigs with a spar forestay, believing that by doing away with the backstays a saving in wind resistance could be made, but none of the owners of these yachts liked or kept the spar forestays and I have given them up. The spar forestay is not an invention of mine, but has been used in central Europe for thirty or forty years and there is probably something in it, although the revolving streamline ones are my invention.

In 1925 I was much interested in schemes for reducing the wind resistance of rigs when closed-hauled. I was working for Starling Burgess at the time, and perhaps because he had been an aeroplane designer we were over wind-resistance conscious. At any rate I was allowed patents on a double luff rig with revolving or oscillating mast [see Fig 3.43]. This scheme was developed principally to do away with exposed wires and on an R boat [shown in Photo 3.24] I believe it did away with about 200 running feet; in other words, there was no headstay, intermediate shrouds, upper shrouds or backstays. The R boat *Live Yankee* was designed to come out with this rig but the regatta committee of the New York Yacht Club, hearing of it, promptly passed the rule prohibiting revolving masts, double luffed sails and similar contrivances.'

Strangely enough, Herreshoff's concept, first conceived as an advanced sail for a boat, was in 1967 converted to an aircraft wing, as depicted in Fig 3.44. A device of this type, called a sail-wing, apparently in honour of the sailor who invented it, was developed at Princeton University. It was subsequently tested in the Langley full-scale wind tunnel to evaluate the aerodynamic characteristics of this novel type of wing (Ref 3.30). It can be seen in Fig 3.44, which shows a typical cross-section of the wing, that the sail construction was developed in an effort to achieve structural simplicity. It consisted of a D-spar leading edge drooped 8°, a wire trailing edge and rigid ribs at the wing tip and root. This framework was covered with a fabric envelope which formed the upper and lower surfaces of the sail-wing. The fabric was tightly stretched by adjustable tension bridle wires attached to the trailing edge. The model was laterally controlled by means of hinged wing tips which effectively caused twisting of the whole wing.

From the experimental investigation a number of points arose which are of some interest, bearing in mind possible applications of this concept in special sailing craft designed for speed. Thus, the lift characteristics, shown in Fig 3.45, for the sail-wing alone were basically the same as those for the complete model. A maximum lift coefficient of 1.5 was obtained at an angle of incidence of about 15°. The sail-wing

Fig 3.44   Sail-wing concept.

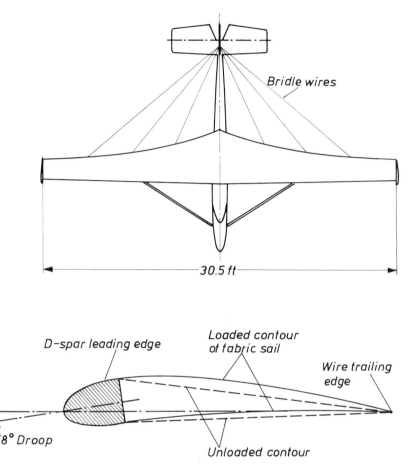

reached a maximum value of L/D of about 28, which is very high by any ordinary thin-sail standard and is about the same as that achieved with smooth, conventional hard wings of approximately the same aspect ratio.

(b) *Padded sails*

The so-called padded sails, to which reference was made in Part 2, Fig 2.73, were conceived along the same line as Herreshoff's double luff sail. As mentioned earlier, rumours about their high efficiency forced the IYRU to introduce a prohibitive rule into the sail measurement instruction. There was little point in this however, as can be seen from the results of fairly simple comparative tests on padded and conventional sails, which are presented in Fig 3.46. Tests were limited in scope and only the lift and drag components were measured. Three different masthead genoas, a conventional sail, a padded sail and finally the envelope of the padded sail, all of

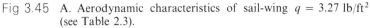

Fig 3.45 A. Aerodynamic characteristics of sail-wing $q = 3.27$ lb/ft^2 (see Table 2.3).

B. Comparison of sail-wing with conventional wings:

Sail-wing AR = 11, c_t/c_r = 0.4, Re = 0.8 × 10^6
Wing 1   AR = 10, c_t/c_r = 0.4, Re = 3.5 × 10^6
Wing 2   AR = 12, c_t/c_r = 0.2, Re = 2.3 × 10^6

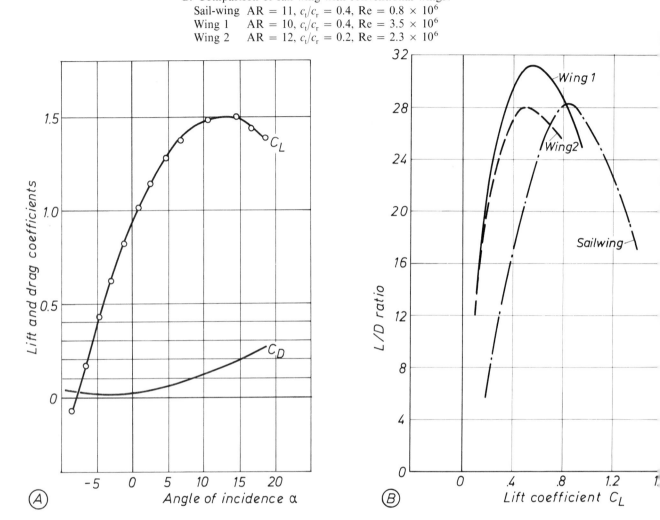

the same area and dimensions, were tested on a rig with the same conventional mainsail set at sheeting angle $\delta_m = 2°$. The height of the headsail was about 7 ft, the angle of heel of the model was 15° and all tests were carried out at the same wind speed of $V = 26.7$ ft/sec.

The sheeting angle of all genoas tested, measured at the foot, was the same ($\delta_f = 7°$), but the tensions applied to the clew in a particular run were initially adjusted so that the sails assumed a shape, with the wind on, that to the practical eye seemed best for the predetermined range of heading angles $(\beta-\lambda) = 17.5°–30°$. At the completion of these adjustments the two components of the total aerodynamic force, namely the lift L and drag D, were recorded. Some verifying tests were also performed at other sheeting angles δ_m and δ_f, but the results did not substantially

Fig 3.46 Comparative test on 'padded' and conventional headsails.
$V = 26.7$ ft/sec.

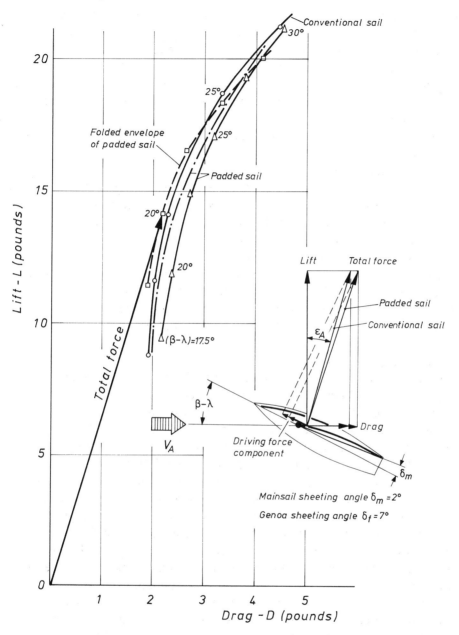

differ, in a relative sense, from those depicted in Fig 3.46 for sheeting angles $\delta_m = 2°$ and $\delta_f = 7°$.

To assess the relative merits of different sails it was assumed, for reasons explained earlier, that if the polar diagram curve L versus D, for a given sail is bodily shifted towards the lower drag relative to the other curve regarded as a yardstick, the sail efficiency is better. This assumption is presented graphically by the sketch in Fig 3.46, where it is seen that a conventional sail has a lower drag than a padded sail for the same lift. So the padded sail offers no better performance to windward than the ordinary, conventional, thin sail supported by hanks; on the contrary, its efficiency is worse. Subsequent tests on the folded envelope of the padded sail, after the inside foam was removed, gave results which are better than that for padded and conventional sails. Why?

The smooth leading edge of the folded envelope wrapping the forestay evidently has better entrance efficiency than the hanked genoa. The influence of the relatively sharp leading edge of the padded sail on its characteristics cannot be assessed readily. However, it became a matter of general experience in aeronautics that the performance of a foil can be drastically impaired by incorrect shape of entry, i.e. the curvature of the foil right behind the leading edge, as discussed earlier in Part 2 in connection with Fig 2.72. For this reason, powerful mathematical methods have been employed in aeronautics in order to find correct shapes of profiles that assure predetermined, desirable flow and pressure distribution.

(c) *Leading edge function*

Amongst a number of geometrical features of the foil cross-section such as the maximum camber/chord ratio, position of the maximum camber and the foil curvature at the leading edge proximity, the last one appears to be at least as important as camber/chord ratio. Figure 3.47A is intended to illustrate the significance of the angle of entry, marked E, which is made by the tangent at the very leading edge of the sail section. Adjusting the luff tension by means of a Cunningham hole, the position of maximum camber and, what is more important, the angle of entry E can be varied while the amount of camber may virtually remain the same. The arrows S and T indicate the variation of the entry angle E depending on whether the sail luff is slack or taut, which of course is controllable within certain limits. So, even if the magnitude of camber is fixed, we are still left with a range of possible shapes which the sail section may have, each having different aerodynamic characteristics.

Sketch B in Fig 3.47 illustrates the physical meaning of the angle of entry E as distinct from the incidence angle α. It shows schematically the path of air particles; as they approach the leading edge of the sail section where the effect of progressively stronger upward circulation velocities, felt as an upwash, causes the particles to rise rapidly at the leading edge. If now the flow is to enter the leading edge smoothly then the nose of the section must curve in some fashion towards the oncoming air particles to meet them gently, in a symmetrical manner. In such a case, the stagnation point S_t will be right on the very edge of the section.

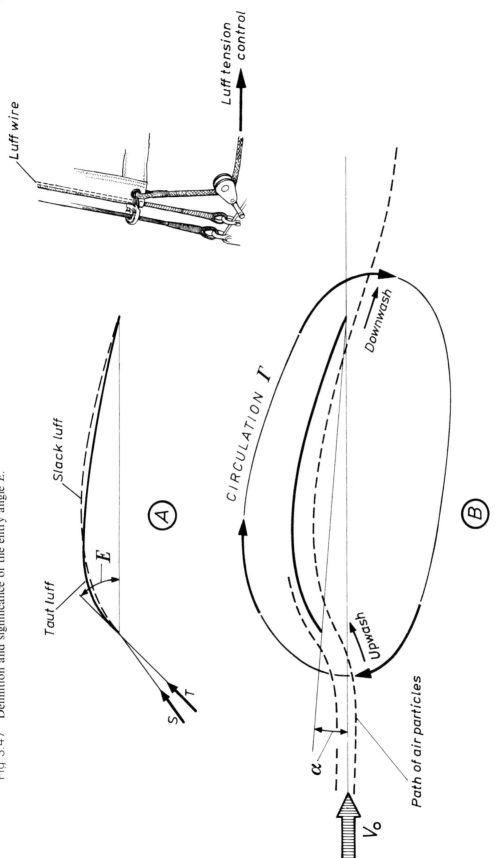

Fig 3.47 Definition and significance of the entry angle *E*.

Luff wire

Luff tension control

Slack luff

Taut luff

E

S

T

A

CIRCULATION Γ

Downwash

Upwash

B

α

Path of air particles

V₀

As pointed out in Part 2 and shown in Fig 2.62, the flow of air on to a section with a sharp leading edge is likely to be smooth, i.e. the two neighbouring streamlines will split at the edge without undue initial shock, only at one particular angle of incidence, which may well be called an ideal or optimum angle. One might therefore expect that for every wind speed there is an optimum combination of sail camber, section curvature at the leading edge and the sail incidence or sheeting angle which produce the best performance.

One more glance at Fig 3.47 and an analysis of Eq 2.12 from Part 2, re-written below for convenience, should substantiate the above statement:

$$\Gamma = f(V_0, c, \alpha) \qquad \text{Eq 2.12 (repeated)}$$

It is evident that the upwash in front of the sail, and therefore the direction of the airstream that enters the leading edge, depends on the strength of circulation Γ, which in turn is a function of wind speed V_0, well ahead of the sail, and the sail incidence angle α. In other words, any section with a sharp leading edge which is required to be a really efficient form should undergo a change in radius of the foil curvature near the leading edge with every change of angle of incidence or wind speed. With the well-rounded entry of thick foils the importance of the leading edge requirement is lessened but not removed.

It is interesting to note that the advantage associated with an adjustable leading edge was discovered as early as 1920. Biplane wings had relatively thin sections because they were biplanes and they suffered from leading edge stall. A cure was found in the form of the so-called droop snoot, i.e. adjustable hinged leading edge flap. Wind tunnel measurements showed that by dropping the nose of a foil the maximum lift coefficient could be increased from about 1.1 to about 1.7 for an optimum snoot deflection of about 30° relative to the chord of the otherwise plain foil section. This reserve of lift was vital in landing conditions when high lift coefficients are required.

A similar desire to secure high lift, and a large rudder force when needed, is behind the concept of the so-called articulated rudder shown in Fig 3.48A. It was designed by H Herreshoff and, as reported by B Devereux Barker III in *Yachting*, was initially employed on a 41 ft Cougar Class yacht. With the modern tendency to have the wetted hull area cut away as much as possible, it is a logical step for designers to develop rudders which although smaller in area are equally efficient. As seen in Fig 3.48A the articulated rudder consists of a fixed centre section with blades forward and aft that turn in the same direction. The forward blade turns half as much as the after blade by means of a connecting linkage. The resulting cambered shape is said to provide excellent steering control with modest rudder drag.

The theory supporting this concept, given by its inventor in his own words, is as follows,

'...the limitation on rudder force with a plain spade rudder is reached when a low pressure (suction) peak near the leading edge may cause the surface air to communicate with the suction side of the rudder blade. This phenomenon is called "ventilation". Use of a skeg in front of the rudder serves to reduce the peak suction pressure and the tendency for ventilation.

Because it effectively resists ventilation, in any application where very high lift (or side force) is required, a cambered shape is preferable to the plain spade or spade-skeg combination.'

The patent pending rudder, shown in a self-explanatory manner in Fig 3.48B and reproduced here from the German Magazine *Die Yacht*, as well as the fin-keel with leading and trailing edge flaps tested on a towing tank model shown in Photo 3.25, manifest the same line of thought. They resemble the high lift devices employed on aeroplane wings during take-off and landing. The flaps make it possible to attain large lift required at low speed without danger of severe flow separation that might otherwise occur.

(d) *What do the tell-tales tell?*

On the assumption that it is advantageous, when close-hauled, to adjust the sail shape so that every section along the sail luff is at its optimum incidence, the practice of using tell-tales near the luff has become quite common. Some devoted helmsmen openly admit that nowadays it is impossible to be competitive without the aid of these woolly yarns. When sailing to windward, the mainsail sheet and course steered should be so adjusted that as many tufts as possible lie smoothly against the fabric of the sail, i.e. without showing any agitation. In this way, it is believed, the leading edge separation on both sides of the sail can be avoided and by doing so the ideal or optimum angle of incidence is attained. Three sketches in Fig 3.49, redrawn here from one of the sailing magazines, illustrate, although inaccurately, this concept. Thus, if the sail incidence is made any smaller (trim too loose in sketch B), a front stagnation point shifts to the leeward side and may cause so-called backwinding (reversal of sail curvature) with separated flow on the windward side. Conversely, if the sail incidence is made greater (trim too tight in sketch C), a stagnation point forms on the concave side with separated flow on the other.

Although most experts agree that '...the early stall indication that yarns give is one of their most important functions', some of them warn of '...the danger of using any set rule for reading the tufts, and advise that each owner should consult with his sailmaker as to their best use, and employ trial and error in interpreting his own streamers.'

The use of tell-tales appears to be still in its infancy and this is well reflected in the recommended positioning of the tufts back from the luff ranging from 6 to 18 in. In this connection it is worth asking what these threads really indicate or detect.

The essential principle has long been employed by aerodynamicists in wind tunnel tests for studying air-flow over wing models. The remarkably detailed investigations, made by McCullough and Gault (Refs 3.31 and 3.32) practically solved the problem of direct correlation between the character of flow over the leeward side of a foil and type of stall caused by flow separation near the leading and/or trailing edge. To the

Fig 3.48 Rudders of variable geometry.

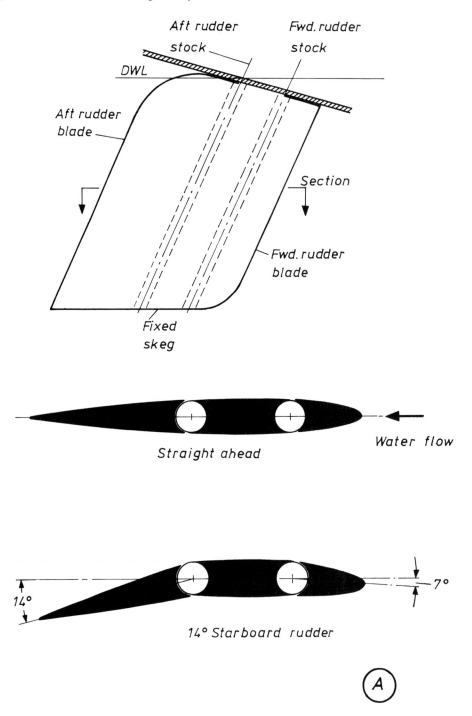

Aft rudder stock

Fwd. rudder stock

DWL

Aft rudder blade

Section

Fwd. rudder blade

Fixed skeg

Straight ahead

Water flow

14°

7°

14° Starboard rudder

(A)

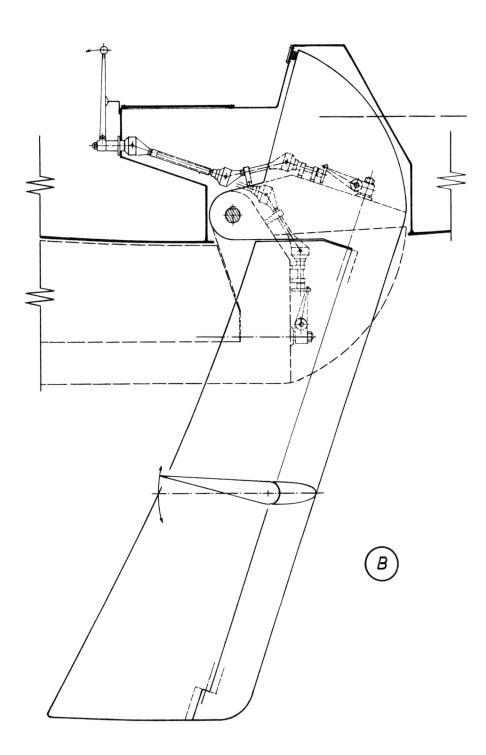

B

Fig 3.49   What do the tell-tales tell?

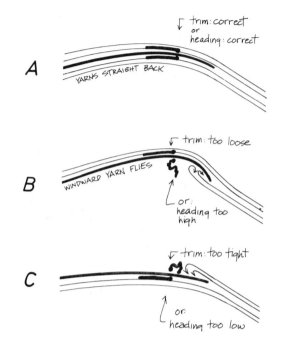

satisfaction of the author, who conducted similar tests on sail-like sections, such as shown for example in Photo 2.25, the McCullough and Gault conclusions are perfectly applicable to flow conditions observed on sails. Accordingly, as discussed in Part 2, the smaller the radius of leading edge curvature, the higher the local velocities, therefore the greater is the local suction peak and hence the more probable is laminar separation near the leading edge. Once separated, the flow passes above the surface of the foil and usually re-attaches further downstream. The mechanism of re-attachment is such that it can be expected, provided the incidence angle is sufficiently low, that the flow will re-attach to the foil surface a short distance behind the leading edge and then will follow the section contour up to the trailing edge without further separation; as illustrated in Photo 2.25A and Fig 2.59B.

When discussing the thin foil stall pattern in Part 2 in connection with Fig 2.57, it was mentioned that the so-called laminar separation bubble plays an important role in determining the behaviour of the boundary layer along the leeward side of the foil and consequently the all-important pressure (suction) distribution. Figure 3.50 illustrates a simplified model of the flow pattern in the presence of a bubble near the leading edge, based on McCullough and Gault's findings (Refs 3.31 and 3.32). Their observations afford a unified interpretation of thin foil stall. An obvious condition to the appearance of the bubble is the existence of an adverse pressure gradient (high suction peak) steep enough to cause laminar separation at point S close to the leading edge. But this condition, although necessary, is not sufficient to initiate the

Photo 3.25   Variable geometry of the fin-keel tested in the Southampton
University towing tank. Below, picture of the flow around
the nose of the foil.
A. Separation occurs at the leading edge.
B. Flow remains attached if the nose is deflected.

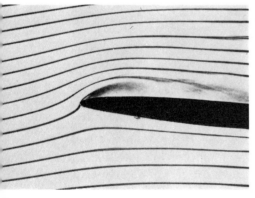

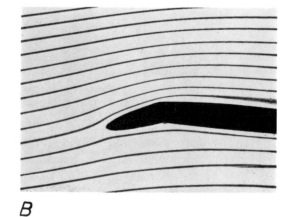

A   B

Fig 3.50 Simplified model of the flow pattern in the presence of a bubble near leading edge of a thin foil. Size of the bubble in vertical dimension is greatly exaggerated for the sake of clarity.

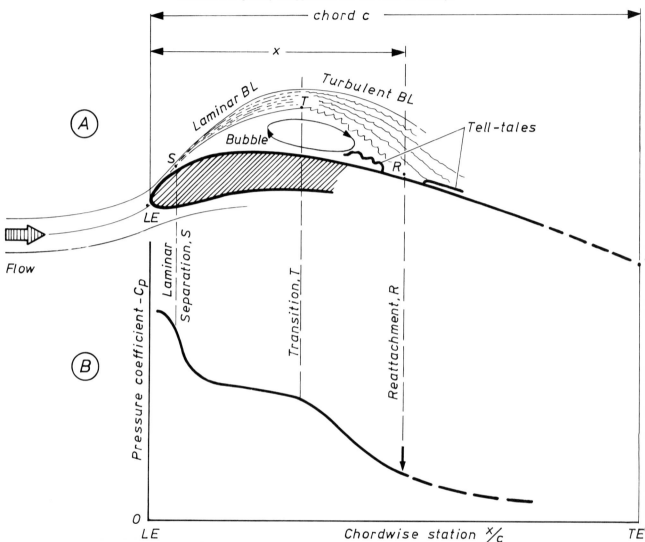

formation of the bubble. If the Reynolds Number is high enough, a case applicable to fast craft, transition from laminar to turbulent flow of the boundary layer will take place ahead of the theoretical laminar separation point S in Fig 3.50A; namely that point at which separation would have occurred if the boundary layer had remained laminar. Under these circumstances, the bubble formation will be precluded. On the other hand, if the Reynolds Number is sufficiently low, a case applicable to light wind or zephyr sailing conditions, the separated laminar boundary layer flow will not change into a turbulent one, hence the separated flow will not re-attach to the foil surface and no bubble will be formed. Thus the bubble formation is possible only for a certain range of Reynolds Numbers, and its character will depend on the foil curvature at the leading edge and incidence angle, all factors finally affecting the pressure distribution.

Referring to Fig 3.50A it appears that the role of the bubble is to facilitate the transition of the boundary layer flow from laminar to turbulent, which is considered to be a prerequisite for re-attachment of the flow to the foil surface. Furthermore, the reversed flow vortex observed inside the bubble attracts the separated flow towards the surface. However, the flow re-attachment is not likely to occur unless sufficient energy is taken from the wind to maintain this reversed flow vortex inside the bubble to compensate for energy dissipation. This explains why, at low Reynolds Numbers or low wind speed, the flow, once separated, will not re-attach; simply because there is not enough energy to be taken out from the low-speed wind to support both the reversed flow vortex and the boundary layer flow against unfavourable pressure gradient.

Now we are in a position to look at the problem of optimum sail camber in light winds from another viewpoint. In general it can be accepted that one should have flat sails in heavy winds and full sails in light winds. However, there is a certain limit to increasing sail camber, and this is determined by the fact that, sooner or later when wind speed decreases, a critical combination of sail camber and low Reynolds Number is reached at which the stall pattern changes rapidly from one type to the other, i.e. from the predominant thin-foil stall to the so-called leading edge stall classified earlier in Part 2. The latter, described as an abrupt flow separation without re-attachment and depicted in Fig 2.59A, is characterized by the complete collapse of the leading edge suction peak accompanied by a rapid and disastrous lost of lift and increase in drag. This type of stall results from the failure of the separated boundary layer flow near the leading edge to re-attach to the foil surface. In such circumstances, the tuft observations will indicate that the flow over the leeward side of the foil is steady at all angles of incidence prior to the stall and will give no warning of any impending change. In other words, the tell-tales may not indicate existence of a localized region of separated flow near the leading edge, such as, for instance, shown in Fig 3.50A. The transformation into the stall pattern, characteristic of full-chord flow separation shown in Fig 2.59A, is seemingly instantaneous. A less cambered sail is more resistant to leading edge stall and therefore more efficient in drifting conditions in which this type of stall is likely to occur.

The critical Reynolds Number, at which the thin-foil type of stall with re-attachment may be replaced by the leading edge type of stall without re-attachment, is believed to be about 6×10^4; the actual value depends on the foil curvature defined not only by the amount of maximum camber but also its distribution along the chord. Besides, the critical Reynolds Number also depends on the turbulence of the oncoming stream, that is, the higher turbulence delays the undesirable laminar leading edge separation; and this is the reason why a so-called turbulator wire placed just in front of the foil leading edge is one of the methods of delaying or

Fig 3.51 Pressure distribution on Gö 801 foil section at the same incidence angle $\alpha = 18°$ and two different Reynolds Numbers (4.2×10^4 and 7.5×10^4) at which two different types of stall occur. The estimated position of re-attachment is shown by downward pointing arrow. This figure is based on Kraemer measurements described in Ref 2.53.

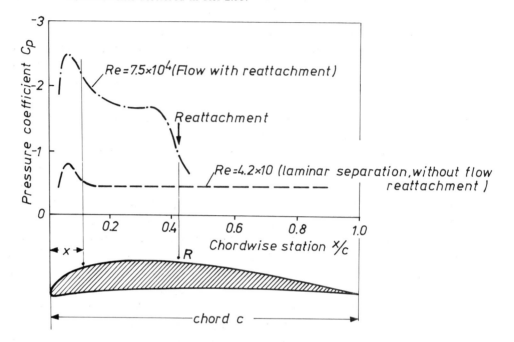

avoiding laminar separation. Sharp leading edge is another means which hastens transition of the flow from laminar to turbulent, and so it is recommended for foils operating in a low range of Reynolds Numbers. The relevant losses in terms of suction, particularly important in the front part of the foil, are well demonstrated in Fig 3.51, which gives the suction distribution over the leeward part of the same foil developed at two different types of flow, one with re-attachment and another without.

Let us now return to the predominant type of thin-foil stall, i.e. partial separation at the leading edge as shown in Fig 3.50A. Below, sketch B illustrates, in a quantitative sense, the chordwise distribution of pressure coefficient C_p (suction) over the leeward side. It can be seen that partial collapse of the suction peak near the leading edge, which usually takes place when the incidence angle is greater than 5–6°, is followed by the appearance of a region of approximately constant pressure which we referred to in Part 2 as a suction plateau. This is indicative of laminar separation in the presence of an underlying bubble. At a certain distance from the separation point S, transition from laminar to turbulent flow takes place, which is marked by the letter T. Transition and the ensuing expansion of turbulent flow were subsequently found to be the necessary conditions to re-establish the flow on the foil surface which occurs somewhere in point R; distinguished as re-attachment in both sketches of Fig 3.50.

The tell-tales can give quite a clear picture as to where the re-attachment actually takes place. The surface tufts located downstream from R indicate relatively steady flow towards the trailing edge while the tufts located upstream from point R indicate the existence of the rough flow, pointing erratically towards the leading edge; and their agitation is the result of reversed flow vortex operating somewhere between points R and LE. Thus, the tell-tales, if properly sited, can detect not only the flow separation, but also, and this is more important, the character of separation. Such a distinction is not a matter of pure semantics, but of essential practical importance. In fact, it is directly linked with the important question—how should the tell-tales be distributed in relation to the sail leading edge to fulfil their function effectively? To grasp the significance of the distance of re-attachment point R from the sail leading edge LE we must consider how the position of re-attachment R is related to the suction distribution over the leeward side of the sail and subsequently to sail efficiency.

According to Ref 3.32, visual observation of tuft behaviour, prior to the appearance of the separation bubble, indicated smooth flow over the leeward side for angles of incidence α up to 4.5°, at which the c_l coefficient was about 0.5. At this point, noticeable intermittent separation of flow near the leading edge occurs. As the angle of incidence is further increased, leading edge separation in the form of a bubble persists and gradually spreads downstream. A relatively large region of separation bubble is first discernible at an angle of incidence of 5°, at which point, as shown in Fig 3.52, it covers approximately the first 8 per cent of the chord c. The thickness or height of the bubble underlying the separated boundary layer is about $\frac{1}{2}$ per cent of the chord c. With increasing incidence, the separated flow region grows in both thickness and chordwise extent x until it covers the entire leeward surface of the foil at an incidence angle $\alpha = 9°$, corresponding approximately to maximum lift coefficient c_l about 0.9. The maximum thickness of the underlying bubble for the boundary layer flow which re-attaches to the surface close to the trailing edge is then about 3.5 per cent of the chord length c. Thick arrows pointing downward illustrate the re-attachment points R, which are shifting gradually towards the trailing edge as the incidence angle increases.

The suction distributions in Fig 3.53, which should be regarded as a supplementary picture to Fig 3.52, clearly show that, after the collapse of the suction peak, the negative pressure is redistributed along the chord into the more or less flattened suction plateau. This extends over the region occupied by the bubble length, as illustrated by thick arrows pointing down and indicating re-attachment position R along the chord. It can be seen that with an increase of incidence angle the suction plateau is lowered and lengthened. Coincident with it is an abrupt increase in drag.

Bearing in mind the knowledge we already have that the driving force in the sail is

Fig 3.52   Thickness of the bubble underlying separated boundary layer. Observations made on NACA 64A006 section are similar to the author's own findings made on sail-like forms. Attention is drawn to the fact that since the model was tested in two-dimensional flow conditions, the incidence angles indicated in this and Fig 3.52 refer to *effective* incidence angles.

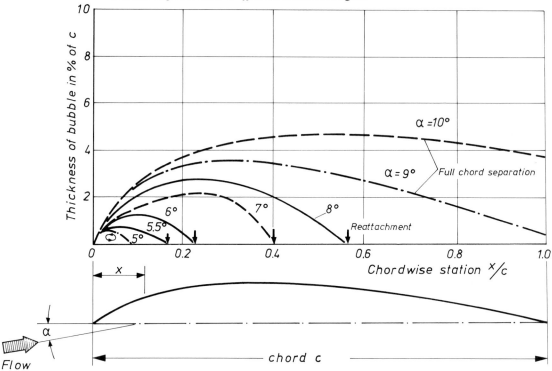

generated in its forward rather than its after part (see Fig 2.77), one may anticipate that there must be a certain optimum distance of the re-attachment point R from the leading edge, beyond which sail efficiency deteriorates rapidly. Evidently, according to the conclusions reached in Part 2, when discussing the practical indications of Fig 2.77 we found that, in close-hauled conditions, the sail should be trimmed in such a way that the leeward side pressures are concentrated close to the leading edge. Otherwise, when those pressures are shifted towards the trailing edge they will tend to give a large drag component. It appears, on the basis of Fig 3.53, that the optimum position of the tell-tales in relation to the leading edge should be such that they are capable of detecting re-attachment somewhere between 5 to 15 per cent of the sail chord. So any standard formula which says, for instance, that '...the yarns should be set back from the luff between 10 and 16 in,' regardless of the sail chord, is not accurate enough for the purpose. In other words, when positioning the tell-tales along the luff of, say, a genoa, the distance D between the luff and the line of yarns set along three to four stations, marked 1, 2, 3, 4, in Fig 3.54, should vary according to the chord length at the station in question. The exact position and eventual

Fig 3.53   Pressure distribution on thin foil at different incidence angles from 5° to 10°. Downward arrows indicate the position of re-attachment.

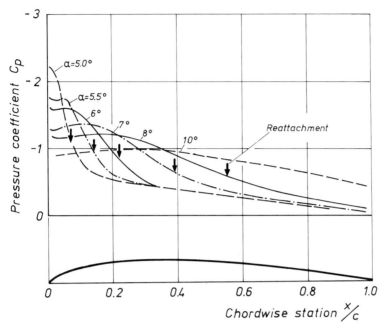

number of yarns in a row can only be established by a trial-and-error routine and personal preference.

In order to find the correct position for the tell-tales on a genoa, it is desirable to have at one station, say, 2 in Fig 3.54 and Photo 3.26, several yarns, one after another, attached chordwise in a line right from the leading edge up to about 20 per cent of the chord length, as suggested by Gentry (Ref 3.33). The general rule in interpreting their behaviour when beating would be:

1. In very strong winds, when incidence angle is small, none of the leeward side yarns should flutter.

2. In gradually decreasing wind speed, when the incidence angle of the sail is larger, the course to windward should be selected in such a way that some of the yarns located at, say, distance 5–10 per cent of chord, may be allowed to flutter. The number of yarns actually twirling will tell us the actual size of the bubble which might be tolerated, bearing in mind the correlation between the bubble size and pressure distribution shown in Figs 3.52 and 3.53. This can only be established by experiments on a particular sail and the last yarn in the row which is allowed to flutter will give an indication as to where the remaining single yarns in station 1, 3 and 4 should be located.

Apart from the aid of tell-tales, correct mainsail tuning can be facilitated by painting a distinct colour-strip going straight aft from the luff, some distance from

Fig 3.54   Essential set-up for tell-tales and tell-tails.

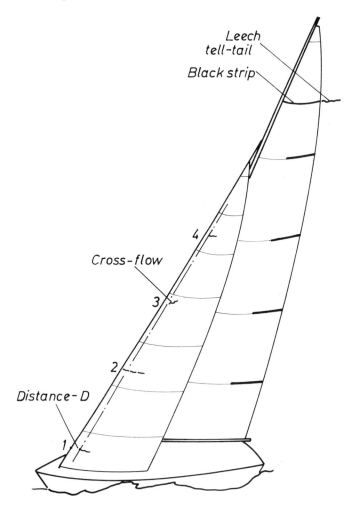

the sail head, as shown in Fig 3.54 or Photo 3.16. Such a strip, together with leech tell-tail also shown in Fig 3.54, will assist in estimating the camber and twist of this important part of the sail where harmful separation, so crucial to boat performance, is most likely to occur first.

Suggested arrangement of tell-tales, tell-tails and coloured strips will certainly help the crew to think of sails in three-dimensions and subsequently change the three-dimensional sail shape according to requirements imposed by wind and course sailed.

An interesting arrangement of tufts, shown in Photo 3.26, was developed by A Gentry. Readers wishing to know more about the practical aspects of this arrangement are recommended to read Ref 3.33.

Referring to Fig 3.54, it may happen, as mentioned earlier in Part 2, that some

Photo 3.26   Mainsail and genoa set up for tell-tales suggested by A Gentry.
Pictures reproduced by kind permission of A Gentry.

yarns on the leeward side of the sail at station 3 do not stream in the chordwise direction but instead manifest a persistent tendency to point obliquely upwards without fluttering. Such a tendency is evidence of the crossflow of the boundary layer to which every swept foil is susceptible, but in the case of headsails this tendency can either be aggravated or alleviated. The reason for it is the inherent difficulty of achieving uniform effective incidence angle along the whole sail height due to the forestay sag. Sketch A in Fig 3.55 shows this in a fairly self-explanatory manner; depending on the amount of sag in the forestay, both the camber and incidence angle of the sail section will vary from, say, 1 to 2, as indicated. Evidently, section 2 is fuller and set at a higher incidence angle α then section 1.

The differences in camber, due to the forestay sag, can be compensated by the sailmaker who may introduce some correction, possibly in the form of an elongated

Fig 3.55   Deleterious effect of forestay sag on pressure distribution.
In sketch C, V–flow at the leading edge of the genoa
V_U–flow caused by pressure gradient existing on the leeward surface of the sail between points 1 and 2
V_R–resulting flow indicated by the yarns attached close to the leading edge between sections a and b.

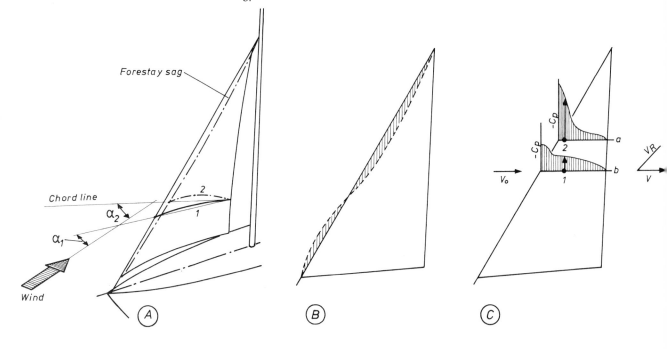

S-shaped distribution of sail cloth surplus along the luff as shown schematically in sketch B. However, the differences in incidence angles α_1 and α_2 cannot be compensated and they will yield different pressure distribution in neighbouring sections as demonstrated in sketch C. Section a, operating at higher incidence angle, may develop the suction distribution depicted by the small graph of C_p in Fig 3.55, while section b, operating at smaller incidence, may develop more favourable pressure with a higher suction peak than section a. Thus, between points 1 and 2 on the sail lee-surface there exists a sharp pressure gradient at right angles to the air flow V indicated by an arrow in a sketch to the right of section a. In other words, the pressure at point 1 is higher (less suction) than at point 2. It has been found that in such circumstances a comparatively small pressure gradient, i.e. small differences in pressure, tend to produce large cross-flow and this is indicated by an arrow V_u (upward flow) in sketch C. As a result, the air particles, which become decelerated in the boundary layer to velocity V from initial velocity V_0 ahead of the sail, will have a tendency to travel in the direction V_R which is the resultant velocity of the two components V and V_u. It thus appears that a really tight forestay is an absolutely essential condition allowing tuning the headsails to their best efficiency, particularly in close-hauled work.

Finally, one should be aware of the fact that even if the sail of a given section shape is trimmed correctly with the help of tell-tales, it does not necessarily mean that the boat will be automatically steered to her best performance, in terms of optimum V_{mg} for given wind speed. Why is that? We already know from Part 1 Fig 1.9D that for every boat there is an optimum apparent course β for given wind speed which produces the best V_{mg}. This particular course β is largely determined by the way the hull resistance builds up with speed V_s and it implies, as demonstrated in Fig 1.9E, a certain optimum sheeting or incidence angle for the sail; no matter whether the sail in question is a headsail or mainsail alone (cat-rigged). Now, let us imagine that the headsail set is too flat for the wind conditions, as drawn for example in sketch C of Fig 3.49, and the helmsman attempts to steer the boat by the tell-tales attached to this sail. It is evident that in such circumstances the helmsman will tend to pinch the boat, which may take her away from the optimum β angle. In exactly the same wind condition, the wool streamers attached to the full sail, shown in sketch B of Fig 3.49, will guide the helmsman to select a rather broader course β and that, again, may take her away from the optimum course to another extreme. It implies that there is only one particular sail curvature, i.e. combination of camber in conjunction with leading edge curvature, which is the ideal one for a given sheeting angle and/or course sailed β. So, beware of tell-tales–they are very sensitive flow indicators but they cannot possibly tell the whole truth as to whether the boat is sailed in a most efficient way.

(5) Sail interaction, slotted foils

'Scientific theories can never be justified or verified. But in spite of this, a hypothesis A can under certain circumstances achieve more than a hypothesis B– perhaps because B is contradicted by certain results or observations, and therefore *falsified* by them.'

K POPPER
Logic of Scientific Discovery

'My child I have been wise, I have never thought about thinking.'

W GOETHE

Although single-slotted or multi-slotted foils, such as are shown for example in Fig 3.56 or Photo 3.28, were employed as a means of propulsion of sailing craft for centuries, no attempt was made to explain their action until they were re-discovered in aeronautics and subsequently used as auxiliary devices to assist in the lifting or landing of heavier-than-air machines. The slotted wing shown in Fig 3.57B is not

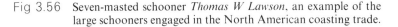

Fig 3.56 Seven-masted schooner *Thomas W Lawson*, an example of the
large schooners engaged in the North American coasting trade.

much different from a jib-mainsail combination. The practical objectives of both
systems, no matter how many slots, flaps or sails are actually incorporated, is
increase in lift and delay of the stall.

As we shall see, in a certain combination of foils a desirable type of flow, and
hence pressure distribution, can be realized which otherwise would be impossible.
The means by which these objectives are achieved always involve the same basic
principles, which are the creation of conditions under which the flow energy, lost
through viscosity (friction) action, is either minimized or restored in an efficient
manner. Thus, as will be demonstrated, slotted foils prevent the separation of the
boundary layer in circumstances where, without slot or slot action, separation would
certainly occur.

(a) *Explanation of slotted foil action*

It was pointed out by A Gentry in Ref 3.33 that '...there are some serious
misunderstandings about the old slot effect explanations in the sailing literature' and
moreover '...all the explanations in the sailing books on the interaction between the
jib and main are wrong'. All one can say in defence of sailing theoreticians is that
these misconceptions concerning jib-mainsail interactions were originally derived
from the most respected and time-honoured, authoritative aerodynamic theories
and faithfully reflected the state of affairs in this field.

'Because,' to quote A M Smith, the author of a splendid paper on *High-Lift
Aerodynamics* (Ref 3.34), 'unlike the birth of Venus, new ideas do not burst forth
fully matured or fully recognized,' a short review of how the concept and theory of
slotted foils developed seems both instructive and desirable to clarify this
controversial issue. The problem is fascinating and topical, since multi-slotted foil

Photo 3.27   The so-called 'segmental sail' made of thin panels of cloth parallel to the leech. It has been claimed by the inventor (G Corbellini) that the panels are self-adjusting so that the sheet need hardly be touched. Reefing is done by removing panels.

The photograph is reproduced with kind permission of *Yachting World*.

Photo 3.27B   Close-up of segmental sail plan arrangement   (*photo by G Corbellini*)

Photo 3.28   Although the multi-slotted ketch configuration is not as good to windward as, say, the sloop or cutter, on other courses it is capable of developing larger driving force.
(*Beken of Cowes.*)

Fig 3.57   Prandtl's slotted foil analogy was largely responsible for the long-cherished theory according to which the flow separation on the foil behind the slat (sketch B) is delayed or avoided by the jet of 'fresh' air that flows through the slot.

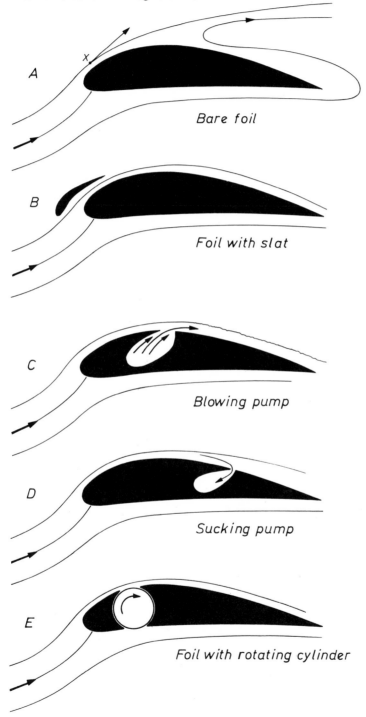

A   *Bare foil*

B   *Foil with slat*

C   *Blowing pump*

D   *Sucking pump*

E   *Foil with rotating cylinder*

configurations still attract the attention of many sailing enthusiasts dreaming about fast sailing machines or more efficient rigs. Photographs 1.19 and 3.27 serve to illustrate some recent developments in this respect.

In about the year 1920 Handley-Page and concurrently G Lachmann conceived, or rather re-discovered, and demonstrated that a foil incorporating a slat in front can, as a system, develop considerably greater maximum lift than a single foil of conventional form. Subsequently, Handley-Page attempted to prove that a combined foil consisting of $n + 1$ elements can produce more lift than one having n elements, and he investigated configurations having up to 8 elements. Figure 3.58A depicts one of his extreme multi-slotted foils, a modified RAF 19 section, set at the incidence angle α of 42°, which corresponded to maximum lift (Ref 3.35). Figure 3.58B shows the lift coefficient curves for the section as it was progressively modified from a one- to a seven-slat configuration. These tests indicate that, with a multiple slot arrangement, an increase in lift can be obtained of two to three times the usual value without a slot. However, the lift/drag ratio L/D drops from about 12 for one slot to about 6 for seven slots. This rather dramatic decrease in L/D ratio, which negates the benefit of having high lift, should be viewed as the result of the quadratic increase of induced drag with lift, as reflected in formula 2.28A.

Incidently, the so-called 'segmental' sail shown in Photo 3.27, developed recently by G Corbellini and reported in Ref 3.36, as well as the seven-masted schooner shown in Fig 3.56, or any other type of multi-sail or multi-slotted foil configuration, are bound to suffer the same kind of setback, in terms of L/D ratio, as the Handley-Page foil shown in Fig 3.58.

The multi-slotted foil is merely a development of the single-slotted foil, so the action of the more complicated system can more easily be understood by considering the effect of the front foil on the following one. Thus, as shown in the rough sketch C of Fig 3.58, each foil imparts a certain downward motion to the airflow, the effect being cumulative. The downwash reached at the fourth slat of a series set for maximum lift is considerable and, according to the implications of Eq 2.28, can be regarded as a measure of the induced drag which in turn determines the resulting L/D ratio.

Numerous tests conducted since Handley-Page on various foil configurations, sails included, have confirmed beyond any argument that a combination of foils operating in tandem or in close proximity, when each foil is subject not only to its own effects but to those of the other foils, is capable of developing very high lift indeed. Unfortunately, the side-effect or by-product of it–a small L/D ratio–limits the application of multi-slotted foil configurations, and sailing people already know that the yawl or ketch configuration, as shown for example in Photo 3.28, is not as good in windward work as the sloop or a single-sailed craft. Therefore, these or similar configurations cannot possibly be used as a propulsion system in any truly fast sailing craft, in which the highest available L/D ratio, as was demonstrated in Part 1, is an absolutely essential criterion of speed potential. Nevertheless, a combination of interacting foils developing large lift can successfully be used on

Fig 3.58   RAF 19 eight slats section with seven slots. Section shown at
incidence angle for maximum lift. Foil of AR = 6 was tested at
Re about 2.5×10^5.

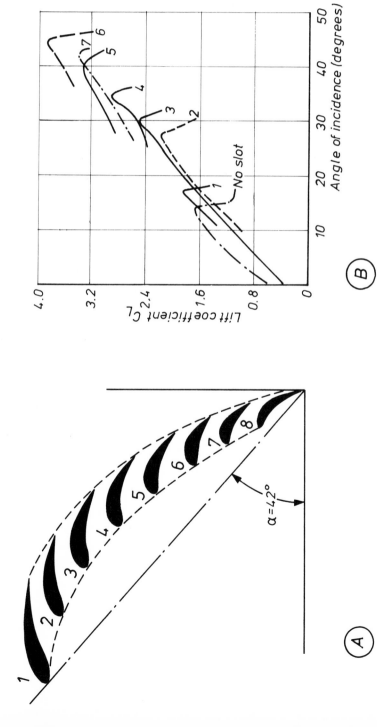

courses other than beating, or in craft which are not primarily designed for speed alone.

So far, so good, but why do slotted foils develop higher lift than a single foil? The famous fluid dynamicist Prandtl was one of the first to give an answer to this question in the book *Applied Hydro-Aeromechanics* published in the early thirties. His theory, illustrated by Fig 3.57, was explained as follows:

'The air coming out of the slot blows into the boundary layer on the top of the wing [see sketch B] and imparts fresh momentum to the particles in it, which have been slowed down by the action of viscosity. Owing to this help, the particles are able to reach the sharp edge without breaking away. A similar action can be obtained by blowing air at great velocities through little nozzles from the interior of the wing into the boundary layer [see sketch C].

Another means of preventing the boundary layer particles from flowing back, is to suck them into the interior of the wing [see sketch D]. This is done by means of a blower, and the air thus transported into the wing is blown off at some place where it cannot do any harm.

Still another method of obtaining the same result is to replace the front edge of the airfoil by a rotating cylinder, or also by putting this cylinder inside the wing [as shown in sketch E]. Experiments made...have shown that airfoils with such a rotating cylinder can be made to have much greater lift coefficients $C_L = 2.43$ with $\alpha = 41.7°$.'

The above explanation and analogies of slotted-foil action appeared to be convincing; Prandtl himself was a well recognized authority in aerodynamics and an expert in boundary layer flow, so his view became a scientific paradigm. With small modifications it was repeated in most subsequent textbooks on the subject. As we shall see however, Prandtl was wrong in his interpretation of the slot effect, but a short excerpt from the sixth edition of the most authoritative monograph on boundary layer flow, published some 35 years after Prandtl's original enunciation, illustrates the fact that erroneous beliefs have an astonishing power of survival.

'An alternative method of preventing separation consists in supplying additional energy to the particles of fluid which are being retarded in the boundary layer. This result can be achieved by discharging fluid from the interior of the body with the aid of a special blower [see again sketch C in Fig 3.57], or by deriving the required energy directly from the main stream. This latter effect can be produced by connecting the retarded region of higher pressure through a slot in the wing. In either case additional energy is imparted to the particles of fluid in the boundary layer near the wall.'

A different explanation, published in fact before Prandtl, was offered in England by Le Page, who conducted a series of wind tunnel experiments on the possibilities of obtaining high lift from foils in tandem with various overlap by utilizing the mutual induction effects (Ref 3.37). One of his tested configurations is shown in Fig 3.59, the

Fig 3.59   Effect of mutual interference between the two foils, set in tandem with small overlap, on pressure distribution on separate foils and while interacting.

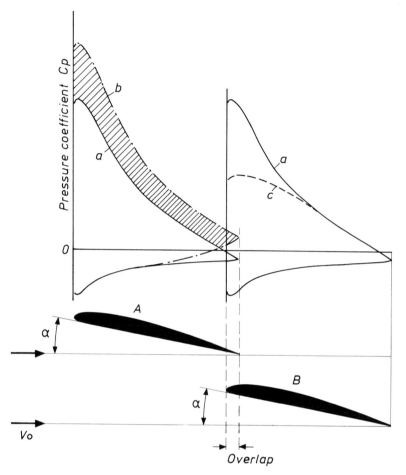

arrangement consisting of two approximately equal foils. Concurrently, a similar theory was developed by Prof Betz of Göttingen. What follows may serve as a rather different explanation of the effects produced by slotting a foil as compared with that given by Prandtl. The wording is close to Betz's version published in Ref 2.32.

The forward foil A, itself set at incidence angle α, would have had a pressure distribution represented by the continuous line *a* on the left pressure diagram. When the other foil B, set at the same incidence α and which by itself would have had approximately the same pressure distribution as foil A, is brought near to the first one, the trailing edge of the forward foil A will be in the region of greater velocity and correspondingly lower pressure produced by the rear foil B. This will have the effect of modifying the pressure distribution curve *a* of the front foil. Since its trailing edge is immersed in a region of flow velocity appreciably higher than that of free stream, the velocity at all points along the foil surface is increased, thus alleviating separation problems or permitting increased lift. In consequence of this alteration in flow conditions the negative pressure or suction over the front foil is more favourable, as indicated by the broken line *b* in Fig 3.59. It is at once obvious that the lift, which is represented by the area enclosed by curve *b*, has been considerably increased in comparison with that enclosed by curve *a*.

Exactly corresponding phenomena occur with the rear foil B. The front foil produces a diminution of velocity in the region of overlap and hence a reduction of suction at the nose of the rear foil. Consequently, the suction peak is much lower and the resulting pressure distribution is given by the broken curve *c* in Fig 3.59. On the whole, the two foils, interacting in the manner described, produce a greater maximum lift than if they were separated; and this increase is to be found entirely at the front foil.

This view on the role of the slot, according to which the flow within the overlap is decelerated, not accelerated, and so opposite to the view of Prandtl, has been corroborated by some other researchers since, as reported in Refs 2.15 and 3.34. Nevertheless in some textbooks, notably *Fluid Dynamic Lift* by S Hoerner, published recently in USA, 1976, both concepts compete for recognition, and accordingly:

'Boundary layer control by means of a slot is based on the concept of injecting momentum into a tired boundary layer. In addition to the supply of momentum to the boundary layer the following mechanism seems to be important for the effectiveness of leading edge slots or slats. Considering airfoil plus slat to be an entity, it is seen [in Fig 3.59], that the peak of the negative pressure distribution is loaded onto the slat. Peak and subsequent positive (unfavorable) pressure gradient on the main part of the airfoil, are thus appreciably reduced. Whatever boundary layer is formed along the upper side of the slat, is carried downstream as a thin sheet between the outer flow and the "jet" of fresh air exiting the slot.'

Looking now at the controversial problem of whether the flow within the overlap of two foils is accelerated or decelerated, one may say that both schools of thought are, in a way, right. The question that has given rise to much controversy would become irrelevant if one agrees on a frame of reference in relation to which the suggested acceleration or deceleration is measured or compared. Thus, as far as the front foil is concerned, the flow at its trailing edge is accelerated due to the action of the rear foil and this induces greater circulation on the front foil, as compared to that which would otherwise develop without the presence of the rear foil. Following Prandtl's analogy, the stream of air coming from the slot has the same effect as a pump sucking the accumulated material of the boundary layer from the surface of the front foil. This favourable effect is similar to the action of the sucking pump, presented earlier in sketch D of Fig 3.57. Of course, this work done by the pump may be viewed as either pumping or suction and is derived from the kinetic energy of the

Fig 3.60   Two interpretations of the interaction effects between the two
foils.
A. Circulation effect.
B. Upwash–downwash effect.

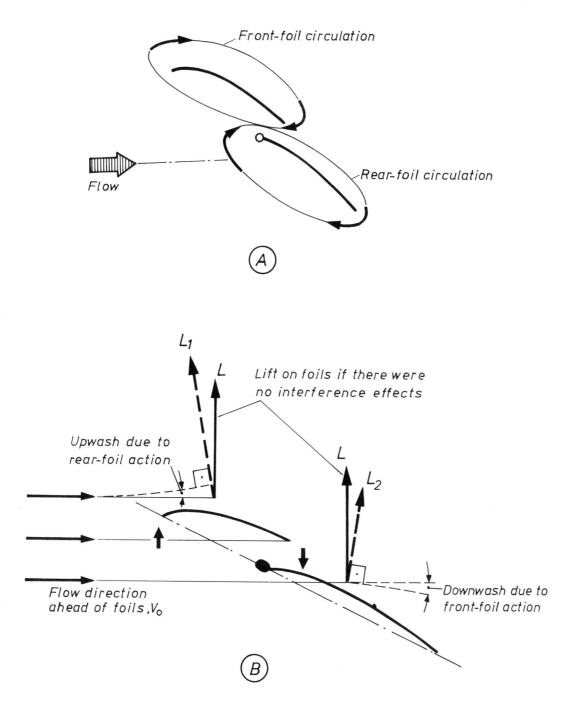

flow passing through the slot. This inevitably implies that the flow round the leading edge of the rear foil must necessarily be decelerated, as compared to the flow velocity which would otherwise occur over the leading edge of the rear foil if there were no front foil.

An interaction between the front and rear foil may also be explained by saying that the circulation round the front foil is increased to the same extent as the circulation round the rear foil is decreased. This is implicitly shown in Fig 3.60A, from which one may infer that in order to satisfy the Kutta-Joukowski condition, discussed in Part 2, the strength of circulation developed by the front foil will automatically be adjusted on account of the higher flow velocity at its trailing edge.

Alternatively, the effectiveness of the two interacting foils can be explained by taking into account the associated upwash and downwash effects shown in sketch B of Fig 3.60. It will be noticed that the forward foil is situated in the upwash of the rear or main foil as indicated by the thick arrow below the front foil. As a result, the incidence angle at which the flow meets the front foil is greater than the incidence angle α. This is in accord with our earlier discussion in Part 2. The air-flow can be traced over the front foil and continued on over the rear foil, where it is found that the angle made by the air-flow meeting the main foil is less than the incidence angle α. In other words, the rear foil is in the region of downwash from the front foil, and the air, having been deflected downwards to some extent already, finds less difficulty in adhering to the surface instead of separating as it otherwise would, or re-attach quickly if separation does occur. The net gain of such an interaction is that the lift L_1 generated on the front foil is higher and more favourably inclined when compared with lift L on the foil without interference effect. Conversely, the lift L_2 developed by the rear foil is comparatively smaller and is inclined unfavourably backwards. In other words, the leading foil carries a heavier load than the rear one.

The powerful effect on the character of the flow exerted by a relatively small auxiliary foil situated in proximity to the leading edge of a much bigger foil is dramatically exposed in Photo 3.29. It will be seen that at an angle of incidence α of about 25°, the full separation and stall is already developed on the main foil alone, while, with the help of a small auxiliary foil, the rear separation just begins at an incidence angle α of about 31°. As a matter of fact, such a configuration reflects in principle the tall-boy type of sail, a short-footed sail set between genoa and mainsail for beating in light winds, or set across the foredeck for running.

A range of positions of the auxiliary foil with respect to the main foil was tested to find out which one of them gave the most substantial gain in terms of maximum lift (Ref 3.38). It was found that the best efficiency was achieved when the auxiliary foil was located (in relation to its trailing edge) 15 per cent of the main foil chord ahead of the leading edge of the main foil and 12 per cent above the main chord line, both chord lines being parallel to each other. In the case of the Clark Y main foil section shape the maximum lift coefficient was about 1.8, which is 40 per cent greater than for the foil alone.

Many problems concerning the interference between a mainsail and a jib were

Photo 3.29   The slat is a powerful anti-stalling device. Tall-boy type of
sail works on the same principle.

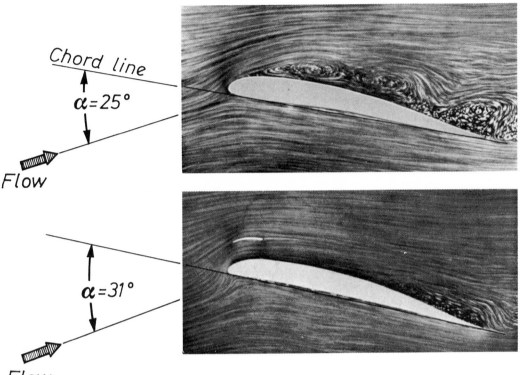

clarified by A Gentry who explained correctly, for the first time, the jib-mainsail interaction effect (Ref 3.33). Figures 3.61 and 3.62, reproduced here by his kind permission, illustrate the flow pattern round the jib and main with and without interference effects and also the corresponding pressure distributions. The pictures of the streamlines around the jib and mainsail or their combination were drawn with the help of the so-called analog field plotter, which is based on an electrical analogy technique (Ref 2.75). The calculated pressure distributions presented in Figs 3.61 and 3.62 perfectly agree, in a qualitative sense, with that in Fig 3.59 based on the Betz theory (Ref 2.23).

The following list describes the major jib-mainsail interaction effects, as classified by Gentry:

(b) *The effects of the mainsail on the jib* (see Fig 3.61)

1. The upwash flow ahead of the mainsail causes the stagnation point on the jib to be shifted around towards the windward side of the sail.
2. The leech of the jib is in a high speed flow region created by the mainsail. The leech velocity on the jib is, therefore, higher than if the jib alone were used.

Fig 3.61   The effect of the mainsail on the flow pattern and jib pressure distribution. (Drawing by kind permission of A Gentry.)

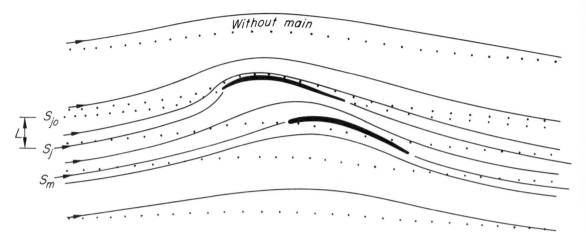

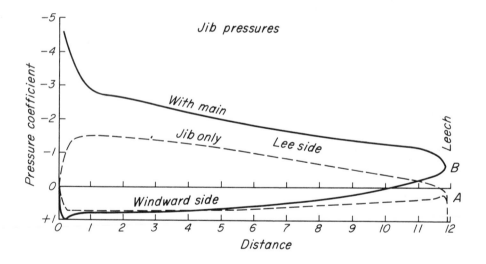

3. Because of the higher leech velocity, velocities along the entire lee surface of the jib are greatly increased when both the jib and main are used, and this contributes to the high efficiency of a jib.
4. The higher lee-surface velocities on the jib mean the jib can be operated at higher angles of attack before the jib lee-side flow will separate and stall.
5. Because of all this, proper trim and shape of the mainsail significantly affect the

Fig 3.62   The effect of the jib on flow pattern and mainsail pressure
distribution. (Drawing by kind permission of A Gentry.)

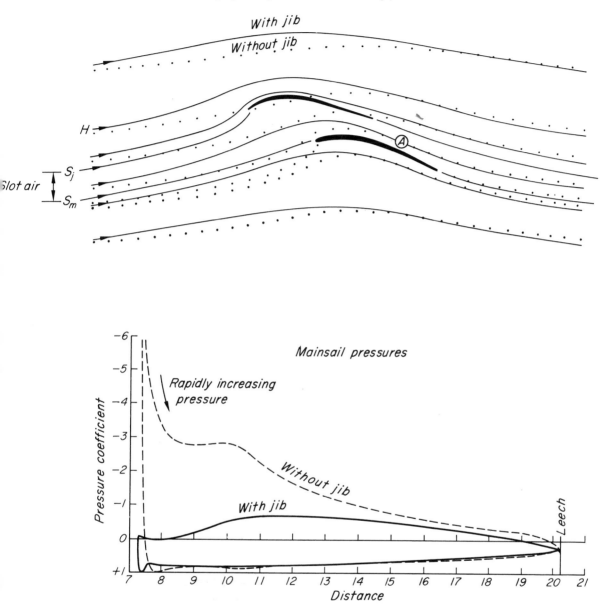

efficiency of the overlapping jib. Anything that causes a velocity reduction in
the region of the leech of the jib, such as some separation on the aft part of the
main, results in a lower driving force contributed by the jib.

6. The trim of the main significantly affects the pointing ability of the boat for it
directly influences the upwash that approaches the luff of the jib.

(c) *The effects of the jib on the mainsail* (see Fig 3.62)
  1. The jib causes the stagnation point on the mainsail to shift around towards the leading edge of the mast (the header effect).
  2. As a result, the peak suction velocities on the forward lee-side of the main are greatly reduced. Since the peak suction velocities are reduced, the adverse pressure gradient is also reduced.
  3. Because of reduced pressure gradients on the mainsail, the possibility of the boundary layer separating and the aerofoil stalling is reduced.
  4. A mainsail can be operated efficiently at higher angles of attack without flow separation and stalling than would be the case with just a mainsail alone. This is caused by a reduction in velocities over the forward-lee part of the mainsail rather than by a speed-up in the flow, which is the popular theory.
  5. As the jib is sheeted in closer to the main, there is a continuing decrease in suction pressure on the lee-side of the main. When the pressures both to windward and leeward sides of the mainsail become equal there is no pressure difference across the sail fabric to maintain the sail shape and the sail begins to flutter.

(d) *Some results of wind tunnel tests on sail interaction*

'As a very nervous new boy to yachting journalism, I crewed for John [Illingworth] in the early days of *The Myth*, and I still remember asking one of the regular crew, "Where does the skipper like this sheet cleated?" The reply was, "Oh, we don't cleat anything in this ship. He trims sheets all the time!"'

B HAYMAN
Yachting World, Editor

Although the significance of the continuous transverse trim of sails was recognized by many top dinghy sailors some years ago, the appreciation of the advantages of having instantaneous athwartships trimming gadgetry by ocean racer crews was, according to a 1973 copy of *Seahorse* magazine, '...the latest go-fast phase'. It was reported that in the 1973 American SORC series many yachts continually varied their sheeting angles with barber-haulers or double sheet leads, as the wind and sea conditions changed. 'Notably', to quote from *Seahorse*:

'the winner of the SORC series, *Muñequita*, used a double sheet system with one led to the conventional track on an $8\frac{1}{2}°$ line, and the other led to the rail at between 12° and 13°. According to her skipper Chuck Schreck: "We barber-haul the genoa and work the mainsail constantly going to weather."

They found while tuning for the SORC that *Muñequita* was very sensitive to the helm and liked neutral balance. The double sheet system gave this degree of

Photo 3.30   A $\frac{1}{4}$ scale Dragon model in the wind tunnel of Southampton
University. From left to right–Lord Brabazon of Tara, the
author, D Phillips-Birt and H Davies–Chairman of the
Advisory Committee for Yacht Research.

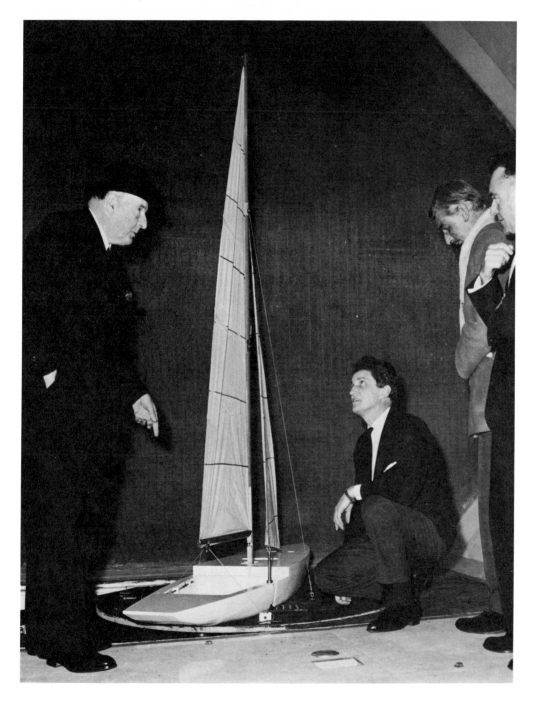

perfection in the constantly varying winds which are a feature of the circuit and, indeed, most race courses. One or even two men were continually trimming and they found that half a knot of speed could be gained in 15 to 20 knots of wind by switching from the inboard to the outboard track.

Even *Salty Goose*, which has multiple genoa tracks at seven, eight, nine and ten degrees, uses a twin sheet system as her skipper, Bob Derecktor, inevitably finds that to get the trim just right he needs to sheet between tracks!'

One may ask whether the difference of half a degree in trimming really matters so much to yacht performance. The answer can relatively easily be given by controlled wind tunnel experiments, and some results presented here may serve as an example. The tests were carried out by the author on a $\frac{1}{4}$ scale model of the Dragon class yacht shown in Photo 3.30 (Ref 2.86). In some tests, the hull above DWL was included; in others, the sails only were used, either singly or together. In some cases, the heeling moments were measured for a number of systematic alterations in sail trim. The tests were confined to zero heel angle only and a relative wind speed V_A of 25 ft/sec (14.8 knots) was applied throughout. This corresponds to force 4 on the Beaufort scale.

No attempt was made to reproduce in the tunnel the counterpart of full-scale sheeting arrangements. Instead, the clew of the genoa and the after end of the main boom were held rigidly by special fittings which were capable of various adjustments. Thus, it was possible to move the clew of the genoa fore-and-aft or up-and-down without altering the sheeting angle, and the clew of the mainsail could be raised or lowered to change the tension in the leech without alteration to the boom angle.

To enable the camber and twist of sails to be defined under various conditions of sheeting, black lines were painted on the sails one foot apart and parallel to the foot. The shape of the sails was recorded using a camera mounted on the tunnel roof and a typical picture yielding mainsail camber and twist is shown in Photo 3.31.

At the time the experiments were planned (1961) very little precise scientific information was available concerning the wind forces experienced by a yacht in a close-hauled attitude. The effects of the numerous adjustments which can be made to the sheets and rigging were almost completely unknown. It was anticipated that the hull itself would contribute something towards the aerodynamic forces and that the magnitude of this contribution would depend upon the relative wind angle and the angle of heel.

The results of the tests with and without the hull showed that the presence of the hull had a marked effect, for not only did it contribute a fairly large drag force but it increased the efficiency of the sails as lift-producing media. In fact, the driving force components were increased due to the presence of the hull. Figure 3.63 shows, conclusively, that there exists considerable interaction between the genoa and the mainsail. In this particular test, the sheeting angles of $\delta_f = 13.9°$ for the genoa and $\delta_m = 5°$ for the mainsail were kept constant and the forces were measured for various heading angles $(\beta-\lambda)$ in the range from 14.5° to 45°. It will be seen that the

Photo 3.31 Model of Dragon rig seen from a bird's eye view.

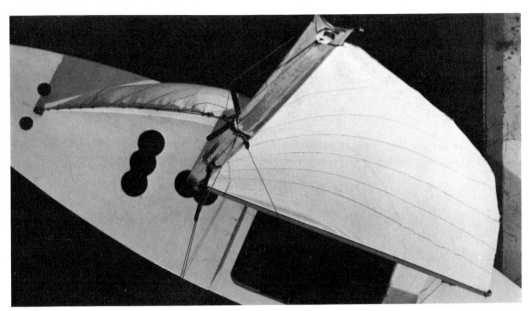

driving force F_x produced by the complete rig is in excess of that produced by the two sails taken separately. No doubt, this result can only be attributed to the slot effect and, as shown in Fig 3.64, the relative positions of both sails are very important indeed.

In another experiment, the mainsail sheeting angle δ_m was kept constant and the genoa sheeting angle δ_f was gradually altered from an initial 7.5° to 22.5°. From the results presented in Fig 3.64 it is clear that, as the heading angle (β-λ) is increased, large differences in the driving force component F_x can be obtained by widening the genoa sheeting angle δ_f; and it was found that the associated changes in either the heeling moment or heeling force are comparatively small.

It can be concluded from Fig 3.63 that the hull has a marked effect on the driving force component F_x, and for the genoa the F_x is increased by approximately 25 per cent, in the range of heading angles (β-λ) from 25° to 30°. This is due to a reduced gap between the sail foot and the sea surface.

As a rule, the genoa, as well as other types of headsail, is a splendid driving sail with its centre of effort (CE) relatively low. As shown in Fig 3.65, at heading angles near 30°, i.e. in the close-hauled condition, for each square foot of sail area the genoa alone produces 30 per cent more drive than the mainsail alone. Moreover, according to Fig 3.66, the genoa contributes 45 per cent less to the heeling moment than the mainsail. From this it is argued that, when a reduction of sail area becomes necessary, because the yacht is heeled too much and making excessive leeway, then it is best to reef the mainsail in preference to using a smaller headsail.

In connection with Fig 3.64, it is perhaps worth noting that the secret of

Fig 3.63   The effect of mainsail, genoa and hull interaction on driving
force component F_x. During test the trim angle of genoa δ_f
= 13.9° and trim angle of mainsail δ_m were kept constant.

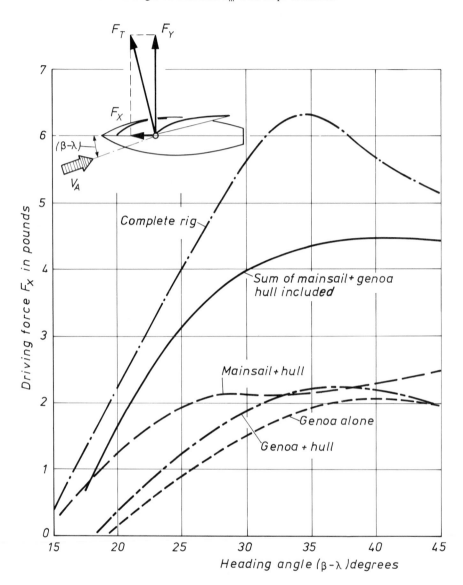

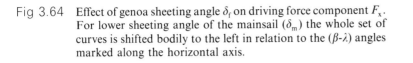

Fig 3.64   Effect of genoa sheeting angle δ_f on driving force component F_x.
For lower sheeting angle of the mainsail (δ_m) the whole set of
curves is shifted bodily to the left in relation to the (β-λ) angles
marked along the horizontal axis.

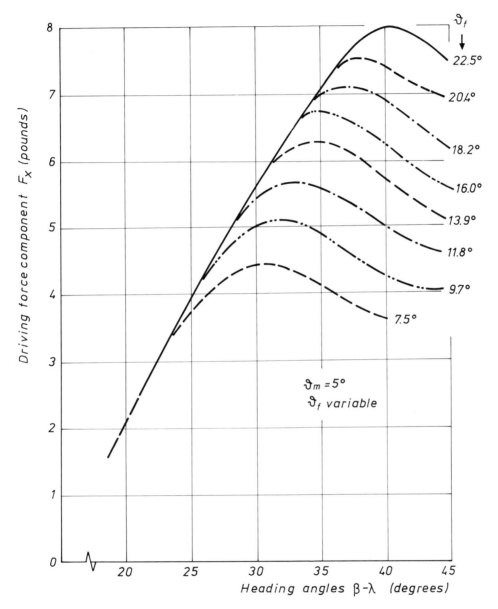

Fig 3.65 Driving force component F_x in pounds per square foot of sail area. Dragon rig.

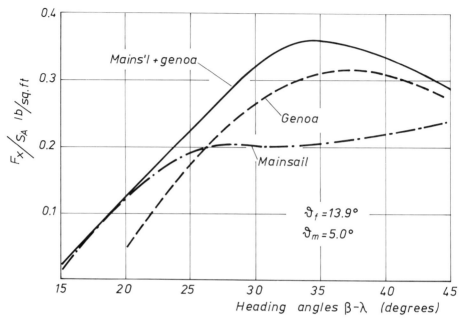

Fig 3.66 Heeling moment per lb of driving force at various heading angles $(\beta-\lambda)$.

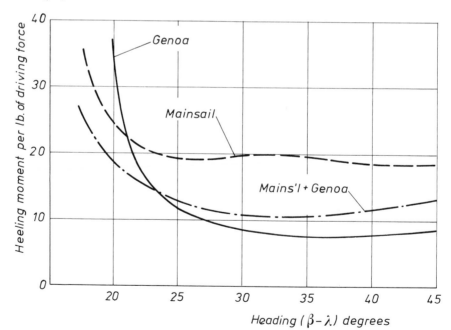

Fig 3.67 Measurement points recorded (C_x values) while attempting to trim the rig to its best performance.

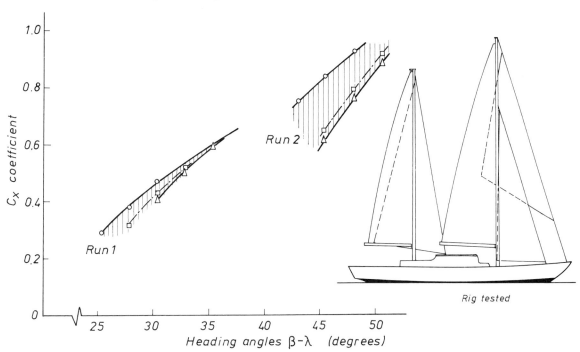

Rig tested

maximizing rig efficiency lies in the selection of the right sheeting angle for a given wind condition, since for any particular heading angle (β-λ) there is an optimum genoa sheeting angle δ_f which produces the largest driving force component. From wind tunnel tests carried out on a more complicated $\frac{1}{16}$ scale model of an 80 ft cruising ketch rig shown in Fig 3.67, it became evident that the correct sheeting angle is more critical and more difficult to find in a close reach than when beating (Ref 3.39).

Two series of experiments, marked Run 1 and Run 2 in Fig 3.67, were carried out at the same wind speed $V_A = 21.73$ ft/sec (12.85 knots) and at an angle of heel of 10°. During each run the trim angles of all the sails involved were adjusted so as to follow the heading angle (β-λ) variation in a systematic way to obtain the best possible sail efficiency. The positions of the clews of all sails set in a particular run were initially adjusted so that the sails assumed a shape and position relative to each other, with the wind on, that to the practical sailor's eye seemed best for the predetermined range of heading angles (β-λ). At the completion of these adjustments the two components of total aerodynamic force, namely the lift L and drag D, were recorded, from which the driving force coefficient C_x was calculated.

The test envelope that subsequently encloses the complete family data illustrates the best attainable driving force coefficient C_x. From Fig 3.67 it will be seen that the scatter areas below the envelopes marked 1 and 2 are different. Data scatter is smaller in close-hauled conditions (Run 1) than that recorded in full and by and close reaching conditions. The noticeable spread of measurement points in Run 2 is not all due to experimental error but rather reveals the difficulties in trimming all the sails to the best advantage. It appears that in the case of the multi-sail ketch configuration with a wide choice of trim angles and clew positions for yankee, staysail, main and mizzen, that in turn affect camber, twist and mutual interaction between the sails, the tuning problem becomes increasingly difficult when the boat bears away from the close-hauled course. It requires a much greater concentration on the helmsman's part and higher trimming ability from the crew.

One may infer therefore that the commonly agreed view that sailing close-hauled on the windward leg is the best test of helmsmanship, may not necessarily be right, provided of course, that in this particular case, by good helmsmanship is meant the art of tuning the sail to the best performance. No doubt, a speedometer can be of great help in overcoming the difficulty in trimming the sails while off the wind; whereas attaining the best V_{mg} is still an art even with a speedometer, unless a yacht also has a V_{mg} meter.

One prominent racing helmsman, who is also a sailmaker, raised the following question concerning sail trimming:

'Behind all the pros and cons of adjustable sheet systems, there lies a fundamental question. Once it has been set up for going to windward, why should a jib sheet have to be adjusted? Or a mainsheet for that matter? Having already allowed for the fact that the wind strength varies constantly, something which a majority of sailors take for granted, there must be a further explanation for helmsmen who play the mainsheet and traveller, mostly by instinct, in a series of reflex actions.'

A very good point indeed which must bother many others! Trying to give an answer to this problem he proposed three reasons: *firstly*–the live nature of the sails, i.e. their stretching properties require compensation, so sheet adjustment is inevitable to get the maximum speed; *secondly*–one must also compensate for the sudden gust of wind by easing the sheet; and *thirdly*–sails of different camber used in different winds also require different sheeting angles.

No doubt he is right, these are the factors which must be taken into account. However, there is also one more fundamental reason which is linked directly with Fig 3.64 and our earlier discussion in Part 1 and in Part 3, section 4d. It has been found, and Figs 1.16 and 1.21 substantiate it, that the optimum course to windward, no matter whether it is measured in terms of apparent course β or true course γ, depends on the rate with which the hull resistance builds up with speed. Consequently, for heavy-displacement yachts, the β and γ angles are smaller than for dinghies. In other words, in the case of keel boats the build-up of resistance with speed is so sharp that it does not pay to sail faster and further off the wind, whereas for a dinghy it does.

In Fig 1.9D, which presents the Finn dinghy characteristics, the optimum apparent course β changes over the range of 24°–36° depending on wind speed, and so the optimum sheeting angle must also change over quite wide limits, as illustrated in Fig 1.9E. For a heavy keel boat this range of optimum β angles is shifted towards lower values of β, and for the 12-Metre the optimum β range is 18°–25°. This implies different ranges of sheeting angles for different types of sailing craft and the data presented in Fig 3.64 must be viewed accordingly. The geometric relationships between the sheeting angle δ_m, heading angle $(\beta-\lambda)$ and incidence angle α is presented in Fig 1.10.

(e) *Tests on genoa overlap*

Some tests have been carried out on a range of sail models of a One-Ton Cup type of rig with varying genoa overlaps in order to determine whether there is an optimum value for close-hauled performance. The model sails, which were of $\frac{1}{6}$ scale (Photo 3.32), are identified in Table 3.14 as follows:

Photo 3.32   The late T Tanner with his $\frac{1}{6}$ scale model of a One Tonner set to test the genoa overlap effect on sail efficiency.

TABLE 3.14

Sail	Description		No of tests (runs)
Mainsail	Aspect ratio	3.0	
Genoa 1	Length of the Genoa foot	1.48 J	12
2	in terms of the fore-	1.58 J	13
3	triangle base, J	1.62 J	24
4		1.67 J	11
6		1.75 J	9
7		1.82 J	18

All the models were tested at a wind speed of 26 ft/sec and at the same angle of heel of 20°. The hull model was designed so that the positions of the clews of the mainsail and the genoa could be moved fore-and-aft, athwartships, and up-and-down. It was also possible to alter the tensions in the luffs of both sails. Each *run* involved a small change in one parameter (adjustment) only and consisted of measuring the lift and drag components as well as the heeling moment over a range of relative wind angles (β-λ) from about 20° to 40°.

The raw data obtained from the wind tunnel was subsequently analysed statistically by means of a computer, using Tanner's criterion; namely, the sail configuration which gives the greatest value of V_s/V_T ratio (where V_s is boat's speed

Fig 3.68   The effect of genoa overlap on boat performance.

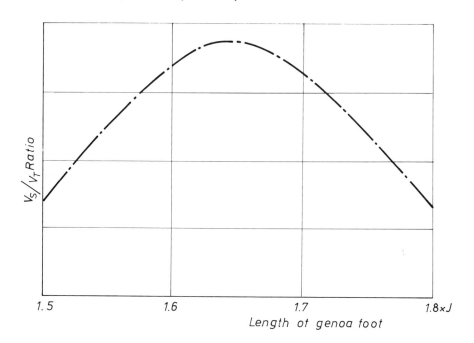

and V_T is true wind velocity) at the chosen heading angle $(\beta\text{-}\lambda)$ is the best configuration (Ref 3.3).

Figure 3.68 presents the best values of V_s/V_T plotted against the actual lengths of the genoa foot in terms of the base of the foretriangle J, for the apparent wind angle $(\beta\text{-}\lambda) = 30°$. It appears that a genoa with a foot length of 1.65 J (65 per cent overlap) will give the best performance in close-hauled conditions. It should be added that, in the processing of the data, a common sail area of 16.83 ft^2 has been used. This was the sail area of the actual foretriangle, i.e. the area of the genoa overlap was not taken into account.

It was also found that greater differences in performance (up to 20 per cent) may result from imperfect adjustment of the sail than by changing its size, and that differences of the same order could be obtained by greater care in the manufacture of the sails.

Porosity of sailcloth is another source of differences in the efficiency of sails which are identical in shape. It was found, for example, that when the tested sail was sprayed with water an immediate rise of 5 per cent in lift force was observed. However, this disappeared within 3 min when the wind was switched on and the sail dried out. Finally, when the sail was sprayed with ethylene glycol, which evaporates slowly, an increase in lift remained until the end of relatively long tests (Ref 3.14).

(6) Downwind rolling

> 'After all, the art of handling ships is finer perhaps, than the art of handling men. And, like all fine arts, it must be based upon a broad, solid sincerity, which like a law of Nature, rules an infinity of different phenomena.'
>
> JOSEPH CONRAD
> *The Mirror of the Sea*

Never will I forget my first downwind rolling experience, which I had in the prime of my sailor's life. It happened in a fresh wind and moderately undulating water. What I remember precisely is a disastrous, fearful feeling of complete helplessness when, quite unexpectedly, I lost control over my dinghy. During a series of rolls of increasing amplitude I saw the boom pointing high in the sky, then a sudden unintentional gybe, vicious broaching-to, and soon afterwards everything was blown flat, and water, cold water, was everywhere. At that very moment I understood the importance of having some knowledge of sailing theory.

There is one thing of which I am quite certain: I do not trust the man who tells me: 'Anyone can drive a yacht downwind.' Rolling and broaching are rather difficult yet fascinating problems, and there is still a lot to learn about the theoretical and practical aspects of these phenomena.

When running before a fresh wind and following sea, the rhythmic rolling and broaching tendency becomes an almost inevitable characteristic of all sailing craft, not just of small racing dinghies. It is not uncommon nowadays to see heavy keel boats involved in a spectacular but rather unpleasant unsteady rolling, as shown in Photo 3.33.

Another example of unsteady motion which is coupled with rolling, and which appears to be exaggerated in some modern yacht designs, is a directional instability which leads occasionally to broaching. In particular it seems to affect those boats with a short fin keel and reduced wetted area of the hull appendages.

According to Ref 3.40,

'The control problem came into sharper focus more recently. A yacht of modest proportions, $29\frac{1}{2}$ ft waterline length, was developed in the testing tank and subsequently built. Tests indicated that the windward performance would be improved with a shorter keel length. Trials of the full-scale yacht confirmed the testing tank's prediction of improved windward performance; however, in downwind sailing with the wind over the stern quarter, the yacht proved unmanageable. In one race it was reported that on the leeward leg the vessel could not be kept on course, and rounded up and broached 33 times in 3 hr!'

These two different types of motion, i.e. rolling and broaching, to which some modern ocean racers are prone, apart from affecting overall performance may also become potentially dangerous. Sailing yachts are normally designed in such a way that they are statically stable and yacht designers are usually content when the boat has a degree of transverse static stability; which is measured by the restoring moment, and determined by the amount of leeward shift of the centre of buoyancy B relative to the centre of gravity G, or the equivalent metacentric height GM shown in Fig 3.69A. This in turn limits the boat's power to carry sail and also her performance.

The Dellenbaugh angle method or the wind pressure coefficient method (Ref 3.41) might be used to check whether a boat will be tender in response or stiff. This rather empirical concept of yacht stability may be justifiable, since the presence of some degree of static stability usually ensures that the sailing craft, after being disturbed, will return towards the equilibrium position in some oscillatory manner. We said 'usually' because, as we will see, it is not always so; a yacht which is statically stable is not necessarily dynamically stable (see Note 3.42). There are both aerodynamic as well as hydrodynamic reasons for dynamic instabilities in yacht behaviour.

(a) *Rolling in still water*

Any yacht which floats freely in still water without sails may be given a rolling motion by the action of external moments or forces which are periodic in character. It could be accomplished by rocking the hull with a halyard when a yacht is moored in harbour, by the crew sallying to and fro across the deck or by wave action.

Let us assume that the disturbing force or moment is suddenly removed when the mast has reached an angle of heel Θ_0 to port (Fig 3.69A). The boat will tend to

Photo 3.33   Rolling severely downwind, dipping the weather gunwale in.

Fig 3.69   Stable rolling motion. The natural motion of such a system will
always be made up of some combination of two elementary
motion patterns:

A. sinusoidal motion which, in a way, represents the variation
of kinetic and potential energy
B. exponential decay which gives a rate of energy dissipation
with time, as represented in Fig 3.69B by the roll decrement
curve.

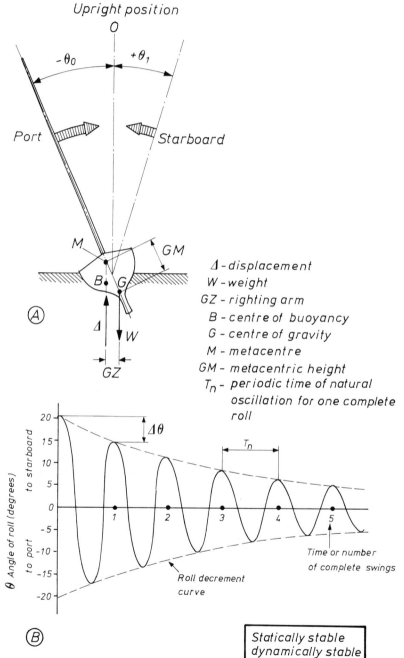

Δ - displacement
W - weight
GZ - righting arm
B - centre of buoyancy
G - centre of gravity
M - metacentre
GM - metacentric height
T_n - periodic time of natural
oscillation for one complete
roll

Statically stable
dynamically stable

return towards the equilibrium position (upright) owing to the action of the righting moment $W \times GZ$. The gravitational potential energy E_p stored in the heeled position, given by the approximate expression:

$$E_p = \Delta GM (1 - \cos \Theta_0)$$

is converted into the kinetic energy of rotary motion. When the mast reaches the upright position and the angular velocity p is at maximum, the kinetic energy accumulated, which is the product of inertia and angular velocity squared, is also at maximum.

$$E_k = \tfrac{1}{2} I_x p^2$$

where I_x is the moment of inertia about a longitudinal axis through the centre of gravity. The hull and the mast, therefore, continue their rotation to starboard. However, not all of this kinetic energy is converted into potential energy as the yacht heels to starboard, a portion is drained away by the work done against the resistance offered by the water. The yacht is therefore brought to rest momentarily at a smaller angle of heel $+\Theta_1$ than that $-\Theta_0$ from which the rolling was started. The cycle of rotary motion begins again and the yacht will perform a series of successive rolls to port and back to starboard, each being less than the previous one until, owing to the damping action of the water, it finally comes to rest in an upright position.

Such damped oscillations are graphically represented in Fig 3.69B by a roll decrement curve. Since the amplitude of roll decays with time, the hull in rolling motion is dynamically stable. The rate at which the rolling dies with time is a measure of the dynamic stability of the hull, and the so-called logarithmic decrement δ, which reflects the rate of amplitude diminution $\Delta\Theta$, may be used as an index of the damping efficiency of the yacht.

The character of the damped rolling of the hull (sail action being ignored), its period and the rate of decay, depend on three fundamental factors:

1. Moment of inertia of the boat; a large moment of inertia serves to increase the periodic time T_n of rolling.
2. Stability of the hull (GM), which affects the oscillation so that a stiff hull (of high stability) performs faster oscillations than a tender one.
3. Damping forces, which are responsible for the gradual extinguishing of rolling motion. They arise as a result of:
   (a) the presence of frictional forces between the wetted surface of the hull and the surrounding water,
   (b) the expenditure of energy in the generation of water waves,
   (c) the dissipation of energy due to the hydrodynamic action of the swinging appendages: fin keel and rudder.

These components of hydrodynamic damping are not equally significant. In the case of a keel boat, the predominant role may be played by the action of the appendages–the fin proper or centreboard and rudder, and also their configuration. Of course, high damping efficiency is desirable, since rolling, apart from bringing discomfort to the crew is also potentially dangerous.

At the moment there is little known about fin keel or centreboard efficiency as damping or anti-rolling devices. However, there is at least a certain theoretical foundation for believing that the modern tendency to reduce the length of the keel and cut down the wetted area, in order to improve the windward performance of the boat, may lead to a reduction in the hydrodynamic damping in rolling. As a matter of fact, the experiments carried out in Southampton University towing tank to investigate the roll-damping characteristics of three keel configurations for a 5.5-Metre Class yacht, showed that the best roll damping was attained with the longest keel. Further work is required on this subject to establish the mechanism of hull and keel damping, their interactions, and to correlate model experiments with theoretical predictions of damping efficiency.

During the experiments, the hull resistance was also measured in order to determine whether or not it was related to damping. Surprisingly, the results indicated that the resistance increase associated with rolling of quite large amplitude was in the order of about 2 per cent only.

Some people believe–let us quote from Ref 3.43–that 'There is little doubt that rolling is caused primarily by the hull balance being lost due to yaw. This is aggravated by bad steering, causing the spinnaker to oscillate in sympathy.' However, the result of a test conducted by the author has proved beyond any doubt that wild rolling may be induced by a sail for an aerodynamic reason. When running downwind, a sail can extract energy from the wind in a self-excited manner by its own periodic motion in such a way that the sail can be regarded as a rolling engine. When studying the self-excited rolling of a yacht, one should focus one's attention on two opposing elements of the rolling motion; namely, the excitation element due to sail action and the dissipation element due to action of the hull and its appendages. The character and magnitude of these two factors determine whether or not, and to what extent, the boat will be able to roll. The process of magnification of rolling amplitude will continue until the rate of wind energy input, due to the sail action, is matched by the rate of dissipation of energy by the damping action of the yacht's underwater part.

(b) Self-excited rolling

Attempting to answer the questions 'why and when' rolling oscillations can be excited aerodynamically, let us assume that a una-rigged yacht is running downwind, as shown in Fig 3.70A. The course sailed β, relative to the apparent wind V_A, is 180° and the angle of incidence α of the sail to the apparent wind direction will be approximately 90°. The total aerodynamic force F_T generated by the sail is more or less steady and acts very nearly along the course sailed. Let us further assume that by some means (it might be wave action or Karman vortex action developing behind the sail, to which we shall refer later) a small rolling motion is induced in the boat's hull.

Fig 3.70 Diagram of forces and windspeeds when running before the wind; without and with rolling motion. Illustration of flow reversal due to rolling (C and D).

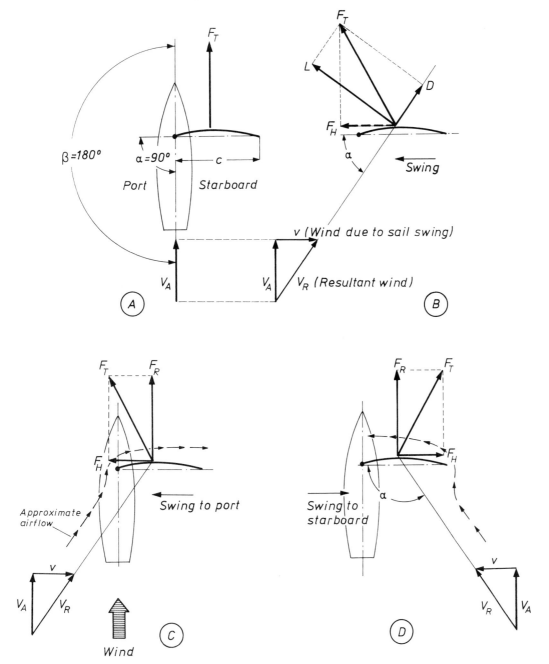

As the sail swings, say, to port, thus acquiring an angular velocity, then the resultant wind, its incidence and aerodynamic force change both in magnitude and direction. This is shown in Fig 3.70B which refers to a narrow, horizontal strip of sail cut at some distance above the axis of roll. The apparent wind V_A is modified by the velocity v induced by the swing. The resultant wind V_R, which is, at any instant, the sum of the two wind vectors V_A and v, will increase in magnitude and the instantaneous angle of incidence α relative to the sail chord c will be less than 90°. As a result of this, the flow pattern round the sail changes radically from that when there is no rolling. A circulation appears, marked in Fig 3.70C by the broken line, which in turn affects the instantaneous aerodynamic force F_T in such a way that its magnitude increases and the force is inclined towards the direction of sail motion. This total aerodynamic force F_T can be resolved into two components, as shown in Fig 3.70C:

1. A driving force F_R acting along the direction of the course.
2. A heeling force F_H acting perpendicular to both the course and the mast, which will tend to increase the angle of rolling (heel angle).

As the heel angle increases, the righting moment due to lateral stability of the hull and damping due to combined action of hull and appendages increasingly oppose the rolling and finally start to return the boat to the upright position. The sail now swings to starboard and the flow pattern is reversed. This is shown in Fig 3.70D. The circulation is opposite to that in the previous swing and the aerodynamic force component F_H is again directed towards the motion pushing the sail to starboard.

Fig 3.71   Recorded oscillation of the $\frac{1}{5}$ scale Finn rig with moderate damping simulating action of appendages (see Photo 3.34). $\beta = 180°$, $\delta_m = 85°$, damping md $= 1.0$.

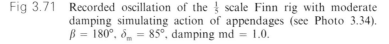

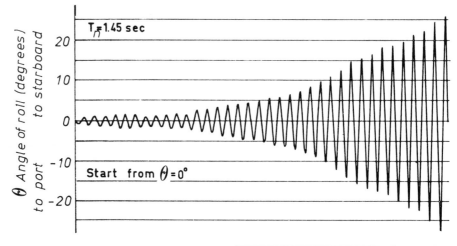

Because of the action of these alternating forces operating in phase with roll velocity, the amplitude of rolling may be magnified progressively.

Figure 3.71 depicts a typical behaviour of the model shown in Photo 3.34, when the rig was set at $\beta = 180°$ simulating downwind sailing and the angle of sail trim $\delta_m = 85°$. It will be seen from it that wild rolling can be induced by a sail for an aerodynamic reason and the rolling amplitude Θ builds up in the absence of any obvious external disturbance.

A $\frac{1}{5}$ scale model of a semi-rigid una-rig Finn-type sail made of Melinex was used for a series of initial tests. The Finn rig was chosen because of its apparent simplicity; there is only a single sail which is rigged on a mast unsupported by any shroud. Moreover, the Finn is well-known as a conspicuous roller and therefore worthy of investigation. Subsequently, further tests were made on a one-eighth scale Dragon rig with spinnaker.

The type of yacht behaviour recorded in Fig 3.71 clearly manifests dynamic instability. Referring also to Fig 3.69, we can say that in both cases the boat is statically stable, since a certain tendency to return to an upright position is maintained; however, in the second case there is a divergency superimposed on the oscillation. So, although the boat is statically stable she is dynamically unstable. Thus, the initial experiments in the wind tunnel have shown that the model responds dynamically according to the prediction based on an analysis of forces presented in Fig 3.70.

Now, it remains to answer the question why the sail, initially in equilibrium, i.e. without any rolling motion, begins to oscillate when wind is switched on? It is known that a flat or cambered plate, immersed in a moving fluid as shown in Fig 3.72, sheds

Fig 3.72   Wake behind the cambered plate or sail. The periodic detach-
              ment of vortices produces a periodic alternating cross force $\pm F$
              on the plate, tending to make it oscillate across the stream.

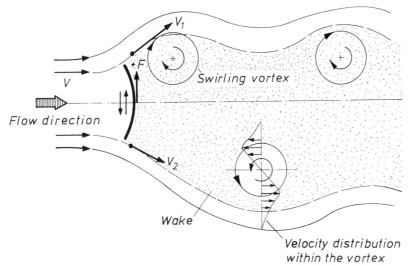

Photo 3.34 The picture to the left shows the Finn dinghy which, while rolling, has already attained the angle of heel close to the point of no-return. To the right, the apparatus used during preliminary tests before it was put into the special tank recessed below the wind tunnel floor in order to reduce possible blockage effect. The apparatus, based on pendulum stability, incorporates:

–an air-bearing support, permitting almost friction-free oscillations about a horizontal axis,

–variable and controllable magnetic damping device,

–a flexure combined with a differential transformer to measure the variation in the drag component D due to rolling,

–a rotary pick-off to measure the amplitude of rolling versus time, up to $\pm 30°$,

–recording facilities, linear recording of angular displacement and drag D versus time.

Photo 3.35 The model of the Finn-rig being self-excited in a manner which is frequently observed in full-scale boats; increasing amplitude of rolling is clearly visible. The model which is free to rotate is capable of reaching a state of the wildest agitation.

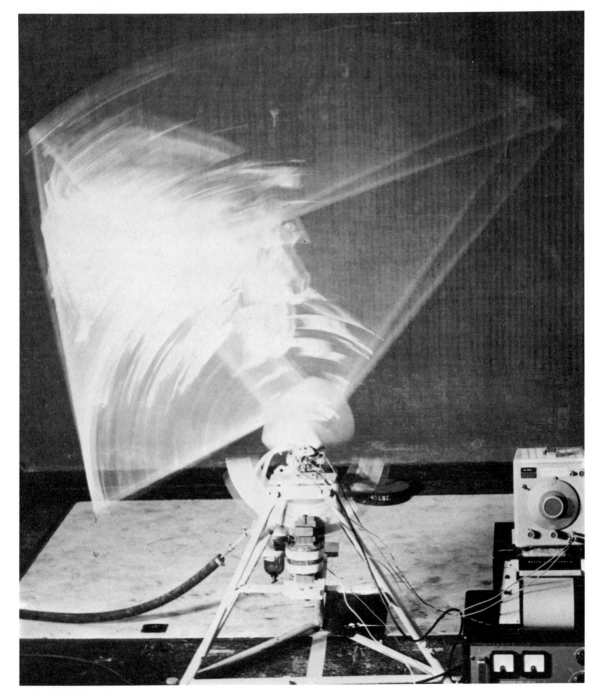

periodically into the wake the well-known Karman vortex trail. This phenomenon, which has been observed by various investigators through the centuries–Leonardo da Vinci, Strouhal, Bernard, Karman and others–is still far from being completely understood. However, the basic mechanism explained by Karman 1911, who made a stability analysis of the vortices being formed in a certain geometrical pattern, is fairly well known, at least for a stationary or non-oscillating body. Each time a vortex is released into the wake an unbalanced transverse force acts on the body, apart from the normal drag component. Whether the surrounding fluid is air or water does not change the basic physical principles. An enlarged picture of this edge vortex being formed is presented in Photo 3.36, which depicts the process of growth of the vortex; small-scale undulations which form a kind of vortex sheet are rolled up and superimposed upon a large-scale circular vortex.

With the vortex swirling in the direction shown in Fig 3.72 and Photo 3.36 there is instantaneously a velocity differential between the opposite edges of the plate, i.e. velocity V_1 at the upper edge of the plate in Fig 3.72 is higher than velocity V_2 at the lower edge. This is because, in this particular instance as illustrated, the flow velocity close to the upper edge of the plate is accelerated due to the presence of the swirling vortex. In accordance with Bernoulli's Principle, this difference in flow velocity must produce a differential static pressure component resulting in a lateral force $+F$ pushing the plate in the direction shown. A short interval of time later, with a succeeding vortex forming close to the opposite edge, a similar interaction between the vortex and flow develops. This gives rise to a lateral force $-F$, acting in the opposite direction to before. Thus, with the formation of alternating vortices, there appears an alternating transverse force which tends to oscillate the plate in a plane perpendicular to the flow direction. A similar situation, but a much more complicated one, can be observed in the case of a rolling sail. Figure 3.71, which presents a record of behaviour of a $\frac{1}{5}$ scale Finn rig, illustrates this point; the system, initially in equilibrium, begins to oscillate, being forced to do so by the Karman vortex trail, which produces an unbalanced transverse force.

The wind tunnel and water channel experiments made by the author suggest that the oscillations identified with vortex shedding at the beginning of the motion of the sail can be classified as forced oscillations. In this case the alternating forces that initiate the oscillation might be regarded as an ignition, which is responsible for initiation of the oscillatory motion. Once the system is set in motion, the alternating forces that amplify and sustain the oscillation are created and controlled by the oscillating rig itself. Since the periodic aerodynamic force is automatically resonant with the natural frequency of the boat, we can distinguish this kind of oscillation from a forced one as self-excited (Ref 3.44).

Rolling is therefore self-excited, drawing its supporting energy from the wind by its own periodic motion. The character of aerodynamic excitation is such that one may say ...*the more the boat rolls, the more she wants to roll*. In general but not always, the more violent the rolling and the greater the amplitude of roll, the higher is the resultant wind V_R (shown in Fig 3.70B) on which, in turn, magnitude of the

Photo 3.36   An enlarged view of the vortex-sheet being developed into
circular vortex.

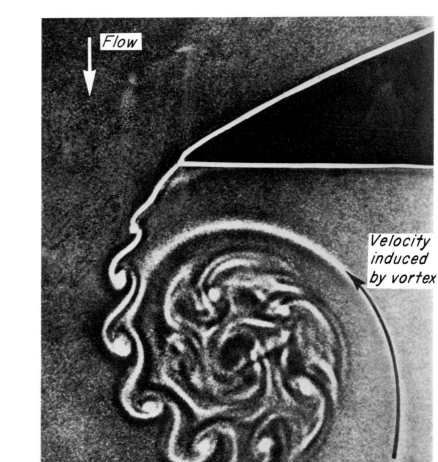

aerodynamic force depends. In this way, self-excited rolling may, in the case of a
dinghy, grow, magnifying the amplitude of heel to the point of no-return.

The modern masthead spinnaker has been cited as the villain responsible for wild
rolling; however, wind tunnel tests prove beyond any doubt that such a motion may
be induced by almost any sail, for aerodynamic reasons, even on a flat sea.
Experience shows that dinghies roll heavily going downwind without spinnakers.

Rolling can also be induced by waves, since the commonly encountered external
force of a rhythmic or periodic nature is that of the waves. The periodic time of

rolling forced by the waves will depend on the relative velocity and course sailed by the yacht to the wave crests, i.e. the frequency of encounter. When the periodic time of a yacht rolling under the action of waves approaches the periodic time of the yacht for natural oscillation in calm water, as described in the previous section, 'Rolling in Still Water', the amplitude of the resultant oscillations increases considerably.

The nature of the forced, synchronous rolling induced by wave action is similar to that when a boat is rolled by the crew running from side to side across the deck at signals timed from the roll. If the waves are high and steep, containing a great deal of energy, large and violent rolling can be built up in a few cycles. Under unfavourable weather conditions, the rolling induced by the waves, or even by wash from a motor boat or ship, can be simultaneously magnified by self-excited aerodynamic forces, to give more or less catastrophic results–broken spars, a knock down, blown-out sails, etc.

Since either aerodynamic or wave forces can cause rolling, it is possible for them to add together or oppose one another. They will tend not to be of the same frequency, and then a condition known as 'beats' will occur. In this case a period of little or no rolling, when the two sets of motion act against one another, is shortly succeeded by an interval of violent rolling when they add. The cycle is then repeated continually.

Referring to the rolling instability of sailing boats induced by aerodynamic forces, one should realize that the basic characteristics of such a self-excited system is of a very complex nature, being affected by a number of parameters, such as the course sailed, the angle of sail trim, the sail twist, the aspect ratio, and so on. However, their effects can be established experimentally using the apparatus as designed and shown in Photo 3.34, which incorporates the essential features of the real system and allows systematic investigation of the most important factors which can be held under close control, measured and compared.

(c) *Results of wind tunnel experiments*

The crucial questions to be answered by the wind tunnel experiments were:

1. In which conditions is the una-rig type, as shown in Photo 3.34, stable in rolling, and in which does it become unstable?
2. What is the relative influence of basic parameters such as:
   angle of heading β_A (in relation to the apparent wind V_A)
   angle of trim of the sail δ_m
   wind velocity V_A
   twist of the sail
   damping action of the hull
   on the rolling behaviour of the rig?

It will be shown that the rolling instability induced by aerodynamic forces can be reduced or eliminated in various ways. Some factors affecting a boat's behaviour and her tendency towards rolling instability can be directly controlled to some extent by the crew; some other factors being predetermined on the designer's desk may be beyond the command of even the best crew (Ref 3.44). More specifically:

(d) *The influence of the course sailed β*

Figures 3.73A–E give examples of rolling oscillations for various angles β from 145°–200°. The tests were performed at constant wind velocity $V_A = 3.05$ m/sec, constant angle of trim of the sail $\delta_m = 85°$ and constant magnetic damping md. At

Fig 3.73   Sketches A–E illustrate recorded rolling amplitudes at various courses β relative to apparent wind V_A. Degree of stability or instability in rolling is given by the index of stability δ. During this series of tests the trim angle δ_m and damping were kept constant. Positive index of stability $+\delta$ indicates that the rig when disturbed will tend to damp rolling motion (deviation-counteracting or equilibrating feedback). Negative index of stability $-\delta$ indicates that the rig, when given small rolling motion, will tend to magnify it (deviation amplifying or disequilibrating feedback).

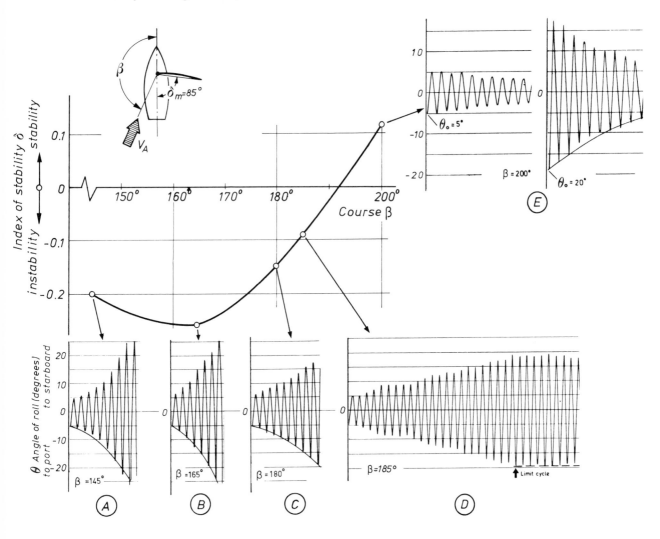

the beginning of each run for the selected V_A the rig was given an initial heel angle $\Theta = -5°$ and then released.

Within the scope of $\beta = 145°–180°$ the recorded oscillations are divergent and the model clearly manifests instability in rolling due to the action of aerodynamic forces. It is apparent that the energy input to the system is not matched by the energy dissipation, limited in a way by the amount of the available positive damping; therefore the amplitude of rolling grows continuously. The degree of instability given by the negative index of stability $-\delta$, being a maximum at β about 165°, decreases when β increases. By setting the sail model in the position of a boat sailed slightly by the lee $\beta = 200°$, the rig becomes dynamically stable and rolling has a definite tendency to die out with time. This behaviour is recorded in Fig 3.73E, which shows the rate of rolling decay when the initial angle of heel was 5° and 20°. When β is greater than 180° the aerodynamic excitation predominates only until a certain amplitude is reached and then energy balance occurs between the self-excitation and the dissipation due to damping. In a condition when damping is capable of balancing the energy input due to sail action, the system reaches a limit cycle steady state motion of finite amplitude. This type of behaviour, shown in Fig 3.73D, might be regarded as a transition from negative to positive stability. The system is unstable at small amplitudes but becomes stable at larger ones, and the rolling oscillations neither grow nor decay with time, being of constant amplitude. The magnitude of amplitude Θ at which the limit cycle is reached, decreases when β increases. Figure 3.73 depicts the relationship between β and the decrement coefficient $\pm \delta$ used as an index of stability.

Such a type of rolling behaviour, as shown in Fig 3.73D, where after a certain period of time during which the system manifests dynamic instability and a limit cycle steady state motion of constant amplitude is reached, is very interesting from a practical point of view. The variations in driving force component F_R, measured in the wind tunnel when rolling downwind, are quite large; up to 60 per cent more driving force F_R can be produced by the sail when it is rolling (amplitude of rolling approaching 30°) than when it is steady. Since the increase in the hydrodynamic resistance due to the rolling motion is less than the associated increase in driving force, so if the rolling amplitude can conveniently be controlled it may, for the purposes of racing, be worth while for the crew to endure the discomforts which are entailed. Such a possibility is quite feasible because experiments have shown that the magnitude of the rolling is determined largely by the sail twist which can in turn be controlled by the kicking strap.

The fact that rolling does not cause deterioration in yacht performance may be substantiated by the following excerpt from the American magazine, *Sail*:

'The eventual winner was *Magic Twanger*, co-skippered by owner Martin Field and Tim Stearn. *Twanger*, a modified PJ 37, took three firsts in the five-race series that had one throw-out. Her most spectacular win was in the deciding final 250-mile long distance race when she survived to win despite wild broaches and a roll that put her windward spreaders into the water.'

It can be seen in Fig 3.73 that the course sailed β has considerable effect on rolling. By applying a technique of sailing by the lee, $\beta = 200°$, the rig becomes dynamically stable and rolling will die out in time. However, sailing by the lee is always considered to be a cardinal sin on the part of a helmsman. Yet, according to wind tunnel findings, it may eliminate rolling; the danger of an unintentional gybe can be excluded by using a combination of fore-guy and preventer, or kicking strap, to effectively lock the mainsail boom.

The following quotation from a sailing magazine may illustrate the point that sailing by the lee could be regarded as a safe routine:–

> 'First Rodney gybed and I sensed that he had gybed a little too early, leaving himself a dead run to offset the tide sweeping round by the leeward mark. We went another 100 yards before gybing and then found that we too had gybed too early; both of us sat there in our boats for about 400 yards, running almost by the lee in a full gale, not risking to do two additional gybes and just praying that we would survive–at least I was and he certainly looked as though he was.'

The author himself experienced a similar situation and, to his surprise, did not capsize in spite of the fact that many others did; it happened well before he was given an opportunity to study this problem in the wind tunnel.

(e) *The influence of trim angle* δ_m

Figures 3.74A to F show records of oscillation for various trim angles of the sail δ_m measured between the boom and the hull centre line. The tests were performed at constant wind velocity $V_A = 3.05$ m/sec, constant $\beta = 180°$ and constant damping.

The rolling instability was most spectacular at an angle of trim $\delta_m = 85°$. By gradually hauling in the mainsheet and decreasing δ_m, the degree of instability was drastically reduced. At $\delta_m = 70°$ the rig reached a kind of neutral stability in rolling. Further pulling in the boom encouraged a positive stability, i.e. the aerodynamic force developed on the sail acted as a suppressor of rolling, producing positive damping. The curve plotted in Fig 3.74 indicates that the damping efficiency of the rig expressed by the positive value of stability coefficient $+\delta$ increases rapidly when the sail trim angle δ_m is reduced below $70°$.

The aerodynamic positive damping is quite profound, particularly when the initial amplitude of rolling Θ is large (Fig 3.74E $\delta_m = \cdot 65°$).

Figure 3.74F shows the rolling oscillation for $\delta_m = 85°$ recorded in the same condition as before, $\beta = 180°$, but the twist of the sail was increased by easing the tension in the kicking strap. One can notice that the rate of increase in the angle of roll at the beginning of the rolling motion is much higher than in the case when the twist was relatively small (Fig 3.74A) and the rig behaves differently. After several swings, during which, as mentioned earlier; the amplitude increases rapidly, the rig reaches a limit cycle with steady state motion of finite amplitude of about $22°$ at which the driving force component F_R is much higher than that recorded in steady conditions when the rig does not oscillate or the oscillations are small.

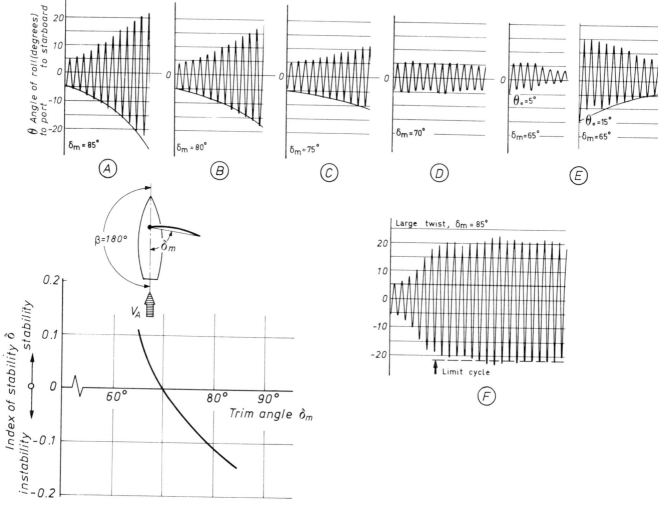

Fig 3.74   Sketches A–F illustrate recorded rolling amplitudes at various trim angles δ_m. During this series of tests the course sailed β and damping were kept constant. Index of stability δ gives the degree of stability or instability in rolling.

(f) *The influence of wind velocity*

Wind velocity influences instability in such a way that heavy rolling may not necessarily occur in very strong winds, but rather in moderate winds, and it can be seen in Fig 3.75 that the index of instability–the negative coefficient $-\delta$–is numerically greater at lower wind velocity. Further details are given in Ref 3.44. This problem appears to be related to the sail area/displacement ratio.

Certainly, carrying more sail area than is prudent when running vastly increases the aerodynamic input coming from a sail–a rolling engine. When coupled with low

Fig 3.75   Variation of stability index δ depending on wind speed V_A.
During tests the course sailed β, sail trim angle $δ_m$ and damping
were kept constant.

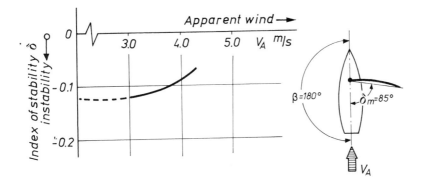

inertia, light-displacement and inefficient hydrodynamic damping, such a combination of factors may stimulate wild and occasionally disastrous rolling. This
point may be illustrated further by an example described by K Adlard Coles in his
book *Heavy Weather Sailing*:

'When we came to race *Cohoe II* in 1952, which was a season of fresh and
strong winds, we found her fast in light or moderate breezes but she proved to
be overmasted and overcanvassed in strong winds, and the world's champion
rhythmic roller. This was partly due to her being designed to carry a lead keel,
but having had an iron one substituted, as lead reached a peak price in the year
she was built.

Accordingly, in consultation with her designer I had the sail plan reduced the
following winter by cutting the mast at the jumpers and cutting the mainsail.
The reduction in sail area was drastic, being equivalent to two reefs...the
alteration greatly improved the yacht. From being a tender boat she became a
stiff one...gone was the rhythmic rolling.'

(g) *The influence of damping*

Figures 3.76A–E present the records of rolling behaviour affected by damping md of
increasing intensity, ranging from 1.0–3.0. During the tests $V_A = 3.05$ m/sec
β = 180° and $δ_m$ = 80° were kept constant. As expected, the combination of
resisted rolling due to the action of damping and self-excitation due to sail action
must produce a different response, depending on the amount of positive damping. It
is demonstrated that the higher the degree of damping, the less rapidly the amplitude
of rolling builds up and the lower is the final amplitude reached in limit cycle steady
state motion.

Thus, positive hydrodynamic damping due to the action of the hull and its
appendages seems to be of essential importance. Figures 3.76D–E suggest that there
is a certain critical damping which makes the system dynamically stable. The

Fig 3.76   Recorded rolling at various damping due to action of appendages. During tests the wind speed V_A, course sailed β and sail trim angle δ_m were kept constant.

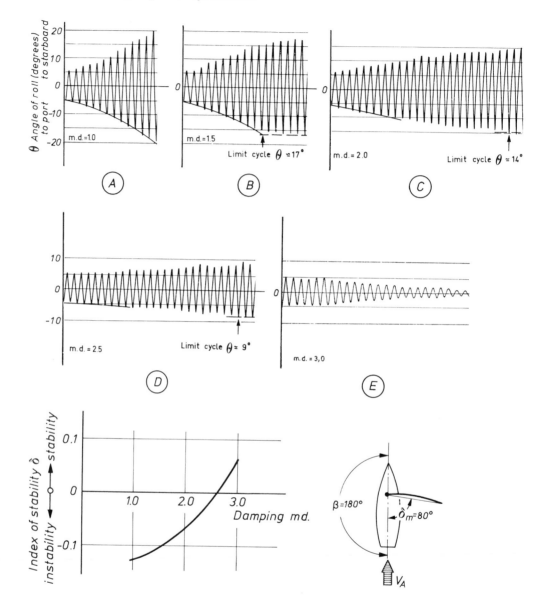

Fig 3.77   Anti-rolling sail configuration.

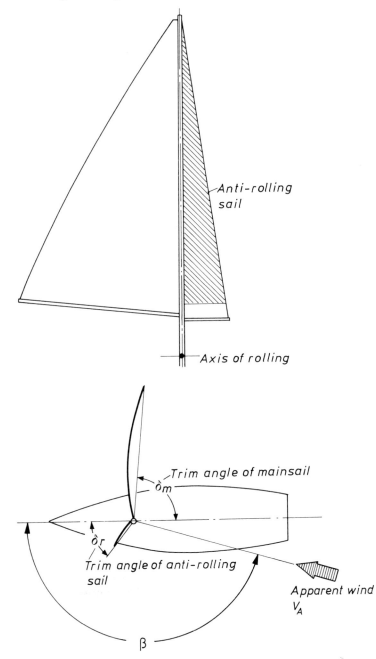

Fig 3.78   Recorded rolling behaviour of the rig shown in Fig 3.77 with anti-rolling sail. During tests the course β, wind speed V_A and trim angle of the mainsail δ_m were kept constant.

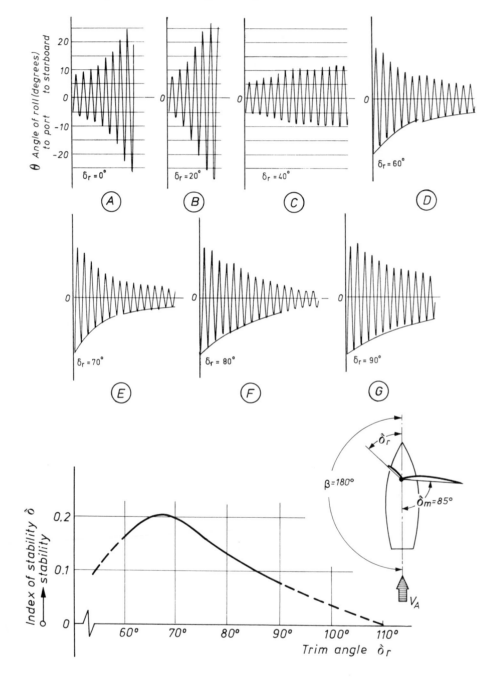

experiments justify the already expressed view that the modern tendency to reduce the wetted surface of the hull by cutting down the area of appendages in order to improve the boat's speed performance may lead to a reduction of hydrodynamic damping below an acceptable minimum imposed by the dynamic stability requirements. It may happen that in some unfavourable conditions the maximum aerodynamic input is far greater than the hydrodynamic damping. In such circumstances, the rolling amplitude will build up into one of those nightmarish affairs that both cruising and regatta racing know only too well.

(h) *Anti-rolling sail*

An attempt was also made to devise an anti-rolling rig which could produce a positive aerodynamic damping. Figure 3.77 shows some details concerning the anti-rolling sail. It is a tall and narrow sail, much shorter in the foot than any headsail would be, and its area is about 20 per cent of that of the mainsail. The tack can be taken to a point on the gunwale or a spreader (strut) on the opposite side to the mainsail. It is essential for the damping efficiency of the anti-rolling sail that there should be no excessive gap between the mast and the 'leech' of the sail, which is attached to the foreside of the mast.

The damping characteristics of the rig depend on the angle of trim of the anti-rolling sail δ_r relative to the centre line of the hull. This is shown in Figs 3.78A–G. When δ_r is greater than 45° and less than 110° the rig becomes dynamically stable even in the absence of damping due to hull action. The tests were performed at constant $V_A = 3.05$ m/sec, $\beta = 180°$ and $\delta_m = 85°$. The damping efficiency of the rig as shown in Fig 3.78 is greatest when the angle of trim of the anti-rolling sail $\delta_r = 65°–70°$.

Experiments with an anti-rolling sail set together with a spinnaker of Dragon type and a mainsail showed the same pattern of behaviour as manifested by the low aspect ratio Finn-type rig. When the wind was switched on, the whole rig stood firm and upright with scarcely any tendency to oscillate. This device could be quite easy to fit to a full-size cruising yacht and there is nothing in the International Offshore Rule to prevent its use while racing.

If full-scale tests could confirm the wind tunnel finding, one can expect that the hazard of being knocked down by a rolling spinnaker, a real danger almost all yachtsmen face, would be greatly reduced.

References and notes to Part 3

3.1   *The test tank as a basis for improving the design of GRP craft* J R Flewitt, PI-RPG Symposium, September 1972.

3.2   *A Method of Determining the Effect of Sail Characteristics on a Yacht's Clase-Hauled Performance and of Comparing the Merits of Different Rigs* P G Spens, SUYR Report No 15.

3.3   *An Approximative Method of Determining Relative Efficiencies of Sail Configurations from Wind Tunnel Tests* T Tanner, SUYR Report No 12.

3.4   *Some experimental studies of the sailing yacht* K S Davidson, TM No 130, Stevens Institute of Technology, 1936. Since Davidson's method several other methods have been developed, among the most notable by H Barkla (Ref 1.1, Part 1). *Methods of estimating yacht performance* and *Estimation of Effects of Sail performance on Yacht Close-Hauled Behaviour* P Crewe, RINA 1964. *A method of Predicting Windward Performance by the Use of a Digital Computer* P F Mills, SUYR Rep 21. *Windward Performance Prediction* J Sainsbury, ACYR Paper No 39.

3.5   *Scale Effects in Sailing Yacht Hydrodynamic Testing* K Kirkmann and D Pedrick, SNAME, November 1974.

3.6   *Towards Better Racing Rules* J K Kerwin, *Sail*, September 1976.

3.7   *Yachtsman's Guide to the Rating Rule* J Johnson and others, Nautical Publishing Company, 1971.

3.8   *Cascade* Ted Jones, *Yacht Racing*.

3.9   *An open letter to Olin Stephens* from J McNamara Jr, *Sail*, February 1973.

3.10 *Stretch Luff Sails* A Farrar, *Yachting World*, March 1969.

3.11 *Hydrodynamics and Aerodynamics of the Sailing Yacht* H Herreshoff, SNAME Paper, November 1964.

3.12 *The Technology of the Design of Sails* J Milgram, First AIAA Symposium on the Aer/Hydr of Sailing, 1969.

3.13 *Sail and Power* Uffa Fox, Scribners and Sons, 1937.

3.14 *The effect of fabric distortions on the shape and aerodynamic characteristics of a stretch luff genoa* M Yendell, SUYR Rep 29; also ICI Fibres Limited–private communications.

3.15 Unpublished papers on Sail Fabrics and Porosity, M J Yendell, Southampton University.

3.16 *The prediction of yacht performance from Tank Tests* W A Crago, RINA 1962.

3.17 *A review of three-dimensional sail Aerodynamics* T Tanner, University of Southampton, 1969.

3.18 *Instability of Sailing Craft-Rolling* C A Marchaj, SUYR Rep No 33.

3.19 *Performance Trials of the 5.5-Metre Class* Yeoman *and comparison of full-scale results with tank model tests* YRC Rep No 1, June 1955, submitted by NPL, Teddington, England.

3.20 Seminar on *Instrument sailing, the measurement of full-scale performance and its relation to predicted performance* SUYR Rep No 22.

3.21 *Sailing Theory and Practice* C A Marchaj, Adlard Coles Ltd and Dodd, Mead and Co.

3.22 *Sail force Coefficients for systematic rig variations* J H Milgram, SNAME Technical Research Rep 10, 1971.

3.23 *The analytical Design of Yacht Sails* J H Milgram, SNAME Transaction, 1968.

3.24 *The wind tunnel balances in the large low speed tunnel at Southampton University* P O A L Davies, R W Dyke, C A Marchaj, AASU Rep 208.

3.25 *The aerodynamic characteristics of a 2/5th scale Finn sail and its efficiency when sailing to windward* C A Marchaj, ACYR Rep No 13.

3.26 *The Application of Lifting Line Theory to an upright Bermudan Mainsail* T Tanner, SUYR Rep 16 and private correspondence on the above subject.

3.27 *The Logic of Modern Physics* P W Bridgman, Macmillan, 1932.

3.28 *Tuning a Racing Yacht* M Fletcher with B Ross, Angus and Robertson, 1972.

3.29 *The Common Sense of Yacht Design* W F Herreshoff, *Rudder*, 1946.

3.30 *Full-Scale Investigation of the Aerodynamic Characteristics of a Model Employing a Sailwing Concept* M Fink, NASA Langley, 1967.

3.31 *Boundary-Layer and Stalling Characteristics of the NACA 63-009 Airfoil Section* D E Gault, NACA Tech Note No 1894, 1949.

3.32 *Boundary-Layer and Stalling Characteristics of the NACA 64 A 006 Airfoil Section* G B McCullough and D Gault, NACA Tech Note No 1923, 1949.

3.33 *How Sails Work* A E Gentry, *Sail* Magazine (in several issues between April and November 1973).

3.34 *High Lift Aerodynamics* AMO Smith, AIAA Paper No 74-939 Los Angeles, California, 1974.

3.35 *The Handley-Page Wing* F Handley-Page, The Aeronautical Journal, June 1921.

3.36 *Segmental Sail* E Corbellini, *Yachting World*, June 1976.

3.37 *Further Experiments on Tandem Aerofoils* W L Le Page, ARC, R and M 886, 1923.

3.38 *Wind Tunnel Tests of a Clark Y Wing with a Narrow Auxiliary Airfoil in Different Positions* F E Weick and M J Bamber, NACA Rep 428.

3.39 *Rig Development Tests of a 1/16.6 scale model of an 80 ft cruising ketch* C A Marchaj, SUYR Rep No 36.

3.40 *Some Further Experimental Studies of the Sailing Yacht* P Spens, P De Saix, P W Brown, SNAME, November 1967.

3.41 *Skene's Elements of Yacht Design* F S Kinney, Dodd, Mead and Co, 1962.

3.42 *The definitions concerning static and dynamic stability.*

'Equilibrium' is a state of balance between opposing forces or moments. The equilibrium of a boat is said to be 'stable' if, after being displaced, the new orientation of forces or moments is such that they tend to bring the boat to her original equilibrium or trimmed attitude. It is 'unstable' if the forces and moments act to increase the initial displacement from this attitude.

'Stability' is a boat property which causes her, when equilibrium is disturbed, to develop forces or moments acting to restore her to the original condition of equilibrium. If the boat possesses instability, she deviates further from her original condition when disturbed.

'Static stability' is the property of a boat which causes her to maintain her steadiness or stability. In a static stability discussion the complete motion is not considered at all and, when a boat is said to be statically stable, it means only that, after being disturbed, the 'static forces and moments' tend to restore the boat to her equilibrium or trimmed state. It is assumed that the accelerations set up are small and inertia forces introduced by oscillating acceleration or deceleration are negligible.

'Dynamic stability' is that property of a boat which causes her to maintain her steadiness or stability only by reason of her motion. This general term is not to be confused with what is known in some quarters as dynamic metacentric stability, involving the righting energy available to bring a heeled boat back to her initial upright or trimmed position. In dynamic stability we consider the motion of a boat system following a disturbance from the equilibrium state, taking into account inertia forces and damping forces, as well as static forces or moments.

A statically stable system may oscillate about the equilibrium condition without ever remaining in it. In such a case the system, although statically stable, may be dynamically unstable.

'Metacentric stability' is that property of a boat by which the action of the buoyancy and weight forces causes her to return to her original position if her equilibrium about a given axis is disturbed. This occurs when the metacentre M lies above the centre of gravity G (see Fig 3.69).

If a boat is stable against a disturbance in heel she has 'transverse metacentric stability'. If the centre of buoyancy B and metacentre M are above the centre of gravity G, the boat is said to have pendulum stability.

3.43 *Do you suffer from spinnaker runs?* Chris Freer, *Yachts and Yachting*, September 1975.

3.44 *Instability of Sailing Craft-Rolling* C A Marchaj, SUYR Rep No 33, 1971.

Appendix

(A) Dimensions and units

The physical phenomena in this book, dealing with the mechanics of fluids and yacht motion, can be described quantitatively in terms of three fundamental dimensions and several other derived quantities. The fundamental dimensions are: force F, length L, and time t; all other physical quantities have dimensions that are derived from a combination of these three. In order to describe the magnitude of each quantity British engineering units are selected–therefore, pounds are units of force

Fig A.1   Comparison of 2 cubes having length of edges L and $2L$. The volume of the larger cube is 8 times the volume of the smaller cube, but its surface area is only 4 times as big as that of the smaller one.

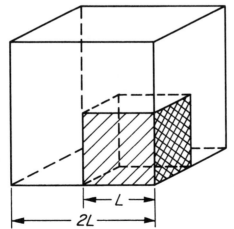

F, miles, feet or inches are units of length L, and seconds or hours are units of time t.

Some of the quantities are described by only one of these basic dimensions; for example, the length of the hull or a distance have dimension L. Volume is proportional to the third power of unit length, L^3. Others require combinations, for example velocity has the dimensions of length per unit time L/t, pressure has the dimensions of force F per unit area L^2, i.e. F/L^2, and so on. Figure A.1, relevant to some problems associated with model testing, demonstrates the relationships between the linear dimension L, area L^2, volume L^3 and the scale of the two cubes.

When two quantities are equated or added we must measure the quantities in the same dimensions. It is impossible to arrive at any sensible conclusion by adding quantities measured or given in two different dimensions.

The table below lists some of the quantities or variables used in the book.

Table of dimensions and units

Symbol	Quantity		Dimensions Brit eng units	Notes
F	Force	F	pound (lb)	
L	Length	L	foot (ft)	
t	Time	t	second (sec)	
A	Area	L^2	foot^2 (ft^2)	
a, g	Acceleration	L/t^2	ft/sec^2	Acceleration due to gravity
Θ	Angular displace- ment (dimensionless)		radian (rad)*	$g = 32.2$ ft/sec^2 $= 9.8$ m/sec^2
ρ	Density	Ft^2/L^4	lb sec^2/ft^4	
m	Mass $\left(\dfrac{W}{g}\right)$	$F/(L/t^2)$	lb/(ft/sec^2)	Unit of mass is called 'slug'
M	Momentum ($m \times V$)	mL/t	lb sec	mass × velocity
p	Pressure	F/L^2	lb/ft^2	
τ	Shearing stress	F/L^2	lb/ft^2	*Note*–Conversion tables
γ	Specific weight	F/L^3	lb/ft^3	and conversion factors,
V	Velocity linear	L/t	ft/sec	facilitating the transfor-
ω	Velocity angular	$1/t$	rad/sec	mation of British pound-
μ	Viscosity	Ft/L^2	lb sec/ft^2	feet-second system into
v	Viscosity (kinematic)	L^2/t	ft^2/sec	kilogram-metre-second system and vice versa,
V	Volume	L^3	ft^3	can be found on pp 683–
W	Weight	F	lb	687

* There are two systems for measuring angles: degrees and radians. It is the second that is frequently used in aerodynamics. To express a plane angle in radians, take a circle of any size and lay off along its circumference an arc equal in length to the radius of the circle R. As shown in Fig A.2 this arc R measures an angle of 1 radian.

Noting that the circumference of a circle is $2\pi r = 6.28\, r$, we see that a complete plane angle around a point, measured in radians, is $2\pi r/r = 2\pi$ rad. So $2\pi = 6.28$ rad are equivalent to $360°$, and

Fig A.2 Definition of radian measure.

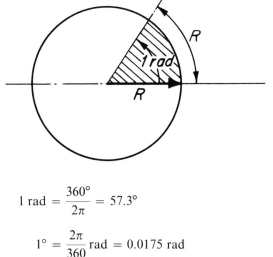

$$1 \text{ rad} = \frac{360°}{2\pi} = 57.3°$$

$$1° = \frac{2\pi}{360} \text{ rad} = 0.0175 \text{ rad}$$

Sometimes physicists prefer to choose mass m as a primary dimension with force F as a derived quantity. This is perhaps due to the fact that 'mass meant as a quantity of matter never changes'. To understand mass as an unchangeable property of all kinds of matter, one should grasp first the meaning of weight. In scientific parlance, weight is the official name for the force that seems to pull all earthly things towards the ground, or more precisely towards the centre of the Earth. Aristotle would say, '...the natural place of things is on the ground, therefore, they try to seek that place'. Whatever it is called, weight is a force that can be defined as:

Force (Weight) = mass × acceleration,

or using symbols as presented in the Table of Dimensions and Units it can be written:

$$F(W) = m \times g$$

The peculiar characteristics of weight as a force is that it is always vertical and unavoidable. Under the action of the pull of gravity, which is more or less constant on the Earth's surface, but changing with height (it is zero at the centre of the Earth, where gravity pulling in every direction would have zero resultant effect), the weight force is proportional to the mass m. The strength of a gravitational field reflected then in acceleration due to gravity g may change. It is, for example, much less on the Moon than on the Earth, but mass m will be the same everywhere, whether on the Moon, Earth or in Space.

Dealing with force equilibrium problems, it is convenient to elect force as a primary dimension. Then, the mass as a derived quantity can be defined:

$$\text{mass} = \text{Force/acceleration}$$
$$m = F/g$$

The unit of mass is thus lb/(ft/sec^2) and is called 'slug' in the British Unit System.

It is customary in aero- and hydrodynamics to call the unit of mass the slug, that a body of weight W has a mass $m = W/g$ slugs. For example, the weight of 1 cubic foot of salt water, so-called specific weight γ_w, $= 64.0$ lb/ft^3. Hence, the mass of this quantity of salt water, so-called mass density ρ_w, can be obtained by dividing the weight per unit volume by acceleration due to gravity g, which is about 32.2 ft/sec^2 $= 9.8$ m/sec^2. Thus mass density of water (salt) $=$

$$\rho_w = \frac{64.0}{32.2} = 1.99 \text{ slugs/ft}^3$$

Sometimes, ρ_w is taken as 2.0 slugs, as a round number.

In an attempt to avoid confusion between the force unit and the mass unit, a new unit for force has been introduced: the 'newton', denoted by N in the universal 'SI' metric system gradually spreading all over the world. One *newton* is the Force F that gives a mass m of 1 kilogram an acceleration of 1 metre/sec^2. If we place a mass m of 1 kg near the Earth, the pull of the force F with which gravity pulls on any mass is:

$$F = m \times g = 1 \times 9.8 = 9.8 \text{ newtons (N)}$$

The strength of the gravitational field is therefore 9.8 newtons per kilogram. The gravitational field on the Moon's surface is much weaker, and 1 kg of mass is pulled with a force of about 1.6 newtons only, i.e. six times less than the Earth's pull. This explains why astronauts visiting the Moon can so easily carry heavy burdens on their shoulders and are capable of making spectacular jumps.

One poundal (pdl), a unit still in use, is a unit for 1 pound of mass being accelerated at 1 foot/sec^2. Recalling that 1 lb $= 0.454$ kg and that 1 ft $= 0.305$ m, we may write poundal $= 0.454$ kg $\times 0.305$ m/sec^2 $= 0.138$ N.

These units, which are used in engineering, can briefly be compared with the newton N as follows:

1 kg-force $= 9.8$ N
1 pdl $= 0.031$ lb-force $\simeq 0.138$ N
1 lb-force $= 0.46$ kg-force $\simeq 4.45$ N
1 Ton (short) $= 2000$ lb-force $\simeq 8900$ N
1 Ton (long) $= 2240$ lb

It is customary in engineering practice, when referring to pounds-force and kilograms-force, to say simply 'pounds' and 'kilograms' although these actually refer to units of mass.

Whenever in practical calculation mass m enters as a unit, one is liable to lose a factor of 32.2 in the British, or 9.8 in the continental system, and end in disaster with an answer 32 or nearly 10 times too big or too small depending on which basic unit has been used, pounds and feet, or kilograms and metres.

Recommended reference book: *Physics for the inquiring mind*, E M Rogers, Princeton University Press, 1960.

(B) Conversion tables

Inches into centimetres (cm)

in	0	1	2	3	4	5	6	7	8	9
	cm	cm	cm	cm	cm	cm.	cm	cm	cm	cm
0	..	2.54	5.08	7.62	10.16	12.70	15.24	17.78	20.32	22.86
10	25.40	27.94	30.48	33.02	35.56	38.10	40.64	43.18	45.72	48.26
20	50.80	53.34	55.88	58.42	60.96	63.50	66.04	68.58	71.12	73.66
30	76.20	78.74	81.28	83.82	86.36	88.90	91.44	93.98	96.52	99.06
40	101.60	104.14	106.68	109.22	111.76	114.30	116.84	119.38	121.92	124.46
50	127.00	129.54	132.08	134.62	137.16	139.70	142.24	144.78	147.32	149.86
60	152.40	154.94	157.48	160.02	162.56	165.10	167.64	170.18	172.72	175.26
70	177.80	180.34	182.88	185.42	187.96	190.50	193.04	195.58	198.12	200.66
80	203.20	205.74	208.28	210.82	213.36	215.90	218.44	220.98	223.52	226.06
90	228.60	231.14	233.68	236.22	238.76	241.30	243.84	246.38	248.92	251.46
100	254.00	256.54	259.08	261.62	264.16	266.70	269.24	271.78	274.32	276.86

Feet into metres (m)

feet	0	1	2	3	4	5	6	7	8	9
	metres	metres	metres	metres	metres	metres	metres	metres	metres	metres
0	..	0.305	0.610	0.914	1.219	1.524	1.829	2.134	2.438	2.743
10	3.048	3.353	3.658	3.962	4.267	4.572	4.877	5.182	5.486	5.791
20	6.096	6.401	6.706	7.010	7.315	7.620	7.925	8.230	8.534	8.839
30	9.144	9.449	9.754	10.058	10.363	10.668	10.973	11.278	11.532	11.887
40	12.192	12.497	12.802	13.106	13.411	13.716	14.021	14.326	14.630	14.935
50	15.240	15.545	15.850	16.154	16.459	16.764	17.069	17.374	17.678	17.983
60	18.288	18.593	18.898	19.202	19.507	19.812	20.117	20.422	20.726	21.061
70	21.336	21.641	21.946	22.250	22.555	22.860	23.165	23.470	23.774	24.079
80	24.384	24.689	24.994	25.298	25.603	25.908	26.213	26.518	26.822	27.127
90	27.432	27.737	28.042	28.346	28.651	28.956	29.261	29.566	29.870	30.175
100	30.480	30.785	31.090	31.394	31.699	32.004	32.309	32.614	32.918	33.223

Square inches into square centimetres (cm²)

sq in	0	1	2	3	4	5	6	7	8	9
	sq cm	sq cm	sq cm	sq cm	sq cm	sq cm	sq cm	sq cm	sq cm	sq cm
0	..	6.45	12.90	19.36	25.81	32.26	38.71	45.16	51.61	58.06
10	64.52	70.97	77.42	83.87	90.32	96.77	103.23	109.68	116.13	122.58
20	129.03	135.48	141.94	148.39	154.84	161.29	167.74	174.19	180.65	187.10
30	193.55	200.00	206.45	212.90	219.35	225.81	232.26	238.71	245.16	251.61
40	258.06	264.52	270.97	277.42	283.87	290.32	296.77	303.23	309.68	316.13
50	322.58	329.03	335.48	341.94	348.39	354.84	361.29	367.74	374.19	380.64
60	387.10	393.55	400.00	406.45	412.90	419.35	425.81	432.26	438.71	445.16
70	451.61	458.06	464.52	470.97	477.42	483.87	490.32	496.77	503.23	509.68
80	516.13	522.58	529.03	535.48	541.93	548.39	554.84	561.29	567.74	574.19
90	580.64	587.10	593.55	600.00	606.45	612.90	619.35	625.81	632.26	638.71
100	645.15	651.61	658.06	664.52	670.97	677.42	683.87	690.32	696.77	703.22

Square feet into square metres (m²)

sq ft	0	1	2	3	4	5	6	7	8	9
	sq metres	sq metres	sq metres	sq metres	sq metres	sq metres	sq metres	sq metres	sq metres	sq metres
0	..	0.0929	0.1858	0.2787	0.3716	0.4645	0.5574	0.6503	0.7432	0.8361
10	0.9290	1.0219	1.1148	1.2077	1.3006	1.3936	1.4365	1.5794	1.6723	1.7652
20	1.8581	1.9510	2.0439	2.1368	2.2297	2.3226	2.4155	2.5084	2.6013	2.6942
30	2.7871	2.8800	2.9729	3.0658	3.1587	3.2516	3.3445	3.4374	3.5303	3.6232
40	3.7161	3.8090	3.9019	3.9948	4.0877	4.1806	4.2735	4.3664	4.4594	4.5523
50	4.6452	4.7381	4.8310	4.9239	5.0168	5.1097	5.2026	5.2955	5.3884	5.4813
60	5.5742	5.6871	5.7600	5.8529	5.9458	6.0387	6.1316	6.2245	6.3174	6.4103
70	6.5032	6.5961	6.6890	6.7819	6.8748	6.9677	7.0606	7.1535	7.2464	7.3393
80	7.4322	7.5252	7.6131	7.7110	7.8039	7.8968	7.9897	8.0826	8.1755	8.2684
90	8.3613	8.4542	8.5471	8.6400	8.7329	8.8258	8.9187	9.0116	9.1045	9.1974
100	9.2903	9.3832	9.4761	9.5690	9.6619	9.7548	9.8477	9.9406	10.0335	10.1264

Pounds into kilograms (kg)

lb	0	1	2	3	4	5	6	7	8	9
	kg	kg	kg	kg	kg	kg	kg	kg	kg	kg
0	..	0.454	0.907	1.361	1.814	2.268	2.722	3.175	3.629	4.082
10	4.536	4.990	5.443	5.897	6.350	6.804	7.257	7.711	8.165	8.618
20	9.072	9.525	9.979	10.433	10.886	11.340	11.793	12.247	12.701	13.154
30	13.608	14.061	14.515	14.969	15.422	15.876	16.329	16.783	17.237	17.690
40	18.144	18.597	19.051	19.504	19.958	20.412	20.865	21.319	21.772	22.226
50	22.680	23.133	23.587	24.040	24.494	24.943	25.401	25.855	26.308	26.762
60	27.216	27.669	28.123	23.576	29.030	29.484	29.937	30.391	30.844	31.298
70	31.752	32.205	32.659	33.112	33.566	34.019	34.473	34.927	35.380	35.834
80	36.287	36.741	37.195	37.648	38.102	38.555	39.009	39.463	39.916	40.370
90	40.823	41.277	41.731	42.184	42.638	43.091	43.545	43.999	44.452	44.906
100	45.359	45.813	46.266	46.720	47.174	47.627	48.081	48.534	48.988	49.442

(C) Tables of conversion factors

Conversion factors for linear velocity

Multiply number of by → to obtain ↓	Centimetres per second	Feet per second	Kilometres per hour	Knots	Metres per second	Miles per hour
Centimetres per second	1	30.480	27.778	51.478	100	44.704
Feet per second	$3.2808 \times (10^{-2})$	1	0.91133	1.6889	3.2808	1.4667
Kilometres per hour	0.036	1.0973	1	1.8532	3.60	1.6093
Knots	$1.9425 \times (10^{-2})$	0.59209	0.53959	1	1.9425	0.86838
Metres per second	0.01	0.30480	0.27778	0.51478	1	0.44704
Miles per hour	$2.2369 \times (10^{-2})$	0.68182	0.62137	1.1516	2.2369	1

Conversion factors for weight

to obtain ／ Multiply number of — by →	Kilo-grams	Pounds avoir-dupois	Kips, thousands of lb	Tons, long	Tons, short	Tons, metric
Kilograms	1	0.45359	453.59	1016.0	907.19	1000
Pounds avoirdupois	2.2046	1	1000	2240	2000	2204.6
Kips, thousands of lb	$2.2046 \times (10^{-3})$	0.001	1	2.240	2	2.2046
Tons, long	$9.8420 \times (10^{-4})$	$4.4643 \times (10^{-4})$	0.44643	1	0.89286	0.98420
Tons, short	$1.1023 \times (10^{-3})$	$5000 \times (10^{-4})$	0.500	1.120	1	1.1023
Tons, metric	0.001	$4.5359 \times (10^{-4})$	0.45359	1.0160	0.90719	1

Conversion factors for length

to obtain ／ Multiply number of — by →	Centi-metres	Feet	Inches	Kilo-metres	Nautical miles	Metres	Geo-graphic miles	Milli-metres
Centimetres	1	30.480	2.5400	10^5	$1.8532 \times (10^5)$	100	$1.6093 \times (10^3)$	0.1
Feet	$3.2808 \times (10^{-2})$	1	$8.3333 \times (10^{-2})$	3280.8	6080.2	3.2808	5280	$3.2808 \times (10^{-3})$
Inches	0.39370	12	1	$3.9370 \times (10^4)$	$7.2962 \times (10^4)$	39.370	$6.3360 \times (10^4)$	$3.9370 \times (10^{-2})$
Kilometres	10^{-5}	$3.0480 \times (10^{-4})$	$2.5400 \times (10^{-5})$	1	1.8532	0.001	1.6093	10^{-6}
Nautical miles		$1.6447 \times (10^{-4})$		0.53959	1	$5.3959 \times (10^{-4})$	0.86839	
Metres	0.01	0.30480	$2.5400 \times (10^{-2})$	1000	1853.2	1	1609.3	0.001
Geographic miles	$6.2137 \times (10^{-6})$	$1.8939 \times (10^{-4})$	$1.5782 (10^{-5})$	0.62137	1.1516	$6.2137 \times (10^{-4})$	1	$6.2137 \times (10^{-7})$
Millimetres	10	304.80	25.400	10^6		1000		1

Conversion factors for volume

to obtain ↓ / by → / Multiply number of →	Cubic feet	Cubic inches	Cubic metres	Gallons (US Liquid)	Litres	Barrels (US Liquid)
Cubic feet	1	$5.7870 \times (10^{-4})$	35.315	0.13368	$3.5315 \times (10^{-2})$	5.6145
Cubic inches	1728	1	$6.1024 \times (10^{-2})$	231.0	61.024	
Cubic metres	$2.8317 \times (10^{-2})$	$1.6387 \times (10^{-5})$	1	$3.7854 \times (10^{-3})$	0.001	0.15899
Gallons (US Liquid)	7.4805	$4.3290 \times (10^{-3})$	264.17	1	0.26417	42.0
Litres	28.317	$1.6387 \times (10^{-2})$	1000.0	3.7855	1	158.99
Barrel (US Liquid)	0.17811		6.2899	$2.3810 \times (10^{-2})$	$6.2889 \times (10^{-3})$	1

(D) Scientific notation

In scientific work a very large or very small number is frequently expressed as a number between 1 and 10 times an integral power of 10. Thus 1,570,000 may be written 1.57×10^6 and 0.000157 may be written 1.57×10^{-4} which also means $1.57/10^4$. One million (1,000,000) can be expressed in short as 10^6. Such a notation has certain advantages. The magnitude of the number is revealed at a glance by the exponent of power. Compare, for instance, the values of coefficients of kinematic viscosity of air and water tabulated in Tables 1 and 2 (Part 2). In recording large numbers the space is saved, a particularly important point in tabulating data, or when inscribing data given in the form of graphs.

(E) The Greek alphabet

A	α	alpha	(ăl'fà)	A	α	
B	β	beta	(bā'tà; bē'tà?)	B	β	
Γ	γ	gamma	(găm'à)	Γ	γ	
Δ	δ	delta	(dĕl'tà)	Δ	δ	
E	ε	epsilon	(ĕp'sĭ lŏn)	E	ε	

Z	ζ	zeta	(zā′tȧ; zē′ta?)	Z	ζ
H	η	eta	(ā′tȧ; ē′tȧ)	H	η
Θ	θ	theta	(thā′tȧ; thē′tȧ)	Θ	ϑ
I	ι	iota	(ī ō′tȧ)	I	ι
K	κ	kappa	(kăp′ȧ)	K	κ
Λ	λ	lambda	(lăm′dȧ)	Λ	λ
M	μ	mu	(mū; mōō; mü)	M	μ
N	ν	nu	(nū; nü)	N	ν
Ξ	ξ	ksi	(zī; keē)	Ξ	ξ
O	ο	omicron	(ŏm′ĭ krŏn; omī′; om′i kron)	O	ο
Π	π	pi	(pī; pē)	Π	π
P	ρ	rho	(rō)	P	ρ
Σ	σ	sigma	(sĭg′mȧ)	Σ	σ
T	τ	tau	(tou)	T	τ
Υ	υ	upsilon	(ūp′sĭ lŏn)	Y	υ
Φ	φ	phi	(fī; fē)	Φ	φ
X	χ	chi	(kī; kē; kē-chē)	X	χ
Ψ	ψ	psi	(sī; psē)	Ψ	ψ
Ω	ω	omega	(ȯ mē′gȧ; ō′mĕ gȧ; ȯ mĕg′ȧ)	Ω	ω

INDEX